Evolution and Escalation

GEERAT J. VERMEIJ

Evolution and Escalation

AN ECOLOGICAL HISTORY

OF LIFE

PRINCETON UNIVERSITY PRESS

PRINCETON, NEW JERSEY

Copyright © 1987 by Princeton University Press
Published by Princeton University Press, 41 William Street,
Princeton, New Jersey 08540
In the United Kingdom: Princeton University Press, Chichester, West Sussex

All Rights Reserved

Library of Congress Cataloging in Publication Data will be
found on the last printed page of this book

ISBN 0-691-08446-7

This book has been composed in Linotron Trump

Princeton University Press books are printed on acid-free paper
and meet the guidelines for permanence and durability of the
Committee on Production Guidelines for Book Longevity of the
Council on Library Resources

Printed in the United States of America

Designed by Laury A. Egan

10 9 8 7 6 5 4 3 2

TO EDITH AND HERMINE

Contents

Contents · viii

Contents · ix

PART FOUR: Escalation, Diversification, and Extinction

Preface

It was a June evening in 1968. I was looking over a small collection of shells that I had gathered earlier that day on a Hawaiian reef. There were heavy cone shells with a slit-like rectangular aperture and a knobby spire that barely protruded above the evenly conical main part of the shell, and prickly *Drupa* shells whose aperture was so obstructed by teeth that there seemed to be almost no room for the soft tissues of the snail. It struck me that these shells were somehow very different from those I had collected the previous January at Curaçao in the West Indies. The Hawaiian shells were more robust, more ruggedly sculptured, and their openings were less accessible to the fingers. I had long been fascinated by shell form, shell function, and geography, but now all these interests came together as I began to think about the possibility that geographical differences in shell shape might reflect subtle but important differences in the ecology and evolutionary history of the molluscs that constructed those elegant shells. I started to ask questions about the powerful crabs, fishes, and other predators that were capable of overcoming the seemingly infallible defenses of these heavily armored fortresses. These questions stimulated a series of observations and experiments, which in turn led to the publication of my first book in 1978. In it, I developed the hypothesis that escalation between shell-breaking predators and armored prey had proceeded further in the tropical Pacific and Indian Oceans than it had in the tropical Atlantic or in the temperate zones, and that these interoceanic patterns in escalation were reflected in the traits of other groups of organisms as well.

It soon became apparent that history played a key role in bringing about the interoceanic contrasts in the level of escalation between marine animals and their biological enemies. With this in mind, I began to probe the fossil record of antipredatory armor and the history of predators. I was amazed to discover that the kinds of armor that are now so common and so highly developed in modern tropical snails are geologically of rather recent origin, and that the familiar shell-breakers of today are also relative newcomers. Armor and the predators that overcome it are only the tip of the iceberg. All sorts of ecological processes and conditions changed over the course of earth history, not just in the sea but also on land and in the air. The fine layering of marine sediments that was characteristic of the earliest phase of the history of animals, for example, came in time to be replaced by a much coarser sedimentary structure, thanks to the evolution of animals that fed, burrowed, and built shelters under the surface of sand and mud on the sea bottom. Over time, organisms played an increasingly important role in the weathering of rocks, and in most environments there was a general increase in the speed of animal locomotion. In short, there was far more to the history of life than a mere parade of long names and dates of fossil plants and animals. The characteristics, ways of life, and biological surroundings of organisms also changed over the course of earth history, with the result that the world became an increasingly risky place in which to live.

To Darwin, this interpretation would perhaps not have been particularly surprising. After all, he expected modern organisms to be better adapted to the conditions of life than were their antecedents. Nevertheless, such an interpretation required careful scrutiny, and important questions remained to be answered. Can we reconcile the fossil record with Darwin's claim of progress? Have plants and animals improved in their methods of survival and reproduction relative to their surroundings, or have the biological surroundings themselves "kept pace" with adaptive improvements? Have any long-term trends in adaptation been obscured or nullified by the great crises that periodically have ravaged life on earth? If the interpretation of adaptive improvement or escalation is upheld by the evidence, what factors stimulated large-scale adaptive change, and is evolution by natural selection at the level of the individual organism sufficient to explain observed patterns? These questions and the research they stimulated prompted me to write this book.

To answer these questions, I found it necessary first to specify what is meant by terms such as adaptive improvement and escalation, and to outline criteria by which adaptive features can be distinguished from incidental consequences and from nonadaptive characteristics. This line of inquiry led to the development of a theory of adaptation that could be applied to fossil as well as to living organisms, and that at the same time would embody predictions about the course of evolution. This theory is set out in the first section of this book.

The main body of the book is one biologist's interpretation of the participants, conditions, processes, and events that together constitute the chronology of life during the last six hundred million years of earth history (the Phanerozoic eon). It is an essay on the nature and dynamics of evolutionary change of organisms and their biological surroundings. The main conclusions are that the capacities of organisms to acquire and retain resources such as food, shelter, and mates have increased episodically through time; that this increase in capacity has made many environments riskier places in which to live; and that species whose competitive, antipredatory, and reproductive capacities are modest have become either extinct or restricted to situations where the rate of supply of available resources is low—such as the deep sea, polar regions, caves, and other dark and cold places—or where enemies have severely constrained the times and places of activity.

In the final section of the book, I examine the reasons for the episodic nature of adaptive escalation. I argue that the major episodes were stimulated by extrinsic events that brought about worldwide increases in productivity and biomass, and that the great crises which periodically decimated life on the planet caused temporary reversals in escalation and set the stage for subsequent, even greater escalation.

This book is written for those who, like me, are fascinated with evolution, but I hope it will also be read by historians of human affairs. There are some obvious parallels between the history of life and the cultural history of man—the episodic pace of change, the effects of crises, and the role of population growth, for example—and the fossil record teaches some sobering lessons about the futility of the seemingly unstoppable arms race between nations.

In an attempt to make the book accessible to readers without a strong background in biology or geology, I have defined most technical terms in the text, provided a geological timetable in the Appendix, and distilled the essential points in summaries at the ends of the chapters. The liberal sprinkling of names of organisms throughout the text is meant for the specialist, but the names are not essential for evaluating the evidence or grasping the conclusions.

Because the fossil record preserves armored animals and their encounters with certain of their enemies more faithfully than it does many other organisms and interactions, shell-bearing animals and their predators are the focus of a large part of the discussion. This bias is inevitably exacerbated by my own research experience. Nevertheless, I have tried to examine other groups as well so as to give the broadest possible picture of adaptive evolution within the limits set by the fossil record and by current knowledge.

To attempt a synthesis such as this means hours in the library, museum, and field. This book could never have been written without the superb libraries and collections of the U.S. National Museum of Natural History in Washington, D.C. I have also made extensive use of the libraries and collections at Yale University in New Haven, and of Lucius G. Eldredge's remarkable collection of reprints in Guam. Many authors have kindly made available to me copies of important manuscripts. Much of the biological fieldwork that stimulated the present inquiry was carried out at the University of Guam Marine Laboratory and on board the RV *Alpha Helix* (perhaps the only ship that ever sailed on the pep-tide). I am deeply grateful to all these institutions and individuals.

My research on molluscs has been very generously supported over the years by the Program of Biological Oceanography of the National Science Foundation. I am also deeply grateful for a book subsidy award from the General Research Board of the University of Maryland.

Many people have helped me enormously by discussing and criticizing ideas, pointing out facts and citations, and showing me specimens and data. Bettina (Elizabeth C.) Dudley has spent untold hours measuring shells, working with me on research projects, reading innumerable books and papers to me, and proofreading all drafts of this manuscript. My wife Edith Zipser has been an unfailingly constructive critic, a collaborator in research, and an enthusiastic field assistant. My graduate students have been a source of intellectual stimulation, and through their own research they have greatly expanded my horizons and knowledge. Jay A. Blundon, Loren Coen, and Roy Kropp were enormously helpful in their preparation of the illustrations.

My deepest gratitude is perhaps reserved for the reviewers of this manuscript. Despite their often severe criticisms, all reviewers displayed a level of honesty, constructive criticism, and scientific integrity that I found both admirable and immensely helpful. The book has, I believe, benefited greatly from the hard work of these people: Jay A. Blundon, Loren Coen, Bettina Dudley, David Jablonski, Egbert Leigh, Jeffrey Levinton, John Sepkoski, Miriam Smyth, Charles Wahle, Thomas Waller, Edith Zipser, and an anonymous reviewer.

Finally, I wish to thank Janet Stern, my copyeditor at Princeton University Press, whose attention to detail and ability to uncover inconsistencies and errors are nothing short of incredible. I might add that she does not live up to her last name; on the contrary, she has been a pleasure to work with.

I dedicate this book to my wife, Edith, and to my daughter, Hermine Elizabeth. They have greatly enriched my life and done everything to help me write this book.

Friday Harbor
2 May 1986

PART ONE

A Theory of Adaptation

CHAPTER ONE

Aptations and Selective Agencies

NATURAL SELECTION AND ADAPTATION

Even to the casual observer, plants and animals seem to be well suited to their places of residence and ways of life. Many showy and fragrant flowers, for example, are shaped in such a way that when a pollinating insect such as a bee comes to fetch nectar, the pollen that is being carried by the insect is brushed onto the stigma while the flower's own pollen is placed on the insect for transport to the next flower. The complexly ridged high-crowned teeth of elephants are well suited for the maceration of tough plant material. In other words, there exists a close fit between an organism's form and behavior on the one hand and its environment and the manner in which essential functions of life are carried out on the other hand. In Darwin's (1872) view, this close correspondence is the outcome of natural selection, a process whereby individuals bearing certain traits have an advantage in survival or reproduction over genetically related individuals in which these traits are either less well expressed or lacking. Through the action of natural selection, traits change in frequency and expression over time, and adaptive evolution takes place.

I shall use Gould and Vrba's (1982) term *aptation* to mean any heritable attribute that enables an individual organism to carry out a specific function or to cope with hazards in the environment, and that confers an advantage in survival or reproduction in that environment. The hazards and challenges of the environment constitute the selective agencies that affect the aptation or that, in the absence of the aptation, cause failure or death. The use of the term *adaptation* is here restricted to the act or process by which an aptation is achieved. (Gould and Vrba also used the word adaptation to mean an attribute that has become elaborated to carry out its original function. An *exaptation*, according to their terminology, is an attribute that, when it evolved, either had no function or had a different function from the one taken on subsequently. Although this distinction between adaptation and exaptation may be useful, the functional history of a given attribute is usually difficult to trace, and I shall not use Gould and Vrba's distinction in this book.)

The fundamental observation on which Darwin's view of evolution by natural selection was founded is that the number of offspring produced by an organism is almost always greater than the number surviving to maturity. Darwin reasoned that this struggle for life continually favors improvements in the ways organisms stay alive and propagate. Because improvement of one species affects many other species (Van Valen, 1973), unilateral improvement is unlikely to go on for long. Instead, many species improve simultaneously, with the result that the struggle for life becomes increasingly intense (Darwin, 1872; R. A. Fisher, 1958). In Van Valen's words, species must keep "running in place" just to stay even with a changing (and usually deteriorating) environment. Species that are unable to cope with the higher risks as the struggle for life intensifies become either extinct or restricted to environments where the risks are lower. According to this interpretation, the history of life is characterized by two simultaneous trends: increasing risks to individuals from potential enemies and increasing incidence and expression of aptations to cope with these risks. Modern organisms may be no better adapted to their biological surroundings than ancient ones were to theirs, but the biological surroundings have themselves become more rigorous within a given habitat. I shall refer to this expectation as the hypothesis of escalation.

Does the record of life on earth bear out Darwin's prediction of escalation? This is the central question of this book. By examining the correspondence among form, habit, and environment in living organisms and applying this information to the fossil record, I shall trace the history of adaptation of plants and animals to their enemies, and the history of the risks and challenges that enemies and nonbiological crises have imposed. Not only the direction of change, but also the timing, rate, and consistency of escalation will be examined critically with all available evidence.

At least three arguments potentially undermine the hypothesis that adaptive escalation between organisms and their enemies has been important in evolution. The first is that enemies—competitors, predators, parasites, and agents of disease—are less important as selective agents than are nonbiological hazards such as storms, floods, extreme heat and cold, scouring by ice and sand, fire, and heavy-metal poisoning. If this argument is valid, species would be expected to track changes in the physical environment more often than they respond adaptively to their biological surroundings. An evaluation of this argument requires that the identity and mode of action of selective agencies be established, and that selective agencies be ranked in order of their effect on particular species. I will attempt such an evaluation later in this chapter.

The second argument is that adaptive change is frequently prevented even though the environment changes. Adaptational stability may prevail because improvement with respect to one kind of environmental hazard often engenders greater susceptibility to other hazards. Rapid body growth, for example, may enable a plant to out-compete a neighbor for light, but this advantage is achieved at the expense of the ability to resist drought and to produce high concentrations of chemicals that protect against attack by fungi and insects. Moreover, the nature and magnitude of natural selection are so variable on a small scale from place to place and through time that mutations which confer a benefit in only one kind of environment cannot easily spread or become fixed in the population as a whole. The contribution of drilling predation to mortality (and probably, therefore, to selection), for example, varies more than fourfold (from 17 to 75%) among populations of the pelecypod *Ctena bella* on Guam, an island only 50 km long (Vermeij, 1980a). This kind of small-scale variation in the environment seems to be the rule rather than the exception (Vermeij, 1982a). Finally, many species do not adapt when conditions change, but instead seek out conditions to which they are already well adapted. When a forest is cut, the native forest birds do not adapt to open fields or towns, but move to patches of woods that remain. Selection, in other words, typically serves to maintain the adaptive status quo and only rarely results in adaptive innovation or improvement. In short, selection is seen as stabilizing the adaptive attributes of individuals.

The third argument, and perhaps the most serious objection, is that natural selection at the level of the individual organism is less important than, and is often antagonistic to, other processes that influence the establishment of aptations and other traits. These processes include genetic drift, in which the frequency of genes is governed by chance, and selection occurring at levels below and above that of the individual organism—among genes, among gametes, among populations, and especially among species (Lewontin, 1970; Arnold and Fristrup, 1982; Vrba and Eldredge, 1984). A given organism may be well adapted to its surroundings, but if those surroundings disappear or if the individual happens to belong to a population that, for reasons unrelated to the presence of aptations, is unusually prone to extinction, the aptations will disappear as well. If selection at these other levels is important and opposes selection among individuals, adaptive improvement and escalation might be little more than temporary trends whose overall significance to the history of life is slight (Gould, 1982).

To evaluate the role in the comings and goings of aptations of processes acting at the level of populations and species, it is necessary to probe the relationship between the susceptibility of populations to speciation and extinction on the one hand and the habits, habitats, and aptations of individuals on the other. This topic will be taken up in the final section of this book.

No investigation of adaptive improvement and escalation can be undertaken without the establishment of criteria by which aptations can be distinguished from incidental consequences, nongenetic effects, and nonadaptive traits. Moreover, it is important to identify and rank in importance the various agencies of selection, and to understand the limitations and compromises that determine how one adaptive pathway comes to be followed instead of another. In short, as G. C. Williams (1966) realized, clearly we need a theory of adaptation—a system of definitions, postulates, observations, and hypotheses about the ways organisms respond to, affect, and are limited by their environment. It must make predictions about the course of evolution, and it must be applicable to fossil as well as to living species. The development of such a theory is the object of the first three chapters of this book.

Skeptics may wonder if a theory of adaptation is really necessary. After all, adaptation has been treated in quantitative terms by many population geneticists, including R. A. Fisher (1958), Dobzhansky (1968), Leigh (1971), Roughgarden (1979), and S. J. Arnold (1983), to name but a few. These formulations treat adaptation in terms of the fitness of genes, that is, the numerical advantage that one allele gains over another from one generation to the next. Unfortunately, it is difficult to measure fitness, especially in fossil species, and the precise genetic mechanism responsible for a given shape or behavior is usually unknown. In practice, therefore, students of adaptation often regard the demonstration that a given trait works in a way that is thought to be useful to the organism as sufficient evidence that the trait is adaptive. Others find evidence of adaptation in statistical departures in the frequencies of genes from values that are predicted by models assuming selective neutrality, even though the function of the genes in question is unknown. There is thus a large gap between theory and practice. The result is that adaptation has become an imprecise concept subject to widely varying interpretations. Some think of almost every trait as being adaptive, whereas others see most traits as inevitable byproducts of a rigid pattern of development. According to the latter interpretation, traits determine whether and how individuals can adapt (Gould, 1977).

Two features set my formulation apart from other treatments of adaptation. The first is my emphasis on the importance of demonstrating the presence and assessing the nature of environmental hazards as selective agencies. These hazards may be weather-related calamities as well as such biological challenges as competition, predation, and disease. The second way in which my formulation differs from other treatments of adaptation is that it is couched in nongenetic terms. Although the heritability of aptations is a fundamental feature of all formulations of adaptation, including mine, I maintain that the frequencies of genes that code for aptations need not be known in order to establish that a given trait is adaptive. Moreover, the coefficients of selection that are familiar from population genetics cannot be calculated in fossils, because neither the frequency of genes nor changes in these frequencies can be known. It is therefore operationally necessary to emphasize phenotypes—the structural and behavioral manifestations of the genes—and the way phenotypes affect survival and reproduction.

THE RECOGNITION OF APTATIONS

The perpetuation of genes from one generation to the next requires that the individuals bearing them survive and reproduce. This means that individuals at all life stages must be able to cope with hazards in

their environment and acquire enough resources to grow and reproduce.

If we are to show that a given trait functions as an aptation with
respect to the survival of an individual in a given environment, three
criteria must be satisfied: (1) the trait must be heritable, (2) the selective agency to which the individual is thought to be adapted must be
shown to be present, and (3) the incidence or expression of the trait
must be greater in individuals that prevail in or learn from an encounter with the selective agency than in individuals that either
have had no encounter with the agency or have been killed or vanquished by it. If the aptation is to become established, its benefits
must outweigh any disadvantage that the trait has with respect to
other environmental challenges (Vermeij, 1982d, 1985).

The first criterion is difficult to satisfy even in living organisms,
and it is impossible to satisfy completely in fossil species. There is
inevitably some variation in all traits, and research on living species
shows that most traits are under at least some genetic control. The
external environment, however, has a profound modifying effect on
the expression of many traits. Body size, even in species that cease to
grow in the adult stage, depends to an important degree on temperature, food supply, and other environmental conditions, as well as on
genetic factors. The extent of allometry (change in shape with increasing size) also depends on environmental as well as genetic factors. In the gastropod *Littorina littorea*, for example, rapid growth
under conditions of ample food causes the shell to grow with a low
rate of allometry into a thin-walled, low-spired configuration,
whereas slow growth in a regime of high population density and low
food availability results in a thicker, higher-spired shell whose shape
changes more rapidly as the shell increases in length (Kemp and Bertness, 1984). The way form varies according to the conditions of
growth may, of course, in itself be adapted and be under genetic control, but this cannot be assumed to be true without additional evidence. If a trait such as size or size-dependent shape changes through
time, a genetic basis for the change cannot be inferred automatically.

The second and third criteria for the recognition of survival-related
aptations deal with the identification and effects of selective agencies. The second criterion requires that individuals possessing a pur-
ported aptation either had nonlethal encounters with the agency of
selection or learned to recognize diagnostic cues from individuals

that had direct encounters. In other words, evidence that the selective agency occurs in the presence of adapted individuals must be available. To satisfy the third criterion, we must show that the incidence or expression of the trait in question is greater in individuals that survived, prevailed in, or learned from encounters than in individuals that were not exposed to the selective agency or were killed or vanquished by it.

It is important to specify what is meant by an encounter. An encounter takes place when an individual senses or becomes affected by a hazard, or when it comes into the sensory field of a potential enemy. Neither the potential victim nor its enemy need be aware of the encounter; indeed, it is often the function of the aptation to prevent detection or knowledge. The key is proximity; the individual must co-occur with a selective agency in order to have an encounter with it.

These criteria embody a view of adaptation that is more restricted than the one advocated by Sih (1985) and many other biologists. I regard aptations as traits with demonstrable current functions, and I believe that selection in favor of these traits is brought about by the action of verifiable agencies. The inability of individuals to cope with hazards indicates a lack of adaptation to those hazards. This inability, which is manifested either by death or by restriction to times and places in which encounters with the hazards do not take place, is a consequence of, rather than an aptation to, environmental hazards. Species may evolve aptations to cope with hazards in the environments to which they have become restricted, but these traits cannot be construed as adaptive responses to the hazards that caused the restriction in the first place (Vermeij, 1985).

Some examples will make these points clear. Mimicry and cryptic coloration can be thought of as antipredatory aptations of the avoidance type. Individuals with these attributes are able to coexist with visually hunting enemies because they either remain undetected or are mistaken for undesirable prey. On the other hand, the restriction of activity to predator-free environments such as the upper reaches of a tropical rocky shore or the undersurfaces of large boulders may not be the result of adaptation, unless the individual learned to recognize cues from previous experience with danger. Instead, ecological restriction may often be a consequence of predation and an indication that the individual in question cannot exist where risks of predation are predictably high.

Although the preceding discussion has dealt with aptations that enable individuals to survive, many traits are aptations in that they enhance the capacity of individuals to reproduce. Attributes that increase the representation of an individual's descendants in the next generation may be regarded as reproduction-related aptations if they are heritable. These aptations include traits that enhance the probability of mating, promote the protection or dispersal of offspring by the parents, and enable the parents to provide resources for the offspring. To show that such traits are adaptive, we must demonstrate that the presence of the traits increases the number or survival of the individual's offspring.

Note that the criteria for the recognition of aptations do not provide a research program for ascertaining how aptations function. This is the province of functional morphology, biomechanics, and physiology. The chief aims of this kind of functional analysis are to assess the absolute capacities of individuals—that is, the rates, duration, and magnitude of activities and processes—and to study how individuals cope with environmental hazards. This research generally calls for experiments in which traits and hazards are varied systematically, or for comparative observations of organisms that respond differently to encounters with a given hazard.

Evolutionary biologists have long known that a given trait is likely to have multiple functions and effects. The demonstration that a trait is beneficial when the individual is exposed to a given environmental hazard does not preclude other functions in other contexts; neither does it guarantee that the environmental hazard under investigation was responsible for the evolution of the trait. In general, hypotheses about the functions and selective basis of aptations are likely to be complementary rather than exclusive.

Effectiveness, Improvement, and Escalation

Hypotheses of improvement and escalation through time are statements about how well individuals cope with their environment. Evaluation of such hypotheses requires not only that aptations be identifiable, but also that an individual's effectiveness in coping with hazards be measurable.

I define an individual's effectiveness in a given environment as the probability that the individual survives or prevails in an encounter with a hazard. This probability can be estimated as either (1) the number of individuals surviving or prevailing during an encounter divided by the total number of individuals exposed to an encounter or (2) the number of successful encounters of individuals divided by the total number of encounters (Vermeij, 1982d). Whichever estimate is used, the effectiveness of individuals is usually assessed at the population level.

As defined here, effectiveness is a property of individuals and not of aptations, but it clearly depends on the nature and expression of aptations. Because an individual's ability to survive or prevail during an encounter with a hazard usually is influenced by more than one aptation, it is typically not possible to determine how much any one aptation contributes to the individual's success. This underscores the important point emphasized by Gould and Lewontin (1979) that aptations cannot be treated atomistically, and that an individual's traits make up an integrated body.

Effectiveness depends on properties of the individual organism as well as on the severity of the environmental challenge. It is therefore a measure of the adaptive gap between the capacity of the individual and the capacity of the environmental hazard. An effectiveness of 1 (or 100%) means that no adaptive gap exists; in other words, none of the encounters between the individual and the hazard is fatal or unsuccessful for the individual. An effectiveness of 0 means that all encounters are unsuccessful from the perspective of the individual.

The concept of an adaptive gap has much in common with the "factor of safety" that is frequently used by engineers in designing a structure. The factor of safety of a structure is the ratio of the load (force) that would cause it to fail (that is, to break, buckle, or irreparably deform) to the maximum load that it must withstand (R. M. Alexander, 1981, 1984). Structures with a high factor of safety are resistant to most expected loads; the adaptive gap between the individual bearing the structures and the expected maximum load would therefore be close to unity. An adaptive gap of 0 implies that the expected maximum load typically exceeds the load sustainable by the organism during an encounter.

Although the factor of safety can be usefully applied to analyses of biological structures (Alexander, 1981, 1984), I prefer the adaptive gap as a measure of effectiveness. Calculation of the factor of safety requires knowledge of the absolute values of forces (or other characteristics) of structures and of agencies impinging on them. Difficult enough to obtain for living individuals, this kind of information is usually beyond the reach of those who study fossils. Estimates of the adaptive gap require only a tally of the successful and unsuccessful encounters regardless of the magnitudes of the capacities of the individual and the environment in which it lives.

Darwin recognized that few, if any, individuals are perfectly suited to their surroundings. There are several reasons why an organism cannot be expected to be perfect, that is, to have an effectiveness of 100% with respect to any one kind of hazard. To begin with, selection distinguishes between available alternatives and favors the ones that confer better survival or more offspring. The favored traits are not necessarily the best that could be designed, but as long as they are better than those belonging to potential rivals, they will be favored. Second, if individuals belong to a species that only recently differentiated from its ancestor, or that colonized a new environment, many adaptive refinements may not yet have appeared. Third, an individual encounters hazards of many kinds, and coping with one kind of hazard may favor traits that are incompatible with traits enabling the individual to cope with other hazards. Aptations are therefore usually compromises.

That organisms fall short of the ideal or optimal fit with the environment does not mean that they are poorly adapted. It means merely that individuals are as good as they have to be to survive and reproduce. In some situations, a sufficient fit may be quite poor, whereas in others the standards may be set very high indeed.

If individuals are generally imperfectly adapted, the potential for improvement is almost always present. This potential depends on three factors: (1) the number of individuals that encounter a given hazard, (2) the effectiveness of individuals in coping with the hazard, and (3) the probability that an individual possesses a mutation that would improve the individual's effectiveness. Maintenance or enhancement of aptations is most likely when the selective agency affects, or "tests," a large proportion of individuals in a population. If only a few individuals have encounters with the selective agency, the benefits conferred by an aptation during the encounter would accrue to only a small minority in the population. Many individuals possessing the beneficial variant gain no advantage from it because they were never placed in a situation where the benefit would be realized. Even if the agency affects most members of a population, its potential for effecting selection is low if most of the individuals that have encounters die or fail, that is, if effectiveness in coping with the hazard is low. Accordingly, two reasonable measures of the potential for the maintenance of survival-related aptations to a given hazard are (1) the number of individuals with successful encounters divided by the total number of individuals in the population and (2) the number of successful encounters divided by the total number of individuals (Vermeij, 1982d). Measurements of the first type can take values between 0 and unity, whereas those of the second type can have any value equal to or greater than 0.

The potential for selection in favor of traits that increase an individual's effectiveness is expected to be generally lower than that for the maintenance of existing aptations. Adaptation generally occurs by the enhancement of an already existing trait, and it is likely to occur only in those individuals that already have the trait to some degree and that are therefore successful in encounters with the selective agency. Because the probability of a favorable mutation arising is small, the greatest potential for the increased expression of aptations exists in very large populations. After all, the law of large numbers suggests that a favorable mutation will be found only if many individuals have successful encounters. For a population of relatively ineffective individuals, this can occur only when the number of individuals encountering the hazard is very large.

Large population size has one additional important effect on the potential for selection. As Kimura (1983) points out, a mutant with only a very small advantage will behave as a neutral mutant in a small population because the effects of random fluctuations in population size overshadow the effects of selection. A mutant will be favored only if its selective advantage over other variants is large. In large populations, however, mutants with a small benefit may still be favored. An important consequence is that selection in any given direction may be carried further in a large population than in a small one, everything else being equal.

Evolutionary trends in adaptation may now be expressed in terms of the effectiveness of individuals in coping with environmental hazards. Improvement takes place when effectiveness increases, that is, when the adaptive gap between an individual's capacity to prevail during an encounter with the hazard and some absolute measure of the severity of the hazard narrows. It is effected either by elaboration of already existing aptations or by a decrease in the severity of the hazard. In the latter case, improvement is a happy coincidence, whereas in the former case it is an adaptive response. Escalation occurs when the hazards become more severe and the aptations to these hazards become better expressed. Whether the adaptive gap widens or narrows as a result of escalation depends on the rates of change in the species and in the environment. If the severity of hazards increases faster than a species is able to adapt, the adaptive gap widens; but if adaptation not only compensates for but also confers improvement relative to the hazards, the adaptive gap narrows.

Escalation implies change both in the species under investigation and in the environment of the species, and it is therefore broader in scope than is the related concept of coevolution. The term *coevolution* (Ehrlich and Raven, 1964) applies to a series of special cases of escalation in which the environmental change is brought about by adaptation of one or more of the enemies or benefactors of the species in question. Strict coevolution implies adaptive reciprocity between two species. As one species evolves aptations to cope with a second species, the second responds by evolving aptations to the first, and so on. A broader concept of coevolution, often called *diffuse coevolution* (Futuyma and Slatkin, 1983), implies adaptive reciprocity between a given species and two or more other species. Adaptive reciprocity is the essential property of coevolution, whereas adaptation to enemies is the cornerstone of escalation. If adaptive reciprocity involves enemies, as may be the case in interactions between predators and retaliating prey, then escalation has a strong coevolutionary component. I suspect that many alleged cases of coevolution, including such classic examples as evolutionary interactions between flowering plants and pollinators or between plants and herbivorous insects, involve a substantial selective influence of enemies and a much smaller component of adaptation to species that serve as sources of food (see below). Many cases of escalation will resemble instances of adaptive reciprocity, and like coevolution they may stimulate the forging of mutually beneficial relationships between species (see Chapter 5); but the evolution of mutualism cannot be taken as evidence of reciprocity. In short, although both Van Valen (1983) and I (1978) have previously used coevolution in the very broad sense of selection due to other species, I now believe that the word coevolution should be restricted to cases of adaptive reciprocity (Vermeij, 1983a).

Adaptation to hazards is only one of three possible consequences of an increase in the risk that the environment poses to individual survival. Alternative consequences are (1) restriction or displacement of individuals to sites where conditions are more compatible with existing traits and (2) death or extinction. Because the potential for selection in favor of existing aptations declines as the adaptive gap widens (that is, as the severity of the challenges increases relative to the ability of individuals to cope with those challenges), and fewer individuals survive to leave offspring, I expect adaptation to be a less common consequence than are ecological restriction and extinction. This would be especially true for small populations and for those whose individuals are already relatively ineffective in coping with existing hazards. In other words, once a species is established, its adaptive characteristics will remain constant more often than they will change.

Improvement and escalation may also occur with respect to reproduction-related aptations. Effectiveness in this case can be defined (1) as the probability that an individual with a given trait either finds a mate or leaves offspring or (2) as the number of offspring left by that individual. Because not every individual reproduces, and some reproducing individuals leave more offspring than others, there is always opportunity for improvement. As in viability-related selection, improvement in one individual may affect the reproductive performance or potential of others, so that escalation between adaptation and the reproductive capacities of competitors can occur. Indeed, as West-Eberhard (1983) points out, there may be fewer constraints on reproduction-related escalation than on the evolution of survival-related aptations.

ENVIRONMENTAL CHALLENGES AND PATHWAYS OF ADAPTATION

Although environmental hazards and the survival-related aptations to them are bewildering in their diversity and interdependence, they can be meaningfully classified into several broad categories. Selective agencies are either nonbiological (extremes of temperature, turbulence, radiation, and so forth) or biological (competition and predation in the broad sense). Individuals have three basic ways of coping with a challenge. First, they may occupy a place where the hazard is minimal, or they may adapt to biological hazards by appeasing, frightening, fooling, or remaining undetected by an enemy. Cryptic or warning coloration, threatening displays, features that make an individual appear large, and the ability to detect danger at a distance are all important avoidance aptations. Learning is also important if it enables individuals to use diagnostic cues in order to minimize contact with danger. The second type of adaptive response is escape, in which a potential victim moves to a safer place after it detects danger. The third response is resistance, the ability to tolerate a hazard as the latter impinges directly on the body. Enemies may be resisted by armor, toxicity, aggression, the ability to inflict injury, autotomy (the voluntary loss of body parts while the rest of the individual escapes), and large size, among other attributes. Weather-related extremes are resisted in either of two ways, by maintaining constant conditions within the body (homeostatic aptations such as endothermy and osmotic regulation), or by acclimation (the achievement of independence of function relative to some factor such as temperature or ionic concentration which varies both inside and outside the body). Large size also can be effective in buffering the individual against fluctuations in the external environment.

An important goal of the study of natural history is the identification and ranking of selective agencies. A full discussion of this topic is far beyond the scope of this book and perhaps of any book. My purpose here is to outline the diversity of ways in which adaptation is brought about, and to show why some selective agencies are apt to be more important than others.

Nonbiological Hazards

Weather-related agencies commonly cause death and are therefore potentially important in effecting selection as well. Data on the mortal impact of weather is becoming available for some populations, but surprisingly little is known about the role of weather in selection. One reason for this is that most weather-related catastrophes cannot be predicted. Studies of weather-induced selection require that the characteristics of a population be known both before and after selection.

Several good examples of weather-related selection come from studies of birds. In his classic paper on the selective mortality of house sparrows (*Passer domesticus*) during a fierce February storm in Providence, Rhode Island, Bumpus (1899) compared the dimensions of birds that he found dead on the ground with those of birds that, though on the ground, revived in his laboratory. He discovered that males had a higher probability of survival than did the generally smaller females, and that birds which were extreme in size or shape were least likely to survive. Fortunately, Bumpus published all his measurements of the birds that he rescued and of those that died, making it possible for later investigators to reanalyze his data statistically (P. R. Grant, 1972; Johnston et al., 1972; O'Donald, 1973; Lande and Arnold, 1983). Although the later interpretations have varied somewhat, Bumpus's conclusions have in the main been verified. Strong correlations between the characters he measured make it uncertain just how the winter storm affected the sparrows differentially. This uncertainty is heightened by the fact that, as Lande and Arnold have pointed out, we know nothing about the birds that survived but did not fall to the ground.

This difficulty was overcome in a second study of selection among house sparrows (Johnston and Fleischer, 1981; Fleischer and Johnston, 1982). Before the severe winter of 1978-1979, house sparrows in Kansas were marked, measured, and released. In the following spring, the measurements of survivors were compared to those of birds that had succumbed during the winter. Surviving males were found to be larger, and surviving females were found to be smaller, than were birds of these sexes before the winter. Sexual size dimorphism consequently became accentuated as a result of size-related selection during the winter. Surviving birds were also found to have relatively shorter limbs than did the birds that evidently perished. That these differences resulted from selection, and not from differential growth during the winter, can be inferred from the fact that only adult birds, which do not grow, were considered in the analysis.

In their elegant work on *Geospiza fortis*, one of the Darwin's finches on the tiny island of Daphne Major in the Galapagos, Boag and Grant (1981) showed that a decline by 85% in the population during a severe drought in 1977 was associated with strong selection in favor of large birds. This selection was correlated with the size of available food. Small birds depend on small soft seeds, which suffered a much greater decline than did the larger and harder seeds that are available only to larger individuals (Abbott et al., 1977).

These studies of birds cannot be interpreted as unequivocal examples of selection being effected only by weather. Instead, it is likely that competition among birds during inclement weather contributed substantially to selection, at least in the house sparrows. Some of the overwinter mortality of the house sparrows probably resulted from fights in which the larger males are typically victorious (Johnston et al., 1972; Johnston and Fleischer, 1981).

Death and selection during inclement conditions are probably influenced by biological agencies in many other instances as well. Food shortages during droughts in the Kalahari Desert provoke potentially mortal fights among oryx males at water holes (W. J. Hamilton et al., 1977). In British Columbia, the limpet *Collisella digitalis* protects itself against long periods of desiccation by attaching itself loosely to rocks by means of a thin mucous film which reduces water loss. Attachment by mucus is weaker than is adhesion by the sole of the foot, with the consequence that the limpet is probably more susceptible to predation by birds and mammals while it is attached by mucus (Breen, 1972). B. W. Kent (1981b) observed that 11% of the population of the rocky-shore gastropod *Ceratostoma foliatum* had become lodged upside down on the adhesive upper (aboral) surfaces of the sea urchin *Strongylocentrotus purpuratus* after storms in Oregon. Waves churned up by the storms probably dislodged the snails and carried them to crevices where the urchins live. If the snails' shells, which render the urchin less conspicuous to enemies, remain attached to the urchin long enough, the snails are likely to die. Other dislodged molluscs may become entrapped among the byssal threads of mussels or in the gullets of predaceous sea anemones such as *Anthopleura xanthogrammica* (Dayton, 1973; Sebens, 1976). Dislodgment due to storms might not be fatal were it not for the presence of biological traps. The susceptibility of corals to toppling by waves during storms depends in part on the presence of boring clionid sponges, which weaken the base of the coral skeleton (Tunnicliffe, 1979; Highsmith, 1980c, 1982; Mitchell-Tapping, 1983). Limpets such as the Chilean *Scurria scurra*, which excavate deep scars on the stipes of kelps, increase the susceptibility of these marine plants to damage by storms (Black, 1976; Vermeij, 1978).

The important point illustrated by these examples is that biological agents of mortality and selection are widespread and important even when physical calamities would appear to be directly responsible for the regulation of population size. Weather does, of course, influence selection, and there are probably situations in which the effects of weather are little influenced by biological factors. Garrity (1984), for example, has elegantly shown by experiments with gastropods on the hot and dry shores of the Pacific coast of Panama that high temperatures and desiccation kill snails, favor various aptations, and prevent animals from being active for much of the day, and that these effects are not necessarily influenced by competition or predation. Even in these gastropods, however, the particular weather-related aptations that are favored are apt to be determined to an important extent by the activities of predators, competitors, parasites, or even beneficial symbionts.

BIOLOGICAL HAZARDS

I believe that biological agencies—competition and predation in the broad sense—are the primary agencies of selection for the vast majority of species. Natural selection in favor of traits that enhance the capacity to acquire and retain resources should affect all species (Van Valen, 1976, 1983). These resources are often in the form of organisms or organic products, and they must be obtained by killing, interfering with, or outdoing other organisms. Resources must be defended by making them unavailable to other organisms or difficult for them to obtain. Even if the resources are inorganic, their quantity and rate of supply are likely to be influenced or controlled by competitors and predators (Van Valen, 1983).

Consumption of one individual or part of an individual by another organism is a very important way in which limited nutritional resources are distributed in a community of species. This predation, which for convenience will here be regarded as encompassing carnivory, herbivory, and parasitism, is potentially an important cause of selection in nearly all species. The chief reasons for this are that most species have predators and most predators are incapable of killing all the prey they encounter. In a survey of studies of predators ranging in size from copepods to lions, I found that only 19% of the interactions between one species of prey and one species of predator were characterized by predation efficiencies (the number of successful attacks divided by the total number of attacks, measured after the prey was detected) of 90% or more (Vermeij, 1982d). Most of the interactions in which predatory efficiency approached 100% involved victims of a particular size class of especially vulnerable individuals. Had the efficiency of predation been calculated for the prey population as a whole, it would almost always have been substantially lower than 100%. In fact, many interactions have been described in which fewer than 10% of the detected prey actually fall victim to the predator. Examples include predation of gastropods by certain intertidal sea stars (Dayton et al., 1977; Fishlin and Phillips, 1980); shell-breaking predation of some well-armored gastropods by calappid crabs (Vermeij, 1982b); and predation of pollinating insects by crab spiders (Morse, 1979, 1981).

The predominant form of antipredatory aptation is governed by the stage of predation in which the efficiency of the chief predators is lowest (Vermeij, 1982d). An act of predation comprises three phases: (1) detection (or searching), (2) pursuit, and (3) subjugation (Holling, 1966; D. Griffiths, 1980). The respective adaptive responses are avoidance or undetectability, escape, and resistance.

Competition occurs when one individual or colony influences the capacity of another to acquire and defend resources. Although competition may lead to the death of one of the contestants, it does not involve consumption of the vanquished party by the victor. Resources for which individuals compete include living space, shelters, nest sites, mates, and food.

A widely held view is that competition does not occur unless the contested resources limit population size or growth. Now, resource limitation may be an appropriate operational criterion for the establishment of competition as an important factor in population regulation, but many individuals compete fiercely for defensible resources that are not limiting to the population as a whole. We need only think of two children fighting over a toy to understand this point. A tug of war over a toy does not imply that toys are a limiting resource for the children, only that the particular toy in question is desired by the contestants. Similarly, adjacent corals may attempt to digest each other's tissues as their colonies approach one another, but such interference usually does not imply that space is limiting to corals, or that competitive interactions influence large-scale patterns of coral distribution (Birkeland et al., 1982; Bradbury and Young, 1982; Sheppard, 1982).

Competition should be an especially potent agency of selection for animals whose tolerance of starvation is low, and for organisms in which fecundity and survival are strongly dependent on body size. The inability to acquire or defend food in a competitive situation leads to little immediate harm in animals with large energy reserves, low metabolic rates, or large bodies. For animals with high metabolic rates or small bodies, however, competitive failure results in starvation and eventually death (Goulden and Hornig, 1980; Sih, 1982). The adverse effects of failure probably are especially acute for warm-blooded animals (endotherms), which can survive during lean periods only by hibernating or aestivating, that is, by reducing metabolic demands, or by migrating (Bakker, 1980). Even if losers survive, they may leave few offspring if their growth rate is significantly curtailed by a lack of food. Rapid growth, which often results in large body size, is commonly associated with large numbers of offspring and is therefore a substantial benefit of competitive success.

One of the important consequences of competition among predators is that there is strong selection in favor of traits that enhance the predator's capacity to detect, pursue, and subdue prey. Dawkins and Krebs (1979) think that the prey also has exerted a powerful evolutionary influence on its predators. I believe, however, that this influence is as a rule much weaker than that of the predators' own enemies. The chief reason why the evolutionary effect of prey on predators is seen to be of secondary importance is that the failure of a predator to kill and eat a particular item of food is rarely fatal or even deleterious to the predator, whereas failure of the prey to detect, escape from, or resist the predator results in the prey's death. This asymmetry in the effects that prey and predator have on each other has been dubbed the "life-dinner principle" by Dawkins and Krebs. If predators *are* adversely affected by their failure to acquire prey, many prey individuals belonging to many species are apt to be responsible. Most predators eat more than one kind of prey, so even if one victim species is taken with a low rate of success, there are usually suitable alternative prey available with which the predator is more successful (Vermeij, 1982d, 1983a).

Many of the structures predators use to acquire and subdue prey are also useful in bouts with competitors or as weapons against predators. Well-studied examples include the claws of crabs (Schäfer, 1954; Sinclair, 1977); the raptorial second pair of maxillipeds of stomatopod crustaceans or mantis shrimps (Caldwell and Dingle, 1975; Kunze, 1981); the beaks of birds; the claws of cats; the fangs of snakes; and even the jaws of some sea urchins (Grünebaum et al., 1978). Even the webs of spiders have a dual function as collectors of prey and as an early-warning system for the detection of enemies.

When potential victims can retaliate against attacking predators and are able to cause harm, they may have a substantial evolutionary influence on the predators. Such retaliation is rather widespread. Examples include attacks with horns or antlers by ungulates on wolves and lions (Mech, 1966; Schaller, 1972); stinging by scorpions and by aculeate Hymenoptera like wasps and bees (Langley, 1981); biting of ant lions by prey ants (Lucas and Brockmann, 1981); and pinching by the claws of crabs or the pedicellariae of sea urchins (Robinson et al., 1970; Glynn, 1976, 1977; Moitoza and Phillips, 1979). Even such passive defenses as spines, toxins, and large size could prove fatal to some careless predators. These retaliatory characteristics probably also help to teach would-be predators not to attack, particularly if associated with a feature that is conspicuous to the attacker before physical contact with the victim is made. Predators and prey may therefore influence one another evolutionarily in some cases, especially when the predators are catholic and the prey protestant.

REPRODUCTION

The selective agencies that effect survival-related aptations also influence the reproductive patterns in species—the size and number of offspring, provision of resources to the offspring by the parent, traits that enable potential mates to recognize and choose each other, the timing and frequency of mating, and the extent of reliance on sexual reproduction relative to asexual propagation. Incisive general reviews of the extensive literature on this topic have been presented by G. C. Williams (1966, 1975) and Stearns (1976, 1984). Rather than treat this subject in detail, I shall merely outline some of the main conclusions.

Reproduction-related traits are governed to an important degree by the risk of mortality. A high risk of juvenile mortality is associated with two contrasting patterns of reproduction: (1) the production of a large number of poorly defended progeny that are often capable of feeding or of making their own food at a very early stage, and (2) the production of a small number of relatively large, well-provisioned progeny that are well protected in an egg or seed or by the parents. Where the risk of death of small juveniles is low (usually in low-energy environments), the few progeny produced are provided with enough resources by the parents to enable the young to grow to a size large enough to begin fending for themselves in their nutrient-poor surroundings. Sexual reproduction seems to be especially favored at times of radical environmental change, which may either come at regular intervals or be unpredictable. Asexual propagation is effective when favorable circumstances exceed in duration the generation time of individuals. The frequency of reproduction is perhaps dictated by the relationship among body size, fecundity, and risk of adult mortality. If the risk of adult mortality is high, selection often favors reproduction once at an early stage (semelparity), whereas if adult mortality is low, multiple episodes of reproduction in a long-lived individual (iteroparity) are apt to be the rule.

THE RANKING OF SELECTIVE AGENCIES

An agency that results in the death of an organism or the loss of capacity of an individual has the potential for effecting selection among individuals. Nevertheless, some of these agencies are apt to be more important than others. I believe it is possible as well as useful to construct a crude ranking of the major selective agencies. In general, those agencies that affect a large number of individuals and species should elicit more aptations and should consequently play a larger role in adaptive evolution than do agencies that affect a minority of individuals and species.

Competition for resources may be the most ubiquitous selective agency, for it affects nearly all individuals in all species. Moreover, it influences not only individual survival, but also the acquisition of mates and therefore the capacity to reproduce sexually. Competition should be especially potent as a selective agency in cases in which failure during an encounter results either in death or in a substantial reduction in the loser's capacity. As I argued in the preceding section, warm-blooded animals and other organisms with high metabolic demands should be especially susceptible to selection due to competitors.

Predation may rank a close second behind competition. All species have predators at some stage in the life cycle, although not every individual is exposed to predators. For many species with the capacity to endure competitive loss without a substantial reduction in essential life functions, predation is probably the most important agency of selection. This may be so, for example, for many relatively inactive marine invertebrates.

Food and physical calamities would seem to play a distinctly secondary role in adaptive evolution. In the case of food, the effect on consumers is generally smaller than the effect that consumers have on their prey. Physical calamities often cause death, but biological intermediaries are probably important in selection during weather-related events.

The idea that enemies are the most important agencies of selection is not new. Darwin (1872), R. A. Fisher (1958), and A. G. Fischer (1960) are among the many biologists who have stated this view explicitly in their writings.

The ranking of agencies of selection is likely to be different from the ranking of agencies of mortality. Competition may lead to the death of one of the contestants, but usually its effects are more subtle. Losers are often prevented from acquiring or retaining some resource, or they are restricted to less favorable parts of their environment by the winners, but they usually survive. Weather, on the other hand, frequently causes death. In fact, mass mortality in populations is often triggered by unusual weather, and the extinction of many species may be due largely to catastrophic changes in the nonbiological environment (Chapter 14). In short, whereas competition might rank first among selective agencies, it ranks last as a cause of death. Weather and predation, then, are the chief causes of death, while competition and predation are probably by far the more important agencies of selection.

The fact that biological agents of selection evolve has important consequences for the adaptive evolution of organisms. As predators and competitors evolve, migrate, and disappear, the traits favored by selection change. The result is a certain evolutionary instability.

Nonbiological agencies vary over time as well, of course, but no truly novel weather probably has appeared during the Phanerozoic (Schopf, 1980), save for infrequent short-term catastrophic events to which most organisms are not adapted. Aptations to a particular weather-related agency therefore do not become obsolete in the long run, whereas aptations to a particular enemy may never again be effective once the enemy has evolved further or disappeared.

The importance of competition, predation, and physical factors goes well beyond their role in imposing selection and causing death. Together with an individual's pattern of development, they also impose limitations and risks that determine the kinds of adaptive pathways available to a given lineage. This is the subject of the next chapter.

CHAPTER TWO

Limits to Adaptation

THE NATURE OF LIMITATION

Adaptation occurs when, as a result of natural selection, a choice is made between alternative heritable traits in an environment that poses hazards of various kinds to individual organisms. A given hazard can elicit any of several forms of aptation, some of which are more compatible with already existing traits and with other environmental hazards than are alternative adaptive pathways.

A central problem in evolutionary biology is to understand the mechanisms and conditions that govern the establishment of one body plan or pathway of adaptation instead of another. Although solutions to this problem will be specific to particular lineages and environments, two kinds of information are always required to discover these solutions: (1) a list of the alternatives, that is, the range of variation that is compatible with the genetic and developmental system of the organism, and (2) the identity, relative importance, and mode of action of the chief agencies of selection, as well as the risks these agencies pose to the establishment and maintenance of given pathways of adaptation. In effect, we wish to know why certain types of organisms do not exist. Have the requisite genetic variants not arisen? Is a given pathway incompatible with the pattern of ontogeny? Are the ecological risks unacceptably high? How does risk vary in nature and magnitude with habitat and on a geographical scale? These are the questions we must answer if we are to understand both how pathways of adaptation are determined and what controls the extent to which adaptive improvement and escalation can occur before limitations and compromises take effect. In this chapter, I shall treat only the more general aspects and implications of these problems. Conflicts, limitations, risks, and compromises that are specific to particular groups or circumstances will be taken up in appropriate sections of subsequent chapters.

GENETIC AND DEVELOPMENTAL LIMITATION

Without heritable variation, a trait cannot change no matter how beneficial a change would be. Evolution therefore depends on heritable changes in the genetic system. Genetic variants arise as chromosomal alterations or as mutations in the DNA sequence either in the germ-line (eggs and sperm) or, in species with clonal or vegetative reproduction (most plants and fungi, and many invertebrates) in somatic cells.

Laboratory experiments have generally shown that almost any character can be selected for; that is, ample genetic variation exists in most species and populations (Lewontin, 1974; Charlesworth et al., 1982; Slatkin, 1983). It is still possible, however, that appropriate variants for particular aptations are unavailable in many situations, particularly in small populations. Even if appropriate mutations were available, the likelihood of their becoming established would be high only if they arose as dominants, that is, as mutants whose favorable effects were expressed in both the homozygous and heterozygous states (J.R.G. Turner, 1983). If mutations arose as recessives and had effects only in homozygous individuals, their benefits would accrue only to a few individuals that possessed the mutation, and their frequency would remain precariously low for many generations.

The likelihood that a favorable mutation arises in a population is enhanced by greater population size, a larger genome (more sites on the DNA molecule or on the chromosomes where change can occur), and a greater number of cells that participate in the creation of the next generation. Clonal organisms can incorporate mutations both in somatic cells and in germ-line cells (Buss, 1983a, b); they thus provide unusual opportunity for the creation and testing of genetic novelties.

It is likely that the rate of mutation is itself under genetic control, so that some parts of the genome are more susceptible to change than are others. A possible, but as yet undocumented, consequence may be that some traits are subject to more frequent genetic change than are others. The basis of such differences is poorly understood, but at present the possibility that genetic variation is directed and non-random cannot be eliminated or dismissed (Leigh, 1973; Layzer, 1980).

Although in principle a genetic change can occur anywhere in the genome and affect any character, in practice only a very small number of traits are affected. In a study of variation within and between species of the tropical American salamander genus *Bolitoglossa*, for example, Alberch (1983) showed that the prefrontal bones are the only bones of the skull that vary in terms of presence or absence. The prefrontals are typically absent in *B. occidentalis* and are occasionally missing in abnormal individuals of other species. The pattern of reduction and loss of limbs in tetrapod vertebrates follows a highly predictable course within groups. Reduction begins in the digits (fingers and toes), then affects the forelimb segments (radius-ulna and tibia-fibula), and finally proceeds to the humerus and femur. In the reduction of fingers and toes or in the phalanges of these digits (in many amphibians, dinosaurs, and horses, for example), the first or fifth toes are the first to disappear (Lande, 1978; Alberch, 1985; Alberch and Gale, 1985). The degree of limb reduction is related to the size (or the number of cells) of the limb bud; the fewer the cells, the greater is the reduction of the limb (Alberch and Gale, 1985).

Evidence from animals suggests that the most likely morphological changes are those that affect events late in ontogeny. The prefrontals of *Bolitoglossa* and the phalanges of vertebrates are late structures whose addition, reduction, or loss would not interfere with other aspects of skeletal development. Incompatibilities resulting from changes early in ontogeny may cause the embryo's death.

Some examples contradict the generalization that changes in early events are usually lethal. The change from right-handed to left-handed coiling in gastropods occurs in maternally inherited genes whose effect is noticeable even before fertilization of the egg; yet this mutation has appeared quite frequently in gastropods and has even given rise to some lineages of characteristically left-handed snails (Vermeij, 1975; Freeman and Lundelius, 1982). In plants, there may be many examples of change early in ontogeny, for development does not depend on cell movement as it does in animals (Buss, 1983a).

Even in plants, however, there is a widespread tendency for a given pathway to persist, and for variation to be concentrated in only a few of the many possible characters. Stebbins (1967) has referred to this as the principle of conservation of organization. In members of the

Asteraceae and related families (composites), for example, the number of florets per head may vary drastically, but the number of ovules per floret (one) is always constant. Innovations and adaptive changes occur in directions that are most compatible with the pattern of development. McNamara (1982) and Alberch (1983) have emphasized this same point in their studies of brachiopods and amphibians, and the high incidence of parallel evolution in closely related lineages gives ample testimony to the generality of this conclusion.

Another important consequence of developmental limitation is that, if environmental conditions cause a change in the rate of development, the morphology of the organism will change in predictable ways. Life in cold water, for example, generally slows the rate of development in salamanders and therefore either directly or indirectly results in a neotenic condition in which a large animal with juvenile characters is able to reproduce (Gould, 1977). The morphology that results from a change in the rate of development may not itself be an aptation to new conditions, but may instead be the inevitable by-product of the way the environment affects development. Two additional examples will further amplify this important point.

Consider first the case of molluscan shell allometry—the change in shape as the shell grows by the incorporation of calcium carbonate at the growing margin. The molluscan shell is essentially a tube whose diameter increases evenly from the apex (oldest part) to the growing margin. This tube is usually coiled about an axis either symmetrically (planispiral coiling) or asymmetrically (conispiral coiling). In the absence of allometry, a given point, or site of calcification, traces a logarithmic curve as the shell grows. Strict logarithmic growth implies that shape remains precisely the same (D. W. Thompson, 1942). Allometry typically results in a departure from a strict logarithmic pattern of growth. There is now good evidence that the rate of growth of the shell is an important determinant of the intensity of allometry in gastropod shells. In principle, there are many ways in which allometry can be achieved, but in practice one type, called doming, largely predominates (Gould, 1968; Vermeij, 1980c). In doming, the straight sides of the right circular cone of the strictly logarithmic conispiral shell become convex outward. A reduction in growth rate intensifies doming (Gould, 1968; Kemp and Bertness, 1984). Similar effects are widespread in other animals bearing accretionary exoskeletons. Among bivalved animals, slow-growing individuals build inflated valves, whereas rapidly growing ones produce flat valves with a smaller increment of volume for each areal unit of shell that is laid down (Coe and Fox, 1942; Rudwick, 1962; Seed, 1968; Newell and Hidu, 1982). Whether growth rate is under genetic or environmental control, the important point is that morphological responses to a change in growth rate are stereotyped, possibly reflecting insurmountable rules of development and geometry.

In a series of elegant experiments with the eastern North American hermit crab *Pagurus longicarpus*, Blackstone (1985) has uncovered several startling ways in which the morphology of the animal responds in predictable and probably inevitable directions to changes in the conditions of growth. He showed that hermit crabs that are raised in high-spired gastropod shells of the genera *Ilyanassa* and *Urosalpinx* have a relatively longer right claw and a broader cephalothorax than do crabs from the same original population that were raised in low-spired shells of the genera *Littorina* and *Neverita*. He proposes that the difference in shape comes about by the stunting of growth under cramped conditions in the high-spired shells. The intensity of allometry of claw length relative to cephalothorax length seems to be greater when growth is slower, either as the crab's age increases or when the crab lives in tight quarters.

These examples illustrate a very widespread phenomenon, that of the norm of reaction: a change in conditions causes a predictable alteration in the phenotype. Norms of reaction are the rule in the dependence of reproductive characteristics on environmentally controlled growth rates (Stearns, 1984) and of rates of activity on temperature. Genes may modify and sometimes partially compensate for deleterious effects of these dependencies (Scholander et al., 1953; Bullock, 1955; Smith-Gill and Berven, 1979), but compensation is rarely complete. Phenotypic responses to a given extrinsic change are stereotyped; they reflect developmental limitation and greatly constrain the availability of adaptive pathways by leaving open only one or two possibilities.

ECOLOGICAL LIMITATION

ENERGY AVAILABILITY AND THE SCOPE OF ADAPTATION

Even if a change in phenotype is consistent with the established course of development, it may be incompatible with an individual's surroundings. Such ecological limitation takes two forms. First, the availability of energy in some environments is so low that individuals can survive only if they have low metabolic rates. Low metabolism, in turn, is generally incompatible with traits requiring high fluxes of energy. The scope of adaptation—that is, the number and potential degree of expression of possible aptations—is thus severely constrained in such low-energy environments. The second form of limitation is imposed by hazards due to weather and enemies. An aptation to one kind of hazard may be inadmissible because the attendant risks from other environmental dangers are unacceptably high.

I believe that the scope of adaptation is greatest in environments where primary production—the construction of living cells from inorganic matter by photosynthesis and chemosynthesis—can take place continuously and at a high rate. Most primary production takes place in the photic zone, where sunlight is available for photosynthesis. Life in environments lacking primary producers depends entirely on transport of nutrients from the photic zone. The rate of renewal of organic resources in environments without primary producers thus depends on the rate of transport from the photic zone and on the rate at which these resources are exploited; it is consequently always lower than the primary productivity of the source area in the photic zone. This means that, on the average, the metabolic rates of organisms in environments lacking primary producers are lower than those of organisms in the photic zone. If this were not so, the supply of organic matter in low-energy environments would be depleted, and life could not be sustained.

A. Clarke (1983) attributes many of the properties of animals in cold polar and deep-sea waters to the need to maintain low metabolic rates in these low-energy environments. Although photosynthesis is possible during the short polar summer despite the cold, it cannot take place during the long winter when there is no sunlight. Animals in winter must therefore conserve energy in order to survive until the summer, when food once again becomes plentiful. Among the characteristics that Clarke believes are manifestations of low metabolic rates are slow body growth, the production of small numbers of large, starvation-resistant offspring, and slow locomotion. These attributes may also be characteristic of organisms in other low-energy environments, such as caves (Culver, 1982) and the upper reaches of rocky seashores (Vermeij, 1972, 1978). Janzen (1974) argues that forests on white-sand, nutrient-depleted soils support only slow-growing trees whose small leaves place severe limits on rates of photosynthesis and transpiration.

Slow growth, in turn, has important implications for individual life span and body form. Animals and plants growing slowly at low temperatures typically have longer individual life spans than do faster-growing relatives in warmer surroundings (Thorson, 1950; Smith-Gill and Berven, 1979; Lyman et al., 1981). Because growth by cell enlargement is less constrained in the cold than is growth by cell division, organisms growing at low temperatures have a higher nuclear DNA content and larger cells (Grime and Mowforth, 1982). This may also be true of other slow-growing organisms. The effect of growth rate on allometry has already been noted in Chapter 1. Slow-growing species can be expected to show more intense allometry than more rapidly growing ones.

Still another example of the effect of slow growth rate on form comes from land plants. Where growth is slow, plants generally have short internodes (stem or branch lengths between adjacent leaves), and the base of each blade tends to be narrow. The petiole of the leaf is typically short. Fast-growing plants generally have long internodes, broad to cordate (heart-shaped) blade bases, and long petioles.

Some of the adaptive limitations that are imposed by conditions inducing slow growth are particularly clear in plants (Givnish and Vermeij, 1976; Orians and Solbrig, 1977). If the availability of water or nutrients in the soil is low, transpiration of water and the upward pull of water from the soil must be curtailed; that is, leaves and other surfaces through which gas exchange takes place must be kept small and relatively impermeable. Leaves are therefore typically small and leathery or waxy and have few stomatal openings. The competitive mechanism that allows plants to shade out rivals from sunlight is greatly reduced in effectiveness in habitats that favor or induce small leaves, because the size of the shadow cast by a leaf increases directly with leaf size and would therefore be very small under a tiny leaf (Horn, 1971). Moreover, the maximum sustainable growth rate of plants is low. This limitation reduces the differential in growth rate between fast-growing and slow-growing plants and therefore substantially reduces the competitive advantage that fast-growing vines would have over slower-growing plants such as herbs, shrubs, and the seedlings of trees. Thus, even if there would be great advantages

to a plant shooting up to the canopy of a tall forest growing on nutrient-depleted soils, competition from established plants as well as
nutrient limitations may prevent vines from gaining a significant
growth advantage over other plants on the forest floor. The conspicuous absence of climbing plants in the coniferous forests of the taiga
belts of northern North America and Eurasia may reflect physiological limitation, but it does not necessarily indicate that competition
for light is an unimportant agency of selection.

This last example illustrates why curtailment of primary production limits the scope of adaptation. Where high metabolic rates can
be sustained, the potential difference between the capacities of different species can be very large, but where metabolic rates are low,
rates of activity and chemical reaction are so constrained that one
species differs by only a small degree from another. In the northern
coniferous forest, the potential difference between the annual
growth rate of a vine and that of a tree seedling is small, whereas in a
moist lowland tropical forest the difference in growth rate, and therefore in competitive ability, is very large. High metabolic rates enable
organisms to achieve a greater degree of specialization in a given direction (along a pathway of adaptation) than do low metabolic rates;
that is, there is greater opportunity for adaptive improvement and escalation in environments that permit (but do not require) high metabolic rates.

Low temperatures probably limit the scope of adaptation in several
ways other than the constraint on metabolic rate. The precipitation
of calcium carbonate in skeletons is limited in cold waters because
this mineral's solubility increases with decreasing temperature.
Many chemical reactions typically double or triple in rate for every
10° C rise in temperature. Enzymes often reduce and sometimes
eliminate this dependence on temperature over a thermal range commonly encountered by individuals of a species (Hochachka and Somero, 1973; A. Clarke, 1983), but compensation is rarely complete.
Consequently, the activities of many animals are reduced in winter,
and Arctic fishes are unable to attain the fast swimming speeds of
many tropical species (Holeton, 1974). Rapid locomotion at low temperatures is possible in some animals, but only under certain very
special circumstances. Among moths, for example, powered flight is
usually possible only when the thoracic temperature exceeds ±30°
C, but in geometrids of the genera *Operopthera* and *Alsophila* flight

has been observed at temperatures close to freezing. Low-temperature flight is made possible through a combination of factors: very low wing-loading (ratio of wing area to body mass ranges from 3.2 to 3.9 mg/cm^2); slow wing-beat frequencies (about 2 hz); and small body mass (10 mg) (Heinrich and Mommsen, 1985). Without these low-energy specializations, flight at low temperatures could not be sustained. Strong dependence of the rate of locomotion on temperature has also been well documented in burrowing pelecypods (Ansell and Trevallion, 1969; McLachlan and Young, 1982); crawling gastropods (B. W. Kent, 1983c); and running ocypodid ghost crabs, chrysomelid beetles, frogs, and lizards (Florey and Hoyle, 1976; Dreisig, 1981; K. Miller, 1982, Avery et al., 1982; Hertz et al., 1982, 1983). Muscle power is also sensitive to temperature. Rome's (1983) work on the frog *Rana pipiens* shows that the shortening velocity (rate of contraction) and the power output of the muscle increase three- to fourfold over the range of 5° to 25° C, and that isometric force (the force produced even though no work is done by a contracting muscle) increases by about 20% between 5° and 15° C and is temperature-independent between 15° and 25° C. This pattern of thermal dependence does not vary according to the temperature at which muscles have been held during an eight-month period preceding the tests. In the carp *Cyprinus carpio*, various mechanisms of acclimation reduce but do not eliminate the thermal dependence of muscular power and swimming speed (Rome et al., 1985). Sensations such as smell, touch, and hearing are also impeded at low temperatures (H. W. Campbell, 1969).

The ecological implications of these effects are profound. Locomotion permitting escape from enemies and the location of food cannot be exploited to the same high degree by ectotherms in the cold as it can be in warm weather. Skeletons built under cold conditions may be generally weaker than those formed in warm waters (Vermeij, 1978). The ability of ectotherms to subdue armored prey and to detect danger or food by long-distance olfaction or hearing may be generally impeded in the cold except under unusual and limited circumstances.

The internal production of body heat (endothermy) is an effective method for circumventing many of the limitations engendered by low temperatures, but it has its own limitations (Pough, 1980). Endothermy is energetically unfeasible for vertebrates weighing less than 3 g and for snake-like or strongly flattened vertebrates in which the ratio of surface area to volume is high. Endotherms cannot tolerate long periods of scarcity of food and water unless they have evolved mechanisms to reduce metabolic rate temporarily during hibernation, torpor, or aestivation.

The storage of energy is another way of overcoming some of the limitations in a low-energy environment. There are no obvious constraints on the energy flux in elastic structures such as ligaments and tendons. The sudden release of energy from these structures makes possible such short-term high-energy activities as jumping and high-speed prey capture (R. M. Alexander, 1968, 1981). Food stored in roots, seeds, or eggs can help sustain high growth rates despite unfavorable ambient conditions. The duration of all these processes is limited by the amount of stored energy and ultimately by the rate at which the fuel can be replaced.

At least one pattern of adaptation ostensibly violates the generalization that defenses are best developed in high-energy situations. Slow-growing land plants and marine algae living in conditions where growth is inherently slow often have high concentrations of tannins and other polyphenols, compounds that have been shown to inhibit herbivory (Janzen, 1974; McKey, 1979; Coley, 1980, 1983; Hay, 1981a, b). Together with such structural deterrents as spines and leathery textures, these compounds confer a long, functional life span to photosynthesizing tissues. Fast-growing plants usually have low concentrations of polyphenols and a higher turnover of photosynthesizing tissues. They are also, typically, more heavily attacked by herbivores. If they are chemically defended, these plants rely on substances that are easily translocated and recycled in the plant. When tissues are lost to herbivory, they can be replaced quickly. The presence of polyphenols is in many ways like the production of heavy mineralized armor in marine invertebrates. Both constrain body growth (A. R. Palmer, 1981, 1983a), although rapid growth is still possible in young stages when polyphenol concentrations and mineralization of the skeleton are slight, and both are associated with tissue durability. Neither polyphenols nor armor, however, are restricted to low-energy environments; they are often found in slow-growing species that coexist with faster-growing forms in high-energy situations, or in slow-growing mature stages of species which, as juveniles, grow very rapidly. Whereas passive defenses such as polyphenols and some kinds of armor may be the only available deterrents in low-energy environments, they as well as many additional forms of protection become available when the potential flux of energy is high.

Curtailment of primary production is not the only factor responsible for low metabolic rates in some species. Avoidance of predators or competitors may also result in low metabolic rates, especially if avoidance entails long periods of immobility. Cryptic coloration, for example, is most effective when animals remain perfectly still in the vicinity of a potential enemy. For animals that are incapable of resisting or escaping from their enemies, the times and sites of activity are greatly restricted by predators and competitors. Many marine gastropods, for example, hide from diurnally hunting predaceous fishes by occupying crevices during the day; they are active only at night or when the tide is rising or falling (Bertness et al., 1981; Garrity and Levings, 1981; Garrity, 1984). The only exceptions are some heavily armored species, such as *Thais melones*, which are always active. The hunting activities of insect-eating birds greatly limit the time that chemically undefended caterpillars can devote to chewing leaves (Heinrich, 1979; Heinrich and Collins, 1983). If predators and competitors restrict, for individuals of a given species, the time available for feeding, that species may be subject to limitations of adaptation similar to those in environments in which primary producers are lacking.

An additional consequence of decreased metabolic rates in environments lacking primary producers may be a reduction in the frequency of selection-producing encounters with other organisms. This is a very poorly studied but important aspect of evolutionary biology. In effect, I suggest that high metabolic rates increase both the scope of adaptation and the opportunity for selection.

Population Dynamics and Adaptive Limitation

An important factor that may affect the potential for individual adaptation, but that acts at the population level, is population size. In Chapter 1, I pointed out that large populations are more likely than small populations to contain individuals with a mutant that improves an individual's effectiveness in coping with hazards. A second effect of population size is more indirect. Almost every population is subjected from time to time to precipitous declines in the number of individuals as a result of storms, disease, and other unusual events. On small islands or in habitats where populations are necessarily small, such fluctuations could easily bring the population down to levels so low that recovery is unlikely before another calamity strikes. Moreover, very small population size promotes inbreeding and could lead to a loss of vigor. Low metabolic rates enable individuals to withstand periods of scarcity and to live at high densities. For any given area, therefore, a population of individuals with low metabolic rates is apt to be larger than one composed of high-energy individuals, all else being equal. Populations that are constrained by a small habitat area may thus generally consist of individuals with low metabolic rates and a narrow scope of adaptation, for populations of high-energy members are prone to rapid extinction. This effect of limited habitat area has long been recognized for freshwater animals and the land biota of islands. Ectothermic lizards and turtles on oceanic islands, for example, often assume the role of mammals (Bakker, 1975, 1980). Freshwater habitats tend to be small relative to marine ones, and most freshwater mammals are small, in striking contrast to their often very large marine counterparts, whereas freshwater reptiles such as crocodiles often attain very large sizes. Bakker (1980) attributes the persistence of large freshwater reptiles to their relatively low metabolic requirements, which permit populations to be large even in habitats of small size.

ECOLOGICAL RISKS

An individual's surroundings influence the availability of pathways of adaptation in another way: weather and enemies create conditions that make certain modes of life highly risky or completely impossible. The importance of enemies as agents that prevent adaptation in all but a few directions has been appreciated by many biologists, including Darwin, but this topic has not been dealt with systematically. In this section I shall treat only the more general aspects, but the theme of adaptive restriction by enemies is one I shall turn to repeatedly in this book. Jeffries and Lawton (1984) have independently emphasized that competition among individuals for what they call "enemy-free space" is a primary agency determining the evolution of different modes of life among coexisting species.

Interesting adaptational incompatibilities arise when selection favors traits that enhance mate recognition by making their bearers attractive or conspicuous. Flowers that depend on animal vectors for pollination are often showy or fragrant. Plants with animal-dispersed seeds attract dispersing agents by their tasty fruits or even by their foliage (Janzen, 1984). Mate attraction in animals often involves distinctive scents and visual or auditory displays. All these characteristics are at odds with aptations against enemies, for the latter may use the conspicuous features as cues for locating victims. Escape or resistance defenses may be especially important for species that evolve conspicuous mate-attracting traits. The adaptational conflict between attractiveness and inconspicuousness should be minor in species whose gametes are fertilized away from the parents or whose male gametes are transported by wind or water to the female. Moreover, there are ways of averting the adaptive incompatibility in species that do possess mate-attracting traits. Endler (1978), for example, points out that some color patterns are cryptic at a distance and conspicuous at closer range. Some animal sounds may work in the same way. Nevertheless, the adaptational incompatibility between competition for mates and inconspicuousness to enemies is probably widespread.

The evolution of many kinds of resistance requires that survivors of attacks by enemies be tolerant of injury and that, if injury is detrimental, the wounds heal quickly. To understand the conditions that are necessary for the evolution of resistance, we must determine the nature and consequences of risk of injury.

Many studies suggest that injury poses a significant risk to individuals with respect to both survival and reproduction. Loss of toes in toads and the tail in small garter snakes results in an increased probability of death (R. D. Clarke, 1972; Willis et al., 1982), probably because of a reduction in locomotor performance. A similar handicap occurs in lizards that use the tail as a balancing organ while running, but in species in which the tail does not function in this way, running speed is actually increased after tail autotomy because of the reduced body weight (Punzo, 1982; Dial and Fitzpatrick, 1984; Daniels, 1985). Tail loss diminishes a lizard's social standing, and therefore it may place the victim at a competitive and reproductive disadvantage (Fox and Rostker, 1982). Loss of the raptorial appendages of stomatopods reduces the ability of these crustaceans to acquire and defend rock cavities for shelter (Berzins and Caldwell, 1983). Reduction in individual growth rate has been documented in tail-regenerating sal-amanders (Maiorana, 1977); siphon-regenerating pelecypods (Peterson and Quammen, 1982); and spine-regenerating sea urchins (Ebert, 1968). Sexual maturation is delayed and fecundity is reduced in sal-amanders regenerating the tail (Maiorana, 1977), and fecundity is re-duced in spionid polychaetes when these worms lose and then regen-erate posterior segments of the body (Zajac, 1985). Despite these effects, autotomy of limbs and tails has evolved in many salaman-ders, lizards, amphisbaenians, snakes, and crustaceans as an effective way of fleeing from and resisting predators and perhaps thwarting competitors (Robinson et al., 1970; H. W. Greene, 1973; Vitt et al., 1974; E. N. Arnold, 1984).

Some injuries may not be detrimental. I already pointed out that running speed is actually increased in some lizards after the tail is au-totomized (Daniels, 1985). Individuals of the gastropod *Conus spon-salis* whose shells were artificially broken to resemble the way they are broken by predaceous crabs in unsuccessful attacks had the same probability of survival after six days as did uninjured controls (Zipser and Vermeij, 1980). Shell repair was effected within twelve days. No difference in compressive shell strength was detected in shells of *Lit-torina irrorata* with and without repaired injuries (Blundon and Ver-meij, 1983). Injured anchovies (genus *Stolephorus*) were actually less likely to be attacked by predaceous jacks (genus *Caranx*) than were uninjured ones, because they did not move in the characteristic way that allowed jacks to detect prey (Major, 1979). It was not determined if injured anchovies had a long-term advantage in survival, but stud-ies by Foster (1985) reveal that injured blue tang (*Acanthurus coeru-leus*) had the same low rate of mortality as did uninjured individuals, and that cleaner fishes (wrasses and gobies) may have prevented the wounds of injured fishes from becoming infected.

Some intriguing experiments suggest that injury by, or even the mere presence of, predators stimulates the production of defensive structures or substances. The rotifer *Brachionis* develops spines with a demonstrated antipredatory function in the presence of the preda-ceous rotifer *Asplanchna* (J. J. Gilbert, 1966). Similar induction of spines has been observed by Harvell (1984) in the bryozoan *Mem-branipora* when predaceous nudibranch gastropods attack the col-ony. Species of the cladoceran crustacean genus *Daphnia* form crests in the presence of predaceous insects. The crests render the potential victims more difficult for the predators to manipulate (J.W.G. Grant and Bailey, 1981; Krueger and Dodson, 1981). Some trees increase the concentration of herbivore-repelling substances in leaves and branches that are attacked by rabbits or damaged artificially (Brant, 1981; Edwards and Wratten, 1982; Tuomi et al., 1984). Injury to the meristem of plants often promotes branching or an increase in the production of flowers (Black, 1976; Simberloff et al., 1978; Inouye, 1982).

Rapid repair of deleterious injuries is important if aptations that enable individuals to resist injury-producing agencies are to evolve. Many authors have commented on the fact that lizard tails are regenerated in a matter of weeks (Congdon et al., 1974; Vitt et al., 1977; Vitt and Ballinger, 1982). In experiments with fifteen species of gastropod from rocky shores in British Columbia, A. R. Palmer (1983a) found that repair of the damaged outer shell lip is especially rapid in shells with a crossed-lamellar microstructure and relatively slow in nacreous shells. Although nacre is a little stronger in most respects than is cross-lamellar structure (Currey, 1977), it also contains a higher percentage of organic matter whose production may be slower than that of the inorganic (calcium carbonate) fraction. If Palmer's results apply to skeletons generally, they suggest that the most rapid skeletal repair is possible in tissues with the lowest organic content, and that speed of repair may, in some cases, be more important in combating injury than the innate strength of the building material.

These examples illustrate how the extent of adaptation may be limited by risks incurred during injury. If injuries are detrimental and if repair or regeneration is slow, the injury-producing agency causes so many individuals to die or falter that only a few survive to contribute to the next generation. Yet if injury is an inevitable prerequisite for the evolution of many kinds of resistance defenses, as it seems to be, selection for such resistance will be weak unless the detrimental consequences of injury can be reduced.

Forbidden Phenotypes

If disallowed variants are often eliminated by ecological agencies, environments in which such agencies play only a minor role should allow the existence of phenotypes that in most other situations are forbidden. It is instructive to examine areas of high primary productivity that, for some reason, lack the usual complement of competitors and predators, for it is in these situations that ecological limitations on form and behavior are most apt to be transcended.

It has long been known that organisms that colonize empty islands or other unoccupied habitats have often diverged from ancestors to such an extent that they transcend previously established limits of form. Myers (1960) called this extralimital evolution and gave as an example the evolution of highly unusual cyprinid fishes in Lake Lanao in the Philippines. Frazzetta (1970) documented the existence of a radically new jaw-suspension mechanism in the Bolyerinae, a small subfamily of boid snakes endemic to the Mascarene Islands in the southwestern Indian Ocean. Another very interesting example comes from the sciaenid fish *Bairdiella icistius*, which in the early 1950s was introduced from the Gulf of California into the Salton Sea of southern California (Whitney, 1961). Among the fish born during the first two years in the Salton Sea, all kinds of abnormalities survived, including blind fish and individuals with branched lateral lines, a reduced maxillary, a curved vertebral column, and an unusual number of anal-fin spines. Sciaenids generally have two (and sometimes one) spines on the anal fin, but *Bairdiella* in the Salton Sea transcended this limitation by having anywhere from one to five spines. I do not know if any of the unusual variants were genetic and if they persisted after the population reached saturation, but the important point is that transcendent morphologies arose as soon as conventional ecological limitations were removed. Ford (1965) described still another example in the British butterfly *Euphydryas aurinia*. When this species increased in numbers during the early 1920s, it exhibited an extraordinary amount of individual variation—far greater than that in the 1800s, before the population declined. By the time numbers stabilized again in 1925, most of the variants had been purged, but the phenotype of individuals was somewhat different from that before the decline. Like the other examples, this one suggests that the removal of the usual constraints allowed the population to increase and variants that ordinarily would have died to persist.

Other radical departures in adaptation are permitted in progenetic species, which reproduce as small juvenilized individuals and lack the adult characters of the ancestor. Gould (1977) points out that progenetic species are notable for their ability to exploit temporarily abundant resources by virtue of their very high fecundity. According to his interpretation, selection for survival-enhancing traits is relaxed under these conditions. Combinations of traits that were disallowed in the ancestral stock may therefore be permitted in progenetic species and perhaps conserved in their usually larger descendants.

If many phenotypes are forbidden under normally prevailing ecological conditions, theoretically independent traits often become highly correlated, and variation in these traits is constrained within well-defined limits. The resulting pattern of covariation may be so pervasive that it is often interpreted as a manifestation of developmental constraints. Students of gastropod shells, for example, have long known that many conceivable shell geometries are rarely or never encountered in nature (Raup, 1961, 1966; Vermeij, 1971; Schindel, 1986). In most environments, shell form is constrained by the requirement that successive whorls (or turns) must touch or overlap. This requirement might be interpreted as a geometrical necessity (or developmental constraint) were it not for the fact that loosely coiled shells occur from time to time. They are permitted in situations where the risk of shell breakage and of pursuit by fleet enemies is low (see Chapter 7). In this case, limits to variation are set by minimal ecological standards, not by geometrical or ontogenetic constraints.

In short, the relaxation of ecological limitations permits traits to persist that would normally be inadmissible. Relaxation occurs either when populations expand following a calamity or when individuals colonize a previously unoccupied or temporarily favorable environment. Even then, of course, permissible traits are constrained by the rules of genetics and development. After all, mutations only modify preexisting phenotypes and do not reconstruct organisms from the ground up, and many fundamental features of development are shielded from the effects of selection by being insensitive to changes over a wide range of conditions (Alberch, 1985). Pathways of adaptation are thus determined by both developmental and ecological circumstances. In combination, these constraints introduce a pervasive conservatism to the evolutionary process.

THE SPATIAL DIMENSION

The ecological limitations I have discussed in the previous section vary considerably according to habitat and geography. The spatial variation in factors that affect the type and expression of aptations must be understood if temporal trends in adaptation are to be recognized and explained.

Primary production—a factor I have identified as potentially important in determining the scope of adaptation—varies spatially according to the availability of energy and nutrients. Land plants and marine flowering plants (mangroves, salt-marsh grasses, and sea grasses) obtain most of their nutrients through roots in the soil, whereas marine and freshwater algae take up nutrients directly from the water. Accordingly, the highest primary productivity on land is expected in areas where the soil is nutrient-rich, as it is in flood plains and volcanic regions. Where the soil has been leached of minerals, as in many lowland tropical rain forests and cold-temperate coniferous woods, rates of plant growth and therefore of primary productivity are rather low (Janzen, 1974). Runoff from the land (especially if the organic components are easily degradable) and upwelling of nutrients from deep waters are the chief stimulants of high primary production in aquatic environments. Rivers discharging sediments are the chief sources of runoff in the sea. Runoff is therefore extensive on the shores of continents and large islands, whereas it is unimportant on low islands (atolls) and in waters far from land (Rex, 1973; Marsh, 1977; Birkeland, 1982). Upwelling is caused either by winds blowing surface waters away from the coast or by the diversion of water currents by coastal bulges or by submarine topographic relief. In any case, nutrient-rich bottom water is forced to the surface, where it stimulates growth of photosynthetic plankton, which in turn supports a large population of animal consumers. Upwelling is most prevalent on the western coasts of continents, especially those of North and South America, Europe, and Africa, but it is also known in the northwest Indian Ocean, southeast Australia, the north coast of South America, the southeast United States, and parts of northern Australia, New Guinea, Indonesia, and the Philippines, as well as in a narrow belt on either side of the equator in the central Pacific Ocean (Sheldon, 1981; Riggs, 1984).

Besides geographical gradients in the availability of nutrients, there are various gradients in primary productivity that arise from the differential availability of light. Increased primary productivity, and therefore a greater scope of adaptation, can be expected in the following directions: from polar (high) to tropical (low) latitudes, from high to low altitudes (mainly because of greater cloud cover at higher elevations), from deep to shallow water, from the understory to the canopy of a forest, and from caves to lit habitats. Availability of water increases from high to low levels on the seashore and from deserts to forests. Scope of adaptation should also increase from cold to warm habitats and along a vertical gradient from deep layers of the sediment to the sediment surface. This latter gradient arises chiefly because locomotion is greatly impeded within the sediment or soil, especially when compared with locomotion in the overlying water or air. Finally, the potential for adaptive specialization is greater in habitats of large size than in small island-like habitats.

The potential for selection in favor of aptations that enhance the competitive and antipredatory merits of individuals is expected to increase along the same gradients as does the scope of adaptation. As discussed earlier in this chapter, the potential for selection increases as the number of interactions between rivals increases relative to the lifetime of the average individual in a population. Where individuals are widely spaced, sedentary, and slow-growing for much or all of their lives, the competitive and predatory interactions that determine who succeeds and who does not are few in number when compared with situations where individuals live at high densities, have the potential for rapid or extensive movement, and grow rapidly.

Evidence pertaining to this conclusion comes chiefly from the incidence of unsuccessful predatory attacks during the resistance phase (Vermeij, 1982d). Frequency of tail loss and regeneration in salamanders and lizards, for example, increases from high to low altitudes (Shaffer, 1978; Ballinger, 1979) and from islands to the mainland (Rand, 1954). Increases in the incidence of repaired shell injuries in gastropods from high to low latitudes have been documented in the species *Littorina littorea, Terebra dislocata,* and *Odostomia impressa,* in the *Hastula cinerea* species complex, and among species of the families Terebridae and Thaididae (Vermeij, 1978, 1982a; Vermeij et al., 1980; W. Miller, 1983). I know of no examples of a reversed gradient.

Geographical and ecological gradients in risk have also been assessed by setting out animals of a particular type in each of several areas. Bertness and his colleagues (1981), for example, cemented empty snail shells to rocks at different levels on the seashore in cold-temperate New England and on the tropical Atlantic and Pacific coasts of Panama. They found that snails disappeared at high rates in

Pacific Panama, at lower rates in Atlantic Panama, and not at all in New England. Disappearance was correlated with the presence of porcupine fishes (*Diodon* spp.), which prey on shell-bearing animals and were observed to wrench the glued shells from the rocks. On the Pacific coast of Panama, the rate of disappearance of the cemented shells declined in an upshore direction. Tagged hermit crabs also showed higher rates of mortality on the Pacific coast than on the Atlantic coast of Panama, and they were less apt to fall victim to predators at high shore levels than lower on the shore (Bertness, 1981a, 1982). Similar studies on terrestrial ants and spiders reveal that the risk of predation on these animals is greater in the tropics than in the north temperate zone (Jeanne, 1979; Rypstra, 1984).

Assessment of risk by monitoring the fate of organisms under different conditions has also been used effectively in studies of grazing on marine plants. Hay and his colleagues (1983), for example, used pieces of sea grass attached to bits of rock at different depths as assays to study the intensity of grazing by fishes and sea urchins. Like Vine (1974) and Steneck (1983b), they found that the rate of grazing declined exponentially with water depth. The risk to algae from grazers therefore decreases sharply in deeper water.

Some ecologists have interpreted the latitudinal gradient in the expression of aptations as reflecting a latitudinal pattern in mortality. At high latitudes, weather is thought to be the chief cause of death, whereas in the tropics biological agents predominate (A. G. Fischer, 1960; Sanders, 1968; Connell, 1972; MacArthur, 1972). Selection by enemies is weak if physical agencies are the main causes of death; it is strong if enemies are the chief source of mortality. An analogous vertical gradient has been described for species on marine shores, where the chief cause of mortality changes from weather in the uppermost zones to competitors or predators lower on the beach (Connell, 1961a, b, 1972; Vermeij, 1972).

This alternative interpretation of the latitudinal gradient rests on two questionable assumptions. The first is that the strength of selection due to a given kind of hazard is proportional to the contribution of that hazard to overall mortality in the population. The second related assumption is that, if one kind of hazard is the main cause of death, selection due to other hazards is minor. As I pointed out in Chapter 1, agencies of selection and agencies of death show a different rank order. Competition and predation may be important selective agencies even for species whose members are apt to die from nonbiological causes. For many species in the upper intertidal zone and polar waters, for example, weather may predominate as a cause of death, but the overall rate of mortality is low (that is, mean expected individual life span is long); yet selection for aptations against enemies may be weak because encounters with enemies occur rarely and because the scope of adaptation in these low-energy environments is small. In contrast, either physical or biological agencies may cause a high rate of mortality among the short-lived individuals of weedy species, but survival-related aptations cannot evolve because the probability of successful encounter with hazards is low. Note, however, that the scope of adaptation of weedy species is actually quite large, for the high fecundity necessary for persistence of the species requires high energy fluxes. There is, in any case, no evidence that weedy species are relatively more numerous at high than at low latitudes. In fact, high-energy weedy species are generally rare in low-energy environments. I conclude that, although patterns of mortality may closely parallel patterns of risk along habitat or geographical gradients, they do not account for observed patterns in the expression of aptations to enemies.

Summary of the Theory of Adaptation

The following definitions and assertions summarize the theory of adaptation that I have outlined in Chapters 1 and 2.

1. An aptation is any attribute that enables an individual organism to carry out a specific function and that therefore confers a selective advantage to that individual with respect to survival or propagation in a given environment.

2. To show that a given trait functions as an aptation with respect to the survival of an individual in a given environment, we must satisfy three criteria: (A) the trait must show heritable variation; (B) the agency that is thought to impose selection must be shown to be present and to be potentially harmful to individuals; and (C) the incidence or expression of the trait must be greater in individuals that survive, prevail in, or become aware of encounters with the agency

than in individuals that either are not exposed to the agency or have been killed or adversely affected by it.

3. The inability of an individual to cope with a hazard is seen as a demonstration that the individual's traits are unsuitable. This inability is expressed either by the death of the individual or as survival at sites where the hazard does not occur.

4. The effectiveness of an individual in coping with a hazard is the probability that the individual survives or prevails when it encounters or becomes aware of the hazard. Effectiveness of reproduction is defined as the probability that an individual leaves at least one offspring in the next generation, or as the number of offspring left by the individual in the next generation.

5. Because organisms generally are not perfectly adapted to their surroundings, the potential for further adaptation is nearly always present. This potential depends on the number of individuals that are exposed to a given hazard, the effectiveness of individuals in coping with the hazard, and the probability that a given individual possesses a mutation enabling that individual to increase its effectiveness.

6. Adaptive improvement takes place when the effectiveness of individuals increases, that is, when the adaptive gap between an individual's capacity to withstand a challenge and some absolute measure of the severity of hazards narrows. Escalation occurs when the frequency and severity of hazards increase, and when aptations to these hazards become better expressed.

7. Because the potential for selection in favor of existing or enhanced aptations declines as the adaptive gap widens, adaptation to increasingly severe hazards should be less common than are ecological restriction and extinction.

8. Biological agencies—competition and predation in the broad sense—are the primary agencies of selection for the vast majority of species. Selection in favor of the acquisition and retention of resources affects all species. Most species have predators, and most predators are incapable of killing all the prey they encounter, chase, or subdue. Weather and food are of secondary selective importance.

9. The scope of adaptation (number and potential expression of possible aptations) is greatest in environments of high primary productivity and warm temperatures and least in low-energy environments where all organic matter is imported from elsewhere or where

temperatures are low. Aptations requiring a high energy flux are generally unfeasible in low-energy environments.

10. Developmental as well as ecological limitations restrict the direction of adaptation and determine which of several possible adaptive pathways is taken by a particular lineage. Adaptive change takes place most readily in situations where the normally prevailing ecological limitations are relaxed. This occurs either when populations expand following a calamity or when individuals colonize a previously unoccupied or temporarily favorable environment.

CHAPTER THREE

Hypotheses and Their Evaluation

in a Historical Science

PREDICTIONS FROM THE THEORY OF ADAPTATION

The view of adaptation that I have adopted in this book leads to a number of predictions about the course of evolution of plants and animals during the Phanerozoic eon. If biological hazards are the chief agencies of selection, and if selection at the level of the individual is more important in the history of aptations than are processes at higher levels, the following hypotheses should be borne out by empirical evidence:

HYPOTHESIS 1. Traits that enhance the competitive and antipredatory capacities of individual organisms have increased in incidence and in degree of expression over the course of time within physically similar habitats. This has occurred for either or both of two reasons, which are embodied in the second and third hypotheses.

HYPOTHESIS 2. The adaptive gap between the capacities of individuals and the environment has decreased over the course of time in physically comparable habitats; that is, more recently evolved individuals are better adapted to their biological surroundings than ancient ones were to theirs. This is the hypothesis of unilateral improvement of species relative to their surroundings.

HYPOTHESIS 3. Biological hazards due to competitors and predators have become more severe over the course of time in physically comparable habitats; that is, the capacities of enemies have increased, as have the risks they pose to potential victims. This is the hypothesis of escalation.

The first hypothesis can be evaluated by ascertaining the incidence and expression of aptations in assemblages of fossil species from successive stratigraphic horizons that represent the same habitat at different times in earth history. The requirement that the assemblages be from the same habitat is important, because the type and extent of adaptation depend strongly on habitat.

Several predictions are subsumed under the first hypothesis. They include (1) a trend toward more rapid exploitation of resources through time; (2) an increased emphasis on characteristics that enable individuals to interfere with or combat competitors; (3) a trend toward reduced detectability and greater awareness of danger in species that rely neither on escape nor on resistance for defense; (4) an increase in the speed and endurance of individuals that rely on escape as their chief defense; and (5) an increase in the development of armor, retaliatory defense, and other ways of preventing subjugation in species whose chief defense is in the resistance phase of an attack.

Evaluation of the second hypothesis requires measurements of the effectiveness of individuals at different times in earth history. For a given environmental severity, individuals belonging to more recently evolved species should have a higher probability of survival or success than individuals of more ancient species. This prediction can be evaluated only when lethal as well as nonlethal (or unsuccessful as well as successful) encounters between individuals and hazards can be recognized. A weaker test would be to compare the probable capacity of individuals with the severity of the environment, as inferred from functional analysis.

If unilateral improvement were a long-term trend, there would be a corresponding adaptive deterioration in many other species. If, for example, predators increased the efficiency of attack on their prey, the prey would become less well adapted to the predators, and the potential for selection in favor of improved antipredatory aptations would decline. I therefore expect unilateral improvement to be at most a short-term trend, whereas escalation should prevail over the long run.

Absolute measures of the capacity of hazards and the performance of individuals are needed for the strongest test of the third hypothesis. If this hypothesis of escalation is true, the capacities of predators and competitors should have increased through time, and this increase should have been compensated for by the corresponding increases in the resistance of individuals. As in Hypothesis 2, a weaker test consists of indirect estimates of the capacities of organisms from different stratigraphic horizons as inferred from morphology and other preserved evidence.

An important prediction from Hypothesis 3 is that regions or environments where the scope of adaptation and the potential for selection are reduced should be refuges for species that have been outcompeted or that can no longer cope with predators in high-energy habitats. Refuge environments include cold, deep, dark, dry, and nutritionally impoverished places, as well as habitats that provide protection against the usual enemies. Species in such environments are expected to have a more ancient character, in terms of enemy-related aptations, than organisms in contemporary habitats where chemical reactions and biological enemies are less constrained.

In a strict sense, hypotheses about improvement and escalation apply only to evolutionary changes within lineages in the same environment, that is, between ancestor and descendant. I see no reason, however, why the hypotheses should not also apply in cases of evolutionary replacement, that is, when one lineage becomes extinct and is replaced by another lineage in the same environment. Selection takes place among genetically related individuals, but the individuals that are favored must compete with, and defend themselves against, many other individuals that often belong to unrelated lineages. In fact, because the direction of phenotypic change is strongly constrained by the pattern of an individual's development, it could be argued that trends in escalation and improvement will be more evident in cases of replacement than in within-lineage evolution. This would be especially so if adaptation to an increasingly rigorous biological environment is a less common evolutionary response by a lineage than are extinction or habitat restriction (Chapter 1).

The hypotheses and predictions of this chapter are predicated on the idea that selection among individuals is the most important evolutionary process. If, however, selection among higher-level entities such as populations or species is important, the three hypotheses about improvement and escalation would be incomplete if not downright wrong, especially if such selection counteracts individual-level selection.

Several arguments have been advanced in favor of the idea that selection among individuals is generally stronger than is selection among higher-level entities. Because individuals have shorter cycle times (generation times) and are more numerous than higher-level entities, selection among them is more frequent, faster, and potentially stronger than is selection among groups (R. A. Fisher, 1958; G. C. Williams, 1966; Charlesworth et al., 1982). Moreover, selection among individuals is amplified by the process of genetic recombination, which accelerates the fixation or elimination of alleles (Stearns, 1986). Recombination does not occur at or above the level of the population, and therefore it cannot amplify selection among groups. Sober (1984) has partially countered these arguments by pointing out that most cases of selection among populations or among species arise through the cumulative effects on individual organisms. One species persists while another does not because individuals of the second species are prevented from surviving or reproducing, whereas individuals of the first species contribute to the next generation by virtue of belonging to a group that was spared the hazard that befell the second species. He would therefore argue that selection among species is the consequence of many individual-level events, and hence that selection among populations or among species cannot be thought of as rare or insignificant.

It is clear, in fact, that selective processes at the level of populations and species have played substantial historical roles (S. M. Stanley, 1975b, 1979; Gould, 1982, 1985; Vrba, 1983; Jablonski et al., 1983). If highly escalated species are especially vulnerable to extinction or are unlikely to undergo lineage splitting (speciation), then higher-level selection is generally antagonistic to individual-level selection and will tend to prevent or even reverse long-term adaptive escalation. Individuals may be very well adapted to their surroundings, but the adaptive traits will contribute little to subsequent evolution if the species bearing them or the surroundings in which the traits are adaptive disappear. Alternatively, selection among higher-level entities may accentuate and augment individual-level selection and therefore stimulate escalation. An increase in the expression of aptations should then be accompanied by an increased representation of highly escalated species in a lineage or in a given environment. This idea will be developed further in Chapter 13.

I believe that the question is not which level of selection has predominated in the course of evolution, but rather when, where, and how individual-level adaptation is influenced by selective processes at higher levels. It is the task of the historian to establish whether particular changes in adaptive traits occur within lineages or by lineage replacement, and to assess how probabilities of extinction and speciation determine and are determined by the presence of individual traits. Explanations based on different levels at which selection acts should be seen as complementary rather than as strictly alternative.

THE NATURE OF HISTORICAL INQUIRY

Hypotheses about evolutionary trends and their explanations are statements about history. They are educated generalizations (postdictions, to be exact) about ancient conditions, events, and organisms, and about their interrelationships that have not yet been discovered.

There are two general classes of hypothesis in historical science. Statements of the first type assert that there has been some specified change over the course of time. Such hypotheses can be tested statistically against the unambiguous null hypothesis that no net change has occurred. Hypotheses of this kind imply nothing about the cause of the change; in fact, random processes may result in highly ordered temporal patterns, as has been clearly demonstrated in computer simulations of the evolution of morphologically distinctive groups (Raup and Gould, 1974).

The second type of hypothesis is one that proposes a mechanism or explanation for an observed pattern. The validity of such a hypothesis is less easily tested, because conceivable explanations may be complementary to rather than incompatible with the one being tested. Only if an explanation is clearly wrong in a given case can it be rejected. One explanation may be favored over another if it is supported by a greater number of independent lines of evidence, or if the underlying theory suggests that other explanations are unlikely. Even then, we have no guarantee that the proposed explanation applies in all cases or to all intervals of time. We cannot assume without evidence that explanations are time-independent. For these reasons, I agree with J.R.G. Turner (1983) that the methodology of testing a given hypothesis against a null hypothesis or against a strict alternative hypothesis is unacceptably restrictive in a historical science like paleontology. The merits of alternative or complementary explanations must be considered, of course, but acceptance of one form of explanation rather than another is probably simplistic. Given sufficient evidence from enough cases, the preferred or most general explanation is the one that is most consistent with the largest number of independently observed facts.

Two categories of data are available for the evaluation of hypotheses about the history of life. The first is the preserved record of fossils and their enclosing sediments. This record is imperfect and selective in what it preserves. Not only are many environmental challenges impossible to recognize and measure, but many forms of aptation which affect color, behavior, or the traits of soft tissues cannot be ascertained in fossils. Some habitats are better represented in preserved sediments than are others, and in any case the available record represents only a tiny fraction of elapsed time. Spatial variation at any given time may confound the determination of temporal trends. Environments change with time at any one location, and even if they remained the same, conditions within environments vary on a small spatial scale. Temporal trends therefore cannot be taken at face value, but instead must be judged against a background of spatially variable environments and in the light of various time-dependent artifacts of preservation.

The second source of data for the evaluation of historical hypotheses is experimental biology and geology, which provide information about the responses of organisms and their environments to the action of biological and physical events and agencies. This information can be judiciously applied to fossils by invoking the principle of uniformitarianism, the assumption that processes and consequences that can be observed and studied experimentally on the earth today are similar to, and therefore models for, conditions in the past. Application of the principle of uniformitarianism is contingent upon the identification of characteristics that are diagnostic of a given process or environment, and that are preservable in the fossil record.

A discussion of methodology and of the nature of data in a historical science like paleontology or evolutionary biology is important, because the available historical record is easily misinterpreted. Biologists often distrust the fossil record, which they perceive to be hopelessly fragmentary and pervaded by poor and inconsistent preservation. Paleontologists, on the other hand, often have too much faith in the record. They infer anatomy, habits, and ecological relationships from inadequate data, and they frequently apply techniques used in the study of modern assemblages to aspects of the fossil record for which such techniques are inappropriate. An appreciation of the fossil record enables us to direct our attention toward problems that can be solved and away from those that cannot be. Perhaps even more important, such an understanding includes the identification of artifacts that may affect conclusions about evolutionary trends in organic diversity and architecture through time. By taking time-dependent artifacts of preservation into account, we can outline temporal trends in form and diversity that would be expected in the absence of any biological influence.

ARTIFACTS AND RESOLUTION IN THE FOSSIL RECORD

The fossil record preserves only a tiny fraction of elapsed time in the form of sediments. This conclusion stems from the observation that the calculated rate of sedimentation (sediment thickness per unit time) in a given stratigraphic section decreases systematically as the time interval over which sediment accumulation is measured increases (Schindel, 1980; Sadler, 1981). The probability that the stratigraphic section contains gaps, which result either from a lack of sedimentation or from erosion, therefore rises as the time interval of measurement becomes longer. In a stratigraphic section that was laid down in a known amount of time t, it is possible to measure the stratigraphic completeness at a given level of temporal resolution t' by dividing the total thickness s of sediment laid down during interval t (doubled to eliminate the effect of compaction) by the mean short-term accumulation rates s'/t of sediments over the time interval t' (Schindel, 1982). That is,

$$C = \frac{s}{t} \Big/ \frac{s'}{t'}. \tag{3.1}$$

The sedimentation rate in the numerator of equation 3.1 is determined directly from the stratigraphic section; the accumulation rate in the denominator is determined empirically from data on sedimentation in modern environments that are similar to that in which the fossil sediments accumulated. Completeness is therefore an estimate of the percentage of time during which sediments accumulated either continuously or with interruptions shorter than t' (the level of temporal resolution). The shorter the t', the smaller will be the estimate of completeness. Calculations with t' equal to 100 years show that stratigraphic sections that were chosen for their relative continuity are still filled with gaps. With very rare exceptions, such as the Neogene sections in the Turkana Basin of Kenya studied by Williamson (1981), completeness at the 100-year level of resolution is less than 10%. Even less time is represented by sediments when t' is shorter. The time that is represented by fossil organisms is still less, because many horizons, even in fossiliferous formations, are devoid of fossils (Dingus and Sadler, 1982; Schindel, 1982).

One way in which gaps in the fossil record affect evolutionary studies is the nature of sampling. No species, community, environment, or characteristic can be traced continuously through time even in a local section. Moreover, even the thinnest slabs of sediment encompass intervals of time that are, with few exceptions, longer than the lifetime of individual organisms. In other words, it is seldom possible to study species or their environments on an "ecological" time scale, the level of resolution to which we are accustomed in studies of living communities. Some temperate freshwater sediments contain varves that represent annual seasonal cycles, but even here it is difficult to study events of less than a year's duration. The problem of temporal resolution is exacerbated by reworking of sediments. Bioturbation causes sediments that are laid down at different times to be mixed into homogeneous packages in which division into time units is impossible. Some modern crustaceans are known to rework to a depth of 3 m below the sediment surface (Pemberton et al., 1976). The tendency for bioturbation to affect progressively deeper levels of sediment over the course of the Phanerozoic has resulted in a decrease in stratigraphic and temporal resolution from the Early Paleozoic to the present (Thayer, 1979, 1983; Larson and Rhoads, 1983).

The reliability of estimates of completeness depends on the applicability of modern short-term rates of sediment accumulation to similar ancient environments. This raises the question of whether sedimentation rates have remained constant over time or whether they have increased. The production of fecal pellets by animals may have increased sedimentation rates in some environments since the Cambrian (Pratt, 1982; A. G. Fischer, 1984), but research on this topic is too preliminary to permit us to draw firm conclusions.

An important way in which bioturbation affects the fossil record differentially through time is by its influence on the quality of preservation. Under normal conditions of bioturbation, the presence of oxygen permits microorganisms to degrade organic remains, so that soft-bodied organisms are rarely preserved. The greater porosity of sediments under conditions of intense reworking by physical and biological agencies allows water to penetrate deeply into the sediment. When chemically reduced sediments containing sulfides are brought to the surface as a result of reworking, the sulfides are oxidized and sulfuric acid is produced, which in turn corrodes the mineral component of skeletons. In the absence of reworking, conditions beneath the sediment surface become anaerobic and are favorable to the preservation of both mineralized and unmineralized remains (Aller, 1982). If bioturbation increased through time, the quality of preservation must have declined generally (Sepkoski, 1979; Thayer, 1983).

Although gaps usually erase or fail to record potential information, they occasionally help in the interpretation of ancient environments. Schindel (1982) has pointed out that the presence of a gap marks a change in the environment from conditions favoring sedimentation to conditions that are not conducive to the deposition of sediments. Gaps therefore record events that are likely to affect organisms. If gaps can be detected in stratigraphic sections, the frequency of at least one kind of environmental perturbation can be compared between sections representing similar habitats, or between different environments.

Despite the potentially finer stratigraphic resolution in geologically older sediments, sampling of ancient rocks is much less satisfactory than it is in younger deposits. Raup (1972, 1976a, b, 1979) has found that the volume and outcrop area of sediments decrease in an approximately exponential fashion backward through time. Several subtle but important time-dependent sampling artifacts in the number and distribution of species arise from the systematic decrease in the availability of preserved sediments of greater age.

Experience shows that the number of species in an inventory of a biota rises as the sample size of individuals increases. The number asymptotically approaches the total number of species as sampling is intensified. For more inclusive categories, like genera and families, the asymptote is reached at much smaller sample sizes. An inventory at the generic or familial level is therefore apt to provide a better estimate of diversity than is an inventory at the level of species. Raup (1979) gave several illustrations of the dependence of sampling on the choice of taxonomic level. Whereas more than 95% of living pelecypod families are known as fossils, only 86% of genera with calcitic shells and 54% of genera with aragonitic shells have a fossil record.

An important consequence of more intensive sampling in younger rocks and sediments is that rare species are more often included. If rare species differ in some respect from common ones, a subtle but important time-dependent bias would be introduced. It is well known to ecologists that large body size is usually associated with rarity. Predators that feed only on other predators and that are therefore at the top of their food chain or food web are also generally thought to be rare when compared to lower-level predators, herbivores, and plants, especially if the top predator is a metabolically active animal (Hutchinson, 1959; Bakker, 1972). On purely statistical grounds, the observed stratigraphic range of rare species is expected to be short, because rare species are recorded from one or at most a few samples (Koch, 1980; Buzas et al., 1982; Koch and Sohl, 1983). This statistical artifact does not imply, however, that rare species are in fact geologically more ephemeral than common ones. To appreciate fully the sampling artifacts that rare species introduce into temporal comparisons, we need a great deal more information on the characteristics of rare living species than we have now.

Another pattern that can be expected from the better sampling of younger rocks is an apparent increase in global diversity—the number of species, genera, and families on a worldwide scale—and an increase in the ratios of genus to family and species to genus. Moreover, because the probability of finding a representative of a group rises with better sampling, the perceived stratigraphic ranges of genera and families are also expected to increase over the course of time. Because such temporal patterns in bias would be difficult to separate from any real increases in diversity through time, it seems prudent to avoid data on diversity and data that are diversity-dependent.

Estimates of diversity are not the only attributes of assemblages that are affected by sampling. Because delicate skeletons, especially those composed of aragonite, are less preservable than sturdy ones, architectural analyses of fossil biotas are potentially affected by time-dependent factors that influence preservation. Delicate sculpture (spines, frills, and scales) may not be preserved even though the organisms bearing such sculpture do fossilize. This situation is encountered frequently in brachiopods. The effect of these artifacts can be reduced in studies of local assemblages by rejecting all assemblages in which aragonitic groups are poorly represented or in which many fossils are recoverable only as molds or casts.

Another class of time-dependent artifacts has been called the "pull of the Recent" by Raup (1979). These artifacts arise because the Recent biota is better sampled and more thoroughly understood in terms of classification than is any fossil biota. Sampling in the living biota has been so thorough that many species and groups we once thought to be extinct have been found to be still living, often in environments that are poorly represented in the fossil record. Striking examples include actinistian fishes (or coelacanths), which are known as fossils up to the Cretaceous, and limpet-like monoplacophoran molluscs, which are unknown as fossils beyond the Devonian. The calculated probability of extinction of a group should therefore fall as one approaches the Recent, because more and more groups that disappear from the fossil record will be found to be still living in various Recent refuges. The diversity (number of groups) should show a corresponding increase through time. Practices in classification also contribute to this trend. The classification of living organisms relies on traits that often cannot be observed in fossils, or that would not have been considered were it not for experience with living representatives. Fossil species belonging to groups with living representatives, therefore, may often be defined more narrowly.

The fossil record is seriously biased by the differences in the ability of various kinds of environment to be preserved. Much of the marine fossil record consists of quiet-water communities on unconsolidated bottoms of fine sand and mud. Reefs—large constructions of skeletons of colonial or aggregated organisms—are also widespread in the fossil record, but typically only the sessile organisms are well preserved in these environments. Hard-bottom communities other than reefs are poorly represented, as are communities on well-sorted sand. Evolutionary events and morphological types that occur in these environments are therefore unlikely to be observed in the fossil record.

The artifacts I have described in the preceding paragraphs are especially troublesome in studies done on a global (worldwide) scale. Such studies combine into a single datum all occurrences of a given group or architectural type, regardless of habitat or geographical provenance. Such data therefore mask shifts in habitat and preclude the possibility of recognizing restrictions or expansions in geographical and ecological distributions. Findings from unusual or rarely preserved habitats are lumped with those from more typical fossil assemblages. Perhaps more important, only one datum is available for any given slice of time, so that any temporal trend that might be exemplified by the data cannot be judged statistically against spatial variation within habitats or within regions. Global analyses can identify global events, but the latter may be difficult to distinguish from preservational biases without more detailed studies on a regional or local scale.

For these reasons, I shall emphasize comparisons between well-preserved local fossil assemblages that are temporally, geographically, and ecologically restricted. In this way, comparisons can be made (1) spatially between habitats, (2) spatially within habitats between regions, (3) temporally within a local stratigraphic sequence, and (4) temporally within habitats. The various patterns can then be compared to patterns from global compilations, and to patterns that would be expected if only the various artifacts and no net biological change were acting.

In practice, of course, the available data fall far short of this ideal. All too often, it is possible only to sketch a broad outline of the global history of a group, because assemblage-level data simply do not exist.

As is true in any science, data in historical science must be gathered systematically and not anecdotally. The literature of paleontology is replete with studies of isolated selected examples in which a phenomenon is especially common or striking. Situations in which the phenomenon is either absent or inconspicuous often remain unreported or are considered to be uninteresting. This kind of selective research and reporting leads to generalizations that are based on the accumulation of isolated examples. Almost every topic I shall treat in this book would be far better understood if data were collected systematically. Only when phenomena are assessed in all available individuals in a population or species, all available species in a community, all available assemblages in a unit of time, and so on, will it become possible to test general hypotheses and to state inductive generalizations. Arguments based on well-chosen examples may stimulate further work, but they are by themselves weak and always subject to refutation by counter-example.

FUNCTIONAL MORPHOLOGY AND ECOLOGICAL INFERENCE IN FOSSILS

The interpretation of agencies of selection and of the evolutionary responses of organisms to them is the essence of the study of adaptation through time. Inferences about the aptations of fossil organisms depend critically on a knowledge of functional morphology—the study of the correspondence between form, function, and environment—and on paleoecology—the study of processes and conditions in ancient environments. The central problem in the study of fossil organisms and environments is to identify diagnostic criteria from studies of living organisms, and then to apply these criteria to fossils. Many aptations and selective agencies are unavailable for study in the fossil record because they leave no interpretable traces or because they are rarely represented. By recognizing these limitations at the outset, we can direct our studies to aptations and processes that do leave traces and whose history can be reconstructed with some confidence. The aim of this section is to outline some agencies and aptations that are most suitable for historical analysis. Details of functional interpretation and inference are left to later chapters.

Although functional studies of living organisms can lead to the recognition of a great variety of aptations in fossil species, little can be inferred about these aptations other than their incidence and degree of expression. It is impossible to measure the effectiveness of adapted individuals or to assess whether there was improvement, escalation, or an adaptational status quo through time if one cannot observe or infer successful as well as unsuccessful encounters with a given hazard. Estimation of the potential for selection in favor of the maintenance of an aptation requires that survivors that were exposed to the hazard can be distinguished from individuals that were either unsuccessful in the encounter or not exposed to the hazard. Inferences of this kind can be drawn only for those types of resistance aptations that are favored by agencies leaving diagnostic marks on affected individuals.

For most organisms, nonfatal encounters with selective agencies are more easily recognized than are fatal encounters. In colonial marine animals that normally live attached to the sea bottom, for example, nonlethal turbulence resulting from storms can be inferred by a change in growth direction after the colony has become detached from its original position. Mortality due to turbulence, however, would be difficult to discern because dislocation of colonies could be the consequence of post-mortem transport as well as of death-inducing buffeting by storms (Glynn, 1974; Jackson, 1983). Smothering by sediments can be inferred if growth ceases temporarily at a single growth band in all parts of the colony simultaneously and then resumes. Mortal smothering is indicated when growth ceases abruptly and the outer surface of the colony remains free of encrustation or boring (Jackson, 1983). Yet because sudden cessation of growth occurs in response to many other exigencies as well, such as ionic stress and sudden cold, it cannot be used as a diagnostic criterion for the recognition of any one kind of hazard.

Competitive interactions and outcomes often can be reconstructed for fossil organisms, especially when the participants are sessile and when the interactions leave preservable traces. Jackson (1983) has summarized the criteria for the recognition of competitive encounters in clonal species (those composed of repeating units or ramets) and in aclonal forms (organisms that do not propagate vegetatively, and that consist of a single module). The best evidence for competition comes from the existence of defensive barriers between adjacent modules of a colony or between adjacent individuals, and from the presence of distortions and changes in growth that are associated with interference from competitors. Instances of overgrowth of one individual by another can be considered to represent competition if the external surface of the underlying subordinate organism can be shown to have been unbored at the time of overgrowth. Destruction of the skeletal surface of dead marine organisms by boring algae begins a few days after death, at least in Recent West Indian corals and bryozoans (Bak et al., 1977; Palumbi and Jackson, 1982). Encrustation of the dead surface by other organisms is therefore distinguishable from growth over the living surface.

Competition between mobile animals can sometimes be deduced

from injuries sustained during fights. Limb loss in some lizards and crustaceans results from fighting (Vitt et al., 1974; Scully, 1979; Berzins and Caldwell, 1983). Contests between horned ungulates can wound or even kill participants (Geist, 1966, 1967, 1978b; W. J. Hamilton et al., 1977; Morse, 1980; Clutton-Brock, 1982). Injury resulting from such encounters may be difficult to distinguish from predator-induced injury, but plausible instances of competitive fighting have been described in fossil crocodiles (Buffetaut, 1983).

With the rare exception of inferences of escape behavior in the form of preserved tracks and trails, predation in the fossil record can be studied only in the subjugation phase, and then only for those forms of subjugation that involve injury to the victim's preservable hard parts. Examples include partial predation of colonial invertebrates, drilling predation, grazing of encrusting coralline algae, infestation of wood by fungi and insects, and shell breakage.

For most preservable forms of predation, only the potential for selection in favor of the maintenance of aptations can be ascertained, because death due to predation is difficult to infer in fossil remains. Nonfatal breakage of molluscan shells, for example, is easy to recognize because it is often preserved as repaired scars. Mortal breakage is more difficult to assign to specific causes. Not only may breakage be of post-mortem origin, but the extent of damage that is necessary to kill the victim varies from species to species of prey, and it must be determined by observing living predators with each of several prey species in the laboratory. In fossils, drilling is probably the only form of predation for which estimates can be made of the effectiveness of prey during the subjugation phase. Unlike other forms of predation, drilling can be unequivocally recognized both when attempts at subjugation are successful (complete drill holes) and when attempts are unsuccessful (incomplete holes).

In summary, the application of studies of the functional morphology and ecology of living organisms to the fossil record is possible only in those instances where aptations are recognizable on the basis of diagnostic features. Where aptations and environmental hazards can be recognized, inference in fossils is usually limited to an evaluation of presence or absence and to an assessment of the incidence and degree of expression of aptations. Effectiveness of adapted individuals is measurable only for some competitive interactions and for predation by shell drillers.

Given these limitations and the many artifacts of preservation in the fossil record, it seems prudent to emphasize aptations and hazards that can be identified in fossils, and to ignore those that cannot be. This distinction explains my choice of topics as I discuss the history of life in the remainder of this book.

PART TWO

The Acquisition of Resources

CHAPTER FOUR

Mechanisms and History

of Competition

Types of Competition

The survival and reproduction of an individual organism depend on the ability of that individual to acquire and defend resources—food, shelter, mates, living space, and the like. If resources were just there for the taking, resource acquisition would be chiefly a matter of unregulated exploitation of the sort that is all too familiar in our own history. Plants would simply hold their leaves in the sunlight, and animals would do little more than open their mouths, casually pick mates, and occupy any convenient living space. Resources are, however, almost always under the direct or indirect control of other organisms. To acquire resources, an individual must always interact with other organisms. Once acquired, a resource must be defended against rivals or be made unavailable to potential usurpers. Competition between individuals is therefore of fundamental importance as a means of obtaining the necessities of life.

If adaptive improvement and escalation are dominant themes in the history of life, there should be abundant evidence that the capacities of organisms for acquiring and retaining resources have generally increased through time in specified habitats. This chapter and the next two will treat this hypothesis by outlining the functional morphology and history of traits that influence the competitive ability of individuals. The cases I discuss in detail have been chosen because the functional morphology is reasonably well understood and the traits can be traced in the fossil record. Moreover, the analysis of predatory methods of molluscivores (Chapter 6) sets the stage for a consideration of animal armor, which is treated later.

Competitive superiority is achievable in various ways, not all of which are compatible with each other. I can think of at least three general mechanisms by which individuals successfully compete for and acquire resources: (1) rapid location and incorporation of resources, so that potential competitors are prevented from acquiring them first; (2) defense of resources that have already been acquired, or appropriation of resources from another individual; and (3) stabilization or enhancement of the rate of supply of the resource.

The first pathway is referred to as either opportunism or opportunistic competition. It involves a high rate of body growth, quick location of the resource, and rapid feeding. In its extreme form, often called r-selection, this type of resource acquisition favors individuals with a short life span, high fecundity, and poorly developed means of defending acquired resources. Species that exemplify these traits are most conspicuously successful in environments where resources are temporarily plentiful but in the long run unpredictable. Periods of scarcity are usually passed in a dormant or inert state such as a seed, cyst, or egg, or sometimes in a slow-growing and more resistant stage, as is the case with many marine algae (MacArthur, 1967; R. S. Miller, 1967; Case and Gilpin, 1974; D. E. Gill, 1974). Fast growth in algae is associated with a delicate leafy or filamentous body (thallus) and a high rate of photosynthesis. Algae of this construction are grazed intensively and rapidly by herbivores, and they characterize habitats or times of the year that are free of herbivores, or that are so frequently disturbed by weather that slower-growing species fail to become established. The latter types of algal species or life stages are characterized by a turf-like or encrusting habit, or by a resistant leathery or mineralized thallus. They are more resistant to grazers and have a lower rate of photosynthesis (Lubchenco and Cubit, 1980; Slocum, 1980; Littler and Littler, 1980, 1984; Littler et al., 1983a, b; Hay, 1981a, b; Sousa et al., 1979). In molluscs, rapid growth is usually associated with a thin-walled delicate shell (Vermeij, 1980c; A. R. Palmer, 1981, 1983b) except in oysters and some other pelecypods, in which rapid growth is accompanied by a spongy thick-walled shell structure (J. G. Carter, 1980).

An unpredictable supply of nutrients may be necessary for the evolution of short-lived opportunists, but it is not the only situation favoring traits that enhance the rate of resource exploitation. If the acquisition of resources entails a high risk for the individual, selection should favor means by which resources are found and incorporated quickly. This would occur, for example, when two or more scavengers attempt to secure the same carcass, or when the pursuit or subjugation of a prey animal places a predator in an especially vulnerable position (Bakker, 1980; Sih, 1982). Rapid resource acquisition and fast growth are also highly beneficial to young animals, which because of their small size are especially susceptible to predators and larger competitors. These forms of adaptation are therefore well suited to reduce the time of greatest susceptibility to enemies. Shell-bearing gastropods provide good examples of this effect. Heavily armored tropical species tend to develop armor only in the adult stage, when growth ceases. As juveniles, gastropods typically have thin shells that grow extremely rapidly. *Strombus gibberulus* in Guam, for example, adds as much as 0.7 mm of shell per day in the spiral direction during the juvenile growth phase (Vermeij and Zipser, 1986). Such rapid growth rates cannot be sustained without a high metabolic rate. Snails, and probably many other animals as well, therefore effectively combine resistance against enemies with high rates of growth and metabolism that might have evolved as risk-minimizing methods of exploitation.

The second pathway of competition can be described as interference or confrontation. It often involves aggressive behavior, in which mechanical or chemical weapons are used by one individual to prevent another from taking a resource. More subtle forms of interference are also common. One plant may shade another's leaves from the light, and a suspension-feeding animal may create food-laden water currents that are so powerful that neighboring individuals are denied access to water-borne nutrients in the immediate vicinity (Buss, 1979). In cases like these, the distinction between exploitative and interference competition becomes a matter of semantics as well as substance. Interference and confrontation are expected to be favored if they also confer resistance against predators, and if the available resources are defensible (Geist, 1978a, b).

The third pathway of competition may be referred to as resource regulation. This usually involves provisioning or culturing food, with the result that nutrients remain available even during times of scarcity. Food hording by rodents, fungal culture by ants, nutrient storage by plants in tubers, maintenance of "gardens" of edible algae by damselfishes and limpets, and hermatypic associations between marine invertebrates and photosynthesizing algal cells are familiar examples. The supply of shelters can also be regulated and stimulated, as is well illustrated by hermit crabs that occupy, fight over, and often enlarge vacant molluscan shells (see Chapter 8). Regulation often involves mutually beneficial associations between a host and a guest organism. Such associations not only offer competitive and defensive advantages to the host, but initially they may be favored by conditions that post high risks for the guest (Vermeij, 1983a; see Chapter 5).

This scheme of classification of competitive pathways is predicated on the assumption that competitive encounters involve only two individuals. Frequently, however, competitive success is achieved when many individuals of one species overwhelm a single individual of another. Dungan (1985) has documented a nice example of this phenomenon in intertidal barnacles from the Gulf of California. The small *Chthamalus anisopoma* is able to crowd out and slow the growth of the larger *Tetraclita stalactifera* despite the latter's more rapid individual growth rate. The success of *Chthamalus* in this interaction therefore depends solely on large numbers and could not have been predicted from any morphological attribute of individual animals. Cases like this are probably very common and, as Dungan points out, seriously complicate the interpretation of competitive interactions and their morphological indicators in the fossil record.

CLONES AND COLONIES

THE ECOLOGY OF SOLITARY AND COLONIAL INVERTEBRATES

Immobile organisms—those which as adults do not move from place to place under their own power—are ideal for studying the history of competition. Not only have they left a good fossil record, but they also lend themselves well to inferences about competition through examination of growth patterns and morphology.

Jackson (1977, 1979, 1983, 1985) has classified immobile organisms according to various characteristics that strongly influence the mode of life, competitive outcome of interactions, and reproductive capacities of these organisms. The first distinction is between aclonal and clonal forms. Aclonal species are those in which each individual is genetically unique. Individuals may form aggregations in which a certain degree of cooperation is possible, but the aggregation always consists of genetically distinct individuals. Examples of aclonal animals include brachiopods, pelecypods, barnacles, stalked crinoids, vermetid gastropods, and many ascidians, corals, and polychaetes. Clonal organisms are those that propagate genetically identical entities (modules or ramets) by means of asexual reproduction. The modules may either remain in contact to form colonies or become separated. In some colonial clonal animals, the modules become specialized to carry out particular functions (feeding, reproduction, defense, and movement), so that the colony is said to be polymorphic. Modules are referred to as polyps in cnidarians and as zooids in ascidians and bryozoans. Sponges (phylum Porifera) have attributes that make them functionally colonial clonal forms, and they shall be treated as such here.

As a general rule, colonial animals prevail competitively over solitary ones in stable hard-bottom communities below the low tide line in most oceans (Jackson, 1977, 1983, 1985). The chief reason for this superiority seems to be that colonies are generally larger in size than solitary animals. They are therefore less likely to be completely overgrown, consumed by predators, or removed by a disturbance. Sponges and cnidarians are especially effective colonial competitors on light-exposed surfaces, whereas sponges and ascidians are the top competitors on shaded surfaces.

Solitary immobile animals are most abundant and often competitively dominant in the rocky intertidal zone, on small or ephemeral substrata such as shells, seaweeds, and sea grasses, and in unconsolidated bottoms (Jackson, 1977, 1983, 1985; P. D. Taylor, 1979). They also do well in areas of periodic upwelling, which greatly enriches the supply of nutrients (Birkeland, 1977; Schoener and Schoener, 1981). The predominance of solitary forms in the intertidal zone is perhaps attributable to the fact that the chief immobile animals in that environment (pelecypods, barnacles, and some ascidians) have a desiccation-resistant outer cover that is absent or poorly developed in most other colonial animals except bryozoans. The scarcity of colonial forms in the other habitats is thought by Jackson to be related to the physical instability or short functional life span of these habitats. The only feasible way to colonize such ephemeral substrates is by completing the life cycle rapidly, and by relying on dispersal stages to find and occupy newly available substrates of the same kind, where these opportunists are not overwhelmed by colonial forms. Consequently, the only species that can successfully colonize ephemeral substrates are those that produce large numbers of offspring. Moreover, reproductive opportunists can exist on ephemeral substrates because their generally minimal emphasis on defensive traits poses no great disadvantage. In studies of mortality of the Australian limpet *Patelloida mufria*, Mapstone and his colleagues (1984) showed that limpets living on the shells of living snails (the usual habitat) were much less likely to be discovered by drilling predaceous snails than were limpets that occurred on the rock beneath. I suspect that shells, grass blades, and other ephemeral or mobile substrata generally provide a haven from bottom-dwelling enemies. It is therefore perhaps not surprising that various limpets in Australia, California, Chile, temperate eastern Asia, Pacific Panama, and tropical West Africa have become specialized to live on the shells of living mobile snails (Hamai, 1937; Brewer, 1975; Vermeij, 1978; B. S. Morton, 1980b).

It is instructive to probe some of the exceptions to the generalizations presented above. Some very large fast-growing pelecypods are able to hold their own in environments where colonial animals would normally prevail. Because of their external shells, these pelecypods can be overgrown by colonial animals without ill effects. In fact, encrustation by ascidians and sponges benefits scallops (Pectinidae), spondylids, mussels (*Mytilus*), and chamids by protecting the pelecypods against predators that either cannot detect or cannot subdue encrusted prey (Bloom, 1975; Vance, 1978; Forester, 1979; Feifarek, 1985). Heavily encrusted pelecypods are common on rocky bottoms below the low water line in many temperate and tropical regions (Paine, 1976; Keough, 1984). Serpulid polychaetes also frequently remain alive as their calcareous tubes are overgrown by colonial animals. Aclonal vermetid gastropods and sabellariid polychaetes often construct persistent colony-like aggregations in which communal mucous nets for feeding are set up. Reef-like mounds made by these animals are common in warm-temperate and tropical waters (Safriel, 1966, 1974, 1975). Still another class of animals whose members are often persistent despite a solitary habit includes forms like tridacnid giant clams and some large sea anemones, which harbor photosynthesizing symbionts in their tissues. I shall treat these animals in Chapter 5.

Although encrustation of empty gastropod shells typically is by solitary animals, some shells occupied by hermit crabs are encrusted by colonial cnidarians, sponges, or bryozoans. These shells are probably retained by the hermit crabs for long periods, especially if their size keeps pace with the growing crab through shell accretion by the encrusters, so that they cannot be regarded as typical ephemeral substrates.

In spite of these interesting apparent exceptions, Jackson's claim that colonial animals out-compete solitary ones in permanent environments, whereas opportunistic solitary forms predominate as early colonizers and in situations of temporary plenty, seems to be a useful generalization. That colonial animals already had a competitive edge over solitary ones in ancient times is indicated by overgrowth patterns among the encrusters of Late Ordovician boulders (M. A. Wilson, 1985) and Silurian brachiopod valves (Liddell and Brett, 1982), as well as by the history of reefs.

THE HISTORY OF REEFS

Colonial animals have a long history probably dating back to Vendian time. Their most spectacular manifestation has been in reefs—large rigid buildups of skeletons that are raised above the sea floor and are capable of resisting turbulence. Excellent reviews of the history of reefs have been given by N. D. Newell (1971), Copper (1974), J. L. Wilson (1975), and Sheehan (1985). Instead of repeating these accounts, I shall remark only on some competition-related attributes of the reef-builders that have predominated at various times during the Phanerozoic.

The first reefs were built by Early Cambrian sponge-like colonial Archaeocyatha (Rowland, 1984). That competition was already important to these animals was demonstrated by Brasier (1976), who found abundant evidence of interference between adjacent modules (cups) in Australian archaeocyathans. Most archaeocyathan colonies were of uniserial organization (see below), in which the modules were loosely arranged. Some colonies had fused walls, but there was no rapid growth in upward and outward directions. After the Early

Cambrian, archaeocyathans ceased to be reef-builders, but they persisted in shaded cavities until their extinction at the end of the Cambrian (Debrenne et al., 1984).

Bryozoans and tabulates appeared during the earliest Ordovician, but they did not form reefs until Early Middle Ordovician time (Scrutton, 1979; Dzik, 1981b). Early bryozoan-dominated reefs were replaced in the later Middle Ordovician by reefs in which stromatoporoid sponges and tabulate and rugosan corals were more important. Evidence from patterns of overgrowth indicates that stromatoporoids were competitively superior to other colonial reef-builders during the Ordovician, Silurian, and Devonian (J. L. Wilson, 1975; Jackson, 1983). Antler-like and palmate colonies, such as those seen in some fast-growing species of the living staghorn-coral genus *Acropora*, were rare or unknown in the Paleozoic (Copper, 1974), and reefs may have extended only up to wave base instead of to the sea surface, as is usually the case today.

The Late Paleozoic marks one of the few intervals of time when some of the architects of reefs were not colonial animals. Although sponges were important in Late Carboniferous and Permian reefs, several buildups consisting of worm-like animals are known from the Early Carboniferous, and fusulinacean foraminifers and cemented articulate brachiopods contributed to the Late Carboniferous and Permian reef frameworks of North America, Europe, and southern Asia (N. D. Newell, 1971; J. L. Wilson, 1975; Burchette and Riding, 1977; Adams, 1984).

When reef-building recommenced in the Anisian stage of the Middle Triassic, reefs were constructed by the same kinds of calcareous sponges and algae as during the Permian, but the large aggregated brachiopods were extinct (Flügel and Stanley, 1984). Scleractinian corals—the primary builders of modern reefs—originated during the Anisian, but they did not become large or numerically dominant on reefs until the Norian stage of the Late Triassic. Jurassic reefs, also dominated by scleractinians, became especially large and widespread late in the period. The Cretaceous marked the second period when aclonal animals—hippuritacean (rudist) pelecypods—were the chief builders of wave-resistant frameworks (Kauffman and Sohl, 1974). After the demise of rudists at the end of the Cretaceous, aclonal animals once again became subordinate to colonial animals, sponges, and coralline algae during the Cenozoic (J. L. Wilson, 1975; Bosence, 1983).

COLONIAL INTEGRATION

Colony forms are classifiable according to the extent of the connection between modules. The simplest colony form is uniserial, in which individual modules are bounded by only one or two other modules. The modules therefore form a linear array, and they may be separated either by a space (phaceloid condition) or by a common imperforate wall (cateniform condition). Typically, the modules in uniserial colonies are large in size and small in number, and there tends to be little polymorphism. Some uniserial colonies are communicate, that is, there are pores and tubes that connect adjacent modules by extensions of tissue. In multiserial colonies, each module is surrounded by more than two other modules. The walls surrounding modules may be complete (cerioid condition) or reduced or perforated. Many scleractinian corals have large parts of the skeletal surface covered by coenosarc (interpolypoidal tissue) and are therefore highly integrated colonies (Jackson, 1985; Coates and Jackson, 1985).

Colonies with high degrees of integration (the modules communicate) generally prevail in competition with simpler colonies in permanent environments. If there is damage to part of the colony, communication among modules enables the colony to repair wounds and regenerate quickly. This is important if the colony is not to be overgrown by competitors that settle on the damaged portion (Bak et al., 1977; Jackson and Palumbi, 1979; Palumbi and Jackson, 1982; Wahle, 1983). Moreover, greater degrees of functional specialization and coordinated feeding and defense are possible in a more integrated colony.

The only groups in which the evolution of colonial integration has been traced are "corals" and bryozoans. The "coral" groups that have been studied are the orders Rugosa, Scleractinia, and Tabulata. Of these, the Rugosa and Scleractinia are true corals in that they belong to the cnidarian class Anthozoa. Many or most Tabulata, however, may be sponges whose form is strikingly convergent with that of corals (Kaźmierczak, 1984). The identities of rare Cambrian coral-like fossils, which are tubular in form and often colonial, are also uncertain (Jell, 1984). For simplicity, however, I shall refer to all of these organisms as corals.

The history of corals shows a gradual increase in the level of colonial integration through time (Coates and Oliver, 1973; Coates and Jackson, 1985). All Cambrian corals were at most cerioid in construction (Jell, 1984). Communicate phaceloid and cerioid rugosans first appeared in the Middle Ordovician and achieved numerical dominance over incommunicate forms during the Late Ordovician. The first highly integrated (astreoid and thamnasteroid) Rugosa appeared in the Early Silurian. Well-integrated Rugosa constitute 8 to 10% of Silurian and 20% of Devonian genera. They were eradicated during the crisis in the Late Devonian but reappeared in the Early Carboniferous, although they never exceeded 15% of genera during the Late Paleozoic. High levels of integration were more common in the Tabulata. The incidence of highly integrated genera rose from 30% in the Middle Ordovician to 65 to 69% during the interval from the Middle Silurian to the Middle Devonian. After the Late Devonian crisis, the incidence fluctuated between 22 and 36% until the Late Permian, after which the Tabulata ceased to be reef-builders. The Scleractinia achieved a higher incidence of well-integrated colonies (polyp walls reduced or absent and coenosteoid condition) than did either group of Paleozoic corals. From an incidence of 30 to 34% in the interval between the Middle Triassic and the Early Jurassic, well-integrated colonial scleractinians rose in frequency to 65% in the Middle Jurassic, 51% in the latest Jurassic, and up to 73% during the Late Cretaceous. The end-Cretaceous crisis reduced the relative number of highly integrated scleractinians, so that only 55% of Paleocene forms fell into this category, but the incidence soon rose to 78% and then 86% in the interval from the late Eocene to the Recent, with the highest incidence being attained during the Early Miocene (86%) (Coates and Jackson, 1985).

A literal reading of this historical record reveals a pattern of rapid increase in integration during the Ordovician, a rather high stable level during the Silurian and Devonian, a low incidence during the Late Paleozoic, and a high incidence in the post-Triassic, especially in the Cenozoic. As Coates and Jackson point out, this pattern corresponds roughly with the development of reefs. The highest incidence of well-integrated corals occurred during times of greatest reef development, whereas the lowest incidence occurred when reefs had been largely destroyed during the great crises. It must be remembered, however, that corals were not always the numerically or competitively dominant builders of reefs. Ordovician, Silurian, and Devonian reefs may have been competitively dominated by stromatoporoids (J. L. Wilson, 1975; Jackson, 1983), and some Permian reefs were evidently built by large coral-like brachiopods. In the Cretaceous, hippuritacean pelecypods were the prime reef-builders. The somewhat lower incidence of highly integrated taxa during the Paleozoic may therefore reflect the generally subordinate status of corals relative to other reef-builders at that time and does not necessarily indicate that the competitively dominant animals of Paleozoic reefs were less capable than their later coral counterparts.

The only other group of colonial immobile animals in which some aspects of the history of integration have been studied is the phylum Bryozoa. Members of the class Stenolaemata, a group that arose during the Early Ordovician and peaked in diversity in the Paleozoic and Mesozoic, showed an increase in colonial integration after the Triassic (Boardman, 1984). Post-Triassic members often developed communication pores, which permit the transfer of nutrients among zooids. These pores are located in thin parts of the feeding zooids' skeletal wall, which in other portions is strongly thickened. Paleozoic stenolaemates lacked communication pores, and before the Early Carboniferous they also lacked differentiation between thin and thick portions of the skeletal wall. Highly polymorphic colonies that are capable of cooperative feeding among zooids (F. K. McKinney, 1984) are confined to the Cheilostomata, an order of gymnolaemates, which had a Late Jurassic origin. Jawed defensive zooids (avicularia) are also confined to this group (Ryland, 1970).

Lidgard's (1985, 1986) studies of budding patterns in cheilostome bryozoans have provided further strong evidence that colonial integration has increased over time, especially during the Cenozoic. The earliest (Late Jurassic and Early Cretaceous) cheilostomes were characterized by uniserial colonies, in which zooids belonging to adjacent series had little or no communication. Multiserial colonies in which communication among zooids of adjacent series became possible evolved during the Late Early Cretaceous, and in turn gave rise to more integrated types with several styles of budding. These include zooidal budding and frontal budding. In zooidal budding, new zooids are formed by the expansion of preexisting zooids and subsequent partitioning. The process of budding takes place more or less independent of expansion of the colony as a whole. This independence of zooidal and colony growth contrasts with the situation in the more primitive cheilostomes that had intrazooidal budding. In this mode of growth, new zooids are formed by the erection of a porous wall within a zooid, followed by expansion of the bud and the formation of a new exterior wall. Growth of the colony is discontinuous and is dependent on the completion of zooids. Zooidal budding has been the dominant mode of growth in cheilostomes since the Late Eocene and is associated with competitive dominance and rapid growth. Frontal budding enables a colony to grow into a mound-like structure of many layers of zooids. Although the zooids of the inner layers become cut off from the outside, they are kept alive by exchanging nutrients with the overlying zooids through communication pores. The mound-like structure made possible by frontal budding enables colonies to grow rapidly over potential competitors, to smother organisms settling on the colony's surface, and to attain a large size. Frontal budding originated during the Late Cretaceous and represents a major advance in colonial integration among cheilostomes (Lidgard, 1985, 1986).

Among colonial invertebrates, then, integration seems generally to have increased during the Mesozoic and especially the Cenozoic. This increase accords well with the expectation that competitive ability has increased over time. Future studies should concentrate on documenting the competitive superiority of well-integrated colonies, and on making more accurate and more detailed reconstructions of the history of integration in skeletonized colonial groups.

COMPETITION IN PLANTS

The most fundamental resources for a photosynthesizing plant are water, light, and inorganic nutrients in the soil. Competition for these resources is therefore apt to be of primary importance, especially for germlings. Characteristics that enhance the uptake of water, increase the interception of incident light, and prevent competitors from gaining access to essential resources generally should be favored. There is evidence that in fact each of these characteristics has been increasingly emphasized by land plants over the course of the Phanerozoic.

The earliest well-characterized vascular land plants were rhyniophytes such as *Cooksonia*, which appeared during the Pridolian stage of the latest Silurian. This group, typified by the Siegenian (Early Devonian) *Rhynia*, was characterized by short stature, leafless stems, and dichotomous branching (that is, with the stem divided into two more or less equal branches, which then divide again, and so on). There was a small primary xylem with narrow tracheids for the conduction of water and a shallow horizontal rhizome for the uptake of nutrients and water from the soil. The Siegenian zosterophyllophyte *Gosslingia* and the Emsian trimerophyte *Psilophyton* showed the beginnings of a more familiar architecture in which the main stem gives off smaller lateral branches. The primary xylem increased in diameter relative to other stem tissues, and the water-conducting tracheids increased in diameter in several lineages (Niklas, 1984, 1985). The oldest leaves (microphylls) appeared in the Siegenian *Baragwanathia*, the oldest lycopsid, and in the contemporaneous lycopsid-like *Drepanophycus* (Chaloner and Sheerin, 1979; D. Edwards, 1980; Niklas, 1982, 1984; Knoll, 1984).

Broad leaves, tree-like stature, and true roots evolved during the Middle Devonian. Leaves with a pronounced broad blade (lamina) appeared first in the lycopsid *Archaeosigillaria* of the Givetian (Late Middle Devonian) stage. By the end of that stage, plants with large fan-shaped leaves (genus *Enigmophyton*) existed in New York State. Although some Emsian (Early Middle Devonian) trimerophytes may have reached a height of 2 m, true trees appeared only during the Givetian, shortly after the Eifelian appearance of the first progymnosperms of the order Aneurophytales. Progymnosperms not only had relatively larger parts of the stem devoted to primary xylem, but they also evolved true roots and the capacity to form secondary xylem by means of a cambium, which enabled plants to increase in girth and therefore in stature as well. During the Givetian, some lycopods (*"Lepidodendropsis"*) also evolved secondary wood and bark. By the end of the Givetian, trees 13 m in height (the progymnosperm *Eospermatopteris*) had evolved (D. Edwards, 1980; Knoll, 1984; Niklas, 1984).

Knoll's (1984) analysis of local floras shows that the rhyniophytes of the Early Devonian became extinct in most lowland environments after the ascendancy of trimerophytes in the Middle Devonian, and that the trimerophytes were in turn replaced by progymnosperms during the Late Devonian. This pattern of replacement is consistent with the hypothesis that the more recently evolved groups competitively displaced the older ones.

The evolution of the seed in certain progymnosperms during the Famennian stage of the latest Devonian (Gillespie et al., 1981) sparked a major diversification of seed plants in the Late Paleozoic. Carboniferous lowlands were dominated by seed-bearing pteridosperms, but free-sporing lycopsids and calamitalian sphenopsids were also important floral elements. As conditions became drier worldwide during the Permian, these groups were replaced in warm lowlands by gymnosperms (conifers, cycads, cycadeoids, and ginkgophytes), whose mechanically stronger stems and smaller, thicker leaves enabled these plants to out-compete the earlier, weaker-stemmed, moisture-adapted forms (Niklas et al., 1980; Tiffney, 1981; Knoll, 1984; Niklas, 1984). Seeds, which allowed fertilization to take place without the medium of water, enabled these plants to become less dependent on moisture for completing the life cycle. In the free-sporing lycopsids and sphenopsids, fertilization requires the presence of water (Knoll, 1984).

The evolution of the flowering plants (angiosperms) in the Barremian stage of the Early Cretaceous initiated another great episode of competitive displacement among plant groups. The broad leaves, water-conducting vessels (an advance over the tracheids of earlier forms), and various reproductive characteristics (see Chapter 5) enabled flowering plants to achieve high rates of water transport and photosynthesis. Until the Albian (Late Early Cretaceous), angiosperms were restricted to warm climates, but by the end of the Cenomanian (Early Late Cretaceous) they had eliminated most caytonialian and czekanowskialian pteridosperms as well as cycadeoid and ginkgophyte gymnosperms from the tropical zones. Angiosperms achieved a worldwide distribution by the end of the Cretaceous (Retallack and Dilcher, 1981).

The rise of angiosperms was important not only because it radically increased the local and global diversity of terrestrial plants, but also because the plants evolved a host of chemical defenses that previously either had not existed or were of very limited taxonomic distribution. Prominent among these substances are the ellagitannins (or hydrolyzable tannins) and alkaloids. Hydrolyzable tannins are perhaps quite specific as binding agents to proteins, especially to digestive proteins of insects, and unlike the condensed tannins they can be recycled by the plant that synthesizes them. They are especially characteristic of the dicotyledonous angiosperm subclasses Hamamaelidae, Dilleniidae, and Rosidae. Alkaloids, which before the Cretaceous had been characteristic only of the lycopsids, are especially common in magnoliid and related angiosperms, and also appear to act as deterrents to herbivorous insects. It is possible that some of these substances also prevent competitors from becoming established nearby (Swain, 1978; Knoll et al., 1979; McKey, 1979; Swain and Cooper-Driver, 1981; Zucker, 1983; Niklas et al., 1983).

Another important and unique feature of angiosperms is the endosperm, a structure formed by double fertilization that nourishes the growing embryo before and shortly after germination from the seed. Being provided with this nourishment at an early stage of growth may enable the young sporophyte to gain a competitive advantage over other germlings.

A competitive mechanism that may not have been fully developed until the evolution of the angiosperms was the liane, or climbing habit, which enables plants to reach the tree canopy (and thus full sunlight) rapidly and without the necessity of producing large, rigid, upright supporting stems. The climbing habit is very common today in warm, moist regions. At least 30% of the flora of Barro Colorado Island, Panama, consists of plants that climb by various means (twining stems, tendrils, roots, hooks), and even in cold-temperate northwestern Europe about 2% of land-plant species are climbers (Schenck, 1892; Croat and Busey, 1975). Although a few ferns (order Filicales), gymnosperms (Gnetales), and lycopsids (Selaginella) are known to climb, most living lianes are angiosperms, and ascent by tendrils or by twining stems around supports is known only in angiosperms (Darwin, 1876; Schenck, 1892). Some Late Carboniferous plants are believed to have had a scrambling habit because their stems are exceedingly slender. These include the sphenopsid Sphenophyllum (Batenburg, 1981) and stems bearing leaves of the fern form genera Neuropteris and Alethopteris (Wnuk and Pfefferkorn, 1984). No evidence of twining or tendril-bearing climbers has yet come to light from the Paleozoic. Possibly the superior water-conducting vessels of angiosperms (Carlquist, 1975; Zimmermann,

1983) have enabled some angiosperms to conduct large quantities of water rapidly, so that the difference in growth rate between highly specialized climbers with very wide vessels and slower-growing trees is very large.

To sum up the vegetative history of land plants, it may be said that the major evolutionary replacements—rhyniophytes by trimerophytes, trimerophytes by progymnosperms, and gymnosperms and pteridosperms by angiosperms—reflect competitive superiority of plants with respect to water extraction and transport and photosynthesis (Knoll, 1984). The greatest competitive improvements occurred during the Devonian and Cretaceous periods.

It is possible that these improvements in plants led to an increase in the world's primary productivity and biomass as well as to higher levels of oxygen in the atmosphere (Benton, 1979a; H. D. Holland, 1984). If so, evolutionary events in plants not only had a profound effect on herbivorous animals (see later in this chapter), but also allowed the scope of adaptation of organisms in general to increase. By increasing the rate at which potential nutrients are produced for animals, plants make high-energy activities feasible. Escalating competition among plants thus stimulated adaptive escalation among animals as well.

METABOLIC RATE AND COMPETITION

The quantity of food eaten by an individual animal depends on the amount of tissue to be nourished as well as on the metabolic rate of the body. The ability to take in and metabolize large amounts of food rapidly requires and permits high energy fluxes. Evolution of high metabolic rates can therefore be viewed as enhancing both the ability to acquire resources rapidly (exploitation) and the scope of adaptation, and consequently as reducing the risk of predation during times of activity. In the African dung beetle *Scarabaeus laevistriatus*, for example, individuals with high thoracic temperatures (up to 42° C) typically prevail over individuals with lower temperatures during skirmishes over balls of elephant dung (Heinrich and Bartholomew, 1979). Endothermy in birds and mammals is associated with parental care for the young, which enables the young to achieve a large size and exploitative competence rapidly. If the young are altricial—that is, ectothermic and incapable of feeding themselves—they are provided with food by the parents and do not have the metabolic expenses borne by precocial young, which are endothermic and capable of feeding themselves at an early stage (Ricklefs, 1979). Altricial young grow about four times faster than precocial young, and they are therefore able to pass rapidly through the stage during which risk of predation is especially high. Buss's (1979) study of Panamanian bryozoans also shows that faster feeding confers a competitive advantage. *Onychocella alula* was able to interfere with and to overgrow *Anthopora tincta* by virtue of its larger, more powerful lophophore (feeding organ) whose strong currents deny food to neighboring *A. tincta.*

A dominant theme in the history of animals should be an increase in metabolic rate. Evidence that this is so comes from several invertebrate groups and from the record of terrestrial vertebrates.

Suspension-Feeders

Nearly every phylum of animals contains representatives that glean food particles from the surrounding water medium. LaBarbera (1984) has given an excellent overview of the mechanisms by which food-laden water is drawn to the animal and particles are isolated and extracted by the animal's filter. Many suspension-feeders, including most Cnidaria and Echinodermata, rely on ambient currents to bring nutrients to them. Marine algae also depend on currents for nutrients. These organisms are therefore passive and do not expend metabolic energy to create currents that direct particles toward and through the filter. Even for those active suspension-feeders that do create such currents, proper orientation of the body so as to take full advantage of ambient currents may be an important additional way of intercepting food particles. The importance of orientation has been well demonstrated in articulate brachiopods (LaBarbera, 1977, 1978, 1981b).

LaBarbera (1984) has pointed out that, although the metabolic cost of feeding is minimal in passive suspension-feeders, there are substantial risks and limitations associated with this mode of life. Animals relying on ambient currents must occupy exposed sites well above the sea floor, for current speeds close to the sea floor are substantially reduced in accordance with Bernoulli's principle. As fluid moves over a surface, its velocity within the boundary layer close to the surface is essentially zero. Passive suspension-feeders must therefore be extremely well protected against predators; if they are not, they are constrained to live in habitats where predators are rare. The ability to create currents enables active suspension-feeders to live deep in the sediment, in sheltered crevices, and in other places where the risk of predation is reduced and water movement is limited. If risks due to predators increased over time, as the hypothesis of escalation predicts, we should expect to see an increased representation of active suspension-feeders and a tendency among passive ones to evolve autotomy, armor, toxicity, and other effective defenses against predators.

In the modern fauna, passive suspension-feeders—mainly crinoid and ophiuroid echinoderms and hydrozoan and anthozoan cnidarians—are especially prominent on reefs where there is substantial water movement. In the Paleozoic, stalked crinozoans evidently were common on soft unconsolidated bottoms as well. Cnidarians as well as crinoids are well known for their defenses. Cnidarians not only have stinging nematocysts, but frequently also long (sweeper) tentacles and skeletal antipredatory defenses. Tropical crinoids that feed while exposed during the day are generally toxic and autotomize and regenerate body parts easily (D. L. Meyer, 1985). Moreover, they are able to move to shelters and are therefore not constrained to remain on exposed sites when danger threatens, as most Paleozoic suspension-feeding echinoderms were.

Two cases of the apparent competitive supremacy of high-energy organisms among suspension-feeding animals will be examined in some detail. The first case involves stalked filter-feeding echinoderms. The most primitive echinoderms—members of the Early Cambrian class Helicoplacoidea, a group of three genera with a spindle-shaped or pear-shaped plated skeleton (theca) surrounding the soft tissues—were characterized by the presence of only three ambulacra (grooves in which food is gathered). A relatively large body thus required only a small number of feeding structures of limited extent. Derived echinoderms, which also originated during the Late Early Cambrian, have five ambulacra. In the lepidocystoids (two Early Cambrian genera), the ambulacra were extended as projections from the ventral (upward-facing) surface. These projections (brachioles) therefore represented an enlargement of the feeding machinery relative to the rest of the body. Brachioles and true arms evolved independently in most lineages of stalked echinoderms (subphylum Pelmatozoa). From the Early Ordovician to the end of the Paleozoic, members of the pelmatozoan superclasses Cystoidea and Crinoidea were common elements on reefs as well as on level bottoms. Most of these groups had large thecae, but the inadunate crinoids (Early Ordovician to Late Permian), which during the Late Permian or Early Triassic gave rise to the modern articulate crinoids, had very small thecae (Paul, 1977; Paul and Smith, 1984). Filter-feeding stalked echinoderms with large thecae thus became extinct at the end of the Permian. Paul (1977) interpreted this evolutionary pattern as reflecting a trend toward greater efficiency of feeding, but I believe it demonstrates a trend toward an increase in both the rate of feeding and the rate of metabolism. The amount of food necessary to sustain animals in the more recently evolved groups is probably greater than the amount necessary to sustain early echinoderms of similar size. That is, the newer forms have less mass to support with a larger feeding apparatus. This might, in fact, reflect reduced feeding efficiency and higher metabolic activity in later forms. Although echinoderms with large and small thecae coexisted during the Paleozoic, those with the larger thecae and presumably lower metabolic rates perished, while the lineage leading to modern crinoids, which have small thecae, survived.

The second case of competitive supremacy among suspension-feeding animals is the general replacement of articulate brachiopods by pelecypods. Although the two groups of bivalved animals arose at about the same time in the earliest Cambrian, the brachiopods were in all stages of the Paleozoic more diversified than were the pelecypods. Brachiopods were found in most marine communities, whereas most pelecypods, especially in the Early and Middle Paleozoic, were confined to near-shore waters (Steele-Petrović, 1979). The Late Permian crisis reduced the global diversity of brachiopod genera by about 73%, whereas it affected pelecypods by only 27% (Nakazawa and Runnegar, 1973; Gould and Calloway, 1980). After the Permian, the articulates diversified modestly during both the Triassic and Jurassic, but thereafter they were a small faunal component found chiefly in cold, deep-water, and shaded environments (Valentine, 1969; Gould and Calloway, 1980; Valentine and Jablonski, 1983a).

Comparative studies of feeding in brachiopods and pelecypods point to the conclusion that, although articulate brachiopods may be energetically more efficient at extracting particles from the water, they are low-energy animals whose ciliary currents in the feeding organ (lophophore) flow at low velocities, and whose orientation by means of adjustor muscles plays an important role in increasing the flow of water through the lophophore (Rudwick, 1970; Steele-Petrovič, 1975, 1979; LaBarbera, 1977, 1978, 1981b; Shumway, 1982; Thayer, 1986). Both the brachiopod lophophore and the pelecypod gill (ctenidium), which are composed of cylindrical filaments provided with lateral cilia, act as pumps; the cilia beat to produce an inhalant current, which bears potential food particles that are then transported by frontal cilia to the mouth. In pelecypods, the gill filaments are attached laterally to each other and by their tips to the mantle or foot. The filaments of articulate brachiopods are free; therefore, ambient currents or metabolic energy are needed to hold them in place against the mantle surface or body wall in order to separate the inhalant stream from the exhalant water during feeding. The interfilamental connections enable pelecypods to produce area-specific current velocities that are three to six times higher than those of brachiopods (LaBarbera, 1981b). The eulamellibranch condition, in which connections between filaments are formed by tissues and not merely by ciliary junctions, represents an extreme specialization of some pelecypods in this direction, for it enables them to draw in water through long narrow siphons and therefore to live well below the sediment surface (S. M. Stanley, 1968). If only weak currents could be produced, the resistance to flow through long siphons would be so high that it might prevent deep-burrowing pelecypods from obtaining food and oxygen from the water above the sediment surface.

In addition to having low metabolic rates, articulate brachiopods that are attached by a muscular pedicle to the substratum are incapable of locomotion, whereas pelecypods attached by a byssus are able to move slowly from place to place and to reattach if dislodged. This difference probably accounts for Thayer's (1985) finding that in Puget Sound the mussel *Modiolus* overgrows and smothers co-occurring brachiopods of the genera *Terebratalia, Laqueus,* and *Hemithiris.*

The fact that living brachiopods have low food requirements does not necessarily mean that fossil brachiopods had a similar physiology, or that pelecypods have always had higher metabolic rates than those of brachiopods. The size of the brachiopod lophophore relative to the volume of the animal might be one indication of a brachiopod's food requirement. The larger the lophophore, the greater the volume of water the animal can pump. Most living articulate brachiopods belong to the order Terebratulida (Silurian to Recent). Compared to most fossil groups, terebratulides have large lophophores. Spiriferides (Silurian to Jurassic) also had large lophophores, but orthides (Cambrian to Permian) and strophomenides (Ordovician to Permian) had relatively small lophophores and therefore presumably even lower metabolic requirements than those of terebratulides.

The case of the brachiopods and pelecypods is similar to that of the filter-feeding echinoderms in that the purportedly superior forms arose well before the subordinate types disappeared or became ecologically restricted. Steele-Petrovič (1979), in fact, suggests that the superiority of pelecypods over brachiopods became most evident at the times of crisis, especially at the end of the Permian, when populations of all kinds of animals, including those of the two groups of bivalved animals, were greatly depleted. According to her interpretation, the higher feeding and respiratory capacities of pelecypods enabled these animals to expand rapidly after the crises, and therefore to preempt territory that might otherwise have been occupied and retained by brachiopods. Moreover, the faster rate of growth of individual pelecypods shortened the stage during which they were at high risk of being preyed upon, which almost always falls most heavily on small individuals.

Gould and Calloway (1980) view the replacement of brachiopods by pelecypods quite differently. They see this as a classic example of "two ships passing in the night," that is, of processes that occur independent of the characteristics of, and interactions between, individuals belonging to the two groups. In a study of the worldwide stratigraphic occurrence of genera, they showed that when the number of brachiopod genera increased, so did the number of pelecypod genera, and when one group declined, so did the other. The only striking difference between the two groups during the Paleozoic was the much greater reduction in the number of genera of brachiopods than that of pelecypods (73% and 27% of the standing diversity, respectively) during the end-Permian crisis. Brachiopods and pelecypods initially diversified at similar rates during the Triassic (Van Valen and

Maiorana, 1985), but after the Jurassic the diversity of brachiopods remained steady or declined while that of pelecypods continued to increase. Gould and Calloway argued that the same factors controlled the generic diversity of both groups, and that the post-Paleozoic predominance of pelecypods is due chiefly to the greater susceptibility of brachiopods during the end-Permian "great dying."

To some extent, this explanation begs the question, for we are led to ask why brachiopods should have been more vulnerable to extinction. The difference in performance cited by Steele-Petrovič may account for some of the numerical contrast between the two groups, but Valentine and Jablonski (1983a) suggest further that the lack of planktonically dispersed larvae among post-Paleozoic articulate brachiopods is causally connected with poor representation of the group in the tropics, where planktonically dispersing groups (including pelecypods) achieve their highest diversity. When poorly dispersing lineages split to form daughter species, they tend to do so as a result of the extinction of geographically intermediate populations. This kind of divergence therefore does not increase the local number of species. In animals with high-dispersing capability, speciation often does result eventually in an increase in local species number. This occurs because, although isolation is still a necessary condition for lineage splitting, the isolates often reestablish contact through larval dispersal when currents or other conditions change (Valentine and Jablonski, 1982, 1983b). When species reestablish geographical overlap, some degree of ecological segregation may ensue. Of course, we do not know if Paleozoic brachiopods also lacked planktonic larvae.

An additional factor that may account for the extensive post-Permian diversification of pelecypods is the great variety of habitats occupied by this group even in the Paleozoic. Although both brachiopods and pelecypods lived on the sediment surface, partially buried in sediment, and attached to hard substrates, there are no brachiopod counterparts to infaunal pelecypods. This fact introduces a complication in comparisons of the two groups, because brachiopods and pelecypods are not ecologically equivalent. A more appropriate comparison might be between brachiopods on the one hand and epifaunal and partially buried pelecypods on the other.

It is clear from this discussion that the replacement of brachiopods by pelecypods was a complex affair. I suspect that brachiopods were generally more susceptible to extinction than were pelecypods because the latter had exploitative and antipredatory advantages due to their ability to grow, reproduce, and feed more rapidly, and to their being less affected by dislodgment. Differences in susceptibility to lineage splitting may also have played a role. Brachiopods were, of course, not eliminated completely after the Permian crisis, and there is, without doubt, substantial ecological overlap between living brachiopods and pelecypods. Like the brachiopods, some pelecypods (especially protobranchs) have become confined in the Recent fauna to deep-water, dark, and cold places where the low metabolic rates that characterize these relicts are compatible with prevailing conditions. Overall, however, the replacement of brachiopods by pelecypods exemplifies a tendency for high-energy animals to displace low-energy forms.

The case of brachiopods and pelecypods illustrates one other important point. An analysis of patterns of diversity through time, such as the one performed by Gould and Calloway, cannot yield information about differences in adaptation or habitat between the groups being compared. Explanations based exclusively on such patterns are therefore incomplete and apt to be misleading. It is wrong to conclude from the high correlation between the diversity of brachiopods and the diversity of pelecypods that representatives of these two groups did not compete. Evidence concerning interactions and the competitive and antipredatory merits of brachiopods and pelecypods must come from studies of the animals themselves and from the environments in which they lived.

HIGH BODY TEMPERATURE

The ability to maintain the body at high temperatures (30 to 42° C) regardless of surrounding temperatures has evolved many times. It is best known in birds and mammals, but it has also evolved in some insects (butterflies, beetles, bees, and dragonflies); fishes (mackerel shark and tuna); reptiles (sea turtles, pythons, and lizards); and even some araceous and nymphaeaceous flowers (Knutson, 1974; McNab, 1983). Endothermy in birds and mammals is achieved through the production of metabolic heat in specialized tissues. Many animals maintain high temperatures by relying on large body size. In these so-called inertial homeotherms, the low ratio of surface to volume prevents metabolic heat from escaping rapidly from the body. Heliophiles such as many reptiles and insects bask in the sun in order to elevate the body temperature. Many other animals exercise locomotor muscles by shivering, so that the activity temperature is raised. The warm-blooded fishes have evolved a remarkable circulatory system (rete mirabile) which keeps the heated blood far beneath the body's surface away from the cool surrounding water.

Endothermy in animals probably first evolved in the Late Paleozoic. The very large size of many Late Carboniferous flying insects strongly suggests the capacity to achieve and maintain high thoracic temperatures. Warm-bloodedness in vertebrates was achieved first by mammal-like reptiles of the order Therapsida either in the Early Late Permian or during the Early Triassic (Bakker, 1975, 1980; McNab, 1978, 1980). Once these therapsids evolved, communities of land vertebrates were always dominated by warm-blooded animals. During the Jurassic and Cretaceous, the dominant land vertebrates were dinosaurs, which were probably inertial homeotherms (Bennett and Dalzell, 1973; Feduccia, 1973; McNab and Auffenberg, 1976; Benton, 1979c; McGowan, 1979; McNab, 1983). Cenozoic land communities are generally dominated by endothermic birds and mammals.

The evolution of high-temperature physiology has numerous implications. It enabled animals to increase endurance, to process fibrous plant food rapidly, to develop rapid means of subduing prey, and to achieve rapid locomotion. Here I shall take up only the history of herbivory in tetrapod vertebrates, reserving a discussion of methods of subjugation of prey until Chapter 6 and the subject of locomotion until Chapter 12.

HISTORY OF HERBIVORY ON LAND

Evidence of herbivory appears early in the terrestrial fossil record. Lesions were present on the surfaces of some plants from the Siegenian (Early Devonian) Rhynie Chert. Because these lesions are associated with growth modifications of surrounding tissues, they are believed to represent herbivory by sap-sucking arthropods, although subsequent fungal attack is also indicated (Kevan et al., 1975). Compounds such as suberin and cutin, which originally may have protected land plants against desiccation, probably also proved useful against herbivores. Suberin, for example, occurs in the plugs of plant wounds. Lignins enabled plants to stand erect in the thin medium of air, and they also provided protection against enemies by forming a cement between the polysaccharide layers of the cell walls. Procyanidin (or condensed) tannins—highly polymerized phenolic compounds that bind tightly with proteins of all sorts—are characteristic of almost all land plants except lycopsids. They are effective against microorganisms that attack the cell wall with proteolitic, hydrolytic, and other degradative enzymes (McKey, 1979; Zucker, 1983). Lycopsids possess alkaloids that also function against microorganisms and arthropods (Swain, 1978; Swain and Cooper-Driver, 1981).

The most important change in herbivory before the evolution of herbivorous vertebrates was probably the evolution of winged insects. The first winged insects are known from the Namurian stage of the Early Late Carboniferous (F. M. Carpenter, 1976; Shear et al., 1984; Carpenter and Burnham, 1985). Among the early winged insects were forms with sucking mouthparts (orders Megasecoptera, Palaeodictyoptera, and Diaphanopterodea) or with chewing mandibles (orders Protorthoptera and Orthoptera). The first body fossils of beetles (Coleoptera of the suborder Arcostemmata) are known from the Permian (Crowson, 1981), but tunnels in wood that are attributable to beetles occur in the Westphalian stage of the Late Carboniferous (Scott and Taylor, 1983). Herbivorous true bugs (Hemiptera) appeared in the Permian.

Before the Stephanian stage of the latest Carboniferous, the only other certain herbivores were a few gastropods (Solem, 1979; Solem and Yochelson, 1979). During the Stephanian, the first herbivorous vertebrates made their appearance in lowland forest habitats in the equatorial zones of Europe and North America (Milner, 1980). The first herbivorous land vertebrates were large (3 to 4 m long) diadectid brachiosaur amphibians and edaphosaurian pelycosaur reptiles. Because they possessed short limbs and heavy-set bodies, these early herbivores were probably confined to the ground and were therefore incapable of exploiting plant food in the forest canopy. That the earliest herbivorous land vertebrates were large is perhaps attributable to the high cellulose content and resistant nature of land-plant tissues. These characteristics had evolved earlier in conjunction with both the evolution of herbivorous arthropods and the requirement to resist desiccation in air. Breakdown of resistant plant tissues by herbivores requires a long residence time in the gut, the action of symbiotic bacteria in the digestive system, or the powerful cutting and shearing action of teeth so that the food is macerated before being swallowed. Among ectothermic animals like amphibians and reptiles, breakdown of plant food is likely to be accomplished only in a large animal, whose digestive system is in absolute terms longer, and whose jaws are stronger, than those of smaller animals. Even today, there are few small ectothermic herbivorous vertebrates. The juveniles of large herbivores typically eat insects and depend on plants only later as they grow larger. The aquatic juvenile tadpole stages of certain anuran amphibians are the only small ectothermic herbivorous vertebrates, but they scrape only rather delicate algae and do not eat the resistant tissues of vascular plants (Wassersug and Hoff, 1979).

Herbivorous land vertebrates during the Early Permian composed the diadectid amphibians, caseid and captorhinid pelycosaurs, and first pareiasaur reptiles. Together with the Late Carboniferous forms, these animals and the predaceous reptiles that fed on them (sphenacodontian pelycosaurs like *Dimetrodon*) constituted what Bakker (1977) has called the first dynasty of terrestrial vertebrates. This dynasty was apparently restricted to warm climates at low latitudes. The larger members of this dynasty became extinct at the end of the Early Permian and were subsequently replaced in the Kazanian or Artinskian stage of the Permian by a latitudinally more wide-ranging assemblage that included herbivorous captorhinids, caseids, pareiasaurs, dinocephalian therapsids, and endothiodont and dicynodont anomodont therapsids. With the extinction of the pelycosaurs, the second dynasty was replaced by the third, comprising dicynodonts and pareiasaurs. This third dynasty persisted until the end of the Permian (E. C. Olson, 1975, 1976, 1983; Bakker, 1977, 1980; Cooper, 1982).

Triassic terrestrial herbivores included various families of cynodont and small bauriamorph therodontian therapsids, rhynchosaurs, trilophosaurid lepidosaurs, aetosaur thecodonts, and various dicynodonts. These animals, together with predaceous therapsids and thecodonts, composed Bakker's fourth dynasty. With the exception of tritylodontid cynodonts, which persisted into the Jurassic of South America, the major groups of Triassic herbivorous vertebrates became extinct during the Norian stage of the Late Triassic and were replaced by the fifth dynasty. This phase was dominated by prosauropod saurischian dinosaurs, which in turn gave way during the Pliensbachian stage of the Early Jurassic to a fauna dominated by saurischian and ornithischian dinosaurs (Bakker, 1977, 1980; Bonaparte, 1982; Cooper, 1982; Tucker and Benton, 1982; Benton, 1983, 1984a, 1985).

Two important changes in herbivory by vertebrates were initiated in the Late Triassic and Early Jurassic. The first was the evolution of high-browsers. Whereas herbivores of the first four dynasties were low-browsers, cropping plants near ground level, animals capable of feeding high off the ground appeared for the first time with the evolution of prosauropod dinosaurs during the Late Triassic. They were long-necked, long-limbed animals up to 10 m in length (Bakker, 1978; Galton, 1985).

The second important change was the evolution of food mastication. All Paleozoic and most Triassic herbivorous vertebrates cropped plants much as turtles do today, by cutting or slicing off pieces with sharp scissor-like teeth. Grinding between teeth was prevented because fore-aft and lateral movement of the lower jaw relative to the upper jaw was not possible. Among Triassic herbivores, only trilophodont and advanced diademodont cynodonts may have been able to masticate plant food in the mouth before swallowing (Bramble, 1978; Weishampel, 1984).

With the rise of the sixth dynasty in the Jurassic, long-necked, long-limbed dinosaurs (cetiosaurid, brachiosaurid, and diplodocid sauropod saurischians and stegosaurid ornithopod ornithischians) became established along with short-necked low-browsers such as camptosaurid and lophosaurid ornithopods and the earliest armored ankylosaurian ornithischians. Mastication by grinding was achieved independently in several ornithopod lineages (some lophodontids and heterodontosaurids) and by multituberculate mammals (Galton, 1983; Norman and Weishampel, 1985).

The high-browsing members of the sixth dynasty became extinct near the end of the Jurassic and were replaced by dinosaurs of the seventh dynasty during the Early Cretaceous. Included in their ranks were mainly low-browsing forms such as ankylosaurians, pachycephalosaurids, and food-grinding iguanodont ornithopods and, later (mid to Late Cretaceous), ceratopsians (horned dinosaurs with slicing dentition) and grinding hadrosaurids (duckbilled dinosaurs) (Krassilov, 1981; Weishampel, 1984).

The end of the Cretaceous brought the seventh dynasty to a close. The stage was then set for the eighth and final dynasty, that dominated by mammals. Although small, rodent-like multituberculates, which extend back at least to the Late Jurassic, persisted through the end-Cretaceous crisis, all large herbivorous dinosaurs became extinct at or slightly after the end of the Cretaceous.

A characteristic of most larger herbivorous mammals of the Cenozoic was their ability to break down and ferment plant tissues either in the caecum, a sac at the junction between the small and large intestine (most perissodactyls, many rodents), or in the forestomach or rumen (most artiodactyls, some primates, macropod marsupials, and tree sloths). Small herbivorous perissodactyls and artiodactyls were at first restricted to a low-fiber diet, because they were unable to process the large volume of food that is needed to extract proteins from plants rich in cellulose and fibers. As animal size increased, larger volumes could be ingested, residence time of the food became longer, and the capacity to eat foods with a higher fiber content increased. Appropriate dental changes, including the evolution of high-crowned molars with flat grinding surfaces, accompanied this increase in size. Perissodactyls, which arose in the Late Paleocene, were the first large mammals to exploit high-fiber foods on a large scale. They were the chief large herbivores in North America and Eurasia during the Eocene. The major diversification of artiodactyls, an order with Early Eocene origins, came in the Oligocene and Miocene. Although ruminant fermentation had already evolved in the Late Eocene, the great diversification of ruminant pecorans (cattle, sheep, goats, giraffes, and deer), tylopods (camels and hippopotamuses), and suid pigs began in the Early Miocene. Horses (perissodactyls of the family Equidae) apparently take the foods with the highest fiber content. Increased molar specialization is especially evident during the Oligocene and later, not just among perissodactyl and artiodactyl ungulates, but also in South American caviomorph rodents and in other indigenous South American herbivorous mammals (Bakker, 1975; Janis, 1976, 1982, 1984a, b; S. D. Webb, 1977; Marshall, 1981).

Besides mammals, herbivores during the Late Mesozoic and Cenozoic have included turtles (Chelonia); lizards (Squamata); various birds (especially species from South America, oceanic islands, and northern coniferous forests); prosobranch and pulmonate gastropods; and vast numbers of insects—beetles (Coeloptera), bugs (Hemiptera), grasshoppers and crickets (Orthoptera), caterpillars (larval Lepidoptera), and ants (Hymenoptera). Ants, lepidopterans, and wood-eating termites (Isoptera) first appeared in the Early Cretaceous (F. M. Carpenter, 1976; Carpenter and Burnham, 1985; Gall and Tiffney, 1983).

In summary, the history of herbivory on land is marked by several steps. The first herbivores were Devonian ground-dwelling arthropods. These were joined successively by winged insects (Middle Carboniferous); ectothermic, low-browsing, ground-dwelling vertebrates (Late Carboniferous); high-browsing vertebrates (Late Triassic); and high-energy vertebrates with a grinding dentition (Late Triassic and especially later in the Mesozoic and Cenozoic). Higher metabolic rates may have enabled tetrapods to process tougher plant tissues and larger volumes of food. Among living mammals, species with low-fiber diets and unspecialized dentition have markedly lower metabolic rates than those of similar-sized ungulates with grinding teeth (McNab, 1983). The Late Mesozoic and Cenozoic emphasis on dental specialization among herbivorous vertebrates therefore seems to be intimately linked with the ascendancy of endothermy and inertial homeothermy.

COMPETITION FOR MATES

In sexual organisms, reproduction depends on the ability to find mates. This requires that individuals detect members of their own species at a distance, and that members of other species are not mistakenly chosen or attracted. Long-distance communication in animals involves the sensory nervous system, whereas in plants communication is effected either haphazardly by wind or water or more reliably by animal vectors. Once a mate has been attracted, it must be retained long enough for mating to take place. In animals, this usually involves fighting or courting, whereas in plants there is usually a nutritional "reward" (nectar or pollen) for the animal intermediary.

Selection in favor of traits that enhance long-distance communication would be strong even if mate acquisition were the only benefit, but these traits also enable individuals to detect enemies. The conspicuousness that allows mates to be detected at a distance has the potential disadvantage of also allowing enemies to locate the individuals easily. Accordingly, competition for mates should be most intense in species that have well-developed senses and resistance or escape capabilities. Similar comments apply to plants that use animals as dispersal agents for their spores or seeds.

Many traits that enhance the ability of individuals to attract mates are not preserved in the fossil record. These include color, patterns of behavior, and distinctive chemical signals. The ability to see, hear, and make sounds can be inferred in fossils. The study of these important preconditions for mate recognition has only just begun, so that a synthesis of findings is premature. I shall confine my attention to a brief summary of the history of two types of structure that have been associated with competition for mates or dispersal agents: flowers and fruits in plants and horns in tetrapod land vertebrates.

FLOWERS AND FRUITS

Animals have probably acted as intermediaries in plant mating since at least the Late Carboniferous, and several large and strongly sculptured pollen types indicative of insect pollination are known from that period. Angiospermy (the condition in which the ovules or developing seeds are enclosed in an ovary) was achieved several times independently in some Permian glossopterid pteridosperms, as well as in some Mesozoic pteridosperms (orders Caytoniales, Nilssonniales, and Czekanowskiales) and cycadeoid gymnosperms (Smart and Hughes, 1973). The first evidence of insect-adapted pollen among true angiosperms dates from the Late Barremian stage of the Early Cretaceous, shortly after the origin of the group (Retallack and Dilcher, 1981). Insects specialized for pollinating flowers (especially Lepidoptera and bees of the hymenopteran suborder Apoidea) originated during the Cretaceous. The first micropterygid moths appeared during the Neocomian stage of the Early Cretaceous at about the same time that the plants they pollinate today (members of the primitive angiosperm family Winteraceae) originated (Thien et al., 1985). Highly specialized insect-pollinated angiosperm families were well established by the Late Cretaceous and remain the dominant elements of Cenozoic floras (Crepet, 1984). Bats, lemurs, birds, and several other specialized vertebrate pollinators evolved during the Cenozoic.

Dispersal of seeds is just as important as pollination. Its chief benefit may be the colonization of habitats where agents of mortality that affect the parent plant are reduced in number. There is mounting evidence that the mortality of seeds and seedlings declines with increasing distance from the parent plant (Augspurger and Kelly, 1984). It is therefore perhaps not surprising that animals have become important agents of dispersal for plants, and that many features of seeds and fruits serve to attract appropriate dispersers.

Competition for animal dispersers among plants may have had a history similar to that of competition for pollinators. In his analysis of fruits and seeds—diaspores, or dispersal units of land plants—Tiffney (1984, 1986) suggests that large animal-dispersed diaspores with a hard inner coat (sclerotesta) and a soft, often fleshy exterior (sarcotesta) were well represented among tropical medullosaceous seed ferns during the Late Carboniferous and Early Permian, and again among Cenozoic angiosperms. Most Permian and Mesozoic floras were dominated by wind-dispersed plants.

It is interesting that animal-mediated mating and dispersal have rarely evolved in the sea, despite the large number of species whose sexually reproducing stages are immobile. Perhaps the most obvious explanation for this absence is that water may be a more reliable long-distance transporter of gametes or propagules than is the wind. Because of its higher density, water can keep small organic particles in suspension for longer periods and carry them a greater distance than can air. This difference between water and air may also explain why suspension-feeding among animals is much less common on land than in the aqueous medium.

HORNS

Combat is, among animals, a very common form of competition for food, mates, and territory. Among the weapons of combat that have evolved, horns and other growths on the head are widespread in vertebrates and beetles (Geist, 1966, 1978a, b, 1983; Janis, 1976, 1982, 1984a, b).

In living mammals, horns and antlers have any of several functions. The most primitive, according to Geist (1966) and Barrette

(1977), is to throw the opponent off balance so that the enlarged tusks (incisor or canine teeth) of the opponent cannot inflict damage to the body. The horned head is moved sideways to strike the opponent's head as the two contestants stand parallel to each other, side by side. More derived uses of horns include (1) catching the opponent's head so that the horns interlock; (2) butting or pushing the opponent's head as the animals stand facing each other; (3) intimidating or displaying to the opponent, but avoiding combat; and (4) defending against predators. The locking function is promoted by a branching architecture or by a sculptured surface on the horns. Butting is associated with a massive skull, broad horns, and often a thick head shield (Schaffer, 1968). Display horns are often large, complex structures that make the animal look large. Defense against predators is enhanced by sharp horns whose tips point away from the base (Packer, 1983).

The earliest horned herbivorous tetrapods were Middle Permian anteosaurid and some tapinocephalid dinocephalian therapsids, which had thick dorsal head shields and other aptations for butting opponents (Barghusen, 1975). Three groups of Cretaceous ornithischian dinosaurs evolved structures for head combat (Hopson, 1977). The earliest were the dome-headed dinosaurs of the family Pachycephalosauridae, which ranged from the Barremian (Early Cretaceous) to the Maastrichtian (latest Cretaceous). Probably no group of land vertebrates evolved ramming devices to such a high degree. In *Pachycephalosaurus*, the ramming dome on top of the head had a bone thickness of at least 22 cm (Galton, 1971). Horns evolved in the Late Cretaceous duck-billed dinosaurs (Hadrosauria) and in the Ceratopsia (Hopson, 1975; Farlow and Dodson, 1975).

The earliest herbivorous mammals with horns were rhinoceros-like perissodactyls of the family Brontotheriidae, also known as titanotheres (S. M. Stanley, 1974; Schoch, 1983). Two additional groups evolved horns during the Oligocene in North America. These were the dicerotherine Rhinocerotidae, a group of perissodactyls, and the Protoceratidae, a family of artiodactyls belonging to the suborder Tylopoda. The protoceratids had already evolved in the Late Eocene, but they did not develop horns until the Oligocene.

A much more extensive development of horns occurred in the Late Early Miocene among members of the artiodactyl suborder Ruminantia. According to Janis's (1982) analysis, horns evolved independ-

ently in at least six families: Giraffidae, Cervidae (deer), Hoplitomer-
ycidae, Dromomerycidae, Antilocapridae (pronghorns), and Bovidae
(sheep, goats, antelopes, cattle). Although some of these families (notably the Antilocapridae) migrated to the New World, the evolution
of horned ruminants was an exclusively Old-World phenomenon in
animals weighing at least 15 to 20 kg. Forms with relatively advanced horn types that are used either in display or in head-to-head
combat are today especially characteristic of large temperate species
living in open country, whereas primitive horn types are found
chiefly in ruminants of tropical forests.

Because it is difficult to measure the effectiveness of fossil animals
in competitive combat, little can be said about whether large modern
herbivorous vertebrates are better able to fight off enemies than were
their antecedents. Nevertheless, it would appear that opportunities
for the evolution of horns and other combat-related structures on the
head were more numerous in the Cenozoic than they were earlier.
Janis (1982) has related the evolution of horns in artiodactyls to the
capability of fermenting cellulose-rich plant food in the digestive
system. Fermentation and the ability to specialize on a diet of fibrous
plant tissues imply high metabolic activity (McNab, 1983). The hypothesis that high metabolic rates are associated with the evolution
of horns and of other combat and display structures on the head is
supported by Hopson's (1977) observation that the presence of display structures implies a keen visual sense, which in turn is linked
with a large well-developed brain. Most dinosaurs had small brains
relative to those of Cenozoic mammals. This was especially true of
Jurassic sauropods, among which no head structures for combat or
display are known. The increased emphasis on display-related head
structures in Late Cenozoic mammals as compared with earlier large
herbivorous land vertebrates is therefore interpreted as a manifestation of higher activity levels in the more recently evolved forms.

Summary

Organisms require resources (food, shelter, mates, and the like) in
order to live and reproduce. Because resources are usually under the
direct or indirect control of other organisms, competition among individuals for resources is a ubiquitous agency of selection. There are

at least three adaptive pathways for achieving a competitive edge over potential rivals: (1) opportunism or exploitation—the rapid location and incorporation of resources, so that rivals are denied first access; (2) interference or confrontation—the defense of resources that have already been acquired, or the appropriation of resources from other individuals; and (3) regulation—the stabilization or enhancement of the rate of supply of resources. Opportunism is favored when the resource supply is unpredictably plentiful and the risk of predation or other harm is high. The two other pathways are apt to be followed when resources are defensible and directly under the control of organisms, and when the resulting traits also enhance antipredatory defense.

The available evidence indicates that the exploitative capacities of organisms have increased over time. In colonial corals and bryozoans, colonial integration—the degree of connection and communication among modules of a colony—has increased through time, especially during the Ordovician and again in the Late Mesozoic and Early Cenozoic. Integration enhances the growth rates, feeding rates, regenerative powers, and competitive abilities of colonies. In vascular land plants, water-conducting and photosynthetic capabilities increased over time, especially in the Devonian and Early Cretaceous. Among marine suspension-feeding animals, there has been a general trend for high-energy active filter-feeders to have replaced low-energy forms. This trend can be seen within the passively suspension-feeding echinoderms and in the replacement of articulate brachiopods by pelecypods. On land, high-energy endothermic and inertial homeothermic vertebrates replaced low-energy ectotherms during the Permian or Triassic. High-temperature physiology made possible the exploitation of resistant plant foods by land vertebrates. Dental specialization for grinding and shearing fibrous plant foods is especially marked in Late Cretaceous dinosaurs and in Oligocene to Recent mammals. That competition for mates has escalated through time is suggested by the evolution of flowers and specialized pollinators during the Cretaceous, and by the repeated evolution of horns and other head structures used by land vertebrates for display and combat. These horns are especially prominent in Cretaceous and Cenozoic tetrapods.

The supremacy of high-energy organisms reflects not only greater adaptation of individuals to competitors, but also a greater capacity to detect, escape from, and resist predators. This aspect will be treated more fully in later chapters.

The historical accounts in this chapter are generally deficient in that they are based mainly on the dates of appearance and disappearance of different adaptive types on a global scale. A deeper understanding of the history of competition requires an analysis of competition-related traits in local assemblages from comparable habitats.

CHAPTER FIVE

Safe Places, Anachronistic Aptations,

and the Recycling of Resources

THE INVASION OF SAFE PLACES
AND ITS CONSEQUENCES

An important prediction stemming from the hypothesis of escalation (Chapter 3) is that, as high-energy species with well-developed aptations against enemies evolve, low-energy species with less well-developed competitive and defensive capabilities and with a small scope of adaptation become restricted to safe places and situations, that is, to environments where enemies are less potent and encounters with enemies are infrequent. Situations that, in the beginning at least, must have served as safe places for species unable to cope adaptively with enemies on and above the sea floor include the infaunal habitat (the environment below the surface of unconsolidated sediments), the endolithic habitat (the environment within rocks and other hard substrates), the deep sea, caves and other cryptic habitats, fresh water, the dry land, and sites in or on enemy-adapted species.

The status of some safe places as museums for the adaptively obsolete may change profoundly as the resident species overcome some of the limitations inherent in a low-energy existence. The dry land, for example, was almost certainly a safe place for marine species during the mid-Paleozoic, but innovations related to air breathing, water conservation, photosynthesis, and independence from water soon enabled organisms on land to form a complex economy in which the scope of adaptation of species became very large. By Permian or Triassic time, in fact, the land was already exporting species back to the sea, so that it could hardly be called a safe place any longer. The deep sea, on the other hand, may always have served as a safe place for adaptively obsolete species, because primary productivity in this vast environment is lacking almost everywhere except at hydrothermal vents (Karl et al., 1980).

The invasion of safe places is of fundamental importance for the history of primary productivity and the recycling of resources. As organisms came to occupy safe places, resources that previously had accumulated there began to be recycled back into the economy of the biosphere. For safe places within the photic zone, recycling led to an increase in primary productivity and a rise in the level of atmospheric oxygen. As a result, escalation was stimulated not only in the safe places themselves, but also in the environments from which the original refugee species came.

In this chapter, I propose to document these assertions. I shall first consider some safe places in which escalation has been modest and then turn to those in which adaptive improvement of species has been so great that evolution in the original source environments has been influenced profoundly.

THE IDENTIFICATION OF SAFE PLACES

To show that a given habitat serves as a safe place for adaptively anachronistic species, we must demonstrate that the residents of that environment are inferior in competition or defense to species living in the biologically more demanding habitats from which they were excluded. Such a demonstration is difficult for several reasons. First, the identity and the aptations of the initial invaders of safe places are often unknown, not only because there is an absence of fossil evidence, but also because the ancestry of species in the safe places cannot be inferred with confidence. In other words, we are often uncertain about which environments acted as the source for the anachronistic species. Second, since data on the effectiveness of individuals in coping with enemies are often unavailable, it is difficult to show that the species in question are actually adaptively anachronistic. Finally, it is quite a challenge to show that adaptively obsolete species did in fact have a broader ecological distribution in the past than they do today.

The concept of a safe place must be carefully distinguished from that of a refuge. A refuge is a place to which a formerly more widely distributed species or group of species has become restricted as a result of climatic or geographical changes, but it need not be (and often is not) a haven for species that are unable to cope with enemies in the environment from which they were excluded. In the marine tropics, the Indo-West-Pacific and Eastern Pacific regions are prime examples of such refuges. Many genera of invertebrates which during the Miocene and Pliocene were widely distributed throughout the tropics have become restricted to these regions after they became locally extinct in the Atlantic (Vermeij and Petuch, 1986; Vermeij, 1986b). These regions have remained stable for so long that escalation between species and their biological surroundings has proceeded very far. Other examples of modern refuges are the marine temperate northeastern Pacific and, for land organisms, many areas of the southern hemisphere.

Deep Water

Few environments make better candidates as safe places for adaptively anachronistic species than the deep sea. Low temperatures and the absence of primary producers place strict upper limits on the rates of most biological processes and therefore on the scope of adaptation. Progressive restriction to the deep sea has been the fate of many previously more widespread architectural types. Some of these will be discussed in more detail later in this book.

Sea urchins with flexible tests, for example, were common throughout the Middle and Late Paleozoic, but since Late Jurassic time they have been restricted to deep and polar waters (see Chapter 14). Stalked crinoids became restricted to waters deeper than 100 m after the Jurassic (Meyer and Macurda, 1977). The shells of deep-sea gastropods and pelecypods are generally thin, delicate, and either smooth or very weakly sculptured, and they have a low incidence of resistence aptations against predators. Narrow and constricted apertures are rare among deep-sea gastropods, and cemented attachment and crenulated valve margins are almost unknown in deep-water pelecypods (Nicol, 1965, 1967; Vermeij, 1978). The deep sea is one of the few environments where loosely coiled gastropods persist as uncemented animals in the modern fauna (Rex and Boss, 1976). Mechanically weak umbilicate pleurotomariid gastropods, which were abundant in shallow waters during the Jurassic (Hickman, 1984), are today confined to waters below the photic zone. Pelecypods of the subclass Protobranchia (or Palaeotaxodonta) are today largely confined to deep and polar waters, but during the Early Ordovician, they were common in warm-water inshore sands and muds (Allen, 1978; Morris, 1978). Growth is slow even in shallow-water protobranchs, and all aspects of their functional morphology point to a low-energy mode of life (Allen, 1978).

For each of these examples, a case can be made that the architectural type that became restricted to deep water is competitively or defensively inferior to types that persisted or evolved subsequently in shallow water. Numerous other examples of apparent restriction to deep water exist for which there is insufficient information to permit an evaluation of the relative adaptive merits of the restricted groups. Limpet-like monoplacophoran molluscs, for example, are found today only in deep water, but during the Early and Middle Paleozoic they were common in warm shallow waters. If the inability of the living *Neopilina* to cling tightly to rocks was a trait of ancient monoplacophorans as well, the restriction of monoplacophorans to deep water is interpretable as another example of anachronistic animals being restricted to low-energy environments. Glypheoid crustaceans, which arose in the Triassic and subsequently gave rise to lobsters, crabs, and anomalans, were widespread in shallow water during the Early Mesozoic, but today they are represented by a single relict deep-water species of *Neoglyphea* (George and Main, 1968; Forest et al., 1976). Little is known about the functional morphology of glypheoids, but this group apparently lacked chelate limbs which in later decapods evolved predatory and combat functions. Anomalodesmatan pelecypods of the superfamily Pholadomyacea were prominent among deep burrowers of the Late Paleozoic and Mesozoic (Runnegar, 1974), but they are confined in the modern fauna to deep water (B. S. Morton, 1981a). I do not know if pholadomyaceans differ adaptively from other deep burrowers or from pelecypods that coexisted with them in shallow waters during the Late Paleozoic and Mesozoic. The deep-water lepadomorph barnacle *Neolepas* is in level of organization very similar to the Triassic *Eolepas* and may represent restriction (Newman, 1979). The deep-water arcoid pelecypod genus *Bathyarca* may be a member of the chiefly Mesozoic family Parallelodontidae (B. S. Morton, 1982b). Again, nothing definite can be said about the adaptive merits of parallelodontids relative to those of other Mesozoic pelecypods or to later arcoids. The same applies to the Abyssochrysidae, a family of deep-water prosobranch gastropods which appears to be related to ecologically wide-ranging Mesozoic Loxonematacea (Houbrick, 1979). If J. H. McLean (1981) is correct in interpreting the deep-water limpet-like *Neomphalus* as a late-surviving member of the Late Paleozoic and Mesozoic superfamily Euomphalacea, a group of open-coiled umbilicate sedentary forms, a plausible case could be made for deep-water restriction of a once more widespread group of animals with shell features that by modern standards are anachronistic.

This anecdotal survey underscores the need for detailed studies of functional morphology not only in deep-sea relicts, but also in their shallow-water ancestors and in those groups that were not relegated to deep waters. The available evidence points to the conclusion that deep-water environments do indeed serve as safe places for adaptively anachronistic groups, but acceptance of this conclusion must be provisional.

Analyses of the composition of fossil marine assemblages have yielded the generalization that new communities arise in near-shore environments, invade deeper offshore habitats, and are eventually restricted to the latter as the inshore environments become occupied by more recently evolved groups. The most detailed evidence comes from the Ordovician (Sepkoski and Sheehan, 1983) and the Cretaceous (Jablonski and Bottjer, 1983), but the pattern is believed to hold in other geological periods as well (Jablonski et al., 1983). Not all groups or communities became restricted in this way, however. Although Late Paleozoic trilobites were greatly reduced in diversity relative to the earlier Paleozoic, they generally did not live in deep offshore environments, but instead occupied shallow inshore habitats (Owens, 1983).

I can find no compelling example of the re-invasion of shallow waters in the photic zone by groups whose ancestors had previously become restricted to the deep sea. Some deep-sea groups are represented by shallow-water species at high latitudes, but generally they have not penetrated into the tropics. The lack of export of species from the deep sea is one more indication that escalation in this environment has not proceeded as far as it has in the photic zone.

In summary, the deep sea is apparently a refuge for many groups of animals that in more ancient times were common in shallow water. The persistence of these groups can be attributed to low temperatures, low food supplies, and low rates of biological activity, as well as to the fact that the deep sea has been far removed from the environments in which the ravages of the mass extinctions were most evident. Among the problems that still need to be solved are the timing of colonization of the deep sea by the refugee groups, the effect of newcomers on the existing deep-sea biota, and the extent of escalation in the deep sea.

CAVES AND INTERSTITIAL ENVIRONMENTS

Like deep waters, caves and other cryptic environments lack primary producers and usually receive low nutrient supplies from the photic zone (Buss and Jackson, 1981; Culver, 1982). Terrestrial, freshwater, and marine animals that are found only in caves or interstitially between sand grains have long been recognized to have a decidedly ancient cast.

Phyletically diverse sponges with a rigid skeleton of silica or calcium carbonate were widespread on open surfaces during the Paleozoic and Mesozoic, but today they are confined to the shaded surfaces of caves and to waters below the photic zone (R.E.H. Reid, 1968; Jackson et al., 1971; Hartman, 1979; Vacelet, 1979, 1981). In the case of *Acanthochaetetes*, this habitat restriction had already taken place by the Albian stage of the Late Early Cretaceous (Reitner and Engeser, 1983). Sponges with rigid skeletons apparently grow slowly and would therefore have been out-competed and overgrown by flexible sponges and hermatypic corals.

Caves are among the few places where articulate brachiopods are still common. Genera such as *Argyrotheca* and *Thecidellina* are abundant on the shaded undersurfaces of plate-like corals in Caribbean and Indo-West-Pacific reefs, but articulates are absent from well-lit open surfaces (Jackson et al., 1971). During the Paleozoic and Triassic, articulates were abundant in all kinds of marine habitat. During the Jurassic, Cretaceous, and Cenozoic, they became ecologically restricted to chalky bottoms and sponge reefs, environments where sedimentation rates are low and the sea bed hardens to a rocklike consistency shortly after deposition of sediment (Fürsich, 1977;

Surlyk and Johansen, 1984). Articulates are low-energy animals that are generally inferior competitors to pelecypods (see Chapter 4).

T. J. Palmer (1982) has documented a post-Triassic restriction of the bryozoan *Stomatopora* and various serpulid polychaetes to cryptic surfaces. These suspension-feeders are easily overgrown by colonial encrusters (Jackson, 1983).

Various primitive Crustacea, some of which have a fossil record extending back to marine deposits of the Late Paleozoic, are today found only in interstitial ground waters and caves in restricted geographical regions of the southern hemisphere (Schram, 1977). These include bathynellacean syncarids and members of the recently discovered class Remipedia (Iliffe et al., 1983, 1984). The reduced eyes, locomotor capacities, and skeletons of these and other cave dwellers strongly suggest that the predators usually confronted by surface dwellers are absent or unimportant in caves (Culver, 1982). Accordingly, caves are ideal places for the persistence of anachronistic species.

Caves probably have not exported species back to lit environments in any appreciable numbers. The types of specialization that characterize most cave dwellers are mainly reductions and losses rather than new features that could be the foundation for adaptive breakthroughs. In this as well as other ways, cryptic environments are similar to the deep sea.

THE INFAUNAL ENVIRONMENT

MECHANISMS AND CONSEQUENCES OF BIOTURBATION

Organisms probably lived on the surface of sediments long before they took up life under the sediment surface—a mode of life that is called infaunal. Early in the Phanerozoic, the infaunal environment must have provided a haven from many kinds of predation. Experiments show that, whereas predators greatly reduce populations at or near the sediment surface, they are relatively ineffective in reducing prey numbers at greater depths in the sediment (Reise, 1978; Woodin, 1978; Blundon and Kennedy, 1982b). This effect is especially marked when the roots of grasses, the tubes of polychaetes, and other obstructions prevent predators from burrowing or excavating for prey (Woodin, 1978; Peterson, 1979, 1982; Brenchley, 1982).

Bioturbation—the disturbance of unconsolidated sediments by organisms—has been very thoroughly studied by Thayer (1979, 1983), whose magnificent work serves as the chief source for the following discussion. Bioturbation has had profound effect not only on infaunal organisms themselves, but also on organisms living at and above the sediment surface.

Animals disturb sediments in at least three ways. The first is through resuspension of fine sediment. Deposit-feeding animals ingest sediment, process it in the gut where food is extracted, and void the rest. Egestion often results in the re-suspension of fine particles of sediment in the immediately overlying water (Rhoads, 1970). This sediment may clog the filtering apparatus of suspension-feeders. Accordingly, the presence of deposit-feeders in fine sediments often results in the elimination of other animals, especially suspension-feeders. Rhoads and Young (1970) have termed this effect *amensalism*.

The second method is movement of sediment from deep to shallow levels by conveyor-belt deposit-feeders. Sediment is taken in deep beneath the surface, processed, and then egested near the surface. This mode of feeding results in sedimentary instability and may lead to the burial of organisms that rest on the sediment surface. Transport of surface sediment to greater depth also occurs and has a similar effect.

The third mechanism, which may be the most important of the three in marine communities on sand, has been termed *bulldozing* by Thayer (1979, 1983). Bulldozing is the pushing aside of sediment with the animal's body, or the manipulation of packages of sediment by appendages. It is done either by deposit-feeders as they process sediment, or by predators and other animals as they move in search of food. Rapid burial as a means of escape from enemies is probably still another form of bulldozing. The sedimentary instability that results from bulldozing causes organisms that lie passively on the surface to sink. This effect is demonstrable even for large oysters (La-Barbera, 1981a), but it is probably especially significant for newly settled larvae (Peterson, 1979).

Perhaps the most important effect of bioturbation is on the porosity and aeration of sediments. In the absence of bioturbation, the depth in the sediment below which oxygen is absent and at which reducing conditions therefore prevail lies only a few millimeters below the sediment surface. This is the case, for example, during the winter months on the east coast of the United States, when water temperatures are barely above freezing. When sediment-disturbing animals are active, oxygen reaches much deeper into the sediment. In Long Island Sound during the summer months, when sea-water temperatures approach 20° C, the top 10 cm of sediment are well aerated because of the activities of deposit-feeders and "bulldozers" (Yingst and Rhoads, 1980; Rhoads and Boyer, 1982). Bacterial growth is apparently stimulated by the bioturbation-induced porosity of the sediment, and it in turn sustains a greater biomass of animals (Yingst and Rhoads, 1980). Aeration in burrows, tunnels, and chambers made by callianassid thalassinidean crustaceans allows these animals to culture bacteria on pieces of dead sea grass and other detritus (Ott et al., 1976; Seilacher, 1977). Such "gardening" further enhances the nutrients that are available to other infaunal animals. Experiments in which fiddler crabs of the genus *Uca* were removed from tracts of salt marsh in Rhode Island show that the above-ground productivity of the grass *Spartina* was reduced by some 45% relative to that of control plots where these crustaceans, which burrow to a depth of 30 cm, had not been removed (Bertness, 1985).

Many animals make permanent burrows in the sediment that do not significantly disturb surrounding sediments. This is especially true of animals that use the burrows only for shelter or escape. Nevertheless, these open vertical burrows still enhance the aeration of the sediment, and therefore they probably stimulate the availability of nutrients to other infaunal animals (Reise and Ax, 1979). Weaver and Schultheis (1983), for example, have described deep-sea vertical burrows that penetrate extremely deeply (214 cm) into the sediment. These burrows are very small in diameter (0.5 mm) and probably have a minimal impact on sediment stability, but they may allow other infaunal animals to live at greater depth than would have been possible if the burrows had not been there.

At the opposite end of the spectrum are organisms that stabilize sediments. They are sedentary or sessile forms that build tubes (polychaetes, amphipods, tanaids) or have roots (sea grasses, mangroves, other flowering plants). Movement through the sediment is impeded by the presence of tubes and roots. Although sediment stabilizers are less affected by bioturbation than are immobile suspension-feeders that do not build tubes, they are nevertheless adversely influenced. Vigorous bioturbators like callianassids inhibit the spread of the sea grass *Thalassia* in spite of the latter's tough rootstocks (Suchaneck, 1983). The only sediment stabilizers that may be able to withstand even the most intense reworking of sediments are the various mangrove trees, but their roots are so far apart that small deposit-feeders are not prevented from going about their business in great numbers in mangrove swamps (Thayer, 1983).

According to Thayer (1983), the extent of bioturbation depends on at least five factors: (1) the size of the individual organism that is reworking the sediment; (2) the depth to which the sediment is penetrated; (3) the distance an individual moves through the sediment per unit time, or the volume of sediment processed per unit time; (4) the abundance of bioturbating individuals per unit area or volume of the sediment; and (5) the nature of the sediment. All these characteristics can, with varying degrees of reliability, be estimated or inferred from fossils and their enclosing sediments. Thayer (1983) has calculated estimates of the amount of sediment that is disturbed or displaced daily by individual bioturbators. He identifies intense bioturbators as those individuals that process more than 10 cm³ of sediment per day.

THE HISTORY OF BIOTURBATION

The history of the occupation of unconsolidated sediments can be traced to the Ediacarian or Vendian Period, the first period of the Phanerozoic. Relatively few animal tracks and trails have been recorded from this period. One of the earliest trails, *Bunyerichnus*, is known from Ediacarian (or Vendian) strata just above the basal Varangerian Tillite in Australia and may represent the track of a creeping mollusc. Most of the known burrows, of which some 20 have been described from the Russian platform (Fedonkin, 1982), were very small (less than 1 mm in diameter) and are believed by Runnegar (1982a) to have been made by nematodes. Typically, these burrows (*Cochlichnus*, for example) are sinuous and therefore suggest a flexing action of the body such as is characteristic of living nematodes. These worms burrow without the aid of circular muscles by bending the body in the dorsoventral plane.

The latest Vendian and Early Cambrian were marked by a great increase in the diversity and depth of burrows. Just when this increase began is unclear, but in southeastern Newfoundland a sharp rise in the abundance and diversity of traces made by infaunal animals preceded the diversification of skeletonized animals (Crimes and Anderson, 1985). In Member 1 near the base of the Chapel Island Formation, three horizontal burrows appeared (*Planolites*, *Buthotrephis*, and *Gordia*). In the overlying Member 2, there were twelve kinds of burrow, including the vertical U-shaped *Arenicolites* and the coelenterate resting trace *Bergaueria*. Two more burrowing types, including the vertical burrow *Skolithos*, appeared in Member 3. Skeletonized organisms appeared on the overlying fourth member of this formation. None of the Early Cambrian burrows extended more than a few centimeters into the sediment, although later in the Cambrian and during the Ordovician the vertical burrow *Skolithos* extended to a depth of 1 m (Miller and Byers, 1984; Pickerill, 1984). At least some of these early burrows, including *Skolithos*, were escape burrows or resting shelters (Thayer, 1983) and therefore did not create extensive disturbance of surrounding sediments.

According to Runnegar (1982a), the Early Cambrian marks the first time when coelomate animals using peristalsis (antagonistic contractions of longitudinal and circular muscles in the body wall) became burrowers. The bioturbators that we infer to have been present during the Cambrian—annelids, protobranch pelecypods, lingulide inarticulate brachiopods, and trilobites—either had a small individual impact or disturbed only the thin layer immediately beneath the sediment surface. In the Early Late Cambrian of western North America, for example, bioturbation affected only the top 0.5 cm of sediment (Sepkoski, 1982c). Some Cambrian traces inferred by Miller and Byers to have been made by deposit-feeders (*Teichichnus*, *Syringomorphia*, *Phycodes*, and so on) may have penetrated to a depth of 20 cm, which by Mesozoic and Cenozoic standards is shallow. Several Cambrian, Ordovician, and Silurian trilobites are thought to have lived infaunally in burrows where they filtered particles from the overlying water, but they were probably sedentary in habit. These trilobites were smooth dorsally, and many had ventral terraced ridges whose asymmetrical profile prevented the animal from sinking further into the sediment. Examples include the Late Cambrian plethopeltid *Stenopilus* and many Ordovician asaphids, illaenids, aulacopleurids, and scutelluids (Bergström, 1973; Stitt, 1976; Schmalfuss, 1978; Westrop, 1983).

By the Middle Ordovician, extensive and relatively deep bioturbation extended to outer-shelf environments that during the Cambrian had remained relatively free of sedimentary disturbance. Some branching tunnels of the trace-fossil genus *Thalassinoides*, which superficially resemble those made by Late Mesozoic and Cenozoic callianassid crustaceans, penetrated to a depth of 1 m below the sediment surface in carbonate sediments of the Late Ordovician of the Great Basin in the western United States (Sheehan and Schiefelbein, 1984).

Thayer (1983) has suggested that a major intensification of bioturbation took place during the Devonian. One group that may have contributed to this increase was the Holothuroidea (sea cucumbers), deposit-feeding members of which may have been present by Devonian time. Support for Thayer's hypothesis comes from sedimentological work by Larson and Rhoads (1983). They compared two sections from the Middle Ordovician Trenton Group of New York (the Upper Denley Limestone, laid down below wave base but within range of storm waves, and the Lower Denley, laid down in quiet waters that were unaffected by storms) and two Early Devonian sections from the Helderberg Group of New York (the Kalkberg Limestone and New Scotland Formation, representing storm-influenced and deeper quieter water, respectively). The mean thickness of individual sedimentary layers (beds) is much smaller in the Ordovician sections (23 mm and 30 mm in the Upper and Lower Denley, respectively) than in the Devonian Kalkberg (39 mm) and New Scotland (80 mm) sections. Much of the primary layering (lamination) of the sediment is preserved in the Ordovician beds, whereas it has been eradicated by bioturbation in the Early Devonian sections. Another indication of greater bioturbation in the Devonian comes from the percentage of the section that consists of thin beds (thickness 0.5 cm or less). The environments subject to disturbance by storms show a decline from 62 to 34% in the percentage of thin beds, whereas the decline in the deeper sediments is even more dramatic, from 50.0 to 6.5% (Larson and Rhoads, 1983). Because no Silurian sections have yet been examined, the precise timing of the intensification of sedimentary reworking is not known. The fact that the characteristically deep-burrowing pelecypod superfamily Lucinacea arose during the Silurian

(Bretsky, 1976) suggests that some of the increase in bioturbation was already underway during that period. Watkins's (1979) finding of extensive bioturbation in outer-shelf environments of the Ludlovian (Late Silurian) of the Welsh Borderland further supports this assertion.

If the mid-Paleozoic was a time of increasing activity in the infaunal environment, the Mesozoic and Cenozoic were even more so. A large-scale episode of infaunalization, including the evolution of bioturbators capable of processing large quantities of sediment per individual, began in the Late Triassic or Early Jurassic and continued through the Cretaceous into the Cenozoic (S. M. Stanley, 1977a; Vermeij, 1977b; Thayer, 1983). Lugworms—polychaetes of the family Arenicolidae—are known from Triassic time onward. They are today the most intensively bioturbating polychaetes (individual reworking rate up to 44 cm^3/da). Unlike other polychaetes (maximum burrowing depth 15 cm) and protobranchs (maximum burrowing depth 4 cm), which evolved during the Early Paleozoic, lugworms burrow very deeply (to at least 30 cm). Irregular sea urchins, which are very important disturbers of modern sediments (reworking rate up to 8520 cm^3/da, depth 15 cm below sediment surface), originated in the Toarcian stage of the Early Jurassic. Holothurians, which burrow to a depth of 180 cm and rework sediment at a rate of at least 2250 cm^3/da, probably also expanded greatly during the Mesozoic (Thayer, 1983). Asteriid sea stars, such as the northeastern Pacific *Pisaster brevispinus*, can excavate pelecypods from sediments as deep as 40 cm (Van Veldhuizen and Phillips, 1978; Sloan and Robinson, 1983). In doing so, they disturb at least 3200 cm^3/da of sediment, according to Thayer's calculations. The family Asteriidae is not known earlier than the Pliensbachian stage of the Early Jurassic. Infaunal gastropods capable of bulldozing existed by the Late Triassic, when the first moon snails (Naticidae) are recorded. With the evolution of the order Rachiglossa, infaunal gastropods diversified on a large scale during the Cretaceous and Cenozoic, together with naticids, tonnaceans, and burrowing opisthobranchs (Sohl, 1964; Taylor et al., 1980; Sohl and Kollmann, 1985). Siphonate pelecypods capable of deep burial had existed since the Late Ordovician and underwent a substantial diversification (in the subclass Anomalodesmata) during the Carboniferous and Permian, but the deepest penetration of sediments by pelecypods took place during the Jurassic and Cretaceous with the diversification of heterodonts (S.M. Stanley, 1968, 1977a). Included in this diversification were the highly mobile deposit-feeding Tellinidae, which arose from a suspension-feeding ancestor such as *Tancredia* in the Late Triassic, and which were well established by the Early Cretaceous (Pohlo, 1982). Although the sediment-reworking capacity of tellinids (up to 630 cm³/da) is low in comparison with that of irregular echinoids, it is much higher than that of the older protobranchs (Thayer, 1983).

The deepest known burrowers in modern seas are thalassinidean decapod crustaceans (ghost shrimps). Pemberton and his colleagues (1976) report that *Axius serratus* in cold-temperate Nova Scotia constructs tunnels to a depth of at least 3 m below the surface of the sediment. Tropical thalassinideans may well exceed this depth, but no accurate measurements are available. Thayer's estimates of the reworking rate (35 cm³/da) are probably highly conservative. The earliest thalassinideans are axiids of the genus *Magila* from the Toarcian stage of the Early Jurassic (Förster, 1977). They differ from their Triassic mecochirid glypheoid ancestors, which probably also burrowed, by possessing well-calcified claws (chelae) on the first two pairs of walking legs—legs which are used in burrowing (Förster, 1977).

Other crustaceans are also important bioturbators. Squilloid stomatopods make burrows up to 4 m deep, and many ocypodid crabs (ghost and fiddler crabs) and homaridean and palinuran lobsters also make large burrows (Farrow, 1971; Reaka and Manning, 1981; Bertness, 1985). It is unknown when stomatopods achieved the infaunal habit, but burrowing crabs are of Cretaceous or possibly Late Jurassic origin (Frey et al., 1984).

Fishes and mammals are also important disturbers of modern marine sediments. Many fishes grub in the sediment for food and consequently rework sediments extensively. The cow-nosed ray (*Rhinopterax bonasus*) is renowned as an agent of destruction of grass beds in lower Chesapeake Bay, where it makes excavations up to 45 cm deep and reworks sediments at a rate of 12,000 cm³/da in search of prey (Orth, 1975). Although fishes with ray-like bodies are known since Devonian time, especially among placoderms and in the Late Paleozoic Petalodontoidea (genus *Janasa*), true rays occur from Late Jurassic time onward (Maisey, 1984). Other burrowing fishes, including many eels and goatfishes, are of Late Cretaceous and Cenozoic extraction.

Sediment-disturbing mammals are an important addition to the bioturbating fauna during the Cenozoic. In Alaska, the walrus *Odobenus rosmarus* makes pits and furrows up to 35 cm in depth as it excavates prey clams (Oliver et al., 1983a). The gray whale (*Eschrichtius robustus*), which occurs with the walrus in the Bering Sea in Alaska, excavates the sediment to a depth of 50 cm in search of its small invertebrate prey (Oliver et al., 1983b). The sea otter (*Enhydra lutris*) also digs for clams in unconsolidated sediments of the eastern Pacific (Hines and Loughlin, 1980). These mammals, all of which have Miocene or Pliocene origins (Lipps and Mitchell, 1976; Ray, 1976; Repenning, 1976a, b), disturb the sediment at rates of 10,000 to 100,000 cm^3/da (Thayer, 1983). The sea cows (order Sirenia), including the dugong (*Dugong dugon*), make furrows in grass beds as they feed on sea grasses (Anderson and Birtles, 1978). Sirenians have an origin in the Eocene (Domning, 1976, 1982).

Thayer (1983) has pointed out that deep penetration and rapid bulldozing require substantial metabolic activity. Even for deep-burrowing suspension-feeders such as siphonate pelecypods, effective pumping of sea water from the surfaces to the gills and back out requires that considerable friction drag must be overcome. Movement through hard-packed sand cannot be effected without considerable expenditure of energy. The increased bioturbation and deeper penetration of sediments through time, especially after the Paleozoic, therefore represent yet another example of the increasing proportion of animals with a higher metabolic rate (Figure 5.1).

In short, the history of sediment-disturbing organisms is one of episodic increases in the capacity of individuals to process and manipulate sediments. Taxa with the most rapid individual processing rates are geologically the most recently differentiated. The result has been a substantial increase in sedimentary instability through time.

CONSEQUENCES OF INCREASED BIOTURBATION

One of the earliest consequences of bioturbation was the change in type and distribution of stromatolites, layered structures of calcium carbonate laid down as a result of photosynthesis by blue-green algae (Cyanobacteria) and possibly other organisms. The greatest abun-

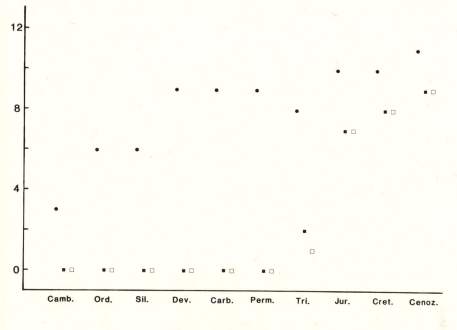

Time Interval

FIGURE 5.1   Global summary of bioturbating animal classes through time. Three groups are plotted: the number of bioturbating classes (circles), the number of classes with deeply bioturbating members (closed squares), and the number of classes with intensively bioturbating members (open squares). Deep bioturbators are defined as those digging 10 cm or more into the sediment. Intensive bioturbators rework the sediment at a per capita rate of 10 cm³/da or higher. Many classes originated long before members became deep burrowers or intensive bioturbators. Only when such members appeared were these classes counted as deep or intensive bioturbators. Data were derived from Thayer's (1983) compilation.

dance and diversity of stromatolites occurred during the Middle Riphean, well before the first fossil appearance of Metazoa in the Vendian. The first group of stromatolites to show a decline in diversity were coniform types such as *Conophyton*. During the Middle Riphean, they constituted 17% of stromatolite taxa in local assemblages. This percentage declined to 8.5% in the Late Riphean and dwindled to 2.5% in the Early Cambrian (Walter and Heys, 1985). Stromatolites of all types showed a noticeable decline after the Late Vendian. Since the Silurian, they have been more or less restricted to subtidal and intertidal environments where grazing and bioturbation are rare (Garrett, 1970; Awramik, 1971). With the advent of bioturbation by metazoans in the Early Cambrian, thrombolites—stromatolitic structures whose primary layering was disturbed during formation—appeared for the first time (Walter and Heys, 1985). Bioturbation is therefore thought to have been important not only in preventing the accumulation of large layered stromatolites, but also in increasing local rates of sedimentation (Pratt, 1982) that are inimical to Cyanobacteria.

The effects of increased bioturbation as well as of increased predation by highly mobile fishes and other predators were especially profound for immobile suspension-feeders living without a means of attachment on unconsolidated bottoms. Throughout the Paleozoic, many brachiopods (notably members of the orders Orthida, Strophomenida, and Productida) lived with one valve lying on or slightly recessed in the sediment but with no means of attachment or locomotion (Rudwick, 1970). Similar modes of life were characteristic of macluritacean and euomphalacean gastropods (Yochelson, 1971; Linsley, 1978b), some oyster-like pelecypods, some tabulate and rugosan corals, and trepostome bryozoans. Epifaunal unattached sedentary echinoderms such as ophiuroids were common during the Paleozoic. All these groups have become either extinct or geographically and ecologically restricted to hard substrates, temperate waters, the deep sea, and predator-impoverished saltwater lakes cut off from the sea (S. M. Stanley, 1977a; Aronson and Harms, 1985). Some large Mesozoic and Cenozoic gryphaeid and ostreid oysters still live unattached on sand and mud as adults, but they were apparently all cemented to objects as juveniles. Scallops (Pectinidae and Propeamussiidae) and other pelecypods may live unattached on the surface of unconsolidated sediments, but they have the capacity to swim when they are threatened. Lunulitiform bryozoans, belonging to several Cretaceous to Recent taxa, are unlike earlier bryozoans on unconsolidated sediments in their being able to propel themselves over the surface (Cook and Chimonides, 1983). Several fungiid and dendrophylliid scleractinian corals typically live on unconsolidated bottoms, but they too are capable of movement, either by attaching themselves to the shell of a motile sipunculan worm or by using tentacles (Goreau and Yonge, 1968; G. A. Gill and Coates, 1977). These examples suggest that the only animals that persisted on the surface of unconsolidated sediments without attachment have the capability of movement.

The history of stalked echinoderms points to the same conclusion. Early Paleozoic stalked echinoderms were attached throughout post-larval life by a small disc (eocrinoids, early crinoids) or by the aboral surface of the theca (edrioasteroids). Crinoids evolved jointed attachment structures (cirri) along part or all of the stalk; these structures enabled the animal to grasp objects temporarily, and to reattach after the original attachment by the disc had come undone. Crinoids in the Ordovician generally lacked cirri, but by the Late Carboniferous some 40% of species had them (C. E. Brett, 1981). Several Middle Paleozoic crinoids also had coiled stems for temporary attachment. A few Late Paleozoic inadunate crinoids and most Late Mesozoic and Cenozoic comatulid articulate crinoids were capable of locomotion (D. L. Meyer, 1985).

For marine organisms in general, the most important effect of an increase in bioturbation through time may well have been an increase in primary productivity. Thayer (1983) and Bertness (1985) have suggested that, by stimulating bacterial growth and by re-suspending fine organic particles into the water, bioturbators recycle nutrients that otherwise would have been buried in the sediment where they would have been unavailable to organisms. The striking rise in the diversity of single-celled, mineralized, photosynthesizing planktonic organisms during the Late Mesozoic is, believes Thayer, causally related to the increase in bioturbation in the Mesozoic.

BIOEROSION AND THE ENDOLITHIC ENVIRONMENT

Another important biological activity that can be studied easily in the fossil record is bioerosion, the mechanical or chemical destruction of rock by organisms. Traces of bioerosion are usually in the form of borings, cavities, and tunnels in shells, corals, and other hard substrata. They are made by organisms that either live in these openings (endolithic organisms) or abrade rock while grazing. The form of the boring is often diagnostic of the various groups of bioeroders, but the identity of borers is not always ascertainable in fossil assemblages. Warme (1975) has given a general account of living bioeroders and their traces. Table 5.1, in which pertinent facts about bioeroders are summarized, is derived from his compilation and from many subsequent studies.

TABLE 5.1   Characteristics and Stratigraphic Ranges of Endolithic Organisms

---

Kingdom Monera

> Phylum Cyanobacteria (blue-green algae); bore to several hundred micrometers into rock; Late Riphean to Recent; examples include *Hormathonema, Hyella, Kyrtuthrix, Mastigocoleus, Plectonema, Scopulonema, Solentia* (Golubic et al., 1975; S. E. Campbell, 1982)

Kingdom Plantae

> Phylum Chlorophyta (green algae); bore to several hundred micrometers into rock; Early Paleozoic to Recent; examples include *Codiolum, Entocladia, Eugomontia, Ostreobium* (Golubic et al., 1975; Kobluk and Risk, 1976, 1977; Highsmith, 1981b)

> Phylum Rhodophyta (red algae); bore to several hundred micrometers into rock; range ? ; examples include *Conchocelis* stage of *Bangia* and *Porphyra* (Golubic et al., 1975)

Kingdom Fungi; bore to a few hundred micrometers into rock; age ? ; examples include *Aspergillus, Penicillium, Cladosporium, Pharcidia* (forming the lichen *Arthropyrenia*) (Rasmussen, 1973; Kohlmeyer and Kohlmeyer, 1979; Kendrick et al., 1982)

Kingdom Animalia

> Phylum Sarcodina, Order Foraminiferida; form shallow pits in shells; *Talpinella*, Late Cretaceous; *Vasiglobulina*, Late Eocene to Early Miocene; probably some Recent genera (Poag, 1971; Baumfalk et al., 1982)

> Phylum Porifera (sponges), Class Demospongiae

>> Family Clionidae; penetrate to 2.5 cm in compact rock, to 8 cm in porous rock; Early Cambrian to Recent (Rützler, 1974, 1975; Highsmith, 1981a; Pickerill and Harland, 1984)

>> Family Spirastrellidae (*Spheciospongia*); penetrate to 10 cm; age ? (Rützler, 1974)

>> Family Adociidae (*Siphonodictyon*); penetrate to 6 cm or more; age ? (Rützler, 1971)

> Phylum Annelida, Class Polychaeta (segmented worms)

>> Family Spionidae; penetrate to 2 cm or less; Early Devonian to Recent (Blake and Evans, 1973)

Table 5.1 (*continued*)

    Family Cirratulidae (*Dodecaceria*); penetrate to 2 cm or less; age ? (J. W. Evans, 1969)

    Family Sabellidae (various genera, including the freshwater *Caobangia*); shallow borers; age ? (M. L. Jones, 1969; Fauchald and Jumars, 1979)

    Family Nereidae (various genera); shallow borers; age ? (Fauchald and Jumars, 1979)

Phylum Crustacea

    Class Cirripedia (barnacles)

        Order Acrothoracica; bore to a few millimeters; Givetian (Middle Devonian) to Recent (Tomlinson, 1969; Rodriguez and Gutschick, 1977)

        Order Thoracica, Suborder Lepadomorpha, Family Lithotryidae; penetrate to 15 cm; Late Cretaceous to Recent (Focke, 1977)

    Class Malacostraca, Order Decapoda

        Family Alpheidae (snapping shrimps) (several species of *Alpheus*); bore to 10 cm; probable Cenozoic derivation (Holthuis, 1980; R. Fischer, 1981)

        Family Callianassidae (some species of *Upogebia*); bore to 6 cm; probable Cenozoic derivation (Kleemann, 1984)

Phylum Ectoprocta (bryozoans), Subphylum Gymnolaemata, Class Ctenostomata; bore to a few millimeters; Late Ordovician to Recent (Pohowsky, 1978)

Phylum Phoronida (*Phoronis* and ichnogenus *Talpina*); penetrate to a few millimeters; Siegenian (Early Devonian) to Recent (Voigt, 1975)

Phylum Sipuncula (Aspidosiphonidae, Phascolosomatidae); penetrate to at least 6 cm; age ? (Otter, 1937; M. E. Rice, 1975; Murina, 1984).

Phylum Mollusca

    Class Gastropoda, Family Coralliophilidae (several coral-dwelling genera); penetrate to at least 6 cm; Late Cretaceous to Recent (Gohar and Soliman, 1963)

    Class Pelecypoda

        Family Arcidae (*Litharca*, Pliocene to Recent); penetrate to 8 cm; family Jurassic to Recent

Table 5.1 (*continued*)

Family Tridacnidae (*Tridacna crocea*, Recent); penetrate to at least 8 cm; family Eocene to Recent (Hamner, 1978)

Family Mytilidae, Subfamily Lithophaginae; penetrate to 10 cm; Late Triassic to Recent (Otter, 1937; Yonge, 1955; B. S. Morton, 1980a; Morton and Scott, 1980; Wilson and Tait, 1984)

Family Gastrochaenidae; penetrate to several centimeters; Middle Jurassic to Recent (J. G. Carter, 1978)

Family Pholadidae; penetrate to 10 cm or more; Jurassic to Recent (Purchon, 1955a)

Family Petricolidae; penetrate to 6 cm; Eocene to Recent (Otter, 1937; Purchon, 1955b)

Family Hiatellidae; penetrate to several centimeters; Jurassic to Recent (J. G. Carter, 1978)

Family Myidae (*Platyodon*, Eocene to Recent); penetrate to 6 cm; family Cretaceous to Recent (Yonge, 1951a)

Family Clavagellidae; penetrate to a few centimeters; Eocene to Recent (B. S. Morton, 1984)

Compared to organisms living on open hard surfaces in the sea, those in the endolithic habitat are relatively protected from most predators and are therefore morphologically less resistant to attack. Rock-boring lithophagine mussels of the family Mytilidae, for example, have notably delicate shells, which contrast strongly with the thick-shelled, often radially ribbed species from open surfaces. Fragility is also the hallmark of the shells of most rock-boring pholadid, gastrochaenid, and petricolid pelecypods, as well as of endolithic coralliophilid gastropods. Being enclosed by rock, these animals are safe from most shell-breaking and shell-drilling predators.

Despite the temptation to compare endolithic organisms with those living on the hard surfaces of marine rocky bottoms, a more appropriate comparison in many instances is between endolithic and infaunal organisms. Many endolithic animals descended from infaunal ancestors rather than from epifaunal ones. The Gastrochaenacea, for example, probably are derived from an infaunal permophorid anomalodesmatan pelecypod stock in the Triassic (J. G.

Carter, 1978). *Platyodon*, a rock-boring myid from the Pacific coast of North America, almost certainly descended from a deep-burrowing infaunal myid (Yonge, 1951a). Even in the Mytilidae, a family of predominantly epifaunal pelecypods, endolithic lineages are probably not derived from surface-dwellers, as Yonge (1955) thought, but from thin-shelled semi-infaunal types that live byssally attached in sand or silt (B. S. Morton, 1980a). Despite assertions that petricolids and hiatellids had epifaunal ancestors (Ansell and Nair, 1969), it is more plausible that their ancestors were mud-dwelling infaunal types. I can think of only two endolithic pelecypods that clearly descended from epifaunal forms. These are *Tridacna crocea*, a member of an otherwise epifaunal byssate genus, and the arcid *Litharca*, which is closely related to the byssate epifaunal *Arca*. Among other endolithic groups, coralliophilid gastropods and lithotryid and acrothoracican barnacles probably also descended from surface-dwellers on rocks, but infaunal ancestry is at least plausible in the polychaetes, sipunculans, phoronids, and alpheid crustaceans. Endolithic sponges often have epifaunal as well as endolithic portions, and at least one species of *Siphonodictyon* (a genus that usually makes large chambers in the skeletons of living corals) has been found in sand (Rützler, 1971, 1974). Unfortunately, no ecological study has been undertaken that compares endolithic and infaunal habitats. Because bulldozing or burrowing predators would generally be deterred from the endolithic environment, I would guess that the endolithic habitat is a substantially safer place than most infaunal environments. Endolithic animals also may be less apt to be unearthed by storms.

Ecologically, bioerosion is almost entirely a marine phenomenon whose effects are most obvious in warm shallow waters. The only freshwater borer of which I am aware is the sabellid polychaete *Caobangia*, which bores into the shells of freshwater gastropods in Southeast Asia (M. L. Jones, 1969). Many freshwater molluscs have deeply eroded shells, but biological agents are apparently not responsible (Golubic et al., 1975). In the sea, the only common borers at high latitudes are algae. Pelecypods, sipunculans, lithotryid barnacles, and adociid sponges are absent, and clionid sponges, spionid polychaetes, and phoronids are rare (Bromley and Hanken, 1981).

A link between productivity and bioerosion has been suggested by Highsmith (1980a). He found that the incidence of boring pelecypods in massive corals was greatest in the highly productive tropical Eastern Pacific, intermediate in the Caribbean Sea and Indian Ocean, and low in the waters surrounding nutrient-poor atolls of the Western and Central Pacific. Comparisons within oceans have confirmed and extended Highsmith's original observations. In the Eastern Pacific, for example, the number of borings per 100 cm^2 in the massive coral *Pavona* is 3 to 4 in the Gulf of Chiriqui, an area without upwelling, and 6.5 to 7 in the seasonally upwelled Bay of Panama and Gulf of Papagayo (Glynn et al., 1983). The reason for this link has not been investigated. Food may be more available for the borers, or the high nutrient content may favor the survival of larvae. It is also unknown if this link extends to temperate zones. I suspect that it does. Shell erosion along the highly productive coasts of California, Peru, Chile, and the Bay of Fundy is strikingly greater than that in the Gulf of Maine and other relatively unproductive parts of the northwest Atlantic (Vermeij, 1978; Bergman et al., 1982).

The first traces of bioerosion in the geological record are tiny borings of possible algal or fungal origin from the Late Riphean or Vendian of Australia (S. E. Campbell, 1982). These borings thus substantially predate the Cambrian explosion of skeleton-bearing and burrowing organisms. In the Early Cambrian strata of Newfoundland, borings 11 to 25 mm deep are recorded from the skeletons of archaeocyathans. These borings, which are referred to the ichnogenus *Trypanites*, are the only borings known from the Early Cambrian. No borings of this form have been recorded from the Late Cambrian (James et al., 1977). The only boring of Late Cambrian age is a vertical cylindrical cavity 2 to 3 mm in diameter, perhaps made in a hard but not quite rock-like substratum (C. E. Brett et al., 1983).

Although borers diversified to at least 4 ichnogenera by the end of the Middle Ordovician (Kobluck et al., 1978), they comprised only 5 to 10% of the sessile fauna of hard-ground communities in the Ordovician, Silurian, Devonian, and Early Carboniferous (T. J. Palmer, 1982). The status of bioeroders in the Late Paleozoic is difficult to judge owing to the rarity of hard-ground communities. By the Triassic, however, borers had clearly increased in importance. Some 45% of species in Triassic hard-ground communities, and 25% of species in similar communities of Jurassic age, are classified by Palmer as borers.

That bioerosion became more important after the Paleozoic is also suggested by the fact that deep borings are known only from the Mesozoic and Cenozoic (Table 5.1, Figure 5.2). Specialized boring pe-

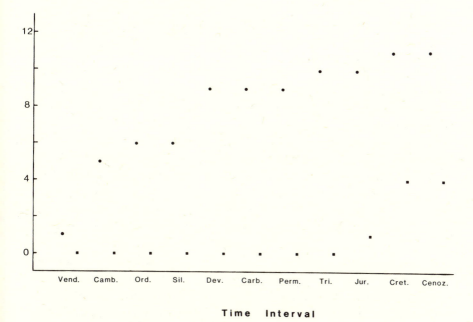

FIGURE 5.2 Global summary of classes of bioeroding endolithic organisms. The number of endolithic classes (circles) is compared to the number of such classes containing deeply boring members (squares), where deep borers are defined as those that penetrate the rock to a depth of 5 cm or more. Excavating animals (chitons, limpets, urchins, and fishes) were excluded. No stratigraphic data are available for sipunculans. Classes were counted as deep borers only when members boring 5 cm or deeper appeared. Data were derived from Warme (1975) and other sources listed in Table 5.1.

lecypods, which excavate holes 2 to 15 cm deep, are known with certainty only from Late Triassic time onward (J. G. Carter, 1978). The Ashgillian (Late Ordovician) pelecypod *Corallidomus* was at most a facultative borer and apparently left no descendants (Pojeta and Palmer, 1976). All endolithic groups originating in the Middle Paleozoic or earlier (algae, fungi, clionid sponges, ctenostome bryozoans, phoronids, polychaetes, and acrothoracican barnacles) penetrate to depths that are at most 2 and 3 cm below the surface of hard rocks, whereas the more recently evolved borers (pelecypods, gastropods, and lithotryid barnacles) bore to depths as great as 15 cm. Nothing seems to be known about the geological history of sipunculans, which in the modern tropical ocean bore to a depth of 10 cm (M. E. Rice, 1975).

Another important form of bioerosion that became common only during the Late Mesozoic and Cenozoic is rock scraping by mobile epifaunal invertebrates and herbivorous fishes (Table 5.2). Although most of the epifaunal rock excavators are herbivores capable of exploiting heavily calcified algae or corals as food (Steneck and Watling, 1982; Steneck, 1983a), a few of the gastropods (notably the Hipponicidae) are detritivores or suspension-feeders. The gastropods and probably the chitons excavate the rock first by secreting acid mucopolysaccharides and carbonic anhydrase from the foot and mantle edge, and then by removing rock particles from the occupied site with the radula (Lindberg and Dwyer, 1983). Protection against dislodgment, especially by predators and competing herbivores, is probably the primary benefit of the formation of home depressions by rock-scraping invertebrates (Vermeij, 1978; Garrity and Levings, 1983; Lindberg and Dwyer, 1983).

The compilation in Table 5.2 shows clearly that, although many of the orders containing rock excavators arose during the Paleozoic and Early Mesozoic, rock scraping is a post-Triassic phenomenon. Scratches attributable to the action of the docoglossan limpet radula have been recorded from the surfaces of shells of Late Jurassic and Early Cretaceous age in England (Akpan et al., 1982; Steneck, 1983a). All chiton and gastropod families containing rock-excavating species arose either during the Cretaceous or the Cenozoic.

Many echinoids are capable of excavating depressions as deep as 10 cm in limestone and even basalt (Focke, 1977; R. Fischer, 1981). No Paleozoic or Triassic echinoid, however, seems to have had the jaw equipment necessary to excavate rock (Kier, 1974; A. B. Smith, 1984). In the oldest true echinoids (the Late Ordovician genera *Aulechinus* and *Ectinechinus*), the jaw (Aristotle's lantern) was broad and consisted of weak teeth. This jaw functioned as a scoop which picked up detritus from the surface of the sediment. Jaws capable of plucking and biting first appeared in the Early Silurian genus *Aptilechinus*. In this and other Paleozoic echinoids, however, the muscles that operated the jaws were attached directly to the inner wall of the test. Although the jaws were capable of some up and down (oral-aboral) movement, the horizontal movement necessary for scraping was very limited in these early sea urchins. The Late Permian cidaroid *Miocidaris* was the first urchin in which the muscles were attached to the perignathic girdle, a series of projections from the inner test wall. It was apparently the only echinoid to have survived the end-

TABLE 5.2 Taxonomy and Geological History of Rock-Excavating Mobile Epifaunal Animals

---

Phylum Mollusca

Class Polyplacophora (chitons): Late Cambrian to Recent (Runnegar et al., 1979)

Order Neoloricata: Early Ordovician to Recent (van Belle, 1977)

Families Ischnochitonidae (Lepidochitoninae), Paleocene to Recent; Chitonidae (Chitoninae, Cretaceous to Recent, and Acanthopleurinae, Eocene to Recent); Cryptoplacidae (Acanthochitoninae), Oligocene to Recent; Mopallidae (Mopallinae), Early Miocene to Recent (data on bioerosion from Otter, 1937; Chelazzi et al., 1983; taxonomy and stratigraphic ranges from van Belle, 1977, 1978)

Class Gastropoda: Late Cambrian to Recent

Order Docoglossa (limpets): Late Triassic to Recent

Families Nacellidae, Late Cretaceous to Recent; Lottiidae, Late Cretaceous to Recent; Acmaeidae, Eocene to Recent (Lindberg and Dwyer, 1983; Lindberg, personal communication)

Order Echinospirida: Late Jurassic to Recent

Families Capulidae, Early Cretaceous to Recent; Calyptraeidae, Early Cretaceous to Recent; Hipponicidae, Late Cretaceous to Recent (Vermeij, 1978; Lindberg and Dwyer, 1983)

Order Rachiglossa: Albian (Late Cretaceous) to Recent

Family Coralliophilidae, Campanian (Late Cretaceous) to Recent (Lindberg and Dwyer, 1983)

Order Basommatophora: Late Jurassic to Recent

Families Siphonariidae (pulmonate limpets), Late Cretaceous to Recent; Trimusculidae (pulmonate limpets), Eocene to Recent (Garrity and Levings, 1983)

Phylum Echinodermata: Early Cambrian to Recent

Class Echinoidea (sea urchins), Late Ordovician to Recent (A. B. Smith, 1984)

Order Diadematoida: Late Triassic to Recent

Family Diadematidae, Late Triassic (Norian) to Recent

Table 5.2 (*continued*)

Order Temnopleuroida: Late Cretaceous to Recent

Family Temnopleuridae, Cenomanian (Late Cretaceous) to Recent

Order Echinoida: Paleocene to Recent

Families Echinometridae, Paleocene to Recent; Strongylocentrotidae, Oligocene to Recent (Kier, 1974; A. B. Smith, 1984)

Phylum Chordata: Middle Ordovician to Recent

Class Osteichthyes (bony fishes): Early Devonian to Recent

Order Perciformes: Paleocene to Recent

Family Scaridae (parrotfishes), Eocene to Recent

Order Tetraodontiformes: Paleocene to Recent

Families Tetraodontidae (smooth puffers), Eocene to Recent; Balistidae (triggerfishes), Oligocene to Recent (Tyler, 1980)

Permian crisis. In *Miocidaris* and all later cidaroids, the protractor and retractor muscles of the lantern had a vertical orientation, so that the jaw was constrained to move chiefly in an up-and-down direction. Consequently, cidaroids continued to feed by plucking and biting in the fashion of their Paleozoic antecedents. The capacity to scrape rocks by lateral movements of the lantern evolved in the euechinoid cohort Echinacea. In these urchins, the muscles operating the lantern came to have more oblique orientations, enabling the jaw to move from side to side as well as vertically. The ultimate development of this jaw apparatus was the camarodont lantern, which evolved first in the late Cretaceous in the Temnopleuridae, although the Oxfordian (Late Jurassic) *Hemicidaris* already approached this condition. Grazing traces (*Gnathichnus pentax*) attributable to scraping echinoids are especially common from Campanian (Late Cretaceous) time onward, but they are known as early as the latest Triassic (Kier, 1974; Bromley, 1975; A. B. Smith, 1984).

Parrotfishes of the family Scaridae are renowned as scrapers of coralline algae and, in the Pacific and Indian Oceans, of living corals (Bakus, 1967; Glynn et al., 1972; Randall, 1974; Glynn and Wellington, 1983). Along with tetraodontid pufferfishes and balistid trigger-fishes, which often break off branches of living coral in order to feed on endolithic organisms, parrotfishes are important agents of bioerosion on modern reefs. All these bioeroding fishes have an Early Cenozoic origin. Scarids and tetraodontids are known from the Eocene, whereas balistids extend back at least to the Oligocene (Tyler, 1980).

The evolution of bioeroding grazers was associated with important changes in the resistance of encrusting calcareous algae to herbivory (Steneck, 1983a, 1985). Early encrusting algae belong to the rhodophyte family Solenoporaceae. This group, which extends from the Cambrian to the Miocene, declined sharply during the Jurassic. Among living encrusting calcareous rhodophytes, the Corallinaceae are the largest group. This family is known from a single Late Carboniferous and Permian genus (*Archaeolithophyllum*) and then again from the Jurassic onward. Members of the Corallinaceae are adapted to withstanding moderate levels of scraping by herbivorous gastropods, echinoids, and fishes. The middle layer of the plant (perithallus), which contains the photosynthesizing cells as well as the reproductive structures (conceptacles), is protected in corallines by an outer layer (epithallus). Corallines such as the tropical *Porolithon* and the cold-temperate *Clathromorphum* have become dependent on grazing in that reproduction cannot occur unless the epithallus is scraped off (Wonders, 1977; Paine, 1980; Steneck, 1982). If the perithallus is damaged, as it often must be by deeply excavating parrotfishes, regeneration can occur rapidly owing to the lateral and vertical translocation of nutrients by means of fusion cells and primary and secondary pits. Solenopores lacked an epithallus as well as fusion cells and primary and secondary pits. Not only were they probably less resistant to scraping, but they were also less capable of repairing superficial damage. Steneck (1983a), in fact, reports that wounds are commonly preserved on Jurassic solenopores, whereas they are rare on contemporaneous corallines. The absence of wounds on corallines is attributed to rapid repair rather than to the lack of injury.

Marine hard-bottom communities were profoundly affected by the Mesozoic diversification of deeply excavating bioeroders. Small mobile predators such as stomatopods and alpheid shrimps, which often shelter in the excavations of echinoids and other borers, may have expanded their populations and diversified as large cavity-formers evolved during the Mesozoic and Cenozoic (Moran and Reaka, 1985).

Cavities of biological origin in modern reefs also provide shelter for a large number of immobile brachiopods, sponges, bryozoans, serpulid polychaetes, and solitary corals (Jackson et al., 1971; Bonem, 1977; T. J. Palmer, 1982). The greater topographic complexity that is created by the presence of excavations made by sea urchins facilitates the settlement of coral larvae (planulae), because the larvae are protected from grazers and other disturbances. This kind of protection is especially important when open surfaces are crowded with erect algae that could smother or shade the coral recruits. (Birkeland and Randall, 1982). The small cavities made by bioeroders in an Early Late Carboniferous framework in Oklahoma were evidently quickly filled with mud (preserved as micrite) and therefore did not serve the same purpose that later cavities did (Bonem, 1977). In short, the Mesozoic increase in the size and depth of excavations made by bioeroding animals can be interpreted as an effect of escalating predation and competition on open rock surfaces and in unconsolidated sediments, but it in turn further stimulated this escalation.

FRESH WATER AND THE DRY LAND

It is hard to imagine the world when fresh water and the dry land were essentially unoccupied by multicellular life, yet this condition seems to have prevailed until some time during the Early Paleozoic. The invasion of non-marine habitats, which had begun by the Silurian at the latest, was important not only because chemical cycles and nutrient availability underwent radical changes (A. G. Fischer, 1984), but also because some descendants of the invaders became secondarily adapted to the sea and profoundly influenced resident species there.

Enormous contrasts exist between fresh water and the dry land. On land, escalation has proceeded to extraordinary levels, and the diversity of life there far exceeds that in the sea. Freshwater environments, on the other hand, have probably always lagged behind the sea in organic diversity and in the incidence and expression of almost every kind of aptation. Despite these striking contrasts, I discuss fresh water and the dry land in the same section because invasion of the land and export of species from the land back to the sea have often occurred by way of freshwater intermediaries.

FRESH WATER

Many processes and adaptive types that are common in the modern ocean are either lacking or poorly developed in freshwater streams and lakes. Only one bioeroder (the polychaete *Caobangia*) is known in fresh water, for example. Drilling predation is unknown in fresh water but very common in the sea. Gastropods with apertural armor (narrowly elongated or toothed aperture) and pelecypods with tightly sealing crenulated valve margins are rare in fresh water, as are colonial immobile animals, conchicoles (animals living in vacated gastropod shells), and rapidly burrowing pelecypods (Vermeij and Covich, 1978; Vermeij and Dudley, 1985). Important marine groups of shell-breaking and shell-extracting predators—sea stars, octopods, gastropods, and stomatopods—are absent from fresh water (Chapter 6). Conversely, several architectural types that were common in the Paleozoic marine environment but are of minor importance in the sea today are still widespread in fresh water. Large armored vertebrates and bivalved crustaceans, which have waned in importance in post-Paleozoic seas, are still well represented in many rivers and lakes, for example. Anaspidid syncarid Crustacea—Australian freshwater relicts of a Late Paleozoic marine group—are today found only in waters where predaceous fishes are rare or absent. They lack the characteristic caridoid escape response of other malacostracan crustaceans, which flip their bodies backward by means of a rapid flick of the tail (Schram and Hessler, 1984).

Another indication that obsolete body plans persist in fresh water comes from bivalved animals. A very common feature of freshwater pelecypods of the family Unionidae is that the shell is strongly thickened anteriorly and relatively thin posteriorly. We interpret this feature as an aptation whose function is to prevent the animal from being exhumed from the sediment by strong currents or waves. A similar weighting method would be equally effective in the ocean, but in marine pelecypods today this form of animal stability is rare. Most marine pelecypods have valves whose thickness is either greatest in the center or uniform throughout. Stability is achieved (or exhumation is prevented) by deep burial, rapid burrowing, or the presence of strong radial and concentric sculpture. Weighting was a common feature in ancient bivalved animals, including many Paleozoic brachiopods (especially Silurian pentameraceans) and Devonian to Jurassic megalodont pelecypods (Skelton, 1978; Bassett, 1984). We suspect that passive stability by anterior shell weighting is still feasible in fresh water, because selection due to predators is relatively weak there, but that weighting is no longer feasible in the sea, where an increased evolutionary emphasis on escape and on aptations related to drilling predation has been pervasive (Vermeij and Dudley, 1985).

Biourbation in fresh water is generally less intense and does not extend as deeply into the sediment as it does in the sea. The bioturbating animals that process the largest daily amount of sediment per individual in the sea (holothurians, sea urchins, thalassinidean and other decapod crustaceans, rays, whales, and seals) are not found in fresh water. The larvae of chironomid dipteran midges are found at depths of as much as 50 cm below the sediment surface in fresh water, and some astacidean crayfishes may make burrows 2 to 3 m below the surface. *Tridactylus*, a minute gryllotalpid mole cricket, excavates hibernation chambers to a depth of 60 cm in the sediment (C. K. Chamberlain, 1975; McCall and Tevesz, 1982). Although these excavations in the sediment are deep, they are chiefly dwelling burrows. Deposit-feeding oligochaetes extend to about 20 cm, and amphipods to about 2 cm, below the surface. M. F. Miller's (1984) analysis of freshwater bioturbators suggests that the much less common occurrence of freshwater bioturbation as compared with bioturbation in the sea has prevailed since the first appearance of freshwater in the Silurian.

The relative biological simplicity of freshwater communities and the apparently low level of escalation among freshwater organisms are difficult to understand. The low concentration of calcium, sodium, and other essential ions in many fresh waters may limit skeletal development, nerve-impulse transmission, and perhaps muscle power in some invertebrates that are unable to maintain high internal concentrations of these ions, but such physiological limitations do not apply either to vertebrates or to freshwater crabs, which are capable of maintaining high concentrations of ions in the body despite the low concentrations in the surrounding water (Hutchinson, 1967). In Chapter 2, I suggested that the small size of freshwater habitats may preclude the evolution and long-term maintenance of populations that consist of metabolically active individuals, for these populations would have to be small and therefore would be highly susceptible to extinction. In other words, the only organisms that are apt to be successful are those with low food requirements. Populations of these individuals could then be large and therefore resistant to extinction during times of crisis. An obvious difficulty with this hypothesis is that the limitation on population size of metabolically active animals would not apply to the numerous species that feed in and affect freshwater communities but that are also found on dry land. Many reptiles, birds, mammals, and insects are important as consumers of freshwater animals, but they spend at least part of their life cycle on land, so that the small size of most freshwater habitats creates no obvious upper limit on their population size or individual metabolic activity. In fact, if these visitors from the land are effective in fresh water, as they seem to be, it may be difficult for exclusively freshwater organisms to adapt to these visitors by evolving resistance defenses. Clearly, the freshwater environment poses many questions but suggests few answers for the evolutionist who is interested in escalation.

The initial colonization of fresh water must have been entirely from the sea. Boucot and Janis (1983) find the first undoubted freshwater armored agnathan fishes in sediments of Late Silurian (Ludlovian) age. Various eurypterids and scorpions, some of very large size, also lived in fresh waters during the Late Silurian, Devonian, and Carboniferous and probably had marine origins (Störmer, 1976, 1977; Rolfe, 1980). Lungfishes (subclass Dipnoi) also colonized fresh waters from the sea early in their Devonian history (Westoll, 1949). Several groups of placoderms, notably the Antiarcha, acquired freshwater habits during the Devonian (Denison, 1978), as did many acanthodians (Denison, 1979), rhipidistians, and pelecypods. Fully aquatic vertebrates with amphibious or terrestrial ancestors are known as early as the Late Carboniferous (Milner, 1980). Since that time, colonization of fresh water has occurred from both the land and the sea.

That fresh waters should be colonized by organisms from the land and the sea is not surprising in view of the low level of adaptive escalation between freshwater species and their enemies, but the frequent colonization of the sea by freshwater species is more puzzling. Freshwater groups including reptiles, fishes, mammals, planktonic insects (skaters), some larval dipterans, and one or two trichopteran caddisfly larvae (Hutchinson, 1967) have all invaded the sea. It is tempting to speculate that the wholesale periodic extinction of large marine reptiles and mammals provided a temporary opportunity for the invasion of less specialized vertebrates from fresh water, for the time intervals immediately following extinction may have been characterized by reduced risks to such colonists. If this were so, invasions from fresh water to the sea should have occurred only when comparable animals were missing from the sea. To my knowledge, investigation of this problem has not been undertaken.

The Invasion of the Dry Land

Until the Silurian, there is no preserved record of organisms that we can confidently regard as having occupied the dry land. For shallow-water marine as well as freshwater organisms, therefore, the land must have provided a place free of competitors and predators.

During the Middle Paleozoic, freshwater environments were probably the chief source of species that colonized the land. A freshwater rather than a marine origin seems likely for the first amphibians in the latest Devonian (Thomson, 1969; Graham et al., 1978); the first land arthropods (Störmer, 1976, 1977; Rolfe, 1980); insects (Kukalova-Peck, 1978); and vascular land plants (Stebbins and Hill, 1980).

The early events in this colonization are not well known. Gray and her colleagues (1982) reported desiccation-resistant trilete plant spores in Caradocian (Late Ordovician) rocks in Libya. They interpret these spores as possible evidence of land plants (though not necessarily vascular plants). The first fossils of vegetative tissues of vascular land plants (*Cooksonia*) are of Ludlovian or Pridolian (Late Silurian) age, but material as old as the Llandoverian (Early Silurian) may also turn out to belong to vascular plants. Tubular fossils of nematophytic construction in Llandoverian and Wenlockian (Middle Silurian) rocks may have affinities to vascular plants or bryophytes (Chaloner and Sheerin, 1979; D. Edwards, 1980; Niklas and Smocovitis, 1983; Thomas, 1984).

The Early Devonian marks the first large-scale invasion of the land by vascular plants. The most rapid rise in diversity came during the Siegenian stage, and a plateau in the diversity of vegetative and spore-bearing fossils was attained in the Emsian stage of the Late Early Devonian (Knoll et al., 1984). By the Givetian (Late Middle Devonian), the first forest trees had evolved.

The first undoubted land animals also date from the latest Silurian or earliest Devonian (Rolfe, 1980). Millipedes—members of the arthropod class Diplopoda—may have been among the earliest land animals, with fossils referred to the genus *Necrogammarus* appearing in the latest Silurian part of the Scottish Old Red Sandstone. Some eurypterids and xiphosurans may also have been capable of short excursions on land. Arachnids, including the extinct order Trigonotarbida, appear together with colembolans (springtails) in the Rhynie Chert, a Scottish deposit of Siegenian (Early Devonian) age (Kevan et al., 1975). The first centipedes (class Chilopoda) and wingless archaeognathan insects are recorded from the Middle Givetian (Middle Devonian) Panther Mountain Formation of New York (Shear et al., 1984).

The Invasion of the Sea from the Land

With the appearance of the land amphibians in the Famennian stage of the Late Devonian and of winged insects in the Namurian stage of the Middle Carboniferous, the broad outline of terrestrial community composition was firmly established. Escalation between land organisms and their enemies had proceeded so far that by Permian time the land began to function as a source of marine species. Although invasion of the sea by land animals often involved freshwater intermediates, many terrestrial groups entered marine communities directly. Most important among these groups are the birds (latest Jurassic to Recent); mangrove trees (Cretaceous to Recent); grass-like angiosperms; and cetaceans (Eocene to Recent) and seals (Miocene to Recent). Man is the latest, and possibly the most important, newcomer to the marine environment.

Species from the land that colonized the sea had a profound impact on the marine biota. Among their descendants are large plankton-eaters (baleen whales and at least three groups of birds); high-level predators of fishes and cephalopods (ichthyosaurs, plesiosaurs, crocodiles, seals, whales, and many diving birds); herbivores capable of grazing tough sea grasses (turtles, sirenian sea cows, and geese); powerful shell-crushers (loggerhead turtles, sea otters); highly effective sediment-stabilizers (sea grasses and mangroves); and the most intensive bioturbators (various marine mammals). With the evolution of angiosperm sea grasses during the Cretaceous, unconsolidated bottoms may have become important sites of primary production for the first time in the Phanerozoic, owing to the fact that sea grasses take up nutrients via the roots, whereas algae must rely on surface absorption which, in the calm waters overlying most unconsolidated bottoms, is slow. In short, escalation in the sea has been influenced to a great extent by invaders from the land.

PARASITISM AND MUTUALISM

Organisms that are highly resistant to predation constitute an important category of safe places for species in high-energy environments. If a well-defended host species lives in an environment where the risk of predation is high, selection in the potential guest species should strongly favor traits that enable the guest to detect and seek out the host. The guest may come to depend nutritionally on the host and establish a parasitic relationship, but the intimate association between host and guest may also become mutually beneficial. As adaptive escalation proceeds, such intimate associations may become increasingly common, and mutualisms that enhance the competitive or defensive capacity of the host may be especially favored (Vermeij, 1983a).

It is impossible at present to test this hypothesis. Indirect support comes from the observation that both the incidence and degree of specialization of marine intimate associations increase in the same directions as the risk and selective potential of predation; that is, they increase from the temperate zones to the tropics, and within the tropics from the Atlantic to the Indo-West-Pacific regions (Vermeij, 1983a).

The subject of mutualism and its impact on other organisms is far too complex and paleontologically too little understood for detailed treatment in this book. Many mutualisms—the association between fungi and the roots of vascular plants, the dispersal of spores and seeds by animal vectors, and the symbiosis between cellulose-digesting prokaryotes and herbivorous vertebrates, for example—can probably be studied in the fossil record, but currently available data are too fragmentary to permit a coherent historical analysis. The association between flowering plants and animal pollinators has already been mentioned briefly in Chapter 4. Although this association dates from at least the Late Carboniferous, it became specialized and common only from the Cretaceous onward. Here I shall treat the so-called hermatypic association between single-celled primary producers and immobile marine invertebrates. This is the only form of mutualism for which an adequately studied fossil record is currently available.

HERMATYPIC ASSOCIATIONS

Partnerships between single-celled photosynthesizers or nitrogen fixers and immobile animals create organisms with formidable competitive ability, even if the animal host is solitary. Cowen (1983) has reviewed the history of these partnerships, which I shall refer to as hermatypic. Nearly all the common immobile animals that live on light-exposed hard surfaces on tropical reefs in shallow waters are hermatypic. Like plants, they are net oxygen producers. In most hermatypes, the algal symbionts are housed in soft tissues which are directly exposed to light, but in the cardiid pelecypod *Corculum* and the brackish-water trapeziid pelecypod *Fluviolanatus*, the shell is constructed so as to transmit light to the underlying mantle tissues containing the symbionts (B. S. Morton, 1982a).

Hermatypic animals apparently owe their competitive success to rapid growth and large size. In skeletonized forms, the photosynthetic activity of the guest cells enhances calcification, so that the skeleton is often massive. Zooxanthellae (symbiotic dinoflagellates) have the opposite effect in certain boring clionid sponges and may in fact promote the excavating powers of these sponges by enhancing the dissolution of calcium carbonate (Vacelet, 1982). Slower-growing ahermatypic animals are typically out-competed by hermatypic relatives in nutrient-poor warm shallow seas. They are found chiefly on shaded surfaces and in deep water. The coral *Tubastrea* constitutes an intriguing exception (Wellington and Trench, 1985). On shallow reefs in the Palau Islands this ahermatype coexists with, and is competitively superior to, a number of hermatypic corals, including *Porites*. Despite its lack of symbionts, this coral grows rapidly. It is rarely attacked by predators, possibly because of chemical deterrents. The fast growth of *Tubastrea* is associated with a delicate skeleton that breaks easily in turbulence, unlike the skeletons of most co-occurring hermatypic species.

Hermatypic animals also are generally more prominent on reefs than are large-bodied algae. The reasons for this are obscure. Hermatypes may be able to exploit food resources (larger planktonic animals) that are unavailable to algae, and many have developed effective competitive and antipredatory structures such as nematocysts (stinging cells), sweeper tentacles, and toxins (Bakus, 1969, 1981).

In the modern ocean, the hermatypic habit is particularly well developed in the tropical Western Pacific and Indian Oceans and somewhat less well in the tropical Western Atlantic, with development in the Eastern Pacific and Eastern Atlantic falling far behind (Vermeij, 1983a). Tridacnid clams, the helioporid coenothecalian coral *Heliopora*, and many scleratinian coral genera have become confined to the Indo-West-Pacific following extinction in the Americas after the Eocene or Miocene (Rosewater, 1965; Heck and McCoy, 1978; Glynn and Wellington, 1983; Colgan, 1984). Didemnid ascidians of the Indo-West-Pacific are anatomically more specialized for culturing the alga *Prochloron* than are American species.

With living hermatypic animals as models, Cowen (1983) and others have sought to recognize hermatypes among fossil groups that have no living relatives. At least four groups in the Late Paleozoic seem to have been hermatypic. Richthofeniacean and lyttoniacean brachiopods and fusulinacean (especially verbeekinid) foraminifers acquired features consistent with a hermatypic interpretation during the Late Carboniferous, whereas the alatoconchid pelecypods appeared first in the Early Permian. All four groups were found at low latitudes in association with reefs and persisted until some time near the close of the Permian. Like some modern large hermatypic foraminifers, the fusulines had large, complexly chambered tests, but their form tended to be cylindrical rather than disc-shaped as in most modern forms. The brachiopods also were often large, with the lower valve either flat and irregular or horn-shaped or coralliform, and always cemented to the sea bottom. The upper valve was thin and delicate, and was evidently covered externally by mantle tissue, which in turn was protected in some species by a grillwork of spines growing from the exposed lip of the lower valve (Rudwick, 1961; Rudwick and Cowen, 1967). Alatoconchids were large pelecypods, up to 1 m long, whose valves had huge dorsoventrally flattened lateral expansions that could have contained photosynthesizing tissues (Yancey and Boyd, 1983).

New hermatypic groups arose during the course of the Mesozoic, but not immediately after the disappearance of the Paleozoic groups. Hippuritacean pelecypods, known as rudists, originated in the Late Jurassic and became gigantic reef-builders during the Cretaceous (Kauffman and Sohl, 1974; Skelton, 1978). Many species had massive lower conical valves and thin, delicate upper valves that either transmitted light or were porous. During the Santonian stage of the Late Cretaceous, some foraminifers became large and probably acquired photosynthesizing symbionts. Rudists and the Cretaceous larger foraminifers disappeared during or shortly after the crisis at the end of the Cretaceous, but the scleratinian stony corals, which appeared in the Middle Triassic but became hermatypic only in the Jurassic (G. D. Stanley, 1979; Cairns and Stanley, 1982), persisted into the Cenozoic era. Larger foraminifers again evolved during the Eocene, but these disappeared at the end of that period. Tridacnid giant clams originated during the Eocene but probably did not become hermatypic until the Miocene, during which stage the modern groups of larger hermatypic foraminifers also became established.

Because all the hermatypes known from the fossil record are reef-associated, the question arises whether pre-Carboniferous reef organisms were also hermatypic. This question must remain unanswered for the present, but I would not be surprised if tabulates and stromatoporoids—groups originating in the Ordovician and continuing as constructors of reef frameworks to the Late Devonian—had algal symbionts. The probable descendants of tabulates and stromatoporoids—the sclerosponges—are known today only in deep water or on shaded surfaces within the reef framework. They grow slowly and lack algal symbionts (Vacelet, 1981). Other groups of sponges with rigid calcified skeletons—the Sphinctozoa and various other Calcarea—were also ecologically more widespread in the Paleozoic and also lack symbionts today (Vacelet, 1981). Nevertheless, the possibility that ancestors of these relicts were hermatypic cannot be discounted.

Even if there are fossil groups that remain unrecognized as hermatypes, the available evidence indicates that the Late Paleozoic and the later Mesozoic and Cenozoic provided highly favorable conditions for the evolution and maintenance of the hermatypic condition. The high correlation in modern seas between the abundance of hermatypic animals and a relatively low availability of water-borne nutrients in warm shallow water further suggests that the Late Paleozoic was a time of low marine tropical productivity. This inference was also drawn by Tappan (1968, 1970, 1982), who has commented on the impoverished Late Paleozoic record of preserved phytoplankton.

The fossil record of hermatypes shows plainly that, despite the apparent competitive superiority of hermatypic animals over immobile animals without photosynthesizing symbionts, hermatypic groups have been unusually vulnerable to extinction during the great biotic crises. As will be seen in Chapter 14, this unusual susceptibility is typical of many organisms with exceptional competitive and defensive capacities.

SUMMARY

Several environments and situations have served as safe places for species whose architecture and mode of life have become adaptively obsolete and anachronistic in high-energy environments in the photic zone, where escalation between species and their enemies has proceeded very far. Such environments include habitats in deep water, in caves and interstitial environments between sand grains, in freshwater lakes and streams, beneath the surface of unconsolidated sediments (infaunal environment), beneath the surface of rocks (endolithic environment), on the dry land, and on or in the bodies of well-defended organisms. Although the invasion of some of these environments (especially deep and dark habitats) had little impact on evolutionary events elsewhere, occupation of most other safe habitats had profound consequences for species in other environments. Burrowing infaunal animals, for example, recycle nutrients that were previously lost in deep layers of unconsolidated sediments; these animals may therefore enhance primary productivity at and above the sea bottom. The increases in burrowing activity during the Early Cambrian, Middle Ordovician, Late Silurian, Early Devonian, and the later Mesozoic and Cenozoic may therefore have stimulated adaptive escalation in the sea. The borings of large endolithic animals provide shelter for a great number of organisms, including predators, which live on the exposed surfaces of shallow-water reefs and sands. Because deeply excavating species are known only from the

Mesozoic and Cenozoic, this stimulative effect of endolithic animals on adaptive escalation in shallow-water marine environments is a relatively recent phenomenon. The dry land originally served as a safe place for marine species during the Middle Paleozoic (Silurian and Devonian), but by the Late Paleozoic it began to export species back to the sea. The marine descendants of land organisms have contributed importantly to marine escalation and include in their ranks deeply excavating sediment-disturbers, powerful predators, sediment-binding sea grasses, and man. Intense predation and competition in high-energy environments may have favored the evolution of associations between poorly defended guest species and well-defended hosts. Some of these associations became mutually beneficial and increased the competitive and defensive capacities of the host. Hermatypic associations between single-celled primary producers and immobile marine animals—associations which evolved during the Late Paleozoic and again during the later Mesozoic and Cenozoic—have been of primary importance in reef-building communities.

In short, the occupation of safe places may have been stimulated by the evolution of increasingly formidable competitors and predators in the high-energy environments of the photic zone, but the descendants of the invaders often reentered high-energy environments and further stimulated adaptive escalation there. This occurred because resources that were previously unavailable to organisms were recycled into the economy of life by species which through time penetrated previously unoccupied environments.

Although the main outline of this scenario appears to be well established, much remains to be learned about the timing of the invasion of safe places and about the adaptive characteristics and ecological origins of the invaders. It is still unclear why escalation in fresh waters has been generally modest, and under which conditions freshwater species have been able to invade the generally more escalated communities in the sea. The history and ecological impact of mutualistic associations is a particularly fruitful area for further research. Even for Recent environments, there is surprisingly little systematically collected information available about the geographical distribution of mutualisms and other intimate associations between species, and about the hypothesis that highly defended species are especially susceptible as targets for poorly defended guests.

CHAPTER SIX

The Predators of Armored Animals:

Functional Morphology and History

Predation is one of the major ways by which animals acquire food. The methods animals use to subdue prey vary almost without end. Whichever method is used, it has a great impact not only on prey species, but also on the way the predator copes with its own enemies. In the presence of competitors and other animals that pose risks to a predator, an increase in the speed with which a victim is subdued and consumed should benefit the predator (Bakker, 1980, 1983). This increase in the speed of subjugation may occur when victims are dismembered rapidly, paralyzed to prevent retaliation, or attacked by groups of predators (Enders, 1975; D. Griffiths, 1980). Instruments for rapid subjugation are frequently useful in combating the predator's own enemies. Slow predatory methods are expected in animals that are themselves highly resistant to attack, and therefore relatively "safe," and in species in low-energy environments where enemies are rare.

One way of expressing the hypothesis of escalation is that individuals are exposed to increasingly greater risks from their biological enemies over the course of earth history. If this hypothesis is true, we should see a temporal trend toward greater reliance on methods of rapid subjugation. To evaluate this hypothesis, we must first survey the methods and capacities of modern predators, and then apply the results of this ecological and functional-morphological work to the predators of the past. The purpose of this chapter, therefore, is to identify the morphological characteristics that enable predators, using specific methods of subjugation, to subdue and consume victims; to discuss the limitations on speed of subjugation and size of victim that each of these methods entails; and to trace the history of molluscivory.

Rather than treat all predators, I have chosen for the following reasons to devote special attention to those that are capable of eating armored prey, especially shell-bearing molluscs. First, not only are some of these predators fairly well understood, but they also illustrate the characteristics of predators in general. Second, since most molluscivores include in their diet many animals other than molluscs, their ecological and evolutionary importance extends well beyond their effect on molluscs. Third, an understanding of molluscivores and how they work is important to the study of the function and history of animal armor. Finally, the animals and their predatory methods are interesting in their own right, and I am not reluctant to dwell on animals that I find fascinating.

METHODS OF SUBJUGATION

Methods of subjugation animals use to subdue shell-bearing victims may be grouped conveniently into five more or less distinct categories: (1) whole-animal ingestion—the victim is swallowed, enveloped, or smothered without damage to the skeleton; (2) insertion and extraction—the enemy penetrates between the valves or plates of the victim's skeleton or enters the aperture of the victim's shell, but it does not damage the skeleton as the flesh is removed; (3) transport—the victim is grasped and then transported to an environment where it could not survive even if the predator were unsuccessful in killing it; (4) pre-ingestive breakage—death is brought about by mechanical destruction of the skeleton before the victim is eaten; and (5) drilling—a small hole is made chemically or mechanically through the victim's skeletal wall without other damage to the skeleton. These five methods will each be discussed in detail below.

WHOLE-ANIMAL INGESTION

Perhaps the most widespread form of predation on molluscs is whole-animal ingestion, in which the prey is swallowed or enveloped whole before digestion begins. Because the prey enters the predator's body intact, it is typically much smaller in size than the predator. Skeletal dismemberment may follow ingestion, as in philinid opisthobranch gastropods and in many fishes and birds, but in most instances it is secondary to the act of swallowing or enveloping in causing the prey's death.

Table 6.1 provides a summary of the living predators that employ whole-animal ingestion as a method of eating molluscs. Most of these predators have very broad diets that include many animals in addition to molluscs. Only the gastropods tend to be specialized to a molluscan diet. Rachiglossan gastropods grasp and envelop the prey by the large foot before swallowing it. The large foot is associated in volutids with a large, wide aperture, but in olivids the foot extends beyond and over the shell, and the aperture is relatively narrow. Accordingly, we can identify no consistent features of the shell that would indicate a predatory habit of swallowing or enveloping hard-shelled prey. A large aperture also occurs in gastropods that ingest whole naked prey such as sea cucumbers (eaten by tonnids) and fishes (eaten by some conids). The large aperture of naticids and cassids is associated with a broad foot that grasps prey in preparation for drilling (Ziegelmeier, 1954; Kohn, 1956; Hughes and Hughes, 1981; Kropp, 1982).

Fishes catch particles of food either by darting forward with open mouth or by suction. Many primitive fishes use the darting method to catch prey. In many post-Paleozoic elasmobranchs as well as in cypriniform and most acanthopterygian (spiny-rayed) bony fishes, however, the upper jaw has become protrusible, so that particles are caught by rapid forward movement of the upper jaw rather than by lunging of the whole body (Moss, 1977; Osse, 1985). Suction is made possible by changes in the volume of the buccal and opercular cavities. When the mouth is closed, these cavities become smaller and a pressure difference is created; thus, when the mouth opens, water is sucked in. Lateral flexibility of the jaw skeleton makes this method of feeding possible. This flexibility, as well as that associated with upper-jaw protrusibility in bony fishes, is incompatible with strong biting and crushing (R. M. Alexander, 1967; Lauder, 1982; Osse, 1985). Swallowing of food is effected by muscles and skeletal elements in the pharyngeal (throat) region. The pharyngeal system in many bony fishes has been adapted to macerate food. In some labrids (wrasses) and freshwater cichlids, for example, the bones in the pharyngeal mill are greatly thickened and provided with molar-like teeth that enable the fish to grind the shells of prey that have been ingested (Liem, 1973; Yamaoka, 1978).

Swallowing whole prey is widespread in birds. Many birds grind food in the gizzard after swallowing, but it is noteworthy that the species with the most powerful crushing devices in the throat are specialized to a diet of stone-like fruits rather than hard-shelled animals. Examples include the turkey (*Meleagris gallopavo*) and the extinct Mauritian dodo (*Graphis cucullatus* (Temple, 1977).

Data on the speed with which individual prey are subdued by whole-animal ingesters are scarce. It takes the sea star *Luidia clathrata* just over a minute to ingest the beach clam *Donax variabilis* (McClintock and Lawrence, 1981), but studies by Christensen (1970) and Massé (1975) on *Astropecten* species show that these sea stars may retain prey in the digestive system for as long as two weeks before egesting it. Birds take hundreds of small prey per day.

The problem with whole-animal ingestion is not the slow speed of predation but the severe limitation on prey size. Even if an item can be swallowed rapidly, the size of individual prey is limited by the size of the mouth opening, and only a few prey can fit into the digestive system at any one time. This limitation is mitigated somewhat in species that break shells after swallowing, but the bulk of the inedible skeleton still must be passed through the digestive system and therefore still "competes" for space with edible flesh.

Predators that ingest their molluscan prey whole have a wide ecological and geographical distribution, but they are especially common and diverse on marine unconsolidated bottoms (sand and mud) and in fresh water (see Table 6.1). All gastropods and sea stars that use this method of predation are found in sand and mud, and most are capable of burrowing. Fishes that consume shell-bearing prey by swallowing them whole and leaving the shell intact in the digestive system also are found mostly on unconsolidated bottoms, although some blenniids, gobiesocids, and cottids are found on rocky shores. Wrasses, grunts, and other fishes that grind the shells in the throat are typically found over hard bottoms.

INSERTION AND EXTRACTION

The chief limitation of whole-animal ingestion as a method of predation is that only relatively small prey animals can be ingested. The

TABLE 6.1 Living Molluscivores Using Whole-Animal Ingestion

Phylum Cnidaria, Class Anthozoa, Order Actinaria: *Anthopleura, Condylactis, Stichodactylus*; feed on molluscs that fall into rock crevices (Dayton, 1973; Sebens, 1976; Minchin, 1983)

Phylum Annelida

Class Polychaeta: Eunicidae, Lumbrinereidae, Nephtyidae, Polynoidae; eat small molluscs, possibly using hardened proteinaceous jaws and eversible pharynges (Fauchald and Jumars, 1979)

Class Hirudinea (leeches): Glossiphoniidae; eat freshwater molluscs (Branson, 1963; Townsend and McCarthy, 1980)

Phylum Mollusca

Class Gastropoda

Order Rachiglossa: Olividae, Volutidae; eat molluscs in sand by enveloping prey in expanded foot (Marcus and Marcus, 1959; Ponder, 1970; Marche-Marchad, 1977; Vermeij, 1978)

Order Toxoglossa: some Conidae; eat molluscs in sand by ingesting prey whole inside digestive tract (Kohn, 1959)

Subclass Opisthobranchia: Retusidae, Aglajidae, Philinidae; eat molluscs on sand or mud; Philinidae crush prey shell with calcified gizzard plates after ingestion (Paine, 1963b; Blair and Seapy, 1972; Rudman, 1972a, b; Rasmussen, 1973; Gosliner, 1980)

Phylum Echinodermata

Class Stelleroidea (sea stars)

Order Paxillosida: Astropectinidae, Luidiidae; swallow molluscs on sand and mud (Mauzey et al., 1968; Christensen, 1970; Massé, 1975; Jangoux, 1982)

Order Forcipulatida: some Asteriidae (*Pycnopodia*); swallow molluscs on sand (Mauzey et al., 1968; Shivji et al., 1983)

Class Ophiuroidea (brittle stars): a few genera (*Ophiura, Ophionotus*); swallow small molluscs on sand and mud (Feder, 1981; Warner, 1982; Fratt and Dearborn, 1984)

Table 6.1 (*continued*)

Phylum Chordata

Class Chondrichthyes, Subclass Elasmobranchii, Order Rajiformes (rays): Rajidae (skates); some swallow whole pelecypods on sand (du Buit, 1978)

Class Osteichthyes (bony fishes), Subclass Actinopterygii

Superorder Elopomorpha

Family Albulidae (bonefishes); swallow whole molluscs first, then crush prey in pharyngeal mill; prey are mainly from sandy habitats (Warmke and Erdman, 1963)

Order Anguilliformes: Anguillidae (eels); swallow whole molluscs on sand (Wenner and Musick, 1975)

Superorder Euteleostei

Order Salmoniformes: Salmonidae (trout); swallow freshwater molluscs (Hart, 1973)

Order Cypriniformes: Cyprinidae (minnows); swallow freshwater molluscs, then crush them in pharyngeal mill (Corbet, 1961; Stein et al., 1975; Heller, 1979)

Order Siluriformes (catfishes): Pangasiidae, Pariidae, Ictaluridae, Mochocidae; swallow whole freshwater and some marine molluscs (Corbet, 1961; Turner and Roberts, 1978)

Order Gadiformes (cod and relatives): Gadidae; swallow whole molluscs on sand and mud (Hart, 1973; Arntz, 1978; Sedberry and Musick, 1978)

Order Gobiesociformes (clingfishes): Gobiesocidae; swallow molluscs on hard bottoms (Hart, 1973; Paine and Palmer, 1978)

Order Lophiiformes: Ogcocephalidae; crush prey in pharyngeal mill after swallowing prey on hard bottoms (Randall, 1967)

Order Beryciformes: Holocentridae (squirrelfishes); swallow molluscs on sand (Hiatt and Strasburg, 1960; Randall, 1967)

Order Perciformes: Mullidae (goatfishes) swallow molluscs over sand; Nototheniidae, Blenniidae, Tripterygiidae swallow whole molluscs, mainly on hard bottoms; Clinidae swallow whole molluscs on hard bottoms; Centrarchidae, Cichlidae swallow freshwa-

Table 6.1 (*continued*)

---

ter molluscs, then crush prey in pharyngeal mill; Branchiostegidae, Pomadasyidae (grunts), Leiognathidae, Sillaginidae (Australian whitings), Labridae (wrasses), Embiotocidae (surfperches) swallow molluscs from hard and unconsolidated bottoms, then crush prey in pharyngeal mill (Randall, 1967; Hobson, 1968; Quast, 1968; De Martini, 1969; Fryer and Iles, 1972; Chao, 1973; Vivien, 1973; Greenwood, 1974; Heller, 1979; Reimchen, 1979; B. Bennett et al., 1983; Lauder, 1983b; B. C. Russell, 1983)

Order Scorpaeniformes: Cottidae (sculpins); swallow prey whole, mainly on hard bottoms (A. L. Rice, 1962; Moore and Moore, 1974)

Class Amphibia

Order Anura: some Ranidae (freshwater frogs) and Hyperoliidae (African terrestrial frogs); swallow prey whole (Elliott and Karunakaran, 1974; Drewes and Roth, 1981)

Order Urodela (salamanders): some Sirenidae and Salamandridae; swallow freshwater molluscs whole (D. E. Gill, 1978; Hanlin, 1978)

Class Aves (birds)

Order Charadriiformes: many Laridae (gulls), Charadriidae (plovers), Scolopacidae (sandpipers); swallow shallow-water molluscs from hard bottoms, sand, and mud (Harris, 1965; A.J.M. Walker, 1972; Burton, 1974; Pettitt, 1975; R. P. Simpson, 1976)

Order Ralliformes (rails): many Rallidae; swallow terrestrial, freshwater, and shallow-water marine molluscs (Moffitt, 1941; Simpson, 1976; Heard, 1982)

Order Strigiformes (owls): some Strigidae; swallow terrestrial gastropods (Mienis, 1971; Zinner, 1978)

Order Passeriformes (perching birds): Motacillidae (pipits); swallow small intertidal gastropods (Gibb, 1956)

Order Anseriformes: many Anatidae (ducks); swallow freshwater and marine molluscs from hard bottoms and sand (Bartonek and Hickey, 1969; Brun, 1971; Pettitt, 1975)

---

other four predatory techniques all enable predators to take relatively larger prey, because part of the food preparation takes place before flesh is ingested into the body.

The first technique that permits predation on larger prey is insertion and extraction. A summary of living predators that are more or less specialized for inserting part of the body into the prey's skeleton or for extracting flesh by way of the aperture is presented in Table 6.2.

Little is known about the specializations that permit turbellarian and nemertean worms to penetrate molluscan prey. The flattened form of turbellarians enables them to attack limpets under the shell. *Stylochus* and related polycladid turbellarians penetrate between the valves of pelecypod prey and slowly eat the soft tissues within.

Insertion is a common predatory technique among gastropods. It is typically done with the proboscis and may involve the injection of a venom or anesthetic into the victim. Pulmonate land snails using insertion typically thrust the entire head-foot complex into the prey's shell. None of these techniques is reflected in the shell shape of the predator, so that the habit of insertion or extraction cannot be inferred in fossils from shell form alone. Some of the thaidids that are specialized for inserting the proboscis under the shells of limpets and chitons and for overturning these prey before the flesh is rasped out have broad apertures; in this way they are indistinguishable from wide-apertured gastropods that ingest their victims whole.

Some buccinacean rachiglossan gastropods, notably members of the busyconine Melongenidae, Buccinidae, and Fasciolariidae, employ the thin, strong edge of the outer shell lip to wedge open the valves of tightly closing pelecypods (Carriker, 1951; H. W. Wells, 1958; Paine, 1963a; Maes, 1967; Nielsen, 1975; B. W. Kent, 1983a, b). The outer lip of the valve-wedging forms protrudes somewhat ventrally, so that the edge of the aperture does not quite lie in a plane as it does in most other crawling gastropods. Because the predator's shell lip is often slightly damaged during attacks on pelecypods, valve-wedging species such as *Sinistrofulgur contrarium* and *Buccinum undatum* frequently have a high incidence of repaired breaks. Nielsen (1975) found a thick-shelled population of *B. undatum* in which half the individuals had such scars, whereas a thin-shelled population, whose individuals probably were unable to wedge pelecypods, lacked scars.

Octopod cephalopods use extraction as one of two principal modes

TABLE 6.2 Living Molluscivores Specialized for Insertion or Extraction

---

Phylum Platyhelminthes (flatworms), Class Turbellaria: *Freemania, Notoplana, Stylochus, Dugesia, Dendrocoelum, Geoplana*; eat terrestrial gastropods, freshwater and marine gastropods, and marine pelecypods (Mead, 1963; Galleni et al., 1980; Phillips and Chiarappa, 1980; Townsend and McCarthy, 1980)

Phylum Nemertea (ribbon worms): *Cerebratulus, Emplectonema, Lineus*; eat marine molluscs on sand and mud; extend eversible proboscis (Kalin, 1984; McDermott and Roe, 1985)

Phylum Mollusca

    Class Gastropoda

      Subclass Prosobranchia

        Order Canalifera: Cymatiidae; insert proboscis and anesthetize prey; eat gastropods and pelecypods on hard bottoms and sand (Houbrick and Fretter, 1969; Vermeij, 1978; J. D. Taylor, 1983)

        Order Rachiglossa: some Muricidae, Thaididae, Buccinindae, Melongenidae, Fasciolariidae, Costellariidae; eat mainly gastropods, but some buccinids and fasciolariids also take pelecypods, chiefly on sand; some melongenids pry open pelecypods by force with shell; some muricids and thaidids take hard-bottom gastropods by prying (Paine, 1963a; Luckens, 1975; Maes and Raeigle, 1975; J. D. Taylor, 1976, 1978, 1983, 1984; Ingham and Zischke, 1977; Vermeij, 1978; Branch, 1979; Fairweather and Underwood 1983; B. W. Kent, 1983a, b; Duncan and Hughes, 1984; Shimek, 1984; Taylor and Reid, 1984)

        Order Toxoglossa: some Conidae; spear and envenomate gastropods under rocks and in sand (Kohn, 1959, 1978; Kohn and Nybakken, 1975)

      Subclass Pulmonata, Order Stylommatophora: Streptaxidae, Haplotrematidae, Rhytididae, Systrophidae, Oleacinidae; insert body into land snail prey (Ingram, 1942; F. X. Williams, 1951; Solem, 1974; Kasigwa et al., 1983; A. Cook, 1985)

    Class Cephalopoda, Order Octopoda: many species of *Octopus*; many gastropods and pelecypods forcibly opened on hard bottoms and on sand (Arnold and Arnold, 1969; Wodinsky, 1969, 1973; Fotheringham, 1974; Nixon, 1979)

Table 6.2 (*continued*)

Phylum Crustacea, Class Malacostraca, Order Decapoda, Infraorder Brachyura (true crabs): many species facultatively extract prey on hard bottoms and on sand; methods described for Cancridae, Portunidae, Grapsidae, Thelfusidae (freshwater species), Xanthidae, Menippidae (Hiatt, 1948; Ebling et al., 1964; Kitching et al., 1966; Shepherd, 1973; Hamajima et al., 1976; Vermeij, 1977a, 1978; M. J. Williams, 1978; Zipser and Vermeij, 1978; Boulding, 1984; Cunningham and Hughes, 1984)

Phylum Chelicerata, Class Arachnida, Order Opiliones (harvestmen): some Trogulidae; extract land snails with chelicerae (Pabst, 1953)

Phylum Uniramia, Class Hexapoda (insects)

> Order Diptera (flies): larval Sciomyzidae; enter shells of freshwater gastropods (C. O. Berg, 1964; Barraclough, 1983)

> Order Coleoptera (beetles): Dytiscidae (freshwater), terrestrial Carabidae (especially the tribe Cychrini), Silphidae, Drilidae, Lampyridae; insert head into shells of land snail prey (Heymons et al., 1927; Heymons and Lengerken, 1932; Schmid, 1933; F. X. Williams, 1951; Mead, 1961; Solem, 1972; A. Greene, 1975; Pollard,1975)

Phylum Echinodermata, Class Stelleroidea (sea stars)

> Order Spinulosida: Solasteridae (*Crossaster*); digest prey on hard bottoms extraorally (Mauzey et al., 1968; Hancock, 1974)

> Order Valvatida: Asterinidae; digest prey on hard bottoms extraorally (Jangoux, 1982)

> Order Forcipulatida: Heliasteridae, Asteriidae; open gastropods and pelecypods by force on hard bottoms and sand (Mauzey et al., 1968; Jangoux, 1982; Shivji et al., 1983; Sloan and Robinson, 1983)

Phylum Chordata

> Class Chondrichthyes, Subclass Elasmobranchii, Order Squaliformes: Scyliorhinidae (dogfishes); extract large marine gastropods on sand (Eales, 1949)

> Class Osteichthyes
> > Order Perciformes: some Cichlidae (freshwater) and Labridae (marine); eat gastropods on hard and sandy bottoms (Fryer and Iles, 1972; Greenwood, 1974; Bertness, 1982)

Table 6.2 (*continued*)

Order Scorpaeniformes: Hexagrammidae (greenlings); extract large marine gastropods on hard bottoms (A. R. Palmer, 1977)

Order Gadiformes: Gadidae (cod); crop siphons of pelecypods (Arntz, 1978)

Order Anguilliformes: Anguillidae (eels); crop siphons of pelecypods (Wenner and Musick, 1975)

Order Pleuronectiformes (flatfishes); crop siphons of pelecypods (Edwards and Steele, 1968; Trevallion et al., 1970; de Groot, 1971; Braber and de Groot, 1973; Peterson and Quammen, 1982)

Class Reptilia, Order Squamata, Suborder Serpentes (snakes): Colubridae (Dipsadinae and Pareinae); extract land snails (Peters, 1960; Savitzky, 1983)

Class Aves (birds)

Order Charadriiformes: Haematopodidae (oystercatchers), Laridae (gulls), Chionididae (sheathbills); overturn and dislodge chitons and limpets; insert bill into pelecypods and gastropods; eat mainly marine molluscs on hard bottoms (Feare, 1971; A.J.M. Walker, 1972; R. P. Simpson, 1976)

Order Gruiformes: Aramidae (limpkins); insert bill into freshwater pelecypods and gastropods (Snyder and Snyder, 1969)

Order Falconiformes: Circidae (kites); grasp foot of freshwater gastropods (Snyder and Snyder, 1969)

Order Ciconiiformes: Ciconiidae (*Anastomus*); insert bill between valves of pelecypods in freshwater habitats (Root, 1963).

Order Passeriformes: Corvidae (*Cassidix*); grasp foot of freshwater gastropods (Snyder and Snyder, 1969)

Class Mammalia, Order Carnivora: Phocidae (*Erignathus*, bearded seal), Odobenidae (*Odobenus*, walrus); attack exposed foot or suck out flesh of permanently gaping Arctic sand-bottom pelecypods (Lowry et al., 1980; Oliver et al., 1983a) of predation. Suckers on their arms are capable of both pulling flesh out of the aperture of gastropods and pulling apart the valves of pelecypods enough to gain access to the soft tissues (Arnold and Arnold, 1969; Wodinsky, 1969; Nixon, 1979). Unfortunately, no measurements of the pulling strength of *Octopus* are currently available.

Although most brachyuran crabs crush or peel shell-bearing prey by placing the shell between apposing surfaces in the scissor-like claws (chelae), most species include in their repertoire of predatory behavior the technique of probing the claws into the aperture of gastropods or into gaps between the valves of pelecypods, and then pulling out the flesh. We have never witnessed extraction without breakage in the specialized shell-peeling Calappidae or in the massive-clawed, shell-crushing Parthenopidae and Carpiliidae (Zipser and Vermeij, 1978; Vermeij, 1982b) and know of no crab that has become specialized as an extractor. Presumably such specialization would involve the development of thin, elongate, and perhaps curved claws that would fit into restricted spaces. The mechanical advantage of compression in such claws is low, however, so that maximum pinching force is limited. Slender claws characterize agile portunids that feed on fish and other highly mobile prey.

In contrast to crabs, insects often have become specialized as molluscan extractors and penetrators. Beetles of several families attack land snails by inserting the long narrow head into the shell, grasping the soft parts with long curved mandibles, and liquifying the flesh (Crowson, 1981; see Table 6.2).

Sea stars are well known as predators that employ the insertion method of predation on molluscs. Most living sea stars are capable of extraoral feeding, in which the stomach is everted so that the prey is digested outside the predator's body. The stomach is capable of penetrating slits as narrow as 0.1 mm (Feder, 1955; Lavoie, 1956). Members of the forcipulate family Asteriidae have combined extraoral digestion with the ability to pull apart the adducted valves of pelecypods by means of four rows of tube feet on each of the five arms (Jangoux, 1982). The use of force allows the stomach to penetrate more deeply into the prey (Christensen, 1957). Estimates of the maximum pull exerted by various large asteriids range from 30.4 N in *Asterias forbesi* (Lavoie, 1956) to 45.0 N in *Pisaster ochraceus* (Feder, 1955) and 54.0 N in *Evasterias troscheli* (Christensen, 1957). There have been no careful studies of the way asteriids create and maintain forces of this magnitude, but Christensen has suggested that these sea stars may have a way of locking the skeleton in place as the longitudinal muscles in the tube feet contract to exert the pull.

Extraction of molluscan flesh by fishes probably occurs widely but remains poorly documented. Bertness (1982) performed laboratory trials that substantiated his suspicion that the labrid wrasse *Thalassoma* on the Atlantic coast of Panama pulls hermit crabs from shells by means of forward-projecting teeth in the jaw. Such incisor-like teeth are widespread among wrasses. Eales (1949) surmised that the opercula and pieces of flesh of *Buccinum* which she found in the digestive system of the dogfish *Scyliorhinus* were acquired by ripping the soft parts out of the shell. Pelecypods with protruding siphons are susceptible to siphon-cropping, a form of predation in which the distal portions of the siphons are bitten off by various fishes. Most of the siphon-cropping fishes are also capable of swallowing whole small pelecypods.

At least two groups of colubrid snakes have become specialized for extracting the soft tissues from the shells of land snails. They do so with slender jaws and curved maxillae that are inserted into the shell in a manner analogous to the way beetles use their mandibles.

The bill of birds is an effective structure for extracting flesh from gastropod shells and for penetrating between the valves of pelecypods. Individual oystercatchers (genus *Haematopus*) seem to be specialized for inserting the bill into pelecypod shells, whereas others hunting nearby hammer their prey with the bill (Drinnan, 1957; Norton-Griffiths, 1967; Heppleston, 1971; Baker, 1974). Insertion would seem to require a strong and slender bill, but extraction probably requires the application of considerable compressive force. In most animals, these requirements are more or less incompatible, but in birds this incompatibility may be less severe. Zusi (1984) has pointed out that even slender-billed birds can exert considerable compressive forces in the bill owing to the flexibility (kinesis) of the upper jaw, which can bend at one or more places along its length. Much work remains to be done on the functional morphology of bird beaks and on the forces that can be exerted by them, but it is already clear that the beak fills the bill of many kinds of feeding techniques.

The foregoing survey shows that, although some predators (gastropods, beetles, sea stars, and some fishes) are more or less specialized to predation by insertion or extraction, few exhibit diagnostic preservable features that are recognizable in fossils. The survey also shows that many predators use the techniques of insertion and extraction in addition to other methods, such as drilling and pre-ingestive breakage.

Predation by insertion and extraction is usually very slow, especially in gastropods, beetles, and sea stars. The melongenid gastropods *Sinistrofulgur contrarium* and *Busycotypus spiratus* require only 15 minutes to an hour to open and consume thin-shelled and gaping pelecypods, but it takes *S. contrarium* 8 to 10 hours to eat a single tightly closing pelecypod such as *Chione* or *Carditamera* (B. W. Kent, 1983a). *Conus pennaceus* in Hawaii takes up to an hour to extract flesh from gastropod victims (Kohn, 1959). This process probably would take substantially longer if the predator did not envenom its prey with a potent neurotoxin that is injected with the spear-like radular tooth. Sea stars take 40 minutes to 2 days or longer to open and consume pelecypods (Jangoux, 1982). *Orthasterias koehleri*, for example, completed its feeding on the tightly closing venerid pelecypod *Humilaria* after 2 days (Mauzey et al., 1968). *Crossaster papposus*, an extraorally feeding species that does not use force to pry open pelecypods, requires hours to open small victims and up to 8 days to subdue and consume large ones (Hancock, 1974). The carabid beetle *Cychrus hemphilli* takes 6 hours to consume an 8-mm land snail of the genus *Allogona*, and as long as 55 hours to take a 14-mm individual (A. Greene, 1975). Trogulid harvestmen take from 1 hour to 2 days to subdue and consume small European land snails (Pabst, 1953).

These slow methods of predation would be unsuitable for predators that are threatened frequently by enemies, but they are feasible in safe habitats or for predators that are themselves well protected against enemies. Insertion and extraction are distributed ecologically in accordance with this expectation. They are the chief methods used on land snails by predaceous worms, pulmonates, terrestrial arthropods, and snakes. I suspect that these predators live in a relatively safe environment—the leaf litter and soil—where movement by animals is limited and disturbance from rivals may be infrequent. Forcipulate sea stars that feed extraorally on molluscs are found chiefly in cold waters on hard as well as on unconsolidated bottoms (Vermeij, 1978; B. A. Menge, 1982). Most marine gastropods that insert or extract are found on unconsolidated bottoms, but some cymatiids and most thaidids are typical of hard-bottom habitats in warm seas. The latter gastropods often have extremely strong shells and as adults therefore may be relatively well protected from predators.

TRANSPORT

Some highly mobile animals are capable of collecting prey and then transporting it to a site where it can be killed at a convenient time or in a way that would pose undue risks to the predator in the prey's own environment. Many octopods, crustaceans, insects, and mammals transport food, but in most instances the prey can survive if the predator is unable to overcome the prey's defenses. When marine or freshwater molluscs are transported to land, however, survival is unlikely even if the predator fails to subdue or eat the prey.

Birds and mammals are the best-known transporters of marine and freshwater molluscs. Larid gulls and corvid crows often bring prey to sites where it can be dropped from a height of several meters onto hard surfaces, thereby shattering the shell (Siegfried, 1977; Ingolfsson and Estrella, 1978; Zach, 1978; B. W. Kent, 1981c). Muskrats (*Ondatra zibethica*), boat-billed storks (*Anastomus lamelligerus*), and other species carry freshwater mussels to the banks of streams and lakes, where the valves eventually gape widely enough for the claws or bill to be inserted (Bovbjerg, 1956; Root, 1963; Snyder and Snyder, 1969). The South American otter *Lutra felina* employs a similar technique on marine molluscs (Castilla and Bahamondes, 1979).

PRE-INGESTIVE BREAKAGE

Predatory methods that involve the destruction of the prey's exoskeleton have evolved in many groups and can often be recognized in fossils. Breakage requires the application of force, which in turn calls for instruments that are capable of withstanding considerable stress. Earlier in this chapter I considered post-ingestive breakage, which occurs in fishes, birds, and some opisthobranchs. This section is devoted to a review of the functional morphology of living predators that break prey shells before the victims are swallowed. A summary of these predators is given in Table 6.3.

Breakage is sometimes the incidental consequence of other methods of predation. *Octopus dofleini*, for example, occasionally breaks pelecypods when drilling them or prying the valves apart (Hartwick et al., 1981). The sea stars *Orthasterias, Asterias,* and *Crossaster* may damage the valves of pelecypods while extracting flesh (Mauzey et al., 1968; Hancock, 1974), and *Meyenaster* sometimes crushes sea urchins (Dayton et al., 1977). In predation on thin-shelled deeply buried pelecypods, the melongenid gastropod *Hemifusus tuba* thrusts its proboscis through the prey's siphon, causing the victim to adduct its valves so vigorously that the ventral edges often break (B. S. Morton, 1985). Hughes and Hughes (1981) report that the helmet shell *Cassis* sometimes crushes the urchin *Diadema* under its heavy shell as the predator pounces upon the victim. In none of these cases has the predator come to rely on breakage as a method of predation.

An unusual form of skeletal breakage occurs in some muricacean gastropods whose outer shell lip is provided with a sharp spine. While feeding on barnacles, Californian species of *Acanthina* first lift the shell off the ground and then bring the spine of the lip down forcefully on the opercular plates of the barnacle from above. This action causes the plates either to break or to separate. If the spine is removed experimentally, the snail drills the victim through the opercular plates. Not only does drilling take 2 to 5 times longer than forced entry, but its probability of success (0.73 for unspined versus 0.94 for spined *Acanthina* attacking *Semibalanus* and *Chthamalus*) is somewhat lower (Sleder, 1981; Perry, 1985). Spine-bearing muricaceans belonging to many lineages also attack pelecypods at the commissure between the valves (Vermeij, 1978; B. W. Kent, 1981b), but it is not known if the spine is used to break the seal between the valves or merely to stabilize the predator as it drills the prey.

The phylum Crustacea has within its ranks some of the most abundant and powerful living shell-breakers. They employ three basic techniques to break shells: crushing, peeling, and pounding. In crushing, the whole shell is squeezed in one or more pulses between apposing surfaces until it shatters. Brachyuran crabs, alpheid shrimps, clawed lobsters, and a few hermit crabs use the chelipeds (modified first pair of walking legs) as instruments of crushing, whereas crayfishes and spiny lobsters crush prey between the molar-like surfaces of the greatly enlarged mandibles. Peeling results from the repeated application of force at the shell edge, which is broken back piece by piece until the retracted edible tissues are exposed. Although most crabs and lobsters will attack the shell edge or lip if the

TABLE 6.3   Living Molluscivores Specialized for Breaking Shells

---

Phylum Mollusca

Class Gastropoda, Subclass Opisthobranchia, Order Nudibranchia: *Dirona*; crush small gastropods in jaws (Robilliard, 1971)

Class Cephalopoda
Order Nautilida: Nautilidae; crush crustaceans and perhaps other animals in calcareous beak (Ward and Wicksten, 1980)

Order Sepiida (cuttlefishes); break crab carapaces in jaw (Hewitt and Watkins, 1980)

Phylum Crustacea, Class Malacostraca

Order Decapoda

Suborder Natantia: Alpheidae (snapping shrimps); crush small molluscs on sand (Beal, 1983)

Suborder Reptantia

Infraorder Anomala: some Paguridae, Diogenidae, Lithodidae (hermit crabs); crush echinoids and small molluscs in claws (Kunze and Anderson, 1979; Schembri, 1982)

Infraorder Astacura: many Astacidae, Cambaridae, and related crayfishes; crush freshwater molluscs in mandibles (Vermeij and Covich, 1978)

Infraorder Palinura: Homaridae, Nephropidae; crush prey in claws on hard bottoms and on sand; Palinuridae crush prey in mandibles, (Randall, 1964; Hollmann, 1969; Vermeij, 1978)

Infraorder Brachyura (crabs): many Cancridae, Matutidae, Portunidae, Xanthidae, Carpiliidae, Menippidae, Grapsidae, Ocypodidae, Parthenopidae; crush prey with claws (Kitching et al., 1966; Vermeij, 1977a, 1978; Zipser and Vermeij, 1978; Boulding, 1984; Cunningham and Hughes, 1984); Calappidae peel sand-dwelling prey with claws (Shoup, 1968; Vermeij, 1982a; Ng and Tan, 1984; Signor, 1985)

Order Stomatopoda: Gonodactylidae and related families; hammer prey with second pair of maxillipeds, mainly on hard bottoms (Kunze, 1981)

Table 6.3 (*continued*)

Phylum Chelicerata

Class Xiphosurida, Order Limulida: Limulidae (horseshoe crabs); crush molluscs in gnathobases, mainly on sand (O. R. Smith, 1953; Botton, 1984)

Class Arachnida, Order Opiliones (harvestmen): some Ischyropsalididae; peel land snails with chelicerae (Martens, 1969)

Phylum Uniramia, Class Hexapoda, Order Coleoptera (beetles): some Carabidae; crush land snails with mandibles (Solem, 1972; A. Greene, 1975; Pollard, 1975; Crowson, 1981)

Phylum Chordata

Class Chondrichthyes (cartilaginous fishes), Subclass Elasmobranchii

Orders Rajiformes (skates) and Myliobatiformes (rays): Rajidae, Myliobatidae, Rhinopteridae, Dasyatidae; crush prey in jaws, mainly on sand (Bigelow and Schroeder, 1953; Randall, 1967; Orth, 1975; Capapé, 1976)

Order Chimaeriformes: Callorhynchidae; crush prey in jaws, mainly on sand

Order Orectolobiformes: Orectolobidae, Triakidae, and allied shark families; crush prey in jaws mainly on sand (Randall, 1964)

Order Heterodontiformes: Heterodontidae (Port Jackson shark); crush molluscs and echinoids in jaws, mainly over reefs (Reif, 1976)

Order Carchariniformes: Carcharinidae (*Galeocerdo*); crush prey in jaws, mainly on sand (Randall, 1964)

Class Osteichthyes

Subclass Dipnoi (lungfishes): Lepidosirenidae; crush freshwater molluscs in jaws (Corbet, 1961)

Subclass Actinopterygii

Order Siluriformes (catfishes): Plotosidae; crush freshwater molluscs in jaws (Turner and Roberts, 1978)

Order Gadiformes: Zoarcidae (wolf eels); crush molluscs from hard bottoms and sand in jaws (Schäfer, 1972; Hart, 1973; Sedberry and Musick, 1978)

Table 6.3 (*continued*)

Order Batrachoidiformes (toadfishes): Batrachoididae; crush molluscs and echinoids from hard bottoms and sand in jaws (Matthews, 1968; Collette and Russo, 1981; Collette, 1983)

Order Cyprinodontiformes: Cyprinodontidae (killifishes); crush freshwater gastropods in jaws (Thomerson, 1969)

Order Anguilliformes: Muraenidae (*Echidna*, moray eels); crush crabs in jaws (Hiatt and Strasburg, 1960; Randall, 1967)

Order Perciformes: Sciaenidae (freshwater and saltwater drums), Percidae (freshwater perches), some Cichlidae (freshwater), Sparidae (porgies), Lethrinidae; crush molluscs in jaws, especially on hard bottoms (Hiatt and Strasburg, 1960; Darnell, 1961; Randall, 1967; Hobson, 1968; Manooch, 1977; M. H. Walker, 1978; B. C. Russell, 1983)

Order Tetraodontiformes: Ostraciidae (trunkfishes), Monacanthidae (filefishes), Balistidae (triggerfishes), Diodontidae, Tetraodontidae, Canthigasteridae (pufferfishes); crush molluscs, echinoids, and corals in jaws, mainly on hard bottoms (Randall, 1967; Vivien, 1973; Bell et al., 1978; Targett, 1978; Tyler, 1980)

Class Reptilia

Order Chelonia (turtles): Cheloniidae (*Caretta*, marine loggerhead turtle), freshwater Carettochelyidae, Kinosternidae, Chelydridae, Emydidae; crush prey in jaws (Snyder and Snyder, 1971; Mahmoud and Klicka, 1979; Pritchard, 1979; Mortimer, 1982; Bramble et al., 1984)

Order Squamata

Suborder Lacertilia (lizards): Teiidae, Varanidae, Iguanidae, Scincidae, Anguidae; crush terrestrial and freshwater molluscs in jaws (Dalrymple, 1979; Rieppel and Labhardt, 1979; Estes and Williams, 1984)

Suborder Amphisbaenia: some Amphisbaenidae; crush land snails in jaws (Pregill, 1984)

Order Crocodylia: many Crocodylidae (caimans and crocodiles); crush molluscs and turtles in jaws (Diefenbach, 1979; Carpenter and Lindsey, 1980; Erickson, 1984; E. R. Meyer, 1984)

Table 6.3 (*continued*)

Class Aves (birds)

Order Charadriiformes: Haematopodidae (oystercatchers); hammer prey with bill on hard bottoms and sand (Drinnan, 1957; Norton-Griffiths, 1967; Baker, 1974; Butler and Kirbyson, 1979)

Order Gruiformes: Aramidae (limpkins); hammer freshwater prey with bill (Snyder and Snyder, 1969)

Order Passeriformes: Turdidae (thrushes) break prey on rocks and other surfaces, mainly on land; Corvidae (crows) hammer with bill (Lamotte, 1950; Clegg, 1972; Shachak et al., 1981)

Class Mammalia

Order Insectivora: Desmanidae (freshwater desmans), Soricidae (terrestrial shrews); crush prey in jaws, sometimes peel (Pollard, 1975; Whitaker and French, 1984)

Order Rodentia: Cricetidae (voles, dormice, etc.), Muridae (mice, rats); crush and peel land snails in jaws (Yom-Tov, 1970; Pollard, 1975; Kasigwa et al., 1983)

Order Carnivora: Mustelidae (*Enhydra*, sea otter) crush molluscs and echinoids in jaw or pound them on hard surfaces, mainly on hard bottoms and sand; Viverridae (mongooses) hurl land snails against hard objects (F. X. Williams, 1951; Calkins, 1978; Hines and Loughlin, 1980; Kasigwa et al., 1983)

---

prey is large or hard to crush, specialization for peeling is apparently restricted to calappid crabs and palinurid spiny lobsters. Pounding is the repeated striking of the prey with a blunt instrument. Gonodactyloid stomatopods do this with the expanded propus segment of the second pair of maxillipeds.

Several features of the crustacean claw are diagnostic for crushing hard-shelled prey (Figure 6.1). The claw is a modified appendage whose most distal segment (the movable finger, or dactyl) closes on an extension (fixed finger) of the penultimate segment, the propus. The muscle that closes the claw is located in the propus behind the pivot of the dactyl and is attached to a planar tendon-like structure (apodeme) within the claw. The apodeme is attached in the interior

CRUSHER

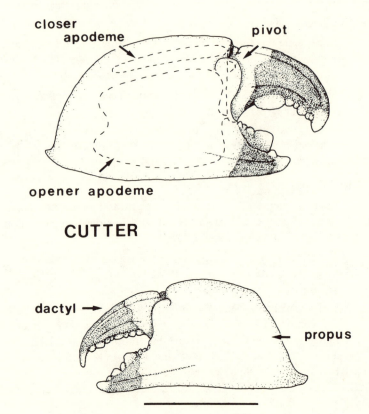

CUTTER

FIGURE 6.1 Crusher and cutter claws of the menippid crab *Menippe mercenaria*. The crusher claw is squatter than the cutter and bears molar-like teeth on the apposing fingers.

of the claw to the proximal end of the dactyl. In crabs and lobsters that are adapted for crushing, the right and left claws differ markedly in size, shape, and dentition. The crusher claw, which is usually on the right side in crabs and on either the right or the left side in homarid lobsters, is thick and stout, has an apodeme of large area, and is equipped with flattened or rounded teeth on the proximal portion of the apposing surfaces. Smaller, sharper, single-cusped teeth adorn the more distal parts of the apposing surfaces. The ends of the fingers are pointed and often cross each other when the claw is closed. The other claw, often called the cutter claw, is usually smaller and more slender and is equipped only with pointed teeth or serrations on the apposing surfaces. Its chief functions are to hold the prey and to tear flesh, whereas the crusher claw holds and crushes the prey. The crusher claw has a high mechanical advantage for delivering compressive force. Mechanical advantage is calculated as the ratio m.a., the distance between the pivot and the point of insertion of the apodeme on the dactyl divided by the distance from the pivot to the tip of the dactyl (Figure 6.1). There is a close correspondence between the degree of development of molar-like teeth and the mechanical advantage of the dactyl of the crusher claw. The claws of the crayfish *Procambarus clarki* (m.a. = 0.22) and the fiddler crab *Uca pugilator* (m.a. = 0.23) lack molars and are not used in feeding. There is a modest development of molars in the claws of the portunid blue crab *Callinectes sapidus* (m.a. 0.23 to 0.29); the cancrid rock crab *Cancer pagurus* (m.a. = 0.33); and the portunid green crab *Carcinus maenas* (m.a. = 0.36). Molar-like teeth are well developed in the American lobster *Homarus americanus* (m.a. = 0.33); the xanthoids *Ozius* (m.a. 0.35 to 0.40), *Eriphia* (m.a. 0.44 to 0.52), *Carpilius* (m.a. 0.45 to 0.58), and *Menippe* (m.a. = 0.50); and the parthenopid *Daldorfia horrida* (m.a. = 0.55) (Schäfer, 1954; Warner and Jones, 1976; Vermeij, 1977a; S. C. Brown et al., 1979; Abele et al., 1981; Elner and Campbell, 1981; Blundon and Kennedy, 1982a; Warner et al., 1982).

Although few measurements of the crushing force of claws exist, those that have been made support the idea that claws of high mechanical advantage and strong molar-like dentition are capable of applying the greatest forces. Direct measurements with strain gauges in the claw show that a large (13 cm wide) *Cancer productus* can muster a force of 178 N (Boulding, 1984), whereas a large (172 mm long) American lobster delivers 256 N in its crusher claw (Elner and Campbell, 1981). *Menippe mercenaria* exerts forces of at least 800 N (Blundon, 1986).

Many crabs and hermit crabs with unspecialized claws are capable of crushing thin or small shells. *Callinectes sapidus* and *Liocarcinus puber* (m.a. = 0.33), for example, crush molluscs even though their claws do not have well-developed molars (Muntz et al., 1965; Blundon and Kennedy, 1982a). Hermit crabs of the genera *Dardanus, Petrochirus, Parapagurus, Eupagurus,* and *Paguristes,* among others, are capable of crushing the shells of small molluscs and the tests of sea urchins in their squat crusher claws, even though these claws do not possess well-developed molars. Almost every clawed crustacean should therefore be viewed as a potential, if unspecialized, crusher.

Calappids have evolved a highly specialized right claw for peeling shells (Shoup, 1968). The dactyl bears a proximal peg-like tooth that projects obliquely downward on the outer face of the claw. This tooth is inserted into the aperture of a gastropod shell; when the claw is closed, the tooth fractures the lip. This process is repeated until either the flesh is exposed or the tooth can no longer be inserted into the opening or break the outer shell wall. Similar but less highly modified claws are seen in a number of xanthids, such as *Lydia*, and in the portunid genera *Thalamita* and *Portunus* (Ng and Tan, 1984).

The pounding technique of stomatopods is best developed in members of the superfamily Gonodactyloidea, in which the base of the penultimate segment of the second pair of maxillipeds is greatly thickened (Caldwell and Dingle, 1975; Kunze, 1981). Snails as well as hermit crabs fall victim to the blows of this hammer (Bertness, 1982). Other stomatopods, such as members of the genera *Oratosquilla* and *Cloridopsis*, are able to break thin-shelled pelecypods by spearing the prey with the sharp tip of the maxilliped (Dingle and Caldwell, 1978). Just as in crabs, therefore, the most specialized method of shell breakage is associated with a highly modified and diagnostic morphology, whereas more modest shell-breaking capacities are found in animals with a more generalized appearance.

Few other arthropods are shell-breakers. The horseshoe crab *Limulus* crushes prey in the claw-like gnathobases on each of the five pairs of walking legs, but the apposing surfaces of the claws are smooth, and only small and thin-shelled prey are taken. A few ischryopsalidid harvestmen have greatly enlarged claw-like chelicerae which are used in peeling the shells of European land snails (Martens, 1969). Many ants, termites, and spiders are well known for their powerful bite and for their ability to crack the armor of insects, but they apparently do not feed on molluscs.

Jawed fishes have evolved the crushing habit numerous times. In rays and skates, the prey is crushed between massive upper and lower jaw plates whose apposing surfaces are composed of fused teeth that form a pavement. A similar dental pavement has evolved in the tetraodontiform pufferfishes (Tyler, 1980). Large puffers are capable of crushing shells of *Thais melones* and *Drupa morum* (A. R. Palmer,

1979; Bertness et al., 1981) that in a testing machine require more than 5 kN of force to break (Vermeij and Currey, 1980). Most other shell-crushing fishes have separate, broad molar-like teeth in the jaws. Reif's (1976) work on the shark *Heterodontus* suggests that crushers with a diet of sea urchins often have strongly sculptured molars, whereas molluscivores tend to have smoother and even more massive teeth.

Reptilian crushers have been little studied, but some are evidently very strong. Many families of turtles contain shell-crushing species with blunt heads and powerful toothless jaws. The loggerhead (*Caretta caretta*) is capable of crushing large thick-shelled molluscs such as *Strombus gigas* and giant clams of the genus *Tridacna*. Many crocodiles, especially those with a blunt snout and large flattened teeth in the back of the jaw, swallow and probably crush molluscs and turtles. Lizards and amphisbaenians that crush molluscs have molar-like teeth. Such teeth are typically found only in adults and are associated with an exceptionally broad diet. Only in the scincid *Tiliqua*, the teiid *Dracaena*, and the amphisbaenid *Amphisbaena ridleyi* are molar-like teeth found in the juvenile as well as in the adult.

The commonest technique used by birds to break shells is hammering with the bill. This behavior is especially well known in oystercatchers. It is remarkable that very few birds use the bill to crush shells, especially because some finches (family Fringillidae) have become highly specialized seed-crushers. The European hawfinch (*Coccothrostes coccothrostes*), for example, is able to crush cherrystones and olive pits that resist compressive forces of at least 300 N (Sims, 1955). The bill in these finches is short and stout and the head is large. Similar modifications would presumably be suitable for crushing shells, but no living bird seems to be specialized in this way.

Many mammals are potentially capable of shell breakage, but it appears that very few have become specialized to this mode of predation. The sea otter (*Enhydra lutris*) crushes molluscs and sea urchins between the heavily enameled molars in the jaws or bashes them against hard objects. The mongoose *Mungos* is reported by Kasigwa and his colleagues (1983) to hurl snails against tree trunks in order to break the victims' shells. As is true in birds, the absence of specialized shell-crushing mammals is surprising in view of the development of multicusped molar teeth in most groups. The ability to crush exceedingly hard nuts, for example, is well developed in sciurid rodents (squirrels), tayassuid artiodactyls (peccaries), and some primates (R. F. Kay, 1981; Kiltie, 1981, 1982). These animals, like the sea otter, have a thick layer of enamel on the molars. The peccary *Tayassu tajacu* is capable of crushing palm nuts which in a testing machine fail under a load of 1350 N, whereas the larger and proportionally more massive-jawed *T. pecari* can crush nuts that fail at 3800 N (Kiltie, 1982). Various primates, including chimpanzees, use tools to break hard tropical nuts (Struhsaker and Leland, 1977; Boesch and Boesch, 1983), but they do not prey on shell-bearing molluscs.

One conclusion we can draw from the study of living shell-breaking predators is that a morphology that appears to be specialized for crushing, peeling, or hammering is not in all cases diagnostic of a specialized diet of shell-bearing animals. Crushing structures in birds and mammals are often used to crush seeds and nuts. Moreover, as Pregill (1984) has emphasized in the case of lizards and amphisbaenians, the development of molariform teeth may permit an animal to take a great variety of foods, ranging from resistant items such as shells to soft-bodied prey such as insects. Similarly, a large, stout beak enables finches to take seeds that vary widely in hardness, whereas slender-billed species are restricted to soft seeds (Abbott et al., 1977). At the other end of the spectrum, a lightly built or small shell can be broken even with relatively unspecialized instruments. Accordingly, any organ that enables an animal to grasp or hold an item of food should be considered potentially capable of breaking a victim's armor.

Available evidence indicates that shell breakage is a rapid method of subjugation. The pufferfish *Diodon hystrix* consumes even very resistant prey within a minute or less (A. R. Palmer, 1979). The sea otter takes up to 90 seconds to break *Saxidomus* clams on rocks but is usually even faster with other prey (Calkins, 1978; Hines and Loughlin, 1980). Easy prey are dispatched by shell-crushing crabs in periods of a few seconds to a few minutes, and even such highly resistant prey as *Conus* and *Drupa* can be consumed and eaten by *Carpilius* and *Daldorfia* in only an hour (P. V. Hamilton, 1976; Hughes and Elner, 1979; M. J. Williams, 1978; Zipser and Vermeij, 1978). These durations are short when compared with those required by gastropods that prey by insertion or drilling.

As a rapid form of predation, breakage is consistent with life in an environment that is risky to the predator. It is therefore not surprising that the ecological distribution of breakage contrasts strongly with that of other predatory techniques. Most families of shell-crushing fishes, crustaceans, and reptiles are found in warm shallow waters, especially on marine hard bottoms. Only a few groups are characteristic of cold waters, and these (homarid lobsters, cancrid crabs, palinurid lobsters of the genera *Jasus* and *Palinurus*, and sea otters) are typically weaker and morphologically less specialized than the warm-water groups (Vermeij, 1977a, 1978; B. W. Kent, 1979). The lobster *Homarus* has a well-developed molar dentition in the crusher claw, but the mechanical advantage and absolute compressive strength of this claw are low in comparison with those of molarized tropical crabs, which moreover are of smaller size. Some species of *Cancer* have stout molarized claws, but again the mechanical advantage, dentition, and strength are modest. Most of the specialized shell-breakers are found on hard bottoms. The only exceptions are species living in permanent burrows in grass beds and mangrove swamps, as well as sand-dwelling calappid crabs. Hammering stomatopods live on reefs and in rubble, whereas spearing types are most typical of sand and mud (Caldwell and Dingle, 1975; Reaka and Manning, 1981). Most of the handful of freshwater shell-breakers have a tropical to warm-temperate distribution. They include crayfishes; some crabs; many fishes (lungfishes, catfishes, cichlids, drums); turtles; teiid lizards (*Dracena* in South America); crocodiles; and some birds and mammals. A few shell-breakers attack land snails, the chief culprits being birds, mammals, and some reptiles and beetles.

DRILLING

Drilling is perhaps the most specialized form of predation on animals with external shells. Although the capacity to drill is not readily apparent from an examination of the predators' hard parts, the results of the activity of drilling predators are readily detectable in the fossil record in the form of shells or tests that bear circular or oval holes whose axes are perpendicular to the shell surface.

Modern drillers are almost entirely molluscan. They belong to at least seven groups of gastropods and to one group of cephalopods.

Some beetles also produce small holes in shells and therefore can be considered to be drillers in the broad sense.

By far the largest groups of living drillers are the moon snails (superfamily Naticacea) and murexes and their allies (superfamily Muricacea). Penetration of the prey shell apparently is effected chemically. The hole made by naticaceans is round and tapers evenly downward. If penetration of the shell is incomplete, the floor of the hole bears an upwardly convex protuberance, or boss. Such a boss is normally not present in incomplete holes that are drilled by muricaceans, although it occasionally occurs in some very deep muricacean holes. The muricacean hole also is more steeply walled and, unlike that of naticaceans, its depth can exceed the diameter (Figure 6.2). The accessory boring organ of muricaceans is located in the foot, whereas that of the naticaceans is located under the tip of the proboscis. Naticaceans envelop the prey in the foot, whereas muricaceans do not (Carriker and Yochelson, 1968; Carriker and Van Zandt, 1972; B. W. Kent, 1985).

The size of the hole is usually an indication of the size of the predator, although other factors, such as blood pressure in the proboscis, also affect the size of the hole (Carriker and Van Zandt, 1972; Wiltse, 1980; Kitchell et al., 1981). In naticaceans, depth of penetration is directly correlated with the hole's diameter, but in muricaceans the ratio of hole depth to diameter is more variable and can attain values as high as 3.0 (B.W. Kent, 1985).

Several interesting variations in drilling have evolved in the Naticacea, the Hipponicacea, and especially the Muricacea. Most drillers are predators that kill their victims, but the muricacean *Genkaimurex* parasitizes scallops of the family Pectinidae by sucking the host's juices via a hole made through the upper (left) valve (Matsukuma, 1977). A similar parasitic habit has evolved in the Capulidae (Matsukuma, 1978). Both *Genkaimurex* and capulids leave traces of their activity in the form of both a circular-to-oval-shaped hole and a scar where the parasite is permanently attached to the host shell.

Another important variation in drilling is boring pelecypods at the shell margin rather than through the shell wall. Edge-drilling is known in many muricacean genera (Matsukuma, 1978; Vermeij, 1978) as well as in the polinicine naticid genus *Polinices* (Vermeij, 1980a). A valve that has been edge-drilled looks as if a circular chip has been removed from the margin (Figure 6.3).

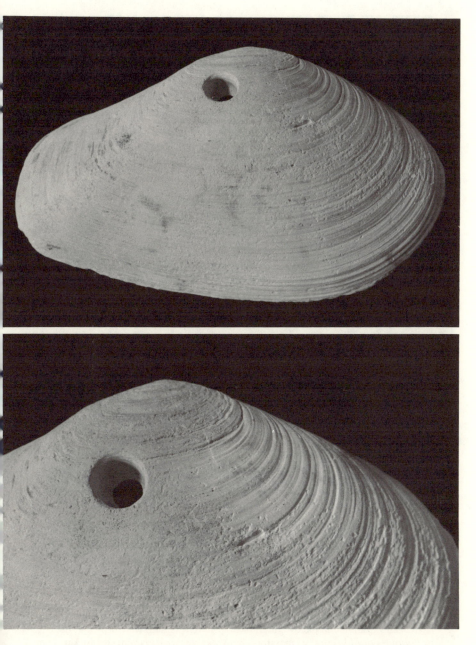

FIGURE 6.2 Two views of a drilled specimen of *Eucrassatella marylandica*, Choptank Formation, Middle Miocene of Maryland. This hole was probably drilled by a muricacean of the genus *Ecphora*.

FIGURE 6.3 Examples of edge-drilled pelecypod valves. *Gafrarium pectinatum* (above) and *Ctena bella* (below), both from Pago Bay, Guam, were drilled at the edge by *Polinices tumidus* in the laboratory. Empty valves with marginal drilling are commonly found in the field at many tropical sites (Vermeij, 1980a).

Helmet shells of the tonnacean family Cassidae have also evolved the drilling habit, but unlike the mollusc-eating Muricacea and Naticacea, they have specialized on echinoderm prey, especially echinoids (Hughes and Hughes, 1981). Little is known about the mechanism of drilling in cassids, but the process is evidently rapid and may involve dissolution by acid.

Drilling in other groups of gastropods and beetles has been very poorly studied. The doridacean nudibranch family Vayssiereidae contains a genus (*Okadaia*) whose members are reported to drill the calcareous tubes of serpulid polychaetes (Young, 1969). At least two genera of pulmonates (*Aegopinella* and *Poiretia*) apparently rasp small holes in the shells of their land snail prey, although this method of penetration is not the only, and perhaps not even the usual, form of predation by these snails (Wächtler, 1927; Mordan, 1977). It has been suggested that drilid beetles occasionally may make holes in shells (von Martens, 1903), but damage of this kind by insects has never been studied. Certainly, the well known capacity of beetles and other insects to penetrate wood and seeds suggests that hole-drilling in shells is not beyond the capacity of predaceous insects.

Octopod cephalopods have also evolved the capacity to drill molluscs. They use the salivary papilla in conjunction with the radula to drill through shells (Nixon, 1979). The holes made by octopods are typically small and irregular in outline. Unlike gastropod drillers, octopods often drill more than one hole into a victim's shell (Arnold and Arnold, 1969; Wodinsky, 1969, 1973; Nixon, 1979).

Drilling is a slow process in naticacean and muricacean gastropods and a moderately slow form of predation in octopods. Species of the thaidid muricacean genus *Nucella* in the Puget Sound region take 2 hours to drill and consume young barnacles (*Balanus glandula*) and 8 hours to eat an adult (Connell, 1970). Other studies of muricaceans, mostly done in temperate waters, reveal durations of 5 to 97 hours to eat limpets, coiled snails, barnacles, mussels, and infaunal clams (J. L. Menge, 1974; Bayne and Scullard, 1978; Garrity and Levings, 1981; Fairweather and Underwood, 1983). The drilling rate of the muricid *Urosalpinx* has been estimated by Carriker and Van Zandt (1972) to be 0.013 mm/h. The naticacean *Neverita duplicata* at 18 to 20° C is a little faster (0.22 mm/h), but it still takes an adult 42 hours to drill and ingest a moderate-sized (36 mm long) *Mya arenaria* clam

(Kitchell et al., 1981). In the Bahamas, the drilling rate of *Octopus vulgaris* was estimated to be 1.25 mm/h (Arnold and Arnold, 1969). At this rate, it takes *O. vulgaris* 30 to 90 minutes to eat various species of gastropod prey (Wodinsky, 1969). The drilling phase of a predatory attack by the helmet shell *Cassis tuberosa* on the sea urchin *Diadema antillarum* in Barbados takes only 10 minutes. Consumption, however, is a lengthier process, taking 1 to 2 hours in both *C. tuberosa* and in the related *Cypraecassis testiculus* (Hughes and Hughes, 1981). An exception to the relatively slow drilling by cassids was found by Kier (1981), who filmed a *Phalium granulatum* drilling and consuming a *Cassidulus* sea urchin in 8 minutes. The short drilling time is probably a consequence of the very thin test wall of *Cassidulus* and other burrowing irregular sea urchins.

Predation by drilling is almost entirely a marine phenomenon. It is unknown in fresh water and practiced by only a few pulmonates and perhaps some beetles on land snails. In the sea, drilling gastropod predators are strikingly more diverse, and they have a greater impact ecologically in the tropics than in the temperate zones (Taylor and Taylor, 1977; Vermeij and Dudley, 1982). Naticaceans are found chiefly on unconsolidated bottoms, although they do on occasion attack clumps of mussels in rubbly areas. Muricaceans, on the other hand, are found on both hard and unconsolidated bottoms, although they are especially diverse on the former. Cassids, which prey on echinoderms, occur in warm-water sandy and rubbly areas. Octopods have a very wide distribution in tropical and warm-temperate seas, but drilling attributable to octopods has not been recorded from cold-temperate areas. Although our knowledge of octopod drilling remains spotty, available data indicate extremely high frequencies of drilling in warm-temperate California, the Mediterranean Sea, and tropical Pacific Panama (Fotheringham, 1974; R. A. Evans, 1980-1981; Ambros and Nelson, 1983; Fawcett, 1984).

The distribution of drilling predators is consistent with the long handling times associated with drilling. Naticaceans, like extracting gastropods, are animals of sand and mud; there they can subdue their prey slowly without undue disturbance while they are buried. Many muricaceans are found on hard bottoms where they would be subject to frequent disturbance by enemies, but their shells typically are extremely resistant to attack by would-be predators. Another advantage of muricaceans in drilling is that only a small part of the animal extends from the shell, so that the feeding predator is no more exposed to enemies than is one that is not drilling. Furthermore, the spine that some muricids and thaidids have on the middle or the anterior edge of the outer lip may serve to stabilize the foothold of the predator while drilling and apparently also aids in the penetration of barnacle prey (Paine, 1966; B. W. Kent, 1981b; Sleder, 1981). It therefore further mitigates the risk of slow predation.

THE HISTORY OF MOLLUSCIVORY

ANCESTRAL AND DERIVED TECHNIQUES

Now that some aspects of the functional morphology of one important class of predators—those which feed on hard-shelled animals, especially molluscs—have been reviewed, we are in a position to evaluate predictions that follow from hypotheses about the history of adaptation to enemies, as set out in Chapter 3.

Whole-animal ingestion is expected to be the most primitive and ancient form of predation on skeleton-bearing animals. This method represents a minor modification of suspension-feeding and deposit-feeding, in which small particles of food are gleaned from the water or sediment. The ingestion of larger, heavier particles is made possible when the mouth is enlarged, and when organs for grasping or sucking are developed. With the evolution of such organs, the way is opened for the development of methods of removing flesh from the prey before ingestion. As a consequence, much larger prey can be attacked and subdued successfully. Insertion, extraction, pre-ingestive breakage, and drilling are therefore expected to be derived methods of predation. Secondary evolution of whole-animal ingestion is, of course, always possible.

I suggested in Chapter 4 that predatory techniques requiring large energy fluxes are generally favored. Not only does rapid acquisition of prey reduce the risks to which the predators are exposed as they subdue and consume prey, but it also enhances the ability of the predators to compete exploitatively and to interfere with their rivals. Envenomation and the application of force are generally associated with rapid subjugation and should therefore be strongly favored as risks from the predators' own enemies increase. Thus, venom-associated extraction and pre-ingestive breakage are expected to have become more common forms of predation over the course of the Phanerozoic, especially in environments where predators are at high risk. Slower methods of predation are feasible for animals that are strongly defended against enemies and in places where high levels of activity cannot be sustained.

Ideally, a historical account is based on an analysis of many local assemblages in each of several habitats at more than one site during each time interval of the Phanerozoic. In practice, this level of precision cannot be achieved. The record of predators is relatively sparse. We must therefore be content for the present with an account based on the global scale. The only smaller-scale studies that are currently feasible are those in which activities of predators are inferred from traces on prey. These studies will be taken up in subsequent chapters after aspects of the functional morphology of prey have been treated.

THE EARLIEST PREDATORS

The early history of predatory metazoan animals is shrouded in mystery. Only two things can be said with certainty about predators during the Vendian (or Ediacarian) period, the first time interval of the Paleozoic. First, if they existed they were soft-bodied, as were their prey. Second, they had no identifiable mouth, feeding appendages, or filtering organs (Seilacher, 1984). There may have been sedentary sea anemones and worms which fed in and around burrows, but the status of these animals as predators cannot be confirmed, and no evidence of predation has yet come to light for any assemblage in the Vendian. Seilacher (1984) has argued persuasively that the animals that previously had been interpreted as hydrozoan and scyphozoan cnidarians (jellyfishes) are in fact a heterogeneous assemblage of burrows of sedentary animals and sandy skeletons.

The Early Cambrian marks the first unequivocal appearance of predaceous animals. In a careful study of the problematic disc-like fossil *Mobergella* from the Early Cambrian of the Baltic region, Bengtson (1968) found evidence of both skeletal breakage and drilling. Repaired breaks are common on *Mobergella* and may indicate that a potentially shell-breaking predator was at work during the Early Cambrian. Bengtson also found several examples of *Mobergella* that were pierced by a single hole, 0.1 mm in diameter, which was oriented with the long axis perpendicular to the shell surface in the same way that drill holes made by living gastropods are.

Other possible indications of predation on skeletonized animals in the Early Cambrian are the presence of several damaged and subsequently repaired trilobites (Pocock, 1974) and the discovery by Alpert and Moore (1975) of trilobite fragments in tubular fillings that are interpreted as the digestive cavities of anthozoans. These fillings were found in vertical cylindrical burrows (*Dolopichnus*) which could have been made by burrowing anemones, but I believe the interpretation of these structures as reflecting predation is equivocal. More convincing is Conway Morris's (1985) finding that at least 20% of the specimens of *Wiwaxia corrugata*, a Middle Cambrian animal from the Burgess Shale of British Columbia, had broken skeletal spines. This proportion of possibly predator-injured individuals might well have been much higher, because not all rows of spines are preserved on most individuals, and because the animal evidently grew by molting, during which an injury could have been repaired without a trace. Several Cambrian arthropod-like animals may have been able to dismember prey. The Early to Middle Cambrian genus *Anomalocaris* had mouthparts that may have served this function and been responsible for several injuries observed on trilobites from the Burgess Shale (Rudkin, 1979; Briggs and Mount, 1982; Whittington and Briggs, 1985). Three genera from the Burgess Shale (*Sidneyia, Naraoia*, and the trilobite *Olenoides*) had pincer-like structures on the thoracic appendages. Fragments of trilobites and hyoliths have been found in the digestive tracts of *Sidneyia* (Bruton, 1981). These animals, none of which was heavily skeletonized (Conway Morris, 1985), represent early stages in the development of grasping structures, but they were not morphologically specialized either for crushing victims or for exerting great force (Manton, 1977). Still another predator known from the Middle Cambrian is the priapulid worm *Ottoia*, whose digestive tract was found to contain intact hyoliths and brachiopods (Conway Morris, 1979). Clearly, there was a substantial diversity of predators and of predatory methods during the Cambrian.

THE REST OF THE PALEOZOIC

Toward the end of the Cambrian, the cephalopods made their first appearance. They began as small straight-shelled plectronoceratids in Chinese strata of probable Franconian age and rapidly diversified to a fauna of some 120 species in strata of latest Cambrian (Trempeauleauian) age (Chen and Teichert, 1983). Nothing is known about the food habits or modes of feeding of these and later Paleozoic cephalopods, but they could have ingested and perhaps even superficially injured armored victims.

Following several episodes of extinction in the Late Cambrian, another large-scale diversification of shell-bearing invertebrates, including predaceous cephalopods, ensued in the Tremadocian and Arenigian stages of the Early Ordovician (Flower, 1976; Crick, 1981). It is possible that early cephalopods were responsible for the frequent repaired damage observed by R. R. Alexander (1986) in Ordovician brachiopods, and for the jagged repaired injuries on Silurian specimens of the gastropod *Euomphalopterus* figured by Peel (1984). However, direct unequivocal evidence is lacking in these cases.

The Early Ordovician also marks the first appearance of stelleroid echinoderms (sea stars and brittle stars); the living forms of these organisms frequently are predators of molluscs. D. B. Blake (1981) believes that the early stelleroids of the Paleozoic fed intraorally and therefore digested their prey piecemeal, and that extraoral feeding had not yet evolved. Claims by R. M. Carter (1968) and others of extraoral feeding in Paleozoic sea stars are based on preservation of pelecypods in close proximity to sea stars, but this evidence cannot be taken to imply predation. Scolecodonts—the calcified jaws of predaceous polychaetes—also appeared for the first time in the Early Ordovician, although the phylum Annelida had a history probably dating back to the Vendian (Kielan-Jaworowska, 1968).

Small circular holes set perpendicular to the valve surface of brachiopods probably indicate drilling predation. They have been sporadically recorded from Late Cambrian inarticulates (R. H. Miller and Sundberg, 1984), and from articulate brachiopods in the Middle and Late Ordovician (Fenton and Fenton, 1931; Cameron, 1967); Silurian (Rohr, 1976); Devonian (Buehler, 1969; Sheehan and Lespérance, 1978); and Carboniferous (Brunton, 1966; Ausich and Gurrola, 1979). The identity of the drillers is unknown, but several authors have noted that drilled articulates of Ordovician and later periods were found in association with platyceratid gastropods (Fenton and Fenton, 1931; Cameron, 1967; Rohr, 1976; Sheehan and Lespérance, 1978). The suggestion by the Fentons that platyceratids were the culprits must therefore not be dismissed. The platyceratids whose mode of life is known with certainty lived attached to crinoids, where they might have fed on nutrients in the host's excretory products (Bowsher, 1955). Living capulid and some muricid gastropods have also evolved sedentary parasitic habits (Matsukuma, 1977, 1978) and may therefore provide models for the evolution of Paleozoic platyceratids.

An important event in the Silurian was the appearance early in that period of acanthodians, the first jawed vertebrates (Denison, 1979). The jaw may have functioned initially as a rigid framework for the suction apparatus of the mouth (Reif, 1982) and therefore may have been part of an improved method of suction. The evolution of the jaw enabled vertebrates, which during the Ordovician were suspension-feeders and deposit-feeders in nutrient-rich waters (Mallatt, 1984), to become particulate feeders, that is, animals capable of ingesting food particles one by one rather than in bulk quantity in a current of water. Eventually the jaw became adapted as an instrument for grasping and holding animal prey. Jaw features that would have enabled the predator to break the shells of molluscs and the armor of jawless (agnathan) vertebrates are unknown in acanthodians, but the potential for their development certainly existed (Gross, 1967).

The eurypterids, a group of chelicerate arthropods with origins in the Llanvirnian stage of the Early Middle Ordovician, must be regarded as potentially capable of shell-breaking predation. During the Late Silurian and Early Devonian, pterygotid eurypterids with massive chelicerae bearing well-worn pointed teeth were associated with, and perhaps preyed upon, armored agnathan fishes (Romer, 1933; Kjellesvig-Waering, 1961). Generally, the teeth were pointed and apparently functioned in slicing and sawing prey. The chelicerae of *Eretopterus* and other pterygotoids had a low mechanical advantage (m.a. = 0.22), comparable to that of the relatively unspecialized crushing claws of *Callinectes* (Selden, 1984). Because the pterygotoids had the most massive chelicerae among the eurypterids, it is tempting to speculate that eurypterids were distinctly less specialized for crushing armored animals than were later chelate decapod crustaceans.

When morphologically specialized shell-breakers first appeared in the Devonian, they did so in several unrelated groups of vertebrates. The earliest to appear were dipnoan lungfishes of the genus *Dipno-*

rhynchus in the Siegenian stage of the Early Devonian. In the Eifelian (Middle Devonian), they were joined by dipterid lungfishes (Westoll, 1949) and by ptyctodontid placoderms (Denison, 1978). A real diversification of potential shell-breakers did not, however, take place until the Frasnian stage of the Late Devonian, when dipnorhynchids and dipterids were joined by ptyctodontid, brachydeirid, pholidosteid, selenosteid, mylostomatid, and leiosteid placoderms (Denison, 1978). All these early fishes vanished after the Famennian, the latest stage of the Devonian. The only groups of possible shell-breakers that arose during the Late Devonian and survived into the Late Paleozoic were helodontid holocephalan and copodontid bradyodont chondrichthyans (Moy-Thomas and Miles, 1971; Zangerl, 1981).

That these Devonian fishes were capable of destroying the shells of brachiopods and molluscs is inferred from the dentition, which usually consisted of blunt teeth or tooth-like growths and sometimes was fused into upper and lower crushing plates. In the succeeding Carboniferous and Permian periods of the Late Paleozoic, fishes with superficially similar dentitions underwent a large diversification (Moy-Thomas and Miles, 1971; Zangerl, 1981). At least 10 families of chondrichthyan and osteichthyan fishes contained probable shell-crushers in the Early Carboniferous, including chondrenchelyids, petalodontids, and chirodontids. By the Late Carboniferous, there were at least 13 families. The Permian caseodontid *Fadenia* and the Permian petalodontid *Janasa* have been found with fragments of brachiopods and other invertebrates in the digestive tract. *Symmorium*, a large (3 m long) cladodont symmoriid shark from the Late Carboniferous, has been implicated by Mapes and Hansen (1984) as a predator of coiled nautiloids. They found a row of punctures and fractures whose shape and size conform to the imprints of the teeth of a large *Symmorium*. R. R. Alexander (1981) has implicated various shark-like fishes of the Early Carboniferous as predators of articulate brachiopods, which bore punctures and healed depressions.

Potentially shell-crushing Crustacea also diversified after the Devonian, but they show no evidence of specialization for shell breakage during the Paleozoic. The order Palaeostomatopoda, whose members were perhaps capable of spearing prey (as are living squilloid stomatopods), was established by the Tournaisian stage of the Early Carboniferous. During the Namurian stage of the Middle Carboniferous, this group was being replaced by the Archaeostomatopoda, among which the Tyrannophontidae were probably also spearers (Schram, 1979, 1982). Malacostracan crustaceans also diversified after their Devonian origin, but nothing in their morphology suggests specialization for shell breakage. Another group of probable crustaceans which may have had raptorial appendages during the Paleozoic, is the class Thylacocephala, including the groups Concavicarida and Conchyliocarida. They are recorded first from the Llandoverian stage of the Early Silurian and persisted until the Late Cretaceous. The Callovian (Middle Jurassic) representatives studied by Secretan (1985) were very heavily armored animals with raptorial appendages. Considerable controversy exists concerning the orientation, mode of life, and systematic placement of these animals (Rolfe, 1985). At this point, little can be said except that the Thylacocephala were probably predators at least during the Mesozoic, and that they were potentially predators during the Paleozoic as well.

In summary, animals with structures that were potentially specialized for shell-crushing are known only from Early Devonian time onward. Not until Frasnian (Late Devonian) time was there a substantial diversity of potential shell-crushers (ten families of fishes). After extinction of the placoderms at the end of the Devonian, a comparably high diversity of shell-crushing fishes was reestablished during the Carboniferous and Permian. If arthropods played a significant role as shell-breakers, it is not reflected in preserved fossil material.

THE MESOZOIC AND CENOZOIC

Although the Late Paleozoic had witnessed an unprecedented diversification of predaceous fishes and crustaceans, an even more profound diversification of predators was to come in the succeeding Mesozoic and Cenozoic eras. The most spectacular among the new predators were drillers and shell-breakers, but representatives of all the predatory methods took part in the diversification, which affected many unrelated clades simultaneously.

One of the groups that evolved specialization to crushing soon after the end-Permian crisis was the molluscan class Cephalopoda. Although this group had been diverse and important since the Late Cambrian, direct evidence for structures enabling cephalopods to crush hard objects is lacking until the Ladinian stage of the Middle Triassic, when the first rhyncholites and conchorhynchs, found in association with the nautiloid *Germanonautilus*, appeared (Gasio-rowski, 1973; Dzik, 1984). These are calcified structures that closely resemble the upper and lower jaws of the living *Nautilus* (Saunders et al., 1978). The upper rhyncholite is a sturdy arrow-shaped struc-ture that acts to dismember the prey by impinging on the cutting edge of the more delicate lower conchorhynch. In the Early to Middle Jurassic (Toarcian to Bajocian stages), rhyncholites also became spe-cialized for the functions of cutting and spearing. Some Jurassic and Cretaceous lytoceratine ammonoids may also have had calcified jaws capable of crushing (Lehmann, 1981a), but most other ammo-noids evidently fed on small organisms (Lehmann, 1981b).

Two Triassic stocks of predators evolved jaws whose morphology suggests a shell-crushing capacity. The placodonts, which appeared early in the period, developed large crushing plates in the jaws and were themselves well armored (von Hüne, 1956). The Middle Trias-sic ichthyosaur family Omphalosauridae possessed broad molar-like teeth (Mazin, 1983). Neither of these groups of marine reptiles is known to have lived beyond the Triassic. A third group of reptiles, the turtles (order Chelonia), originated during the Late Triassic. Some of these early turtles may already have been marine. The food habits of Triassic turtles are unknown, and specialization to a hard-shelled diet seems unlikely, but any slow-moving animal with rela-tively powerful jaws must be considered a potential shell-breaker.

Most of the potentially shell-crushing fishes of the Paleozoic were wiped out during or before the crisis at the end of the Permian, but several fish stocks evolved the capacity again during the Triassic. Pycnodontiform and semionotiform fishes belonging to the Actin-opterygii (ray-finned fishes) appeared in the Late Triassic and may al-ready have had the ability to eat shell-bearing invertebrates at that time (Thurmond, 1974; Böss, 1982).

The Jurassic period marks the appearance of many groups that to-day contain molluscivores. Because almost no functional morpho-logical work has been done on these early predators, we do not know when the various predatory specializations evolved. Ignorance of these developments is most unfortunate, for this was one of the crit-ical times during earth history when large-scale changes were afoot.

Although sea stars arose in the Early Ordovician and may have been important predators throughout the rest of the Paleozoic, spe-cies capable of pulling apart pelecypods may not have arisen until the

Jurassic. D. B. Blake (1981) believes that suckered tube feet, which are necessary for exerting pull, may not have been present in any Paleozoic stelleroid. The family Asteriidae, which contains the only living sea stars capable of pulling apart the valves of pelecypods and brachiopods, is known first from the Pliensbachian stage of the Early Jurassic.

The record of octopod cephalopods may begin in the Callovian stage of the Middle Jurassic (Fischer and Riou, 1982), but it is not clear when the drilling habit evolved in this group. Drill holes attributed to octopods have been described from the Pliocene of Albenga, Italy (Robba and Ostinelli, 1975), but I suspect that the drilling habit is substantially more ancient.

Gastropods may have been predators during the Paleozoic, but the great diversification leading to the highly varied modern fauna of predaceous gastropods did not commence until the later Mesozoic. Unfortunately, the method of feeding in gastropods cannot be inferred from shell form. Although nearly all marine predaceous gastropods have a siphonal canal or notch at the anterior end of the aperture, this feature is found also in herbivorous strombids and cypraeid cowries and in deposit-feeding aporrhaids, and it is absent in predaceous naticids and terrestial pulmonates. The history of predaceous gastropods must therefore be traced by following extant families back through time. This, too, has proven to be difficult, for many of the families of the Rachiglossa are not well demarcated, and little work on lines of descent is currently available. Despite these problems, it is abundantly clear that the Jurassic and especially the Cretaceous were times of unprecedented diversification of predaceous gastropods (Sohl, 1964; J. D. Taylor et al., 1980, 1983).

The drilling Naticidae may have been among the first of the predaceous Mesozoic gastropods. Drill holes similar to those made by living naticids have been recorded from the Late Triassic St. Cassian beds in northern Italy (Fürsich and Wendt, 1977; Fürsich and Jablonski, 1984). However, since no drill holes attributable to naticids are known from the Jurassic (P. H. Fischer, 1962), it is possible that the Triassic drill holes were made by animals other than naticids. Undoubted naticid holes appeared in the Blackdown Greensand of England, a deposit of Albian (Late Early Cretaceous) age (J. D. Taylor et al., 1983). Muricaceans also evolved the drilling habit by Albian time (J. D. Taylor et al., 1980). Tonnaceans may already have been present in the slightly older Aptian stage of the Early Cretaceous, but drill holes made by cassids in sea urchins are not known earlier than the Late Eocene stage of the Cenozoic (Sohl, 1969). Capulids are recorded questionably from Early Cretaceous time onward, but species drilling and living on molluscs appeared first during the Campanian stage of the Late Cretaceous (Hayami and Kanie, 1980). Drilling thus evolved independently in several Cretaceous lineages.

Gastropods using insertion or extraction as predatory modes may have been present by the Early Cretaceous. The first group to appear was the family Buccinidae (Barremian stage), followed by the Fasciolariidae (Aptian) and Melongenidae (Albian). These and other primitive Rachiglossa apparently originated in the cold-water boreal realm of the Early Cretaceous and did not invade the tropical Tethyan realm until Albian time (J. D. Taylor et al., 1980; Kollmann, 1982). Such a geographical origin would be consistent with the observation that insertion and extraction are slow methods of predation. Groups that developed various methods of speeding up insertion and extraction had later origins in warm-water areas. The order Toxoglossa, whose members envenom their prey, differentiated during the Campanian and perhaps earlier in the Cretaceous, and molluscivorous species of *Cylinder* (family Conidae) were present by the Miocene. The lip-wedging technique used by the melongenids *Busycon* and *Sinistrofulgur* evolved during the Middle Pliocene on the Atlantic Coastal Plain of the southern United States (B. W. Kent, 1986).

Gastropods that ingest their victims whole also took part in the Cretaceous diversification. The Volutidae had evolved by the Cenomanian (Early Late Cretaceous), whereas the Olividae appeared during the Campanian (J. D. Taylor et al., 1980).

The decapod Crustacea, which are among the most important shell-breakers in modern seas, began their great diversification in the Triassic (Schram, 1982). The first groups to appear in the Early Triassic were the clawed homaridean lobsters (genus *Protoclytiopsis*) and penaeid shrimps (genus *Antrimpos*). Next came the clawless spiny lobsters (Palinuroidea), whose earliest genus (*Palaeopalinurus*) dates from the Norian stage of the Late Triassic (Förster, 1967, 1973, 1984). Palinurans of modern aspect are known first from the Kimmeridgian stage of the Late Jurassic (genus *Astacodes*). Just when the shell-breaking habit evolved in this group is uncertain, however, for mandibles and other structures that would permit us to infer feeding habits are not preserved in spiny lobsters (George and Main, 1968).

Brachyuran crabs (genus *Eocarcinus*) differentiated in the Pliensbachian stage of the Early Jurassic. Members of the superfamily Dromiacea, whose modern representatives do not break shells, appeared in the Bajocian stage of the Middle Jurassic. The calappid subfamily Necrocarcininae originated from a dromiacean stock during the Hauterivian stage of the Early Cretaceous, but unlike the shell-peeling Calappinae of the Cenozoic, necrocarcinines had unspecialized claws of equal size on the right and left sides of the body (Förster, 1968; Wright and Collins, 1972). Other Cretaceous crabs also lacked specialized shell-peeling or shell-crushing dentition in the claws. The Carpiliidae, for example, were represented during the Albian and Late Cretaceous by the genus *Caloxanthus*, whose equal-sized right and left claws had a row of simple pointed teeth on the apposing fingers (Wright and Collins, 1972). During the Cenozoic, this family gave rise to very highly specialized shell-crushers.

The first brachyurans with molar-like claw dentition date from the Paleocene. By the Middle Eocene, many genera of shell-crushing crabs, including carpiliids and portunids, had become established. Shell-peeling calappines are known as body fossils from the Oligocene onward, but the presence of characteristically peeled shells suggests that these crabs already existed in the Eocene. Cancrids and the massive-clawed *Carpilius* were established by Miocene time (Guinot, 1968; Nations, 1975).

Stomatopod Crustacea are another important group of shell-breakers with Mesozoic roots. They are first known from the latest Jurassic (family Sculdidae). Primitively, they were probably spearers rather than hammerers or smashers. The fossil record of stomatopods is exceedingly poor (Holthuis and Manning, 1969; Förster, 1982), and feeding appendages have not been preserved. Förster's fragmentary evidence suggests that hammering lysiosquillids already existed during the Maastrichtian stage of the Late Cretaceous. I suspect that typical hammering gonodactyloids appeared after the Cretaceous.

The later Mesozoic was also a time of diversification of shell-breaking fishes. Several potentially shell-crushing hybodontid and steinbachodid sharks are known from the Triassic (Reif, 1980). Rays and skates (batoids) and heterodontid sharks are known from Early

Jurassic time onward, although some of the modern families of molluscivorous rays did not arise until the Cenomanian stage of the Early Late Cretaceous (skates of the family Rajidae and stingrays of the family Dasyatidae). Ptychodontid sharks, whose teeth were coalesced into crushing plates as in living myliobatid eagle rays, were shell-crushers throughout the Cretaceous and early Paleocene. Scyliorhinid dogfishes are of Late Jurassic extraction (Kauffman, 1972; MacLeod and Slaughter, 1980; MacLeod, 1982; Maisey, 1984). Semionotiform bony fishes continued to the Late Cretaceous and pycnodonts to the Middle Eocene (Thurmond, 1974; Böss, 1982).

If the Jurassic and Cretaceous were times that were favorable to the evolution of shell-breaking fishes, the Early Cenozoic was even more so, not only because of the diversification of shell-crushing rays, but also because of the explosive expansion in the diversity of actinopterygian (spiny-rayed) bony fishes. All the major living families of shell-crushing rays—myliobatid eagle rays and rhinopterid cow-nosed rays—were established no later than the Middle Eocene, as was the carcharhinid shark genus *Galeocerdo* (Maisey, 1984). The Acanthopterygii originated in the Cenomanian stage of the Early Late Cretaceous, but most of the early families, including the Holocentridae (squirrelfishes), belonged to the order Beryciformes and probably lacked the capacity to break shells (Patterson, 1964). The drums (sciaenids), among the earliest shell-breaking acanthopterygians, were first recorded from the Maastrichtian stage of the latest Cretaceous. Ariid catfishes arose in fresh water in South America at about the same time. During the Paleocene, such important acanthopterygians as the wrasses (Labridae), flatfishes (Pleuronectiformes), and porgies (Sparidae) appeared. These were joined in the Early Eocene by pufferfishes (Diodontidae and Tetraodontidae), trunkfishes (Ostraciidae), and, somewhat later, the grunts (Pomadasyidae). Only a few shell-breaking fish families with a fossil record appeared later than the Eocene. The triggerfishes (Balistidae) are known from the Oligocene onward, and the toadfishes (Batrachoididae) are recorded from the Miocene (Sepkoski, 1982a; Tyler, 1980). Some of these dates of origin may be revised backward in time when more data become available, because the fossil record of marine bony fishes is very spotty. Nevertheless, the phylogenetic scheme outlined by Lauder and Liem (1983) strongly suggests that the great diversification of the Acanthopterygii in general and the shell-breakers in particular is a post-Cretaceous phenomenon.

Reptiles, birds, and mammals also acquired the capacity to break shells in the Cretaceous and Cenozoic. Certain marine mosasaurid lizards, notably those of the subfamilies Globidentinae and Platecarpinae, were common shell-breakers in the Late Cretaceous and have been implicated in causing damage to ammonoid cephalopods (Kauffman and Kesling, 1960). Fossil terrestrial lizards with molarized posterior teeth include the glyphanodontine teiid *Adamisaurus* from the Late Cretaceous of East Asia, two genera (*Pseudeumeces* and *Dracaenosaurus*) of Lacertidae from the Paleogene (Paleocene to Oligocene) of Europe, glyptosaurine anguids from the Late Cretaceous to Oligocene, and the amphisbaenid *Oligodontosaurus* from the Paleocene of Wyoming (Estes and Williams, 1984). The loggerhead turtle genus *Caretta* extends back to the Late Cretaceous, but most shell-crushing turtle families are of Cenozoic origin (Pritchard, 1979; Bramble et al., 1984). Among birds, the large order Charadriiformes (which includes shorebirds as well as gulls and terns) originated in the Cretaceous, but shell-hammering and extracting oystercatchers of the family Haematopodidae are known only from Pliocene time onward (Olson and Steadman, 1978). From the Charadriiformes arose the ducks of the order Anseriformes, which became differentiated in the Early Oligocene (Olson and Feduccia, 1980). Marine shell-breaking mammals are no older than the Neogene. The sea otter *Enhydra lutris* is not more than 2 million years old, although other otters (family Lutridae) may have had the shell-breaking capacity during the Miocene and Early Pliocene (Repenning, 1976a). Seals also arose in the Miocene, but they probably did not achieve molluscivory until the Late Miocene (Repenning, 1976b).

SUMMARY AND CONCLUSIONS

Although this chapter has highlighted only the predators of molluscs, many of its main points are likely to apply to predators in general. A great deal remains to be learned about the functional morphology of living as well as fossil predators, but available information permits us to draw the following conclusions.

1. Whole-animal ingestion (swallowing or enveloping the prey whole before digestion begins) is the predatory technique requiring the least degree of anatomical specialization. Predators using this method are restricted to relatively small prey.

2. Preparation of the prey outside the predator's body permits larger victims to be subdued and consumed. The most important techniques of external preparation of the prey are insertion and extraction, pre-ingestive breakage, and drilling.

3. Insertion, extraction, and drilling are generally slow methods of predation that are used chiefly by predators that live in relatively safe habitats or are themselves highly resistant to enemies. All these methods have probably existed since the Early Cambrian, but insertion and extraction accompanied by the use of force (as in sea stars and octopods) evolved only during the Jurassic, whereas most drilling predators arose during the Cretaceous and Cenozoic.

4. Pre-ingestive skeletal breakage is a rapid method of predation that is especially common in environments where predators are at high risk to their own enemies. Shell breakage is known from the Early Cambrian onward, but predators with obvious morphological specializations for breaking shell-bearing prey are recorded first only in the Late Silurian or Early Devonian. As Figure 6.4 shows, a very large increase in the diversity of marine families of specialized shell-breaking predators occurred during the Late Cretaceous and Early Cenozoic. This increase is of a far greater magnitude than is the increase in the diversity of marine families generally (Sepkoski, 1979, 1981, 1984; Van Valen and Maiorana, 1985).

The developing emphasis on breakage and force-assisted extraction over the course of the Phanerozoic is in broad agreement with the expectation that predatory methods requiring large energy fluxes are favored as risks to the predators themselves increase. The great abundance and impact of drilling predators in the Cretaceous and Cenozoic is less easily reconciled with expectation. Muricacean drillers are heavily armored gastropods and therefore are perhaps less exposed to risks than are other molluscivores, but this cannot be said for naticaceans, whose shells are relatively weak. The evolution of the drilling habit therefore remains an intriguing problem.

Studies of the teeth and jaws of fossil and living sharks show that

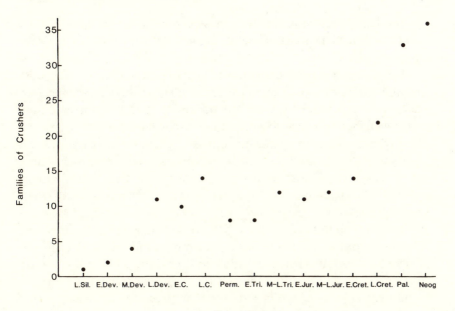

FIGURE 6.4 Number of marine families specialized for predation by shell breakage. Groups considered include eurypterid and crustacean arthropods, cephalopod molluscs, and vertebrates. A plateau in diversity lasted from the Late Devonian to the beginning of the Late Cretaceous and was followed by a steep rise of a factor of 2.5. This Late Mesozoic and Cenozoic increase was substantially greater than that for marine families generally.

the increased emphasis on high-energy modes of predation over time is not limited to animals that prey on hard-shelled invertebrates. Most Paleozoic sharks were characterized by sharply pointed teeth, each with one or more cusps, which were effective in grasping and holding fishes and other swimming prey. Predation by gouging or wrenching large pieces from prey that are larger in size than the sharks themselves has evolved in several squaliform and most lamniform and carchariniform sharks of the later Mesozoic and Cenozoic. In addition to sharply pointed teeth used for grasping and holding the prey, these sharks have also evolved blade-like teeth, often with serrated edges, which slice deep into the flesh of large fishes and marine mammals. Gouging and slicing are aided by vigorous shaking of the head. Predation, in short, has become less passive over time and has placed animals of increasing size at risk (Moss, 1977; Zangerl, 1981).

The rather generalized account offered in this chapter should be accepted with caution. In modern seas, the greatest degree of specialization to predation by breakage occurs in hard-bottom predators such as fishes and crustaceans. Although the hard parts of such predators are more apt to be preserved than the less robust parts of predators using other methods, the hard-bottom environments in which the predators are most common are poorly represented in the fossil record. The apparent increase in diversity and specialization of shell-breakers could therefore be an artifact of better sampling of more recent taxa. I do not believe this, but the fact that this bias cannot be dismissed out of hand underscores the importance of tracing the history of predation within specified environments. One way around this problem is to examine the prey for diagnostic traces of predation. The next several chapters will examine predation from the prey's perspective.

PART THREE

The Evolution of Armor

and Locomotion

CHAPTER SEVEN

The Functional Morphology and

History of Gastropod Armor

THE FUNCTIONS OF ARMOR

In the preceding chapter, I tried to show that insertion, breakage, and drilling have become increasingly important as methods of subjugation of molluscs over the course of the Phanerozoic eon. The time has now come to trace the evolution of the prey, and to ask how the predators influenced the morphology of molluscs and other skeletonized animals.

Few characteristics of organisms are as well preserved and as available for study in the fossil record as is armor. It therefore seems appropriate to begin an inquiry into trends in antipredatory adaptation with an account of the functional morphology and history of armor.

I shall use the term armor to mean a hard, unyielding outside structure that surrounds softer tissues partly or completely. It provides a passive defense which an enemy must break, dissolve, penetrate, or disarticulate. For the purposes of this discussion, I shall generally exclude from the category of armor those features that have a retaliatory function—horns, claws, sharp movable spines, and the like.

Potential victims may benefit from armor in either the detection or the subjugation phase of an enemy's attack. Armor may reduce the likelihood of detection by keeping chemical or other diagnostic cues hidden within an impermeable wall. Evidence for this comes from the studies of Carriker and Van Zandt (1972), who showed that the oyster drill *Urosalpinx* is unable to locate oyster prey when the valves of the oyster are kept artificially sealed. B. S. Morton (1985) found that the melongenid gastropod *Hemifusus tuba* was able to detect and attack tightly closing pelecypods successfully only if blood leaked from the prey through a break in the seal where the valve margins had been damaged. Withdrawal of the cirri into the test enables barnacles to avoid detection by predators (A. R. Palmer et al., 1982). Armor provides protection against enemies during the subjugation phase by making the prey large, difficult to manipulate or to swallow, and hard to remove.

I shall begin the treatment of armor with the gastropods, a group of molluscs whose armor is both fairly well understood and often exceptionally well developed. Many of the principles I shall explore in this chapter apply to other groups as well; thus, this chapter serves as an introduction to the more general aspects of armor, as well as exploring univalved molluscs in detail.

THE CHARACTERISTICS OF GASTROPOD SHELL ARMOR

SHELL SIZE AND THICKNESS

Large size and a thick-walled shell are perhaps the most widespread characteristics of armor in shell-bearing gastropods. Almost every study of the subjugation phase of predation has shown that large, thick shells are more resistant than small thin ones. A shell's strength increases approximately as the square of the linear dimension (Ar et al., 1979; Vermeij and Currey, 1980), and the time required by naticid moon snails to drill through the wall of prey shells increases linearly with shell thickness (Kitchell et al., 1981). Longer handling time and greater shell strength mean that the probability of failure of an attack increases and that the armor is therefore more effective.

Several lines of evidence support the idea that large size and a thick shell wall impart resistance to shells. On the reefs of Guam, gastropod shells occupied by hermit crabs show higher frequencies of lethal breakage due to predation on the original gastropod builders when they belong to thin-walled species (shell wall less than 2.5 mm thick) than when they are thick-walled (Vermeij, 1979a). Incidences of repaired shell injuries usually rise as shell length increases, especially in species from the Late Cretaceous onward (Raffaelli, 1978; Vermeij et al., 1980, 1981; Vermeij, 1982a, b; Vermeij and Dudley, 1982). This rise occurs not only because larger shells belong to older individuals which have therefore been exposed to potential predators for a longer time, but also because the probability that a shell-breaking predator is successful in killing its victim falls dramatically with increasing prey size. This effect has been well documented for crabs in attacks on species of *Littorina* and *Nucella* (Hughes and Elner, 1979; Elner and Raffaelli, 1980; Lawton and Hughes, 1985) and for *Calappa hepatica* in attacks on species of *Strombus* (Vermeij, 1982b). Exceptions occur in high-spired species of *Rhinoclavis* which, when attacked by *Calappa*, are better able to resist peeling at smaller sizes, because the crab's claws cannot fit into the aperture of small individuals. The ratio of incomplete to total number of drill holes (that is, the effectiveness of individuals against drilling in the subjugation phase) rises sharply with increasing shell length in an Indonesian population of the pelecypod *Anadara granosa*, which is heavily preyed upon by the muricid *Bedeva blosvillei* (Vermeij, 1980b), and in the northeastern Pacific mussel *Mytilus californianus*, which is subject to predation by various species of *Nucella* (Dayton 1971; Suchaneck, 1978).

The only method of subjugation for which a small shell may be consistently advantageous to the prey is the dropping of prey by birds onto a hard surface. The impact of a shell as it strikes the ground is proportional to the mass of the shell, whereas strength is proportional to only the two-thirds power of the mass (that is, to the square of the linear dimension). Large shells are therefore more apt to break than small ones (Siegfried, 1977; B. W. Kent, 1981c). This exception is not likely to be important to most marine gastropods, however. Because the hard ground on which shells are dropped is typically out of the water, prey that survive the impact usually die from desiccation or temperature stress. In other words, selection may be weak or absent at the subjugation (dropping) phase of the attack because most or all victims die, whether by breakage upon impact or by being left in a physically hostile environment (Vermeij, 1983c).

The advantage of large size during earlier phases of a predatory attack is much less consistent. Because predators are often capable of learning, a small inconspicuous shell with a tiny morsel of digestable tissue inside may be quite unattractive to a prospective predator and may therefore be avoided or rejected by the predator during the detection-recognition phase. This advantage of small size would accrue whether the animal is armored or unarmored, so that small size by itself should not be thought of as an adaptive attribute of *armor*.

THE LIMPET FORM

Several architectural types, each with its own possibilities and limitations, have appeared again and again in the course of gastropod evolution. Because these types cut across taxonomic lines, I shall discuss gastropod armor by morphological theme rather than by group.

Limpets are characterized by a cap-shaped or ear-shaped shell which covers the soft parts dorsally. Because the mantle cannot withdraw from the shell margin to any marked degree, the ventral surface of the limpet is poorly protected when the animal is removed from the hard substrate on which it lives. Prevention of dislodgment is, not surprisingly, a recurrent theme in limpets, especially in the many groups that live on rocky seashores and in rivers. In such environments, tenacity is favored both by turbulence of the water and by prying predators. *Acmaea mitra* from the Pacific coast of North America, for example, resists the sea star *Pisaster ochraceus* by clinging firmly to rocks (Margolin, 1964). Tenacity in some marine limpets is enhanced by the ability of the animal to excavate a depression or home scar, so that the rim of the shell is wedged snugly where it cannot be dislodged by agencies on the rock surface. In experiments with various limpets on the Pacific coast of Panama, Garrity (1984) and Garrity and Levings (1983) have shown that limpets on home scars have a higher probability of survival than limpets that do not live on home scars, because the former are more resistant to predaceous fishes as well as to high temperatures and desiccation. Deep-sea limpets, including the monoplacophoran *Neopilina*, apparently do not cling tightly to hard surfaces (S. M. Stanley, 1982a; Hickman, 1983).

Lowell (1987) has argued that in many limpet species an upper limit on shell strength is imposed by the force of tenacity. Once a limpet is dislodged, its shell is of little protective value regardless of its strength. Accordingly, the force required to break the shell need not greatly exceed the force required to dislodge the animal. This limitation does not apply to coiled gastropods, whose shell may still serve a protective function after dislodgment.

Ways in which limpet shells may be strengthened include the development of radially oriented ribs or folds and thickening of the shell rim (Lowell, 1987). A low profile (flat, low-conic form) prevents apical breakage of the type inflicted by some predaceous crabs (Bulkley, 1968; Chapin, 1968) and further enables the limpet to avoid strong water currents that could dislodge the animal. Current velocities adjacent to the rock surface are much lower than those further away from the surface.

THE COILED FORM AND RETRACTABILITY

Shell coiling enables gastropods to withdraw the foot and other soft parts into the shell away from the growing edge when danger threatens. Withdrawal has obvious benefits. In the first place, it helps the animal to avoid detection by enemies. Watanabe (1983), for example, found that the Californian sea star *Pisaster giganteus* was less successful in locating *Tegula montereyi*, a species capable of very deep withdrawal, than in locating any of three other species of *Tegula* that withdrew less deeply. In the second place, enemies attacking their victim by way of the aperture may be thwarted from reaching the soft parts, because retractability is usually associated with a small aperture. In laboratory trials at Guam, for example, I have found that the crab *Calappa hepatica* frequently removes a half whorl or more of the shell of *Terebra affinis* and then stops, even though the foot has not been reached. Probably the shell opening becomes so small that the crab's claws cannot gain a purchase on the shell in order to continue peeling. Predation is therefore usually unsuccessful (Vermeij, 1982b). In certain extremely high-spired needle-shaped species, such as *T. triseriata*, the aperture is so small that *Calappa* probably cannot even begin peeling (Signor, 1985).

Perhaps the most important evolutionary consequence of retractability of soft tissues from the shell's growing margin is that the tissues are not harmed even if an enemy manages to damage the shell's lip (Figure 7.1). Injuries of the tissues not only require repair and regeneration, but also allow blood and other diagnostic cues to be released into the snail's environment where they could be detected by potential enemies. The snail could therefore die as the eventual result of the attack, even though the original damage was only superficial and not itself mortal. Our study in Guam of *Conus sponsalis* whose lips had been artificially broken (Zipser and Vermeij, 1980)

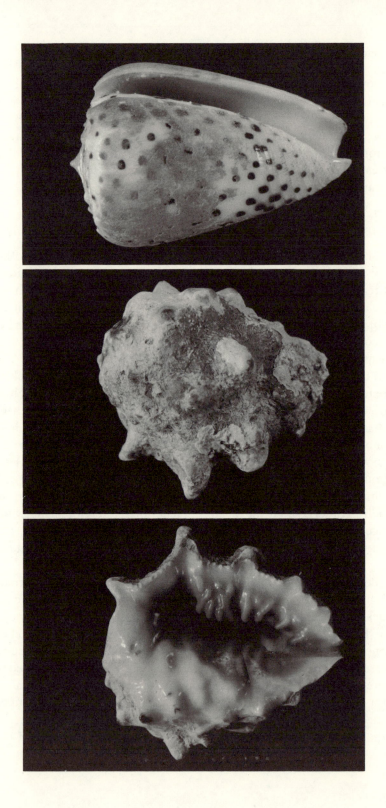

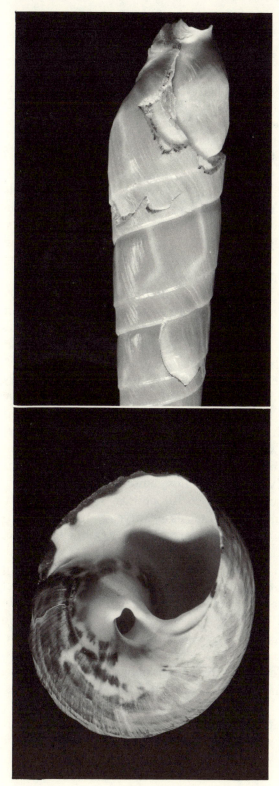

FIGURE 7.1 Some architectural types among Recent marine coiled gastropods. Upper left: Apertural view of *Drupa morum*, collected in July 1981 at Pagan, northern Mariana Islands. Upper center: Dorsal view of same individual of *D. morum*. This species combines most of the traits associated with great shell strength and impregnability—a thick wall, compact globose shape with short spire, narrow aperture bordered by strongly thickened and denticulate outer lip, and knobby external sculpture. Upper right: *Conus pulicarius*, collected in July 1981 at Pago Bay, Guam. This sand-dwelling species has a thick shell, narrowly elongated aperture, and short knobby spire. Lower left: *Cittarium pica*, collected in August 1969 at Fort Randolph, Panama (Caribbean coast). The broad, thin-lipped aperture and wide, deep umbilicus of this species suggest that the shell has a less sturdy build than that of either *Conus* or *Drupa*. Lower right: *Terebra dimidiata*, collected in May 1979 at Luminau Reef, Guam. This is a typical high-spired sand-dweller. Like many members of its genus, this specimen is marked with conspicuous scars resulting from the repair of unsuccessful shell-breaking attacks by the shell-peeling crab *Calappa*.

supports the conclusion that injury to the shell's lip poses no detectable short-term risks for snails. The low risk associated with sublethal shell injury means that there are few ecological limitations on the evolution of strong passive armor in coiled marine gastropods.

Although little has been published about the retractability of gastropods, experienced shell collectors know from their attempts to clean shells that withdrawal of the foot is especially deep in shells with a low expansion rate. Some auger shells (family Terebridae) can withdraw the foot to an angular distance of at least 720° (two whorls) behind the growing edge (B. A. Miller, 1975; Vermeij et al., 1980), and the columellar muscle, which is responsible for retraction, may be up to 4.5 whorls long in *Terebra dimidiata* (Signor and Kat, 1984).

Many groups have achieved deep withdrawal of the tissues by evolving a shell in which the long narrow tube is loosely coiled or secondarily uncoiled. In snails with disjunct (or evolute) whorls, the generating curve is nowhere in contact with earlier whorls, so that the parameter D is always greater than zero. Both evolute and secondarily uncoiled shells are associated with a sessile or sedentary mode of life. Vermetids and the turritellid genus *Vermicularia* have irregularly coiled or loosely twisted shells that are attached at their apical end by cement. Living gastropods with disjunct whorls are found in association with stinging cnidarians (some Architectonicidae and Epitoniidae) or live in very deep water in lakes and oceans (some baicaliids, valvatids, hydrobiids, epitoniids, and trichotropids). Uncoiled siliquariids and coralliophilids live embedded in sponges and corals, and minute caecids and orygoceratids live interstitially in sediments, under stones, or on plants (Rex and Boss, 1976; Vermeij, 1983c). Disjunct coiling may prevent the shell of some mud-dwelling species from sinking into the sediment, because the weight of the shell is spread over a relatively large area. This snowshoe effect (Thayer, 1975a) has been suggested as an explanation for the form of some Paleozoic euomphalaceans (Yochelson, 1971; Peel, 1975). No hypothesis has been proposed to explain the disjunct coiling of certain prosobranch land snails.

Evolute and uncoiled shells are structurally unsuited for resisting breakage, especially if the shell as a whole is subjected to compression. Because the whorls are not in contact, there is no internal buttressing, so that compressive strength is determined solely by the thickness of the tube wall. When the apical and axial portions of the generating curve touch the previously formed whorl, some degree of internal buttressing becomes possible. Deep withdrawal of the foot can be combined with internal buttressing in either of two contrasting types of shell architecture; a low-spired shell with a wide umbilicus or a turreted high-spired shell.

The umbilicate condition lessens, but does not eliminate, the structural weaknesses of the evolute shell. I have seen crabs use the walls of the umbilicus of the trochid *Cittarium pica* as convenient places for the claws to gain a purchase as pressure is applied in breaking the shell (Vermeij, 1976). On the other hand, the presence of an umbilicus enables a retractable gastropod to build a shell with a low center of gravity—a feature that is advantageous in turbulence (Vermeij, 1973a, b). In some minute ground-dwelling endodontid land snails from Pacific islands, and in marine anemone-associated architectonicids, the umbilicus serves as a site for attaching eggs or for brooding young (Bandel, 1976; Solem, 1976). Aside from such examples of probable secondary functions of the umbilicus, this attribute is generally, I suspect, a geometrical consequence of selection favoring deep withdrawal. Systematic surveys of retractability in relation to shell form are needed to evaluate this claim, especially in land snails, in which umbilici are especially common.

Signor (1982c) has pointed out that the presence of an umbilicus is usually incompatible with burrowing in sand or mud. Naticid moon snails are the only umbilicate gastropods that I have seen burrowing, but their shells (including the umbilical region) are covered by extensions of the foot during burrowing. Forward movement in the sediment would be impeded by an umbilicus, because the presence of this feature makes for a large frontal area, and because particles of sediment become entrapped in the umbilicus and must therefore be transported along with the shell as the animal burrows.

Although the structural weakness imposed by an umbilicus can be overcome by the formation of a callus closing the basal cavity, the umbilicus disappears geometrically only when the axial edge of the generating curve lies in contact with the previous whorl along its entire length. For gastropods with the ability to withdraw the foot deeply into the shell, this requires a tall spire (high T) with a small apical angle. Although high-spired shells are still susceptible to enemies that can slice the shell perpendicular to the axis of coiling, they are internally well buttressed and have a solid axial region. They are also compatible with a burrowing mode of life.

Turreted shells provide an effective defense against lip-peeling predators such as calappid crabs. Laboratory studies in Guam, for example, showed that only 9% of shell-breaking attacks by *Calappa hepatica* on *Terebra affinis*, 16% of such attacks on *Rhinoclavis aspera*, and 25% of such attacks on *R. fasciata* were successful (Vermeij, 1982b). Signor's (1985) parallel work with *C. hepatica* in New Guinea showed that shell-breaking attacks were successful 57% of the time on compact terebrids such as *T. affinis* and *T. undulata* and 10% of the time on such slender species as *T. columellaris* and *T. kilburni*. Slender species were also more apt to remain unscathed (94%) when confined with *Calappa* than were the more robust species (82%). The shell-peeling tooth of the crab presumably could not be inserted into the extremely small aperture of the slender species (Signor, 1985).

Studies of the North Atlantic thaidid *Nucella lapillus* confirm and illustrate the importance of a high-spired, small-apertured shell as an effective defense against crabs that attack their victims either by breaking back the lip or by grasping the foot without damage to the shell. Where *N. lapillus* coexists with crabs (*Carcinus maenas* and *Cancer* spp.), usually on sheltered shores or in channels below the low tide mark, the shell is thick and high-spired, shows a fairly high frequency of repair, is resistant to compression, and is immune to crab attack at a small size. Individuals living on wave-exposed rocky shores where crabs are rare have thin shells with a low spire, a broad aperture, a low frequency of repair, and low resistance to compression. Retractability has not been measured in these contrasting morphs, but experiments show that crabs are less successful in grasping the foot of the tall form than that of the squat, large-apertured form (Kitching et al., 1966; Gibson, 1970; Hughes and Elner, 1979; Currey and Hughes, 1982).

THE COMPACT SHELL AND APERTURAL BARRIERS

Although a long, slender shell might be difficult to manipulate in a jaw or claw (Signor, 1985), it can easily be severed perpendicular to the axis of coiling. In environments where shell-crushers predominate over shell-peelers, therefore, a turreted shell might not provide effective armor. Instead, a compact shell with a low spire might be favored. In such a shell, however, deep withdrawal of the foot is impossible. The development of a crush-resistant compact shell should therefore be accompanied by modifications at the opening to prevent lip damage and entry by enemies (Figure 7.1). Once evolved, such apertural barriers would also be effective in turreted shells.

Before discussing the modifications of the aperture, I shall review the features of compact low-spired shells that enable them to resist compression. We have investigated the compressive strength of shells by squeezing them between two flat metal plates, with the aperture placed against one plate and the dorsal wall of the body whorl against the other (Vermeij and Currey, 1980). Short-spired thaidids (those with the apical half-angle greater than 45°) are more resistant than higher-spired species of the same shell weight. A tropical thaidid weighing 2.8 to 3.0 g has a mean compressive strength of 2280 N if it is short-spired and only 810 N if it is high-spired. Just how strong some compact tropical gastropod shells can be is illustrated by the data in Table 7.1.

The low spire of these resistant species is not, of course, the only attribute that confers great compressive resistance. The apertural modifications that restrict entry may also impart resistance against crushing or peeling. These modifications include a narrowly elongated aperture, an aperture occluded by teeth or folds protruding from the margins, a periodically or terminally thickened apertural edge that prevents marginal breakage, and a strong inflexible operculum that tightly seals the aperture when the foot is retracted.

Narrowly elongated apertures (defined as those whose length to width ratio S exceeds 2.5) prevent entry by at least some predators. In tropical Pacific species of *Conus*, *Mitra*, and *Imbricaria*, whose S may exceed 9.0 (Vermeij, 1981), the aperture is often at most 2 mm wide, so that crabs cannot gain a purchase on the outer lip. This effect was nicely demonstrated by Bertness and Cunningham (1981), who found that the xanthoids *Eriphia squamata* and *Ozius verreauxii* were unable to insert their claws into the narrow aperture of Panamanian columbellids. Even when damage was inflicted to the shell, it was superficial. The only way for crabs to gain access to the flesh is to crush the shell outright.

Apertural barriers in the form of teeth and folds have evolved many times in the Gastropoda. In the East African pulmonate land snail ge-

TABLE 7.1 **Compressive Strength of Some Adult Gastropod Shells from Panama and Guam**[a]

Locality and Species	Length	Mass	Strength
	(mm)	(g)	(N)
Pacific Panama			
Thais melones	41.3	29.60	9810
T. triangularis	20.7	2.54	2610
T. kiosquiformis	33.4	3.97	2090
Acanthina brevidentata	25.6	3.01	1300
Guam			
Drupa ricinus	22.7	3.75	2950
D. arachnoides	22.3	3.30	3220
D. morum	31.2	11.70	5530
Morula granulata	29.4	5.25	3170
M. uva	16.7	1.01	1150
Drupella elata	40.7	9.56	5300
Thais aculeata	36.8	6.63	4480
Cronia fiscella	19.8	1.00	970

[a] Data are based on Vermeij and Currey's (1980) study of thaidids and on subsequently collected shells.

nus *Gulella*, they are effective in preventing the streptaxid pulmonate *Edentulina* from inserting its body into the prey aperture (F. X. Williams, 1951; Kasigwa et al., 1983). Böttger (1935) and Solem (1972) have reviewed several indirect lines of evidence that suggest that predaceous beetles are also thwarted by the apertural barriers of many land snails. Apertural constriction is not, of course, a foolproof form of protection. Several molluscivorous species of *Conus* and the thaidid *Nassa* commonly or exclusively prey on narrow-apertured or aperture-occluded species of *Conus* and *Cypraea* by inserting the proboscis into the prey's body by way of the aperture (Kohn, 1959; Kohn and Nybakken, 1975; Vermeij, 1978).

Thickening of the outer lip is clearly associated with greater strength of the shell. Its prime function seems to be as a first line of defense. The thickened adult lip of *Cerithium*, *Strombus*, *Drupa*, and many other well known tropical marine genera often remains intact after a crab such as *Calappa*, *Carpilius*, *Eriphia*, or *Daldorfia* attempts to break the shell of a resistant victim (Vermeij, 1978, 1982b; Zipser and Vermeij, 1978; Bertness and Cunningham, 1981). The only indication that an attack has taken place is a grazed lip or some superficial damage to the surface sculpture. Consequently, the frequency of repaired scars of thick-lipped species is substantially lower than it is for thin-lipped species in similar habitats (Vermeij, 1982b; Vermeij and Dudley, 1982). Thickenings (varices) located at intervals of 180° to 270° around the spire are effective in stopping breaks that are initiated at the aperture from reaching the retracted soft parts. In a study of attacks by *Calappa* on varix-bearing cerithiids of the genus *Rhinoclavis*, I found that all unsuccessful peels ended at the first varix behind the lip (Vermeij, 1982b).

Intuition suggests that tightly fitting inflexible opercula should be effective in preventing entrance into the aperture. Gibson (1970) showed that *Nucella lapillus* with an operculum were more resistant to attack by the green crab *Carcinus maenas* than were individuals in which the flexible operculum had been removed surgically. I suspect that the high incidence of repaired scars on the shells of naticine moon snails, which possess an inflexible calcareous operculum, results from the inability of shell-peeling or extracting enemies to break back the lip beyond the position of the closed operculum. Polinicine moon snails, which have a more flexible "horny" operculum, show a much lower incidence of shell repair. Certainly the shell of moon snails is generally thin and the aperture is broad, so that breakage of the shell would not seem to be substantially impeded by features other than the calcareous operculum. Predation on land snails by streptaxid gastropods is apparently inhibited by the presence of an operculum. In East Africa, the cyclophoracean *Maizania* and the pomatiasid *Tropidophora* are invulnerable to attack by *Edentulina*, which subdues its victims by thrusting its body into the aperture (F. X. Williams, 1951; Kasigwa et al., 1983). Both *Maizania* and *Tropidophora* possess calcareous opercula. The water beetle *Dytiscus* is apparently rarely successful in attacking the operculate snail *Viviparus*, whereas it easily dispatches inoperculate pulmonates such as lymnaeids (Kessel, 1938).

Some gastropods have developed the operculum into an offensive weapon. The slender, curved, often serrated operculum of strombids is used not only in locomotion, but also in stabbing and pushing against enemies. The effectiveness of this behavior has not been investigated, but the surprisingly low success rate (16%) of attacks by *Calappa hepatica* on fragile-shelled juvenile *Strombus gibberulus* perhaps indicates that the snail was able to thwart the crab before shell damage was inflicted (Vermeij, 1982b).

SCULPTURE

Strong external sculpture is still another defense evolved by many gastropods, especially compact species. Ribs, tubercles, spines, and flanges may have any of several functions. Strength is imparted by many kinds of regularly repeated elements of spiral ornament (oriented perpendicular to the apertural edge) and by collabral ornament (oriented parallel to the apertural edge), especially when the sculptural elements cause the shell wall as a whole to be pleated. Tubercles and spines can enhance a shell's strength by providing only a small number of points where the crushing surfaces of a predator's masticating organ can contact the shell. Moreover, these points are the thickest parts of the shell, and they concentrate the predator's crushing force to a very small area where exceedingly high pressures are created (A. R. Palmer, 1979). Many tubercles, spines, and flanges increase the apparent size of the victim, so that it is either difficult for the predator to ingest or mistakenly passed over by the predator as being unsuitably large. Apically directed spines may render the victims difficult or hazardous to swallow or manipulate (Vermeij, 1974a; A. R. Palmer, 1979). The function of granules, beads, delicate spiral riblets, and other fine-scale sculpture is poorly understood and may have little to do with defense.

A poorly documented but possibly important function of sculpture is to render the animal inconspicuous or undesirable because of the organisms that settle on the sculptured surface. Encrusting animals and coralline algae typically settle in cracks or depressions (Bernstein and Jung, 1979; Strathmann and Branscomb, 1979; Strathmann et al., 1981; Schmitt et al., 1983). A surface marked by ribs, spines, nodes, and grooves is therefore ideal for the settlement and survival of such organisms. Some encrusters evidently reduce the detectability of their host. This has been demonstrated for encrusted chamid pelecypods by Vance (1978), who showed that the sea star *Pisaster giganteus* was more successful in locating individuals from which ascidians and other encrusters had been artificially removed than it was in locating encrusted individuals. Feifarek (1986) has shown experimentally that the long spines of the West Indian thorny oyster *Spondylus americanus* do not protect the shell mechanically against attack by predaceous gastropods and shell-breakers, but instead promote the settlement and growth of encrusting sponges, which do seem to provide protection. Specimens from which spines as well as sponges had been removed were more apt to die and were less likely to reacquire a sponge coating than were individuals from which only the sponges had been removed.

Encrustation by coralline algae is the rule in most tropical reef snails. The corallines may render gastropods relatively inconspicuous to visually hunting predators and apparently also protect the shell against boring sponges and other bioeroders. The fact that nearly all reef snails that are susceptible to coralline encrustation are strongly sculptured may point to a causal connection between encrustation and sculpture, but this interesting possibility has not yet been put to an experimental test.

Not all instances of encrustation enhance the defenses of gastropods, and sculpture may therefore be a hindrance rather than a benefit. Settlement of the barnacle *Megabalanus californicus* in holes drilled by *Octopus* in the otherwise smooth shells of the trochid gastropod *Norrisia norrisi* from California causes a reduction in escape speed from enemies and a greater susceptibility to dislodgment by waves from the kelp host on which the snail typically lives. Encrustation also increases the likelihood of death to various bottom-dwelling predators (Schmitt et al., 1983).

It is clear that our understanding of sculpture, encrustation, and their relationship to defense at the three stages of a predatory attack is rudimentary and anecdotal at present. All that can be said is that sculpture attracts encrusters only in species living on hard substrata or on the surface of sands and muds, for encrustation rarely occurs in species that burrow into unconsolidated bottoms.

THE EFFECTIVENESS OF GASTROPOD ARMOR

I pointed out in Chapter 1 that it is impossible to measure the effectiveness of a given purportedly adaptive feature because that feature cannot be divorced from other characteristics of the individual.

The merits of a narrowly elongated aperture are difficult to study separately from the effects of shell size, thickness, and sculpture. Two ways exist to circumvent this problem. One is the experimental modification of the feature in question without affecting the other characteristics. Comparison of the fate of these experimental individuals with the fate of controls should provide insight into the contribution of the trait to the individual's effectiveness in coping with a given hazard. The second, less powerful approach is to make judicious comparisons among variants or species that differ substantially only in the expression of the trait under study. A difficulty with this comparative approach is that the entities being compared almost always differ in ways besides those for which the entities were chosen.

Despite methodological limitations and the paucity of data, the effectiveness of shells as armor against various kinds of predators is an important quantity. Estimates of effectiveness have been made in some common sand-dwelling gastropods that are exposed in the laboratory to predation by shell-peeling calappid crabs (Vermeij, 1982b). An indirect method is available which, despite various sources of error, is useful as a first approximation in the estimation of effectiveness in the field. It requires that in a natural population of snails all lethal and nonlethal attacks that test the strength of the shell be counted, and that the lethal attacks be distinguishable from the nonlethal ones. This can be done if (1) all attacks result in some shell damage, which in the case of nonlethal attacks is repaired subsequently; (2) shells or their crushed remains stay in the vicinity of the living population so that they can be recovered reliably and without a significant loss; and (3) post-mortem breakage is minor or at least measurable. These conditions are most adequately satisfied in gastropods living on sheltered sandy or muddy bottoms where post-mortem transport and post-mortem destruction by borers and shell-breaking agents are unimportant. The effectiveness of the shells against breakage can then be estimated as the number of nonlethal breaks (recorded as the number of scars left when the lip was repaired) divided by the total number (repaired as well as lethal) of attacks. Repairs must be counted in both the living and the "dead" populations. To evaluate the post-mortem artifact of breakage, it is necessary to examine the condition of shells that have been drilled by a gastropod. Muricacean and naticacean gastropod drillers attack living molluscs and seem to avoid shells that are either empty or inhabited by secondary shell-dwellers such as hermit crabs. If a drilled shell is found to have been broken to such an extent that a snail would have been killed, a post-mortem cause of breakage can reasonably be inferred. Accordingly, a minimum estimate of post-mortem "lethal" breakage can be calculated as the frequency f_p of "lethal" breakage in drilled shells. The corrected frequency of lethal breakage F_b is then given by the equation

$$F_b = (n_b/n)(1 - f_p), \tag{7.1}$$

where n_b is the number of shells with breakage extensive enough to have killed the snail and n is the number of "dead" shells. The correction f_p can be applied only in populations in which drilling is a significant agency of mortality. In collections of "dead" shells from protected shallow-water tropical sites, I have found the post-mortem artifact of breakage to be less than 10% in most instances (Vermeij, 1982b).

In Table 7.2 I have gathered together all available values of the effectiveness of gastropod shells against known and unknown agencies of breakage. Three observations about the data in Table 7.2 are important in the present discussion. The first is that the estimates of effectiveness from field collections agree remarkably well with estimates from the laboratory. Small strombids from the Western Pacific, for example, have poorly armored shells despite the thickened adult lip, and in both the field and the laboratory they have an effectiveness of 0 to 0.30 against shell-breaking predators. The second observation is that most other tropical species have relatively well-armored shells that possess one or more of the features I highlighted in the discussion of aptations against breakage. High-spired shells and the narrow-apertured shells of large species of *Conus* are especially effective as armor. The third important observation is that, as with most ecological data, there is considerable variation in effectiveness between populations. Some of this variation is doubtless due to error, but spatial variation seems to be characteristic of most components of natural selection.

All the data in Table 7.2 refer to the effectiveness of individuals against such shell-breaking predators as crabs and fishes. Because these predators usually are capable of moving much faster than gastropods, escape is an ineffective form of snail defense against these enemies. Escape is effective only when the snail releases its hold from a rock and falls among stones or rubble on the bottom. This

TABLE 7.2 Effectiveness of Gastropods against Shell-Breaking Predators[a]

Locality and Species	N[b]	Frequency	Defenses[c]
Pago Bay, Guam, 1979			
Conus sponsalis	78	0.40	nap, lsp
C. coronatus	20	0.40	nap, lsp
C. flavidus	23	0.61	nap, lsp
C. miliaris	13	0.46	nap, lsp
C. lividus	12	0.67	nap, lsp
C. rattus	67	0.18	nap
C. miles	19	0.58	nap, lsp
Rhinoclavis aspera	14	0.50	v, hsp
Terebra affinis	24	0.83	hsp
Pago Bay, Guam, 1981			
Natica gualteriana	15	0.73	co
Cerithium columna	11	0.18	v, hsp
Majuro Atoll, Marshall Islands, 1978			
Conus sponsalis	13	0.46	nap, lsp
Cocos Island, Guam, 1979			
Conus sponsalis	23	0.65	nap, lsp
Rhinoclavis fasciata	121	0.76	v, hsp
Strombus gibberulus	19	0.21	tl
Gun Beach, Guam, 1979			
Conus sponsalis	11	0.27	nap, lsp
Tagbilaran, Bohol, Philippines, 1979			
Clypeomorus bifasciatus	25	0.92	tl, v
Nassarius globosus	20	0.10	tl
Ravao Island, Papua New Guinea, 1979			
Clypeomorus bifasciatus	18	0.82	tl, v
West Coast of Boear, Aru Islands, 1979			
Strombus urceus	10	0.20	tl
Polinices tumidus	11	0.09	
Rhinoclavis vertagus	43	0.54	v, hsp
Kinabohutan, Sulawesi, 1979			
Nerita undata	24	0.63	tl, co
Tumon Bay, Guam, 1979			
Rhinoclavis aspera	18	0.78	hsp, v
R. fasciata	28	0.61	hsp, v
Strombus gibberulus	43	0.14	tl
Bangi Island, Guam, 1979			
Rhinoclavis fasciata	21	0.38	hsp, v
Strombus gibberulus	10	0.10	tl

Table 7.2 (*continued*)

Alupang Island, Guam, 1979			
Strombus gibberulus	89	0.01	tl
Wom Village, Papua New Guinea, 1979			
Strombus labiatus	21	0.05	tl
Nassarius luridus	11	0.54	
N. quadrasi	10	0.30	tl
Bulla ampulla	13	0.07	
Pujada Bay, Mindanao, Philippines, 1979			
Strombus labiatus	11	0.27	tl
Nassarius distortus	17	0.88	tl
Hebra subspinosa	13	0.78	tl
Venado Beach, Panama, 1978			
Nassarius dentifer	15	0.73	tl
N. luteostoma	10	0.40	tl
N. pagodus	26	0.42	
N. versicolor	10	0.80	
Cancellaria jayana	13	0.62	
Terebra spp.	33	0.85	hsp
Paitilla Point, Panama, 1975			
Anachis spp.	16	0.44	nap
Turritella banksi	11	0.45	hsp

[a] Samples were collected on various expeditions to the tropical Pacific.

[b] Effectiveness was measured as the number of repaired breaks in living and "dead" shells divided by the total number of breaks (N) (repairs as well as lethal breaks). All samples in which the total number of breaks exceeded ten were included.

[c] Potential defenses are a narrow aperture (nap), low spire (lsp), very high spire (hsp), varices (v), calcareous operculum (co), and thick lip (tl).

form of passive escape apparently works for the hydrobiid *Nymphophilus* against the fish *Cichlasoma* in certain freshwater ponds in Coahuila, Mexico (D. W. Taylor, 1966), and perhaps for some intertidal nerites (Vermeij, 1978). Active escape becomes feasible when the predators are relatively slow (Vermeij, 1978). Sea stars and predaceous gastropods often elicit highly effective escape responses in their gastropod prey, whereas resistance defenses to these slow enemies are perhaps more the exception than the rule (Ansell, 1969; Vermeij, 1978; Schmitt, 1982).

A striking illustration of this assertion is provided by data on the effectiveness of gastropods against drilling predators (Table 7.3). Only 6.4% of the 47 gastropod species that have been systematically surveyed, and for which the available samples contained 10 or more drill holes, have an effectiveness of 0.30 or greater. The 3 naticacean species show an exceptionally high incidence of incomplete drill holes. In fact, 2 of the 3 species with an effectiveness of 3.0 or greater are naticaceans, and the third naticacean surveyed has an effectiveness of 0.26. I do not know what accounts for the high effectiveness in these gastropods, which are themselves drillers, but the fact that naticids often attack and kill each other (J. D. Taylor, 1970; Adegoke and Tevesz, 1974; Kitchell et al., 1981) may have something to do with it. With these few exceptions, however, it is apparent that gastropods do not have well-developed resistance defenses against drillers in their shells.

The History of Gastropod Armor

Papp and his colleagues (1947) may have been the first to suggest that the history of shell-bearing gastropods was marked by an increase in the incidence of breakage-resistant armor in the Mesozoic era. They were impressed with the high incidence of thickened shell lips in the gastropods of the Miocene of Austria, and they also discussed the potential role of crabs, lobsters, and hermit crabs in the evolution of gastropod armor. Now that good documentation for the antipredatory function of various shell traits is available, we are in a good position to trace the history of gastropod armor, and to assess whether the apparent increase in the power and diversity of shell-breaking predators (Chapter 6) could have been responsible for the increased emphasis on armor in gastropods through time. The account I offer in this section is an extension of earlier work (Vermeij, 1977b, 1983c).

Drilling is also widely regarded as a geologically young method of predation. Carriker and Yochelson (1968) thought that bona fide examples of drilling are confined to the Late Mesozoic and Cenozoic, and P. H. Fischer (1962) and Sohl (1969) made preliminary attempts to trace the history of the chief families of gastropods that drill prey. Although subsequent work has shown that drilling has existed as a method of predation since the Early Cambrian, the conclusion that it

TABLE 7.3   Effectiveness of Gastropods against Drilling Predation

Locality and Species	Effectiveness[a]	Reference
Recent, Western Australia		Black (1978)
Patelloida alticostata	0.120	
Boear, Aru Islands		Vermeij
Rhinoclavis vertagus	0	
Guam (three sites)		Vermeij
Rhinoclavis aspera	0	
R. fasciata	0	
Venado Beach, Panama		Vermeij
Nassarius dentifer	0	
N. luteostomus	0	
Early Pliocene, Gatun Formation, Panama		Vermeij
Cancellaria epistomifera	0	
Antillophos candei	0.048	
Strombina lessepsiana	0.170	
Architectonica nobilis	0.071	
Eupleura thompsoni	0.290	
Middle Miocene, Choptank Formation, Maryland		Dudley and Dudley (1980)
Turritella plebeia	0	
Middle Eocene, France		J. D. Taylor
Mesalia regularis	0	(1970)
Rimella fissurella	0	
Euspira lamellata	0.530	
Ampullina patula	0.360	
Pteropurpura tricarinata	0	
Middle Eocene, France		P. H. Fischer
Bayania lactea	0.091	(1966)
Mesalia multisulcata	0.150	
Turritella carinifera (adults)	0.300	
T. funiculosa mitis	0	
Omalaxis bifrons	0.056	
Architectonica canaliculata	0	
Diastoma costellata	0	
Rimella fissurella	0	
Ampullina patula	0.210	
Natica sp.	0.240	
Clavalithes sp.	0.130	

Table 7.3 (*continued*)

Ancilla buccinoides	0.140	
A. olivula	0.180	
"Olivella" sp.	0	
Volutilithes cithara	0	
Marginella ovulata	0	
Turridae spp.	0.110	
Hastula plicatula	0.180	
Eocene, Nigeria		Adegoke and
Architectonica bendeica	0.250	Tevesz (1974)
Bonnelitia decorata	0.070	
B. amekiensis	0.070	
Clavatula sp.	0.090	
Eupleurotoma sp. 1	0	
E. sp. 2	0.063	
Mesalia amekiensis	0.088	
Mitrella sp.	0	
Natica amekiensis	0.260	
Pleurotoma sp.	0	
Rocellaria unidigitata	0.160	
Surcula africana	0	

[a] Effectiveness, defined as the number of incomplete drill holes divided by the total number of drill holes (complete as well as incomplete), was calculated for each sample in which the total number of drill holes was ten or more.

became ecologically and evolutionarily important only in the Late Mesozoic still stands. In this section, I shall review the history of drilling from the prey gastropod's perspective.

SLUGS

The origins and early history of univalved molluscs remain shrouded in mystery, but there is little doubt that shell-bearing molluscs descended from a small crawling animal whose dorsal wall was impregnated with or covered by unconnected calcareous spicules (Stasek, 1972). Animals of this type are represented today by members of the class Aplacophora, which live either interstitially among grains of sediment or in association with hydroids. These habitats appear to represent safe places in which primitively naked molluscs and other poorly defended animals have persisted.

That there are other situations in which gastropods without shells are successful is indicated by the frequent reduction and secondary loss of the shell in the attainment of the slug condition. This trend has been especially common in the classes Opisthobranchia and Pulmonata, but it has also occurred in one clade (*Buchanania*) of fissurellid keyhole limpets and in the neritacean family Titiscaniidae (Runham and Hunter, 1970; Solem, 1978; J. H. McLean, 1984b; Tillier, 1984). The timing of shell loss in these groups cannot be established with any degree of certainty, but it is clear from ancestor-descendant relationships and from the times of origin of shell-bearing ancestors that most or all of the secondarily slug-like gastropods of today originated during the Mesozoic or Cenozoic. There may have been Paleozoic slugs as well, of course, but the fact that shell loss has occurred in many groups despite the evolution of a high diversity of armor-piercing predators in the Mesozoic and Cenozoic cannot be overlooked. Slugs must have either come to rely on defenses other than armor (toxicity or escape, for example) or evolved in situations where shell-breaching predators are rare.

Solem (1978) has pointed out that most land slugs are found in moist mountainous regions, perhaps in areas where calcium carbonate is scarce. The production of a substantial shell might therefore be slower than in environments richer in calcium carbonate. Moreover, the danger of desiccation is probably minimal in moist mountainous regions, so that at least one selective agency that would have favored the maintenance of an impermeable shell is removed. The shell that could be produced might be so thin that it would offer little protection against predators. If, as Shaffer (1978) has claimed for salamanders, predation on gastropods is less important at higher altitudes than in lowlands, selection in favor of maintaining a protective shell would be weakened even further. However, none of these factors—inavailability of calcium carbonate, lack of desiccation, and minor role of predators—is sufficient to bring about shell reduction or loss, even if they do increase the opportunity for this to occur. This point is further supported by the observation that shell loss is virtually unknown among freshwater gastropods despite the exceedingly low concentrations of calcium carbonate, lack of desiccation, and dearth of predators in many lakes and streams. The only family of shell-less fresh water gastropods that is known to me is the Acochlidiidae, a taxonomically isolated group of small opisthobranchs from fast tropical streams (Hutchinson, 1967).

There is some evidence that slugs have well-developed chemical defenses. Most land slugs secrete copious mucus, making manipulation by beetles and perhaps other predators difficult (A. Greene, 1975). Marine nudibranchs, aplysiaceans, and onchidiaceans rely either on toxic secretions or on the incorporation of toxins and stinging nematocysts from their prey. Many marine opisthobranchs and the Chilean fissurellid *Buchanania* have intimate associations with chemically well defended tunicates, sponges, and cnidarians.

Escape has been important in naked cephalopods (Chapter 10), but its role in gastropods remains unclear. Shell loss in pelagic heteropod mesogastropods and pteropods may have been associated with an increase in swimming speed, and parallel trends of shell reduction in opisthobranchs might have begun in burrowing animals in which rapid locomotion was emphasized (Brace, 1977), but no comparative data on speeds of shell-bearing and slug-like opisthobranchs exist. In the present state of knowledge, these are little more than idle speculations based on comparative functional studies of a few living forms. An important fact bearing on the evolution of shell loss is that no slug-like rachiglossan gastropods are known, despite the very large number of relatively rapid burrowers and crawlers in this group. I regard the whole subject of shell reduction and loss in gastropods as being poorly understood and full of unanswered questions.

LIMPETS

A likely forerunner of the coiled univalve shell is a cap-shaped, spoon-shaped, or plate-like shell that covers the dorsal surface but not the ventral surface of a crawling mollusc. This limpet form was well established in the Early Cambrian. As in the case of slugs, however, the limpet form has been developed secondarily numerous times (Table 7.4). An appreciation of the conditions in which the limpet form is successful today could, as with slugs, provide insights into the situation of the earliest shell-bearing molluscs.

Broadly speaking, limpets are found chiefly in habitats and regions where predators either exert little selection or elicit primary defenses other than shell armor. They are especially prominent in the intertidal zones at temperate latitudes, where they often make up 25% or

TABLE 7.4   Taxonomy and Distribution of Limpet-like Molluscs

---

Class Monoplacophora

Order Helcionellida or Cyrtonellida: Scenellidae, Atdabanian (Early Cambrian) to Early Devonian (Runnegar and Jell, 1976)

Order Tryblidiida: Middle Cambrian to Recent; includes deep-water *Neopilina* and relatives (Runnegar and Jell, 1976)

Class Gastropoda

Order Docoglossa, Superfamily Patellacea: Late Triassic to Recent; many intertidal limpets; many lineages from Late Cretaceous time on capable of making home depressions

Order Lepetellida: Recent; deep-water forms (Golikov and Starobogatov, 1975)

Order "Archaeogastropoda"

Superfamily Bellerophontacea: Pterothecidae, Middle Ordovician to Middle Devonian

Superfamily Fissurellacea (keyhole limpets): Middle Triassic to Recent (J. H. McLean, 1984a)

Superfamily Neomphalacea: Recent; deep-sea rift limpets (J. H. McLean, 1981)

Superfamily Pleurotomariacea: Haliotidae (abalones), Miocene to Recent

Superfamily Platyceratacea: Middle Ordovician to Late Permian; some limpet-like forms such as *Praenatica* (Late Silurian to Early Devonian), many clinging to echinoderms (Bowsher, 1955)

Superfamily Trochacea: Trochidae, Stomatellinae, Norian (Late Triassic) to Recent

Order Neritopsida or Neritacea

Superfamily Cocculinacea: Miocene to Recent; deep-sea limpets

Superfamily Neritacea: Neritidae (*Pileolus*, Early Jurassic to Eocene, marine; *Septaria*, Recent, freshwater); Phenacolepadidae, Recent, marine (Fretter, 1984)

Order "Mesogastropoda"

Superfamily Rissoacea: Hydrobiidae, *Lacunopsis fischerpietti*, Recent, freshwater (Davis, 1979)

Table 7.4 (*continued*)

Superfamily Hipponicacea: Hipponicidae, Late Cretaceous to Recent; make home depressions

Superfamily Calyptraeacea: ? Metaptomatidae, Silurian to Permian; Capulidae and Calyptraeidae, Early Cretaceous to Recent; many make home depressions

Superfamily Eulimacea: Eulimidae, *Thyca*, Recent, on asteroids

Order Rachiglossa
Superfamily Muricacea: Thaididae (*Concholepas*), Coralliophilidae (*Quoyula*), both Recent, marine

Order Notaspidea: Umbraculidae, Recent

Order Cephalaspidea: Smaragdinellidae, Recent

Order Basommatophora
Superfamily Siphonariacea (Pulmonate marine limpets), Late Cretaceous to Recent

Superfamily Ellobiacea: Otinidae (J. E. Morton, 1955)

---

more of the resident open-surface gastropod species, and in the deep sea (Vermeij, 1973a; Hickman, 1983). The limpet form has also evolved independently several times in freshwater prosobranchs and pulmonates (Harry, 1964). Where they coexist with predaceous sea stars, gastropods, octopods, fishes, birds, and mammals on rocky shores, limpets tend to rely on escape, camouflage, or such resistance defenses as copious mucus, tight adhesion to the rock, and a behavior in which the repellent mantle edge extends over the shell's exterior. Strong water movements may have contributed to selection in favor of the limpet form in many rocky-shore and freshwater groups, but it is unlikely to have done so in the deep sea. Even in shallow waters, turbulence is unlikely to be the chief agency favoring high tenacity. Wave action at temperate latitudes tends to be greater than in the tropics, yet tropical Eastern Pacific limpets have tenacities that, on average, are more than twice as high as those of temperate North Pacific limpets (Lowell, 1987), and species that excavate deep home scars from which dislodgment is difficult are confined to tropical and warm-temperate waters.

Armor in limpets is also best developed in warm-water species. Tropical Eastern Pacific limpets are, on average, more than twice as strong at the shell margin as temperate species of similar size (Lowell, 1987). Territorial limpets in which the anterior shell rim is strongly thickened (enabling the animal to push competitors out of the territory) are found in warm-temperate California and South Africa (Stimson, 1970, 1973; Branch, 1976).

With this perspective on the distribution of Recent limpets, we may expect limpets to have declined relative to coiled forms over the course of time in high-energy environments. As Figure 7.2 shows, the fossil record generally accords with this expectation. Inspection of Table 5.2 shows further that limpet families that contain species capable of forming home depressions are known only from Early Cretaceous time onward. Aptations against dislodging agencies have therefore become increasingly well represented among the more recently evolved limpet groups.

Many ancient groups that included coiled forms among their ranks are today represented only by limpet-like forms. The bilaterally symmetrical Bellerophontacea, for example, constituted a large Paleozoic group whose last coiled representatives became extinct early in the Triassic (Yochelson and Yin, 1985). Today, the superfamily is represented only by the Fissurellidae (keyhole limpets). Similarly, the Euomphalacea, a superfamily of sedentary hyperstrophic gastropods with coiled members extending to the Late Cretaceous (Weeksiidae), is perhaps represented today by a single highly derived limpet-like form, the so-called rift limpet *Neomphalus* from the hydrothermal vents in the deep Pacific (J. H. McLean, 1981). The Patellacea may also be the sole (albeit ecologically and geographically widespread and diversified) remnants of a lineage of coiled gastropods (Golikov and Starobogatov, 1975).

UMBILICI AND OPEN COILING

Umbilicate shells have undergone a decline over the course of the Phanerozoic (Figures 7.2 and 7.3). The incidence of umbilicate species in local assemblages (Figure 7.3) was always 25% or higher during the Paleozoic and Triassic, whereas from Middle Jurassic (Bathonian) time onward, umbilicate species constituted 20% or less of local warm-water faunas (Vermeij, 1975, 1983c; Cain, 1977). Incidences of 10% or lower are known only after the Eocene epoch of

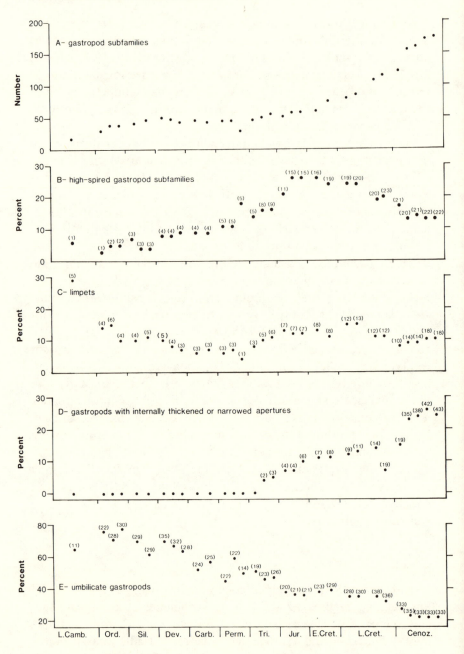

FIGURE 7.2 (*left*) Subfamilial diversity of gastropods with various predation-related shell traits through the Phanerozoic. Numbers refer to the diversity of subfamilies in which the attribute in question has evolved. A subfamily was not counted until at least one genus with the attribute in question appeared. The numbers and percentages of subfamilies will doubtless change with advances in gastropod systematics, but the patterns displayed parallel those seen in local assemblages (Figure 7.3) and are therefore likely to be robust. They include a temporal decrease in limpet-like and umbilicate gastropods, a temporal increase in groups with a narrow or thick-lipped aperture, and a Mesozoic peak in high-spired groups. The data were compiled from more than 50 papers and monographs on gastropod systematics. Univalved Monoplacophora have also been included in this compilation. Only bottom-dwelling groups with at least some marine members are plotted.

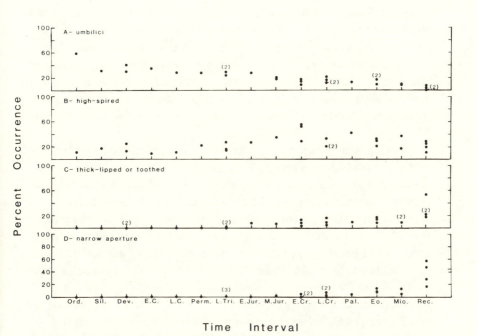

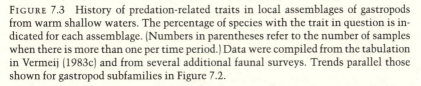

FIGURE 7.3 History of predation-related traits in local assemblages of gastropods from warm shallow waters. The percentage of species with the trait in question is indicated for each assemblage. (Numbers in parentheses refer to the number of samples when there is more than one per time period.) Data were compiled from the tabulation in Vermeij (1983c) and from several additional faunal surveys. Trends parallel those shown for gastropod subfamilies in Figure 7.2.

the Cenozoic. The only modern environments in which umbilicate species constitute more than 10% of the local assemblage are the surfaces of seaweeds, sandy and muddy bottoms of freshwater lakes, and surfaces on land. My data on local faunas do not show a decline in umbilicate taxa during the Devonian, but Signor and Brett (1984) have shown that bellerophontacean genera do have a decreasing incidence of umbilici during this period. In my compilation of umbilicate subfamilies (Figure 7.2), declines are evident in the Early Carboniferous, Early Jurassic, and Early Cenozoic.

Gastropods with evolute or disjunct coiling are very rare in the modern fauna and are known only from environments where shell-breaking predators are rare (Vermeij, 1983c). Rex and Boss (1976) tabulated 15 Recent evolute gastropod species, of which 3 are small land snails from islands, 4 are from deep, cold freshwater lakes, and 8 are deep-water marine gastropods from unconsolidated bottoms. Some of the marine species may be associated with nematocyst-bearing cnidarians. The incidence of evolute gastropods was also low in the geological past. Incidences of 5% or higher are known only from Paleozoic local faunas, whereas no Mesozoic or Cenozoic value exceeds 3%.

Several gastropods have become secondarily uncoiled in such a way that the whorls do not touch. In the modern fauna, shells of this type occur on algae (Caecidae), interstitially in freshwater mud (Orygoceratidae), in marine sponges (Siliquariidae), or cemented to hard surfaces (Vermetidae and the turritellid *Vermicularia*). With the exception of the cemented forms, these gastropods occupy environments where shell breakage is unimportant as a selective agency. Little is known about the geological history of loosely and irregularly coiled shells. Some 30 genera of loosely coiled gastropods are known from Paleozoic rocks (Peel, 1975; Vermeij, 1983c). In addition, Burchette and Riding (1977) believe that some tubular cemented fossils of Carboniferous age were gastropods, but the possibility that they were polychaetes cannot yet be rejected.

HIGH-SPIRED SHELLS

One of the benefits of the high-spired shell is that it protects the soft tissues from being reached by animals that attack by way of the aperture. This benefit has been apparent since at least Late Carboniferous time. In a survey of trochiform gastropods from the latter half of the Late Carboniferous of Texas, we found that the incidence of repaired injuries was positively correlated with spire height (Schindel et al., 1982; Figure 7.4).

In the modern fauna, high-spired shells (defined as those with an apical half-angle less than 10°) are best represented and most highly turreted in the Indo-West-Pacific region. Temperate assemblages show a very low incidence (5% of species or lower). Families such as the Terebridae, Turritellidae, and Cerithiidae, which in the tropics contain exceptionally slender species, are represented in cooler waters only by squat species (Vermeij et al., 1980). Hard-bottom representatives of turreted families are also more squat in shape than species in sand and mud (Vermeij, 1973b; Houbrick, 1978). Turreted shells are also common in fresh water and on land, especially in tropical regions (Cain, 1977, 1978a, b).

Given the antipredatory merits of turreted shells, it is not surprising that their incidence has increased through time. In local assemblages of warm-water species from sand and mud, pre-Jurassic incidences are 30% or lower, whereas the incidences in the Middle Jurassic (Bathonian) and later ages are usually higher, ranging from 20% to more than 50% (Figure 7.3). An analysis of families (Figure 7.2), shows that those of turreted gastropods also became better represented from the Paleozoic to the Jurassic, but after achieving a maximum relative diversity during the interval spanning the Middle Jurassic to Early Cretaceous, they have declined in relative numbers. I have no ready explanation for this trend.

APERTURAL BARRIERS AND SHELL SCULPTURE

As would be expected of strong armor, thickened lips and narrowly elongated apertures are of comparatively recent origin in gastropods and are best represented today in warm shallow seas. Very few Paleozoic gastropods can be described as having thickened lips (some Late Ordovician and Silurian members of the Craspedostomatidae had somewhat thickened, expanded outer lips), and none had occluding teeth or a narrowly elongated aperture. The percentage of species with thick-lipped or narrow apertures was 2% or less in local Triassic assemblages (Vermeij et al., 1982) and rarely exceeded 10% in later Mesozoic assemblages. It then rose through the Cenozoic to a peak

FIGURE 7.4   Representative gastropod shell types from the Late Carboniferous of Texas. A: *Worthenia gabulata*, a rather high-spired trochiform gastropod; Wayland Shale. B: *Macrocheilus* sp., a relatively high-spired smooth gastropod; Wayland Shale. C: *Trepospira* sp., a low-spired form; Wayland Shale. D: *Amphiscapha subrugosa*, a shell with flat base and widely umbilicate apical side, somewhat resembling living freshwater planorbids; Wayland Shale. E: *Bellerophon stevensiana*, a compact planispiral shell; Finis Shale.

in the Recent. This trend is seen both in local assemblages (Figure 7.3) and in a global analysis of gastropod families (Figure 7.2).

In the modern fauna, the highest incidence and greatest expression of thick lips and narrow apertures are found in the Indo-West-Pacific region. Faunas in the tropical Eastern Pacific and Atlantic show progressively lower incidences and less spectacular development of these apertural types (Vermeij, 1974a, b, 1978). These interoceanic differences are evident within many families, including the Conidae, Thaididae, and Cypraeidae (Vermeij, 1979b). The percentage of thick-lipped or narrow-apertured species is reduced to Mesozoic levels (10% or less) on cold-temperate marine shores. This is due not only to the fact that many families with modified apertures are overwhelmingly tropical in distribution (Conidae, Mitridae, Costellariidae, Marginellidae, Olividae, Cassidae, Cypraeidae, Eratoidae, Strombidae, Neritidae, and Ellobiidae), but also to the occurrence of mainly thin-lipped or broad-apertured species at higher latitudes within normally narrow-apertured families (Thaididae, Buccinidae, Columbellidae, Olividae). Only two genera among freshwater snails (the marginellid *Rivomarginella* and the lymnaeid *Acella*) are known to have narrowly elongated apertures, and only one (*Rivomarginella*) can be said to have a thickened lip (Vermeij and Covich, 1978). The incidence of thick-lipped and narrow apertures in fresh water is thus comparable to that in the sea during the Paleozoic. Narrowly elongated apertures are also rare in land snails (some Oleacinidae), but species with the aperture occluded by teeth make up perhaps one fourth of the more than 20,000 known land snail species and belong to unrelated lineages in many families, including the Endodontidae, Charopidae, Corillidae, Urocoptidae, Camaenidae, Polygyridae, Helicidae, Pupillidae, Enidae, and Helicinidae (Solem, 1972).

The ascendancy of gastropods with thick-lipped or narrow apertures in warm seas is attributable both to within-lineage evolution and to unusually high rates of diversification of families characterized by these apertural types. Internally thickened lips arose independently in archaeogastropod groups (Trochacea during the Triassic) as well as in the Neritacea (Triassic), Strombacea (Late Triassic), Cypraeacea (latest Jurassic), Tonnacea (Cretaceous), Rachiglossa and Toxoglossa (Cretaceous), and Opisthobranchia and Pulmonata (Jurassic and Cretaceous). Narrowly elongated apertures evolved repeatedly in at least 9 of the 15 subfamilies of Turridae during the Late Cretaceous and Cenozoic. That families with thick-lipped or narrow apertures have undergone exceptionally high rates of diversification is suggested by the large-scale Neogene increases within Conidae, Cypraeidae, Mitridae, and Costellariidae, especially when these prolific groups are compared with the thin-lipped Volutidae, dorsanine Nassariidae, and Naticidae.

While I have not made a systematic survey of the distribution of strong sculpture through the Phanerozoic, we know that spiny gastropods were present by at least Ordovician time. In the modern fauna, spines and tubercles are found chiefly in species living on hard or unconsolidated surfaces. Sculpture in burrowing forms is limited chiefly to collabral ribs (those parallel to the outer lip).

The capacity to modify the shell's interior by resorbing calcium carbonate from the inner shell walls may have played an important role in the evolution of strong sculpture and narrowly elongated apertures in some groups (Vermeij, 1977b). Most prosobranch gastropods resorb the superficial layer of shell material before the deposition of a new shell layer on the axial (or left) side of the aperture can take place (Signor, 1982b). The capacity to resorb a substantial volume of shell material has, however, evolved only in the Neritacea, the Rachiglossa, the Toxoglossa (family Conidae), and the pulmonate Ellobiacea. Resorption results in the enlargement of the shell cavity as partitions are made thinner or removed. *Conus lividus*, for example, removes about 25% of its internal shell volume without reducing the thickness of the protective outer shell wall or spire (Kohn et al., 1979). In *Nerita* and its allies, resorption of the inner whorls produces a large space that accommodates a sizable visceral mass which no longer has the spiral configuration characteristic of most other gastropods. Without resorption, *Conus*, *Nerita*, *Olivella*, and other genera in which large-scale resorption occurs would have a very small or strongly compressed visceral mass. Resorption may therefore have permitted the evolution of the narrowly elongated or strongly toothed aperture that characterizes these genera today. In many rachiglossans, resorption also permits the development of large sculptural features in a low-spired shell. Without the capacity to remove ribs, spines, or tubercles as they are encroached upon by the axial margin of the aperture during spiral growth, the gastropod would have to grow in such a way that the sculptural features would be either buried under a very thick glaze or retained on the external shell surface behind the aperture, where they are not overgrown by the newly forming whorl. In the latter case, strong sculpture could develop only in a shell with a relatively tall spire. This kind of limitation is observable in trochacean gastropods such as *Tectus*. The large spines of muricid rachiglossans can occur on a compact low-spired shell because they are eliminated by the left mantle margin as the gastropod grows (Carriker, 1972). Resorption therefore represents an important mechanism of shell modification that eliminates some of the adaptive compromises inherent in an external skeleton built by accretion at one end.

The geological history of opercula is difficult to reconstruct, because only calcareous opercula are preserved as fossils, and because their association with identifiable shells is often tentative or nonexistent. The first calcareous opercula, belonging to the maclurita-cean genera *Maclurites*, *Pelliseria*, and *Teiichispira*, are of Late Early Ordovician age (Yochelson, 1979, 1984a). Several other Paleozoic calcareous opercula are known, including members of the Oriostomatidae (Late Silurian to Early Devonian) and the Omphalocirridae (Early to Late Middle Devonian) (Linsley, 1978a; J. H. McLean, 1981). The superfamily Neritacea, which is characterized by the presence of a calcareous operculum in most families, originated with the Naticopsidae during the Devonian, and had calcareous opercula by at least Early Carboniferous time. Liotiids and cyclostrematids (belonging to the superfamily Trochacea) are families with calcareous opercula that originated during the Late Permian or Mesozoic. Other gastropods with calcareous opercula, all with Mesozoic origins, include naticine naticids, turbinids, phasianellids, barleiids, and the freshwater pilids and bithyniids. Inadequate as they are, these dates of origin suggest that the deployment of calcareous opercula came substantially earlier than that of thick lips, narrowly elongated apertures, and apertural dentition.

Calcareous opercula are found only in gastropods whose apertures are circular to broadly oval in shape. They may occur in some species whose apertures are occluded by teeth, as in many Neritidae, but no gastropod with a longitudinally elongate aperture has a calcareous operculum.

At least one other group of probable ancient molluscs, the Hyolitha, had calcareous opercula (Marek and Yochelson, 1976; Yochelson, 1984b). This class of tubular fossils is known from the earliest Cambrian to the Late Permian, but it was especially prominent in the Cambrian. The presence of calcareous opercula in many members of two of the three orders of hyoliths further strengthens the view that opercula provided an ancient method of preventing entry by predators.

At the opposite end of the spectrum are gastropods that have secondarily lost the operculum. That opercular loss should have occurred in many limpets and in families with secondarily internal shells (Ovulidae and Cypraeidae, for example) is perhaps not surprising, but loss of the operculum in groups with a coiled external shell is more difficult to explain. In the neritacean land snail families Proserpinidae and Cereidae, the absence of an operculum is correlated with the presence of folds and teeth that obstruct the aperture (Bishop, 1980; F. G. Thompson, 1980). Many rachiglossan families that are found predominantly on marine unconsolidated bottoms may have undergone a similar functional switch, a slit-like or toothed aperture substituting for the operculum. This may explain the absence of the operculum in costellariids; mitrids; cancellariids; most olivids; many turrids (especially members of the Mitrolumninae, Daphnellinae, Mangeliinae, and Clathurellinae); some Cassidae and Volutomitridae; and all shell-bearing cephalaspidean opisthobranchs except some Acteonidae and Retusidae. Chemical defense may have compensated for the loss of the operculum in broad-apertured tonnids, ficids, and volutids, but no experimental work on such defenses has been carried out. Another possibility is that obstruction by an operculum is less effective in gastropods from unconsolidated bottoms than in those from hard-bottom environments, because the operculum might interfere with locomotion in sand and mud, or because shell-entering agents are less common beneath the sediment surface. Opercular loss in pulmonates was probably associated with colonization of terrestrial and freshwater environments, which at the time of invasion were probably relatively free of predators. Only two genera of pulmonates, both belonging to the salt-marsh and brackish-water family Amphibolidae, have retained the primitive operculum (Solem et al., 1984). The evolution of land predators that are specialized to extract gastropods via the aperture has probably been responsible for the widespread evolution of apertural dentition in pulmonates, as well as for the development of secondarily operculum-like structures. In the pulmonate family Clausiliidae, a flap closing off the aperture probably differentiated from the palatal wall of the shell during the Cenozoic (Nordsieck, 1982), whereas in two genera of charopids from New Caledonia (*Pararhytida* and *Rhytidopsis*) the distal end of the tail has become hardened with collagenous fibers, effectively sealing the aperture when the foot is withdrawn into the shell (Solem et al., 1984).

The History of Traces of Predation

Breakage

Although the morphology of an organism can suggest or be consistent with an adaptive explanation, especially if living models are available, a historical account based solely on changes in architecture through time is inherently incomplete because it fails to incorporate independent evidence that the selective agency believed to be responsible for the adaptation existed. In the case of shell-breaking and drilling predation, such independent evidence is available in the form of diagnostic traces left on the victim by the predators (Vermeij, 1983b, c). The most reliable indication that a given shell was capable of resisting breakage is the presence of a scar, or repaired break. The injury could, of course, have been caused by any of several agencies— predation by a shell-breaking predator, grinding by rocks, competition with rivals that push or retaliate, and even prey that attempt to break the predator's shell—but available evidence indicates that predation is the most important cause of nonlethal injuries in most species. If physical agencies were important in causing nonlethal damage, scars should be common in snails from turbulent freshwater and marine habitats that are exposed to constant heavy wave action and strong currents and in which shell-breaking predators are rare. This expectation is not met, however. Cold-temperate wave-exposed thaidids, tropical wave-exposed planaxids, and temperate pleurocerids from fast-flowing streams all show extremely low incidences of repaired shell injuries despite the high potential for dislodgment and physical abuse of the shell. High-spired terebrids from warm-water sands, by contrast, often show extremely high incidences of scars (up to 13 in one shell of *Terebra gouldi* from Hawaii) despite the absence of turbulence (Vermeij, 1983c).

I argued in Chapters 1 and 2 that resistance against shell-breaking predators cannot evolve unless individuals in the population survive shell-breaking attacks. The potential for selection in favor of the evolution and maintenance of breakage-resistant traits rises as the incidence of unsuccessful shell breakage increases. If shell-breaking predation became relatively more common over the course of the Phanerozoic, as is suggested by the architectural history of gastropods and by the history of molluscivores, we would expect the incidence of unsuccessful shell breakage also to have increased through time. This hypothesis is testable in thin-lipped gastropods, in which nonlethal breakage is preserved as scars. (Thick-lipped species are less reliable because many unsuccessful attacks leave the lip unscathed and it therefore does not require shell repair.)

Before we test this hypothesis, it is prudent to review the somewhat complicated interpretation of shell repair and its incidence in a sample of shells. The incidence of repair (number of scars per shell) should increase if any of three factors also increase: (1) the relative abundance of shell-breaking predators (that is, the rate or probability of encounter between predator and potential prey); (2) the relative strength of the prey (that is, the probability that the predator will fail to cause death during an attack); and (3) the age of the prey (that is, the time during which a given individual has been exposed to potential attackers). The extent to which each of these factors contributes to an observed incidence of repair cannot be ascertained without intimate knowledge of the abundance and strength of predators; the size structure of the predator and prey populations; and the growth rate, age, and age-related probability of encounter of the prey. A high incidence of repair means that many individuals in the population were exposed to unsuccessful attacks because predators were common, prey were resistant, or prey were old. A low incidence, which means that unsuccessful injury-producing attacks were rare, may have any of several causes. First, shell-breaking predators may have been rare, that is, there were few encounters between predator and prey. Second, predators may have been extremely strong relative to the prey. Even if predators had been very common, the incidence of repair would be low if the probability of successful attack during an encounter was high. Support for the latter interpretation would be provided by the frequent occurrence of lethally broken shells. Finally, a low incidence of repair might be due to the predominance of alternative defenses, such as toxicity or the ability to escape. No matter which of these interpretations is correct, selection in favor of resistance of the shell is weak whenever the incidence of scars is low (Schindel et al., 1982; Vermeij, 1982b, d; D. G. Reid, 1986).

We tested the hypothesis that repaired injuries became more common through time by evaluating the incidence of scars in well-preserved gastropod assemblages from shallow-water unconsolidated bottoms representing five different times in the Phanerozoic: the latter half of the Late Carboniferous of Texas (Schindel et al., 1982), the Late Triassic of Italy (Vermeij et al., 1982), the Maastrichtian stage of the Late Cretaceous in the southeastern United States (Vermeij and Dudley, 1982), the Late Miocene or Early Pliocene of Panama, and the Recent of the tropical Eastern and Western Pacific (Vermeij, 1982b). Because we have samples from several localities for all time intervals except the Late Triassic, spatial variation can be taken into account in the evaluation of temporal trends.

The results show that there was a large increase in the incidence of scars between the Late Triassic and Late Cretaceous. This increase may be demonstrated statistically in several ways. The basic data consist of the incidence of repair in each sample of each species for which 10 or more individuals were available. In the first comparison, we calculated the median incidence of repair for each species and asked, by means of the non-parametric Jonckheere Test, whether a temporal trend was detectable. For the five time intervals taken together, we were indeed able to detect an increase in repair at the 0.002 level of significance (Vermeij et al., 1981). This level would have been even higher had we eliminated thick-lipped (mainly Cretaceous, Miocene, and Recent) species from our analysis, for thick-lipped species have significantly lower incidences of repair than thin-lipped species occurring with them. Further analysis showed that the increase occurred almost entirely in the Mesozoic. No difference existed between Late Carboniferous and Late Triassic species, or among those from the Late Cretaceous, Miocene, and Recent. Our findings of low frequencies of repair in the Late Triassic are corroborated by Yin and Yochelson (1983), who found no instances of repair in any of the Late Triassic gastropods from China that they examined.

The second comparison (Table 7.5) involves assemblages. For each assemblage, I determined the proportion of species in which the incidence of repair was 0.25 or higher. Comparison of the eight pre-Jurassic assemblages with the nine post-Jurassic assemblages shows that the proportion of heavily scarred species rose sharply during the Mesozoic ($p < 0.01$, Mann-Whitney U-Test). In addition to the data

TABLE 7.5 Percentage of Species in which the Frequency of Shell Repair Exceeds 0.25[a]

Assemblage	%[b]	N[c]
Late Carboniferous		
Grindstone Creek	25	12
Wolf Mountain Shale	20	5
Colony Creek Shale 1	40	5
Colony Creek Shale 2	25	8
Finis Shale 1	0	7
Finis Shale 2	14	7
Wayland Shale 1	57	7
Late Triassic		
St. Cassian beds, Costalaresc	18	11
Late Cretaceous		
Ripley Formation	63	24
Late Miocene		
Vienna Basin, Austria	14	7
Pliocene		
Gatun Formation, Panama	63	19
Albenga, Italy	40	15
Recent		
Venado Beach, Panama	61	18
Tumon Bay, Guam	78	9
Wom, Papua New Guinea	23	13
Pujada Bay, Mindanao	58	12
Dodinga Bay, Halmahera	100	7

[a] For each assemblage, species for which sample size exceeded ten individuals were considered. The data are calculated from values given by Papp and colleagues (1947), Robba and Ostinelli (1975), Vermeij and colleagues (1981, 1982), Schindel and colleagues (1982), Vermeij (1982b), and Vermeij and Dudley (1982).

[b] % Percentage of species in which the frequency of repair is 0.25 or higher.

[c] N Number of species.

summarized in our 1981 paper, I also included in this comparison the data of Robba and Ostinelli (1975) for Pliocene gastropods of Albenga, Italy, and figures given by Papp and his colleagues (1947) for the Late Miocene of Austria. These published data are not strictly comparable to ours, for they express the incidence of scars as the number of scarred individuals divided by the total number of individuals, rather than as the number of scars divided by the total number of individuals as in our data. The published data are therefore more conservative than ours and would therefore bias the results of the temporal comparison against the expected trend. Even so, the trend is revealed by the data.

A more detailed analysis of temporal trends was possible in our material from the Late Carboniferous of Texas (Table 7.6). Although some strata, such as the Wayland Shale, showed conspicuously higher than average frequencies of repair, we were unable to detect any consistent trends through time (Schindel et al., 1982).

I have also examined the history of the incidence of scars in two families of toxoglossan gastropods. In the Terebridae (auger shells), no temporal change in the frequency of scars was detectable from the Middle Eocene to the Pleistocene, but the Eocene and Oligocene were represented by only 7 samples of 5 species (Vermeij et al., 1980). There was an increase in the incidence of repair in European species of Conidae from the Eocene to the Miocene. The median incidence of scars in 8 samples from the Eocene was 0.11, whereas that in 16 samples from the Miocene was 0.29 (Table 7.7). This difference is significant at the 0.05 probability level by the Mann-Whitney U-Test.

There is also an indication that the absolute strength of shell-breaking predators has increased since the Jurassic. In our Late Carboniferous and Late Triassic samples, the incidence of scars rises with increasing shell length in only 57% and 43% of the species, respectively, whereas in assemblages from the Miocene and Recent it does so in 67% and 80%, respectively. Shell size and frequency of scars are expected to increase together for either or both of two reasons. First, the likelihood that a given predator will be successful in its attack declines as shell length and shell thickness increase. Second, a larger shell has coexisted with predators for a longer time than has a small shell, and it has therefore accumulated more scars. Some shells, however, may become so large that an attack by a predator will not cause damage and will therefore not be recorded as a scar. We

TABLE 7.6 Incidences of Repair in Relation to Time in the Late Carboniferous of Texas[a]

Species	Incidence of Shell Repair[b]							
	Gr[c]	WM	CC1	CC2	Fi1	Fi2	Wa1	Wa2
Glabrocingulum grayvillense	0	0.110	0.087	0.140	0.140	—	0.095	0.170
Ananias sp.	—	0.100	0	0.260	0.059	0.350	1.000	—
Goniasma sp.	0.029	—	—	—	0	0.071	—	—
Worthenia tabulata	0.400	0.620	0.360	—	—	—	1.270	0.310
Meekospira sp.	0	—	—	0.050	—	—	0.071	0.170
Macrocheilus sp.	0.050	—	—	—	0.130	0	—	—
Amphiscapha catilloides	0.170	—	—	0.250	0.091	0.025	—	—
A. subrugosa	—	0.200	—	—	—	—	0.310	—
Trepospira sp.	—	0	—	0	—	—	0.059	0
Phymatopleura spp.	—	—	0.670	0.190	—	0.150	1.030	0.570
Bellerophon stevensiana	—	—	—	0	0.053	0	—	—

[a] The data were collected in connection with the study of Schindel, Vermeij, and Zipser (1982).

[b] The incidence of repair (number of scars divided by number of shells per sample) is given for all species that occurred in at least two strata and for which samples of ten individuals or more were available. Strata are listed from left to right in ascending order.

[c] Gr Grindstone Creek; WM Wolf Mountain Shale; CC1 and CC2 Colony Creek Shale (two localities); Fi1 and Fi2 Finis Shale (two localities); Wa1 and Wa2 Wayland Shale (two localities).

interpret our data (Figure 7.5) as weak support for the hypothesis that pre-Jurassic predators were more apt to attempt shell breakage on small victims and were too small or too weak to attack larger shells, whereas post-Jurassic predators were large or strong enough to leave scars on larger shells (Vermeij et al., 1981).

DRILLING

Drilling is the only method of predation for which fossils can be used to estimate mortality and antipredatory effectiveness in the

TABLE 7.7   Frequencies of Breakage-Induced Shell Repair in European Fossil *Conus*[a]

Age and Species	Locality	Number	Frequency
Early Eocene			
C. deperditus	Vaudancourt	13	0.07
C. edwardsi	Bracklesham	60	0.12
Middle Eocene			
C. deperditus	Grignon	35	0.08
	Fresville	13	0.23
	Parnes	12	0.42
	Damery-Marne	11	0.18
	Croix Blanche	25	0.08
C. cresnensis	Le Ruel	16	0.06
Miocene			
C. mercati	Touraine	38	0.34
	Manthelan	14	0.29
	Pontlevoy	54	0.04
C. antediluvianus	Gram	69	0.83
	Wienerbecken	13	0.31
	Stazzano	12	0.33
C. tarbellianus	Sanbriguez	34	0.44
C. dujardini	Vöslau	38	0.29
	Pontlevoy	38	0.23
	Winterswijk	183	0.17
C. sp.	Cacela Velha	43	0.02
C. ixion	Saucats	26	0.31
	L'Argillierte	22	0.22
C. moravicus	Vöslau	77	0.11
C. pseudoponderosus	Manthelan	34	0.18
C. sp.	Winterswijk	15	0.20

[a] Data are based on collections at the British Museum of Natural History and the Institut Royal des Sciences Naturelles de Belgique.

past. However, several important qualifications and difficulties must be considered before the available data can be interpreted meaningfully. If we are to obtain an unbiased estimate of the contribution of drilling to overall mortality, we must record all deaths due to drilling. Since many shells are broken after death, some shells that were

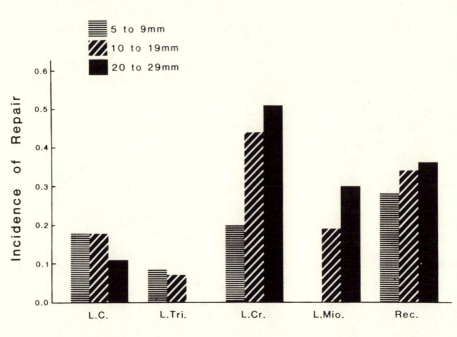

FIGURE 7.5 Relationship between the frequency of shell repair (number of scars divided by number of shells) and shell size in local assemblages during the Phanerozoic. The frequency of repair was calculated by lumping all shells in a given size class regardless of species. Larger shells are generally more resistant to lethal breakage by predators and have a greater mean age (and thus a longer exposure time to predators) than do smaller shells. The expected higher frequency of repair in larger shells is realized in post-Jurassic but not in pre-Jurassic assemblages. This suggests that pre-Jurassic predators may have been weaker than later ones. Alternatively, it is possible that pre-Jurassic gastropods were smaller (and less longevous) than post-Jurassic gastropods, and that there has been improvement by gastropods in coping with shell-breaking enemies.

drilled are lost from the sample. A more important problem is that the incidence of drilling is usually calculated from a sample of otherwise intact shells. If breakage is an important cause of death, however, the incidence of drilling will be seriously overestimated by this technique (Vermeij, 1980a). In the case of drilling by octopods, complete drill holes do not necessarily imply that the victim died as a result of drilling. In their study of the Californian trochid *Norrisia norrisi*, Schmitt and his colleagues (1983) found that many living snails had multiple complete holes drilled by *Octopus*. Kitchell and her colleagues (1986) have shown that complete naticid holes in which the ratio of the inner to the outer diameter is less than 0.5 are too small to permit the proboscis to reach the soft parts of the prey. Except in some terebrids and naticids, such nonfunctional complete drill holes by naticids are extremely rare.

Still another problem arises in the study of effectiveness. Because most drilling predators are relatively slow-moving, many gastropods have evolved effective methods of escape. For most species, therefore, the subjugation phase tends to be successful; that is, the pursuit phase rather than the subjugation phase is the stage of the predatory attack in which the gastropod is most effective. Shell features with a clear function against drillers are thus relatively rare, and a temporal pattern in their incidence or degree of development may not exist even though drilling may have become more important through time as a cause of death. Instead, evolution may have resulted in more effective methods of escape, which cannot be inferred from data on the frequency of drilling (Vermeij, 1978, 1982d).

Drilling by gastropods is most important as a cause of death for gastropod prey in warm oceans, especially in species from unconsolidated muddy and sandy bottoms. In turritellids, which live sedentarily in sand and mud, the mean incidence of drilling in temperate species (0.11) is about one third that in tropical species (0.32) (Dudley and Vermeij, 1978). Temperate intertidal gastropods from rocky shores are attacked only rarely by muricaceans, and even in the tropics predation by these gastropods is confined chiefly to small-bodied species (those with maximum dimension less than 10 mm) and rarely accounts for more than 20% of mortality (J. D. Taylor, 1978, 1984; Vermeij, 1979a). My unpublished data show that drilling from all sources accounts for only 1.5 to 29.9% of the mortality of gastropods from rocky shores on the Pacific coast of Panama, 0 to 14.2% of the mortality of hard-bottom species in Atlantic tropical America, and 0 to 12.0% of the mortality on rocky shores of the tropical Western Pacific. The incidence of drilling is usually higher in gastropods from adjacent sandy or muddy shores. The contribution of drilling to overall mortality of gastropods in these habitats varies from 5.9 to 30.3% in the Eastern Pacific and 5.0 and 38.0% in the tropical Western Pacific and Indian Oceans.

Only one study has been undertaken to investigate the relationship between the incidence of mortal drilling and water depth of the habitat on the continental shelf. Sander and Lalli (1982) found in Barbados that the incidence of drilling increased in five genera, and remained constant but high in two others, along a depth gradient from 125 to 225 m.

The fossil record of drill holes in gastropods begins in the Albian stage of the Late Early Cretaceous (J. D. Taylor et al., 1983). The available data, summarized in Figure 7.6, show that the incidence of drilling was low throughout the Cretaceous and variable but generally higher in the Cenozoic. By the Eocene, the contribution of drilling to overall gastropod mortality had already reached modern levels. This pattern is evident not only at the level of whole assemblages, but also in particular families such as the Turritellidae (Dudley and Vermeij, 1978) and Terebridae (Vermeij et al., 1980). Drilling in members of the molluscan class Scaphopoda (tusk shells) shows a similar tem-

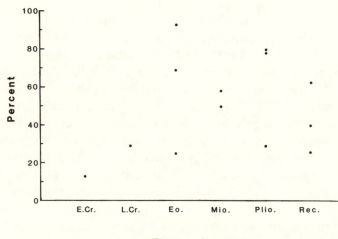

FIGURE 7.6 Drilling as a cause of gastropod death during the Phanerozoic. For each local assemblage, indicated by a dot, the percentage of species in which the frequency of drilling (number of drilled shells divided by total number of shells) exceeds 0.10 is plotted in relation to temporal occurrence. All assemblages were from warm shallow waters on unconsolidated bottoms. Only species in which the number of shells in the sample exceeded ten were considered. Calculations were based on data given by P. H. Fischer (1966), J. D. Taylor (1970), Adegoke and Tevesz (1974), Hoffman et al. (1974), Robba and Ostinelli (1975), Martinell and Marquina (1978), Sander and Lalli (1982), Vermeij (1982b), Vermeij and Dudley (1982), Taylor et al. (1983), and Hoffman and Martinell (1984).

poral pattern. It accounted for 0 to 13% of the mortality of scaphopod species in the Ripley Formation (Early Maastrichtian Stage of the Late Cretaceous of the southeastern United States), 0 to 15% of the mortality of species from the Eocene and Oligocene of the Gulf Coastal Plain, and 10 to 61% of the mortality of Recent species (Yochelson et al., 1983).

Too little information is currently available to permit evaluation of the hypothesis that shells have become more effective against drilling since the Late Cretaceous. Careful studies of trends within lineages and between assemblages are needed to determine whether the ratio of incomplete to total drill holes has risen through time.

The distribution of drilling by octopods is still poorly understood, but there is no doubt that octopod drilling is a major factor in gastropod mortality on some warm-temperate and tropical shores. In southern California, for example, octopod drill holes are found in 89% of the dead shells of the cowry *Cypraea spadicea* (R. A. Evans, 1980-1981), and R. A. Wells (1980) reports that 82% of the shells of the limpet *Collisella limatula* show evidence of having been pried off or drilled by *Octopus*. Drilling by *Octopus* accounts for 35 to 92% of the mortality of the intertidal trochid *Tegula funebralis* in southern California, but for only 3 to 12% of the mortality of this species in the cooler waters of central and northern California (Fawcett, 1984). Sand-dwelling gastropods such as *Olivella biplicata* also show high incidences of mortal drilling (more than 60%) in some populations in southern California (D. C. Edwards, 1969; Fotheringham, 1974). There is also evidence of octopod drilling in the Mediterranean (Ambros and Nelson, 1983) and on the Pacific coast of Panama, but I have not seen it in the tropical Western Pacific.

The only fossil assemblage in which *Octopus* drilling has been recognized is that from the Pliocene of Albenga, Italy. Of the 9 species showing predation by *Octopus*, only one turrid has an incidence greater than 20%, and 8 have incidences of 10% or less (Robba and Ostinelli, 1975).

SUMMARY AND CONCLUSIONS

The gastropod shell can be an exceptionally well-armored structure that functions effectively during either the detection or the subjugation phase of a predatory attack. Characteristics that enhance resistance against shell-breakers include large size, thick outer wall, deep withdrawal of the body, narrowly elongated aperture, teeth and other barriers in the aperture, thick outer lip, and strong sculpture. The limpet form, in which withdrawal of the soft parts is impossible, constitutes effective armor only if it is accompanied by tenacity of the foot. Umbilicate and open-coiled shells permit deep withdrawal, but they are mechanically weak and are therefore susceptible to crushing. The high-spired or turreted form is also associated with deep withdrawal of the foot and is highly effective against shell-peeling crabs, but it may be less effective against shell-crushers. Compact shells with apertural barriers and a thick lip are effective against predators that extract, crush, or peel their prey.

The capacity to withdraw the foot is seen as the principal characteristic that enabled coiled gastropods to emphasize armor. Failed attacks may leave the shell injured, but the retracted soft parts are typically unscathed. The victim therefore survives and is able to repair the damage.

Data on shell form and on traces left by predators on shells show that breakage and drilling became more common during the later Mesozoic and Cenozoic than they were in more ancient times. Traits that prevent access to the aperture, strengthen the shell, and enable the foot to withdraw deeply into the shell increased in incidence and in degree of expression over time in assemblages of warm-water gastropods from unconsolidated bottoms, whereas traits associated with mechanical weakness, unobstructed openings, and limited retractability declined. Modern levels of predation may have become established as early as the Eocene for some groups, but for others they were achieved in the Miocene. The incidence of repaired breakage-induced shell injuries remained more or less constant from the Late Carboniferous to the Late Triassic, rose from the Late Triassic to the Late Cretaceous, and thereafter again reached a plateau. These trends, together with the increase in the diversity and power of shell-breaking predators during the Mesozoic and Cenozoic (Chapter 6), suggest a pattern of escalation between armored gastropods and their shell-breaking enemies. This escalation was conspicuous during some intervals of the Mesozoic and Cenozoic but imperceptible during others (the Late Paleozoic and Early Mesozoic, for example).

Much remains to be learned about the tempo and extent of escalation between gastropods and their enemies. Well-preserved Jurassic and Early Cretaceous assemblages of gastropods are critically needed to provide us with data for studies of the incidence of repair, and to further a more detailed understanding of architectural changes during the Mesozoic. Estimates of the incidence of incomplete as well as complete drill holes are necessary if we are to evaluate the effectiveness of shells in resisting drilling predators since the Early Cretaceous, when drill holes on gastropods first appeared.

Gastropods are not, of course, the only (or necessarily even the best) group in which to investigate these matters. In subsequent chapters, I shall explore predator-prey escalation and its consequences in several other groups of skeletonized animals.

CHAPTER EIGHT

Opportunistic Armor: The Evolution

of the Conchicolous Habit

The Conchicolous Habit

If aptations are beneficial under a broad range of conditions, they tend to evolve independently in many clades. As the account in Chapter 7 demonstrates, the same architectural types of shell armor have appeared over and over again within the class Gastropoda. Given that gastropod shells are highly effective as predation-resistant fortresses, the fact that many animal groups have come to rely secondarily on gastropod shells for defense should come as no surprise. The repeated evolution of the conchicolous habit—the use by other animals of shells as residences after the original builders have died—is impressive testimony to the widespread reliance on and effectiveness of shells as protection. It is therefore of interest to investigate which groups have adopted the conchicolous habit, under which conditions the habit evolved, and at which times in the geological past the various groups became conchicolous.

The occupation of empty shells is casual and facultative in some groups, but it is specialized and more or less obligate in others, notably in the hermit crabs of the decapod crustacean superfamily Paguroidea (see Table 8.1). The use of the shell also varies from group to group. Shells are used as immobile shelters by representatives of the groups that contain only a few conchicoles. The Brazilian sarcophagid fly *Malacophagula neotropica*, for example, pupates in empty land snail shells that are cemented tightly to hard surfaces (Bequaert, 1925). Some jumping spiders in Madagascar and Europe use land snail shells suspended by silk threads from bushes as retreats and as nurseries for spiderlings (Decary, 1926; Fage, 1926; André, 1932; Mi-

TABLE 8.1   Survey of Conchicoles

Phylum Sipuncula: *Aspidosiphon, Phascolion,Golfingia*; mobile, marine; in gastropod shells and other domiciles (Hendrix, 1975; Hylleberg, 1975; Murina, 1984)

Phylum Echiura: *Lissmyema*; immobile, marine; in pelecypod valves (Ditadi, 1976)

Phylum Mollusca, Class Cephalopoda, Order Octopoda: *Octopus joubini* and related species; immobile while in shell, marine; in gastropod shells and pelecypod valves (Mather, 1982; R. B. McLean, 1983)

Phylum Crustacea, Class Malacostraca

Order Amphipoda, Superfamily Corophioidea: species of *Photis* and *Pterunciola* and members of the Corophiidae Siphonoecetinae; mobile, marine; in gastropod and scaphopod shells, many also in self-constructed tubes (Gauthier, 1941; Just, 1977, 1983; J. W. Carter, 1982)

Order Tanaidacea, Suborder Apseudomorpha: Pagurapseudidae; mobile, marine; in gastropod shells (Gutu, 1981; Sieg, 1983)

Order Decapoda

Infraorder Anomala, Superfamily Paguroidea (hermit crabs): mobile, mainly marine, but some terrestrial (*Coenobita*); in gastropod shells, worm tubes, and other objects, some in pelecypod valves (*Porcellanopagurus*) (McLaughlin, 1983; Scully, 1983)

Infraorder Brachyura (crabs): Dromiidae (*Hypoconcha* and *Conchoecetes*), Dorippidae (many genera); mobile, marine; under pelecypod valves (André, 1937)

Phylum Uniramia, Class Hexapoda

Order Diptera
Family Sciomyzidae Tetanocerinae: immobile, freshwater, in gastropod shells (C. O. Berg, 1964)

Family Sarcophagidae: *Malacophagula*; immobile, terrestrial; in gastropod shells (Bequaert, 1925)

Order Coleoptera: Drilidae (some species); immobile, terrestrial (F. X. Williams, 1951)

Order Hymenoptera: Apidae (*Osmia*); immobile, terrestrial (Bar, 1978)

Table 8.1 (*continued*)

Phylum Chelicerata, Class Arachnida, Order Araneae (spiders): Clubionidae (*Olios coenobita*), Salticidae (*Hyllus, Pellenes*), Argyronetidae (*Argyroneta*); immobile, terrestrial or freshwater (Decary, 1926; Fage, 1926; André, 1932; Mikulska, 1961)

Phylum Chordata, Class Osteichthyes, Order Perciformes

Family Cichlidae: *Lamprologus* spp., *Pseudotropheus livingstoni* group; immobile, freshwater (Fryer and Iles, 1972; Ribbink et al., 1983; E. C. Taylor, 1983)

Family Blenniidae: *Gobiosoma, Chasmodes*; immobile, marine (Breder, 1950; Dahlberg and Conyers, 1973; R. R. Phillips, 1977)

Family Serranidae: *Diplectrum*; marine, immobile (R. B. McLean, 1983)

---

kulska, 1961). Some solitary bees of the genus *Osmia* in the Old World build nests in empty land snail shells (Bar, 1978), and European trogulid harvestmen lay eggs in empty shells (Pabst, 1953). All these groups have only a few conchicolous representatives in their ranks. Conchicoles capable of locomotion while protected by the shell belong to groups in which the conchicolous habit is common and often highly developed. Besides the hermit crabs, these conchicoles include tanaidacean and amphipod crustaceans, as well as worms of the phylum Sipuncula (Table 8.1).

Inspection of a shell frequently reveals whether it has been occupied by a conchicole. The columellar (inner) lip is often scratched, the ventral surface near the aperture is often characteristically eroded from having been dragged along the ground, and the outer lip often bears an incision where a hermit crab's claw has rubbed against the shell (Boekschoten, 1966; Baluk and Radwański, 1984). Encrusting organisms often congregate around the aperture. If encrusting organisms have extended the shell at the aperture, the presence of a conchicole can be inferred with confidence, for such extension has never been observed in living gastropods despite the widespread occurrence of encrusters on living gastropods (Douvillé, 1931; H. W.

Wells, 1969). The assertion that the gastropod shell can be enlarged by encrusting bryozoans while the gastropod is still alive (Lecointre, 1929; Roger and Buge, 1947) is founded on erroneous interpretations of fossil shells. Hermit crabs of the genus *Coenobita*, which occupy the upper fringes of the shore as well as nearby habitats on land in the tropics, resorb the inner walls of the shells they inhabit, so that the shell cavity becomes enlarged and loses its spiral configuration (Figure 8.1). Amphipod, sipunculan, and tanaid conchicoles build a tube that may extend some distance beyond the aperture of the shell. Shells that are not immediately occupied after the gastropod's death are usually buried rapidly and do not become encrusted or extensively bored (Stachowitsch, 1977, 1980).

BENEFITS AND RISKS OF THE CONCHICOLOUS HABIT

I suspect that predation is the primary agency responsible for the evolution of the conchicolous habit, and that armor is the primary protection provided by the shell. Without a shell, a hermit crab with its soft abdomen and exposed clutch of eggs would be a highly vulnerable animal unless, as in some secondarily crab-like species that do not live in shells, the abdomen is hardened and tucked under the cephalothorax.

Secondary shell-dwellers fall victim to many of the same shell-breaking and shell-entering predators as do gastropods. Although many sea stars, shell-entering gastropods, and perhaps all shell-drilling gastropods are specialized to molluscan prey and therefore shun hermit crabs and other conchicoles, most fishes, crabs, stomatopods, lobsters, birds, and mammals that eat gastropods also consume other shell-bearing animals, including conchicoles. The characteristics that enhance the effectiveness of armor in gastropods should consequently provide similar benefits to later occupants of gastropod shells. Where there is strong selection in favor of armor in gastropods—at low latitudes and in shallow marine waters, for example—there should also be strong selection for conchicoles to recognize, acquire, and retain fortified shells.

This expectation is fulfilled by some tropical Indo-West-Pacific hermit crabs of the genera *Dardanus*, *Calcinus*, and *Trizopagurus*, which live at or below the low tide line in environments where gas-

FIGURE 8.1   Modifications made to shells by the hermit crab *Coenobita*. *Coenobita*-occupied *Nerita plicata* (upper right) and *Strombus gibberulus* (lower right) differ from the shells of living gastropods (upper and lower left, respectively) in having excavated columellar (inner) and outer lips, and shell walls that are, as a whole, thinner. Nothing is known about the mechanism or possible functions of these highly characteristic modifications. All shells are from Pago Bay, Guam.

tropod armor is especially well developed. These crabs have become specialized to occupy shells of conids, cypraeids, thaidids, strombids, mitrids, and other families with narrowly elongated, obstructed, or slit-like apertures. Bertness (1981a, 1982) has shown that tropical hermit crabs from the lower intertidal zone in Panama generally choose heavy, well-armored shells when given a choice among shells with various combinations of traits.

Armor is not, of course, the only benefit of the shell to a conchicole; neither is it always the attribute of the shell that is most important when an occupant is presented with an array of potential alternative residences that vary in size and shape. In fact, the size and geometry of the shell influence nearly every aspect of life of the secondary occupant. Bertness (1981a, b, c) has carefully studied how several species of hermit crab in Panama choose among shells. A light-weight shell with a spacious interior, such as *Nerita*, enables a hermit crab to grow rapidly and to carry a large clutch of eggs, but it constitutes a poor defense against shell-breaking and shell-entering enemies. A heavy shell with a small or narrow opening has a relatively small internal volume that severely limits the size of the egg clutch as well as the rate of body growth, but it does provide effective armor. Such a shell is also typically ineffective in conferring resistance to desiccation and thermal stress on the high shore, for it is incapable of storing large amounts of water during periods of exposure to air and sun. In short, secondary occupants usually cannot expect to find shells in which defense, growth, fecundity, competitive ability, and resistance to physical stresses are enhanced simultaneously. Instead, adaptational incompatibilities arise which must be resolved according to the pool of available shells and the ranking of the various agencies of selection (Bertness, 1981c). On the higher reaches of the shore, for example, hermit crabs are more apt to prefer shells with a relatively large internal volume than those that are heavily armored with small internal volume, because desiccation and thermal stress may outrank predation in importance. In some environments, predation may be so relaxed that withdrawal by the hermit crab into the shell need not be complete. This is seen in *Pagurus hirsutiusculus*, a species of the intertidal zone of the cold-temperate northeastern Pacific that is frequently found in shells too small for the body to be withdrawn into completely. Terrestrial hermit crabs of the genus *Coenobita* are also relatively large for their shells, but they, like some large marine hermit crabs, compensate for the inability to withdraw by an aggressive disposition and the modification of the larger claw into an operculum that guards the aperture when the crab is retracted into the shell.

Although shells can provide effective passive protection to conchicoles, three processes probably make them less reliable for secondary occupants than for the molluscs that built them: weakening of the shell by borers, irreparable shell breakage, and shell exchange between individual conchicoles. An important difference between shell-bearing molluscs and conchicoles is that boring and nonlethal breakage can be repaired or counteracted by molluscs but generally not by secondary occupants. Molluscs retain the same shell throughout life, whereas hermit crabs and at least some sipunculans, spiders, and amphipods, and possibly other conchicoles as well, change from a smaller shell into a larger one as they grow (Fage, 1926; Murina, 1984). Shells whose original molluscan builders have died deteriorate with time and therefore become unsuitable even for the competitively most subordinate among conchicoles. Shell exchange between hermit crabs, however, poses potentially substantial risks of detection by enemies and makes the conchicoles temporarily vulnerable to attack as the abdomen or other unprotected part is exposed while one shell is vacated in favor of another. We know little about the extent of these risks, or about how the risks vary geographically and with habitat, but one way of assessing these risks would be to estimate the proportion of a living population of hermit crabs that occur without shells. The only published report of homeless hermit crabs is that of Samuelsen (1970), who found them in western Norway. I have never seen hermit crabs living outside shells in the wild anywhere in the tropics. These observations suggest that the shell provides vital protection to hermit crabs, especially in warmer regions, and that selection in favor of characteristics that reduce the frequency or danger of shell exchange would be strong in many situations.

THE EFFECTS OF SHELL SUPPLY

Specialization to life in shells by conchicoles probably could not evolve if the shell supply were low or unreliable, no matter how great the advantage of protection by a shell, or how low the risks of shell exchange to the individual. The dynamics of shell supply have not been investigated thoroughly, but many studies give the strong impression that populations of hermit crabs vary in size according to the availability of shells. Spight (1977, 1981), for example, tracked populations of hermit crabs near Friday Harbor, Washington, for many years, and he showed that the size of the population depends directly on the availability of the most common shell, that of *Nucella lamellosa*. The population dynamics of conchicoles, as well as the potential for the evolution of the shell-dwelling habit, are thus expected to depend on the mortality schedule and abundance of molluscs.

I would guess that predation on gastropods provides a far more reliable source of shells than do such catastrophic agencies as storms and floods. Although rates of predation decline with a decrease in temperature, and are therefore apt to vary with the seasons, at least in temperate regions, predators need to eat regularly as long as they remain active. Empty shells therefore become available at a relatively constant, or at least predictable, rate instead of episodically. That hermit crabs rely on predation by molluscivores for the acquisition of new shells has been well established for shallow-water species in northern Florida (R. B. McLean, 1974; Wilber and Herrnkind, 1982, 1984; Gilchrist, 1984). The hermit crabs are attracted chemically to, and wait near, sites where predaceous gastropods and even crabs are attacking prey gastropods. In this way, some hermit crabs obtain clean, fresh, high-quality shells on a regular schedule, whereas other hermit crabs in the population benefit by taking the shells discarded by crabs that acquire the shells of the newly killed victims. Predators that offer this kind of dependable supply potentially include sea stars, gastropods, octopods, crabs, and fishes. Predators that leave the shell intact and do not also prey upon hermit crabs would seem to be the most reliable sources of shells. Most warm shallow-water marine habitats support an abundance of the appropriate predators.

Physical calamities are important causes of death for gastropods in many parts of the world, but they are unlikely to benefit conchicoles. Because catastrophic events tend to occur at infrequent intervals, shells become available in great numbers at certain times but not at others. Shells that become empty as a result of catastrophe-induced mortality are therefore often lost before potential secondary occupants, which themselves were probably affected by the catastrophe, can secure them, so that the pool of shells entering the population of conchicoles is actually very small.

From these considerations, I expect that environments in which predation is the chief cause of death of gastropods are suitable for the evolution and maintenance of the conchicolous habit, whereas environments in which episodic calamities are the chief agencies of death cannot support populations of specialized secondary shell-dwellers, even if shells would offer significant protection to potential conchicoles. Unfortunately, we still know little about causes of death and their distribution along geographical and ecological gradients, but the rarity of conchicoles in certain habitats and regions is consistent with my impression that shell supply is erratic in these places.

Fresh water constitutes one environment where conchicoles are conspicuously rare. The only shell-dwellers I know of from fresh water are two genera of cichlid fishes (*Lamprologus* in Lake Tanganyika and *Pseudotropheus* in Lake Malawi), the European spider *Argyroneta*, and members of the snail-eating fly family Sciomyzidae. The shell is apparently always used as a retreat or as a brood chamber and is not carried about. Moreover, the use of shells is facultative for the spider and the flies. No hermit crabs, tanaids, or sipunculans are known in fresh water, and none of the freshwater amphipods are known to occupy discarded shells. The rarity of conchicoles in fresh water carries over to some large bodies of brackish water, notably the Baltic Sea and the Danish inner fjords (Rasmussen, 1973).

I suspect that the near absence of conchicoles in fresh water is attributable in part to an erratic supply of shells. Predaceous gastropods, sea stars, and octopods are lacking in fresh water. Snails and sea stars that prey on molluscs do occur in the inner Danish waters and western Baltic, but they prey chiefly on pelecypods and are considered to be minor predators of gastropods (Rasmussen, 1973). At least in rivers, much of the mortality of gastropods is caused by floods (Davis, 1979), but this seems unlikely in lakes.

The secondary shell-dwelling habit in three marine groups (tanaids, amphipods, and sipunculans) probably arose in animals that already built portable tubes. In fresh water, caddisfly larvae of the insect order Trichoptera also build portable cases, which in the family

Helicopsychidae closely resemble dextrally coiled snail shells in shape, but no caddisfly larvae are known to be conchicoles. One wonders if the erratic supply of molluscan shells accounts for this lack of specialization.

An alternative hypothesis is that the shells of freshwater snails are generally so flimsy that they would offer little protection to conchicoles. I reject this hypothesis on two grounds. In the first place, shells in some regions (Southeast Asia, southern North America, and the Rift lakes of Africa) are notably sturdy, yet secondary shell-dwellers are no better developed in these regions than in cold-temperate areas where thin-shelled pulmonates prevail. Second, caddisfly larvae gain significant protection against fishes even though their cases are generally weaker than the shells of most freshwater gastropods. The cases capable of withstanding the highest compressive loads (*Odontocerum albicorne* from Sweden, 5.09 N, and *Helicopsyche* sp. from Ontario, 5.30 N) are much weaker than the shells of the pulmonate *Helisoma* (10.00 N) and the prosobranch *Goniobasis* (52.00 N). Even the weakest freshwater snail shells (*Physa* from Ohio, 2.94 N) are still stronger than most caddisfly cases (Otto and Svensson, 1980; Stein et al., 1984; D. D. Williams et al., 1983). Moreover, caddisfly cases are typically open at both ends, so that it is impossible to achieve the kind of tight seal that can exist in gastropod shells.

Polar regions also seem to lack secondary shell-dwellers. The paguroids that do live in the Arctic and Antarctic are members of the family Lithodidae which have become secondarily crab-like and have lost the shell-dwelling habit (Yaldwyn, 1965). Shell-dwelling amphipods, tanaids, and sipunculans are also unknown from polar waters. Whether this absence is due to an erratic shell supply or whether there is little advantage in living in the relatively poorly armored shells of cold-water gastropods has not yet been ascertained.

IMPROVING THE SHELL RESOURCE

Once the shell-dwelling habit has evolved, there is likely to be strong selection for improving the effectiveness and lengthening the functional "life span" of shells. After all, shells deteriorate if they are not repaired, just as houses do, and there may be substantial risks in moving to more suitably sized shells as the conchicole grows.

It is largely through intimate association with other animals that hermit crabs and shell-dwelling sipunculans have modified the shell resource. Sea anemones of the genus *Calliactis* and related forms have been shown to provide protection to hermit crabs against predaceous crabs and octopods (D. M. Ross, 1971, 1974; R. B. McLean, 1983). Hydroids such as *Hydractinia* and *Podocoryne* may confer a similar benefit to many temperate hermit crabs, and various solitary scleractinian corals, including *Heteropsammia*, may provide protection to the associated shell-dwelling sipunculan *Aspidosiphon* in the tropical Western Pacific. Association with nematocyst-bearing sea anemones enables some hermit crabs to exploit large, fragile shells that would otherwise offer little protection. In fact, the single published survey of anemone-bearing shells occupied by hermit crabs (Cutress et al., 1970) shows that sturdy shells such as *Phyllonotus* are much less likely to be carrying anemones than are the more delicate shells of *Fasciolaria*.

Many associates of hermit crabs enlarge the shells they encrust and therefore enable the crabs to stay in the same shelter for an extended period of growth. Risks associated with shell exchange, as well as problems brought on by an erratic shell supply, are therefore potentially alleviated by these shell-building encrusters. The most remarkable example of shell enlargement has been described by Dunn and her colleagues (1981). After settling on a small gastropod shell inhabited by a deep-water species of *Parapagurus*, the actiniid sea anemone *Stylobates* (which was originally described as a mollusc) enlarges the shell in a spiral direction by laying down a golden-yellow chitinous base around the crab. Hydroids, zoanthids, sponges, and bryozoans are also capable of enlarging the shells of hermit crabs (H. W. Wells, 1969; Provenzano, 1971; Gordon, 1972; Miyake, 1978; Stachowitsch, 1980; Muirhead et al., 1986).

Shell enlargement by cnidarians and sponges seems to be a phenomenon of deep-water and mid- to high-latitude environments where the supply of shells is probably low. Shell-enlarging bryozoans are common on small hermit crabs in the low intertidal and adjacent subtidal zones in New Zealand, the Gulf of Mexico, the Bay of Panama, and West Africa (P. L. Cook, 1968; Gordon, 1972). These areas are all more or less highly productive. I have never seen shell-enlarging encrusters on shallow-water hermit crabs in the West Indies, the islands of the tropical Western Pacific, or the relatively unproductive

Gulf of Chiriqui in the Eastern Pacific; nor am I aware of any instances of secondary shell enlargement among conchicolous tanaids, amphipods, or sipunculans.

THE HISTORY OF THE CONCHICOLOUS HABIT

If gastropod armor increased in strength and incidence through the course of the Phanerozoic, as I suggested in Chapter 7, the advantages of living in empty gastropod shells may also have become greater over time. In other words, the shell-dwelling habit is perhaps rather young geologically. Although the record of conchicolous groups is quite poor, it conforms to this expectation (Table 8.1). Hermit crabs, which are the most numerous and most highly specialized of living conchicoles, are known only from Late Jurassic time onward (Glaessner, 1969). The order Tanaidacea apparently originated in the Carboniferous or earlier, but the suborder Apseudomorpha, to which the shell-dwelling family Pagurapseudidae belongs, perhaps arose during the Early Mesozoic (Sieg, 1983). Amphipods may have evolved in the Triassic or Late Paleozoic, but the Corophioidea (the superfamily to which conchicolous amphipods belong) is probably no older than Early Cenozoic age, if patterns of present-day geographical distribution can be trusted to provide an indication of time of origin (Bousfield, 1983). The only group of living conchicoles that may have originated during the Paleozoic is the phylum Sipuncula. No sipunculan body fossils have yet been recognized, but Gill and Coates (1977) infer the presence of shell-dwelling sipunculans during the Late Cretaceous from associations between shells and solitary corals. These associations closely resemble those between solitary Scleractinia and the sipunculan *Aspidosiphon* from the Miocene Atlantic and the Recent Eastern Pacific and Indo-West-Pacific. It is noteworthy that genera containing shell-dwelling sipunculans and amphipods also contain species that do not live in shells. Sipunculans of the genus *Phascolion*, for example, may be obligate shell-dwellers or live in the empty tests of deep-sea benthic foraminifers (Gooday, 1984). Several species of the amphipod family Siphonocoetidae, and most members of the Photidae, build tubes but do not occupy shells (Sars, 1895; Enequist, 1949). These observations raise the possibility that the conchicolous habit is quite recent among sipunculans and amphipods.

Several examples of heavily encrusted shells whose pattern of encrustation and superficial erosion suggests occupation by conchicoles are known as fossils. They include shells of the Ashgillian (latest Ordovician) gastropods *Spiroecus* and *Lophospira* encrusted by the bryozoans *Monotrypa* and *Diplotrypa* (McNamara, 1978), the Middle Devonian gastropod *Palaeozygopleura* encrusted by the coral *Pleurodictyum* (Brett and Cottrell, 1982), the Late Bathonian (Middle Jurassic) gastropod *Ataphrus* encrusted by the cyclostome bryozoan *Atractosoecia* (Palmer and Hancock, 1973), and Early Miocene to Recent gastropod shells encrusted by the hermit-associated bryozoan *Hippoporidra* (Taylor and Cook, 1981). Because hermit crabs are no older than the Jurassic, the Paleozoic examples cannot have involved these crustaceans as hosts. It is clear from these examples that the conchicolous habit is ancient. They tantalize us with the possibility that shell-dwellers were already tampering with the shell resource during the Paleozoic, but we know nothing about either the abundance of these shell-dwellers or their nature.

Evolutionary Effects of Conchicoles on Shell-Builders

A potentially important aspect of the history of animal armor is the effect that the evolution of conchicoles may have had on the evolution of the molluscs that provided the shell resource. I suggest that the diversification of conchicoles was both a consequence of and a further stimulus for the evolution of shell-entering and shell-breaking predators and of molluscan armor.

Perhaps the most immediate effect of the evolution of conchicoles was to increase the total number of individual animals living in molluscan shells. The chief reason for this increase is that most conchicoles, including hermit crabs, do not prey upon the animals that provide them with shells, but instead acquire shells either from each other or when the shells are vacant. If they did acquire shells by predation, they would ultimately destroy the source of the shells. Just how important this effect is can be seen from the data in Table 8.2. At several shallow-water sites in Guam, I found that shells occupied by conchicoles (hermit crabs, tanaids, and amphipods) constituted 20 to 77% of the total number of animals found to be carrying gastropod

TABLE 8.2   The Relative Abundance of Conchicoles and Living
Gastropods in Guam[a]

Site and Habitat	Conchicoles[b]	
Pago Bay, inshore reef flat with some sand	77%	(548)
Pago Bay, rubble and boulder zone	77%	(498)
Pago Bay, blades of *Enhalus acoroides*	20%	(143)
Pago Bay, algal ridge	23%	(58)
Pugua Patch Reef, depth 12 m	27%	(153)

[a] All gastropod shells with animals living inside were collected by hand during the summer of 1984 at each of several sites representing a range of inshore habitats at Guam, Mariana Islands. The sample from Pugua Patch Reef was collected by R. K. Kropp and J. Dominguez.

[b] Percentages given are for shells that were occupied by conchicoles (hermit crabs, tanaids, and amphipods). Numbers in parentheses refer to the total number of shells in each sample.

shells. Conchicoles far outnumbered living gastropods in many inshore sites on reef flats. Secondary occupation therefore effectively prolongs the functional life span of shells, although estimates of how long a given shell remains in circulation after the death of the original builder are not currently available.

This increase in the total number of shell-dwelling animals and in the life span of shells has represented an expansion of the food resource for the predators that are specialized to remove food from shells. The existence of effective molluscan armor and of shell-entering and shell-breaking predators may have stimulated the evolution of the conchicolous habit, but the molluscs and the predators themselves have been affected by this evolution (Vermeij, 1977b). Because the shells of conchicoles are generally weaker and in poorer condition than those of living gastropods, assaults by predators on the shells of conchicoles may have a higher probability of success than attacks on gastropods whose shells are of the same size, shape, and species. Predators that take conchicoles and molluscs alike may therefore be more apt to attack well-armored or oversized prey than are predators that encounter only shells occupied by the molluscs that built them, because the apparently well-protected shell might turn out to be inhabited by a secondary occupant and to be mechanically weakened. Attacks on equivalently armored shells of living gastropods would usually be unsuccessful, but these attacks test the armor of shells and are therefore important for the maintenance and enhancement of armor. If this speculation is correct, one may envision a kind of diffuse coevolution between primary and secondary shell-dwellers, the consequence of which is an escalation between shell-bearing prey and their specialized shell-entering and shell-breaking predators.

The evolution of the conchicolous habit illustrates a common historical phenomenon. The widespread adoption of a given form of adaptation or way of life may stimulate the evolution of enemies that reduce the effectiveness of some of the individuals bearing the aptation. Further evolutionary and ecological adjustments follow, because benefits and risks keep changing. This kind of instability has long been recognized in cases of animal mimicry. If many animals come to resemble a dangerous or poisonous species, the effectiveness of the resemblance may decline, and the species whose color is being mimicked will also be affected (J.R.G. Turner, 1981). These examples reinforce the important point that a given level of expression of an aptation is not a fail-safe indication of the effectiveness of individuals in coping with the hazards that caused selection in favor of the aptation, and they underscore the importance of measuring not merely the morphological attributes, but also, whenever possible, the performance of individuals under known conditions.

SUMMARY

The frequent evolution of conchicoles—animals that live in the abandoned shells of molluscs—offers powerful testimony to the effectiveness and evolutionary importance of molluscan armor as protection against predators. Most of the conchicolous groups, including hermit crabs, are known or inferred to have evolved after the Paleozoic, but some shells that suggest occupation by conchicoles following the death of the original gastropod builder are recorded from rocks as old as the Ordovician. Adoption of the conchicolous habit requires not only that shells offer effective protection, but also that the supply of shells be high and predictable. I suggest that predation on molluscs is a far more reliable agency of supply than is mortality from nonbiological causes such as storms or floods. By increasing the number of shell-inhabiting animals and the life span of shells, conchicoles may have had important evolutionary effects on shell-entering and shell-breaking predators, as well as on the molluscan shell-builders themselves, and they may have contributed to the post-Paleozoic escalation between predators and their shell-bearing prey.

This chapter is more speculative than the other chapters in this book. Despite the vast amount of excellent research that has been done on conchicoles, especially on hermit crabs, fundamental questions remain unanswered—questions about the effectiveness of the shell as armor for conchicoles, the nature and rate of supply of shells, the influence of conchicoles on shell supply, the extent to which the functional life span of individual shells is prolonged by conchicoles, and the evolution of animal groups that contain conchicoles.

CHAPTER NINE

Locomotion and the Evolution of Movement in Gastropods

GENERAL FEATURES OF LOCOMOTION

BENEFITS OF LOCOMOTION

Armor is a common theme in the history of both gastropods and the animals that secondarily occupy vacant gastropod shells, but it is by no means the only form of adaptation to enemies. Locomotion—the movement of an individual from one place to another under its own power—is both a potentially effective defense against mobile enemies and a means of increasing the rate of encounter with resources. Improvement in locomotor ability therefore enhances the competitive as well as the antipredatory capacities of an individual animal. In groups that are capable of at least some self-generated movement, an evolutionary pathway leading to rapid locomotion would therefore seem to be more satisfactory than a pathway leading to the elaboration of passive armor, because the latter confers protection against enemies without enhancing the capacity to locate resources. The purpose of this chapter is to outline some properties of moving animals and to apply these principles to gastropods. Generally thought of as slow-moving animals, gastropods might at first glance seem to be poor candidates to use for an investigation of the history of locomotion, but they show a diversity of locomotor methods and are amenable to experimentation.

Mechanics of Locomotion

Locomotor performance is not simply a matter of high speed. The component of movement that is most important to an individual depends on the proximity and capacity of pursuing enemies, as well as on the environment and on the mobility of resources. Rapid acceleration permits an animal to escape from a nearby enemy, whereas high velocity is effective when an individual is being pursued in a relatively obstacle-free environment. Maneuverability—the capacity to change direction—is important in moving through cluttered surroundings and in confusing a pursuer. Finally, escape might be ineffective if the distance the animal can travel is small. Endurance is therefore an important component of locomotion for many animals.

A potential risk of movement is that enemies with well-developed sensory systems may be able to detect moving prey more easily than they can detect resting ones. They may be able to see movement or to sense pressure changes in the air or water as animals move through these media. For swimmers and burrowers, this means that the surrounding medium should be disturbed as little as possible. Minimal disturbance of air and water can be achieved by streamlining the moving body, a method that has the additional advantages of permitting high velocity and a more efficient use of metabolic energy. If enemies rely heavily on vision or hearing, fleeing potential victims must be visually camouflaged and move away as quietly as possible. These requirements can be achieved through particular color patterns, light-footedness, and other characteristics that have implications for the animal's mode of life far beyond the necessity for fleetness.

To understand how the various components of locomotion can be improved, we must consider briefly the mechanics of locomotion. Once the complex relationships between a moving body's shape, size, and velocity are understood in relation to the forces acting on the body, criteria for the recognition of locomotion-enhancing morphology can be applied to fossils.

As a body moves through a fluid, particles of fluid must make way for the body and fill in behind to replace the body. The forces that impinge on a moving body depend on properties of both the fluid medium and the body. Important properties of the fluid include its density (mass per unit volume) and viscosity. Let us define viscosity here. Consider two plates of area S that are separated by distance D and that are moving parallel to one another at velocity u. The force (F) needed to overcome viscosity is then

$$F = \mu S u / D. \tag{9.1}$$

The constant μ varies according to the fluid and the temperature. It has a value of 1.8×10^{-5} Nsm^{-2} for air and 1.0×10^{-3} Nsm^{-2} for water, both at 20°C. For a body of length l, the dimensionless Reynolds Number (RN) is defined as

$$RN = (\rho/\mu)ul. \tag{9.2}$$

The ratio ρ/μ has a value of about 7×10^{4} sm^{-2} for air and 10^{6} sm^{-2} for water at 20°C and at atmospheric pressure. Movement of fluid around a body depends on the RN. When the RN is less than 1.0, which occurs either at very low velocities or in very small bodies, there is a velocity gradient of the fluid outward from the body. At the surface of the body, the fluid's velocity is approximately equal to the velocity of the body, whereas far away from the body the fluid is stationary. As the RN increases to values between 2.10^{5} and 2.10^{6}, flow becomes laminar, that is, particles of fluid move in smooth paths around the body in a pattern that could be described as "low commotion." At higher RNs, flow changes abruptly to turbulent, in which particles of fluid trace irregular paths around the body (R. M. Alexander, 1968, 1977b).

Three types of force impinge on bodies moving through a fluid medium: drag, acceleration reaction force, and lift. Drag—the backward-acting resistance to forward motion—is influenced by the shape, size, and velocity of the body, but the nature of this relationship depends on the body's RN. At an RN of less than 1.0 drag D is given by

$$D = k\mu ul, \tag{9.3}$$

where the constant k varies according to the shape of the body. Small bodies moving at very low velocities are therefore affected by drag that depends on the body's length rather than on its area. At higher

RNs, drag is proportional to some area of the body—its surface area (or wettable area) S_w, its frontal area S_f (the projected area of the body perpendicular to the direction of motion), or the plan area S_p (the full-scale area projected beneath the body). It is convenient to regard drag as having two components: friction drag (skin friction of the resistance due to the viscosity of the fluid in the boundary layer immediately adjacent to the body's surface) and pressure drag (the resistance created by the wake, which is due to the replacement of the moving body by the fluid). The friction drag D_f for bodies moving with laminar flow through a fluid is given by

$$D_f = \rho S_w u^2 C_{df}/2, \tag{9.4}$$

where C_{df} is the coefficient of friction drag. The coefficient C_{df} in laminar flow is given by

$$C_{df} = 1.3 \text{RN}^{-\frac{1}{2}}. \tag{9.5}$$

If frontal area instead of surface area is used, friction drag is given by

$$D_f = \rho S_f u^2 (S_w/S_f) C_{df}/2. \tag{9.6}$$

Pressure drag D_P is given by an equation similar in form to equation (9.4):

$$D_P = \rho u^2 S C_{dp}/2. \tag{9.7}$$

The pressure drag coefficient C_{dp} is relatively independent of the RN over a wide range of RN (1.10^2 to 2.10^5); it depends instead on the shape of the body, in particular on the body's fineness ratio (maximum length divided by maximum diameter). A long, slender body with a high fineness ratio has a relatively high friction drag but a low pressure drag. Pressure drag is least in streamlined bodies, which move in such a way that the particles of the fluid trace smooth paths as they pass by. For streamlined bodies, the pressure drag coefficient C_{dp} normally has values less than 0.10 (R. M. Alexander, 1977b).

The second type of force impinging on bodies moving through a fluid medium is acceleration reaction, the resistance to changes in the velocity of a body. It is the force required to increase the kinetic energy of the fluid medium, and to push the fluid from in front of the body to around and behind the body (Daniel, 1984). The acceleration reaction G is defined as

$$G = -\alpha\rho V(du/dt), \qquad (9.8)$$

where α is the added-mass coefficient, ρ is the density of the fluid medium, V is the volume of the body, and du/dt is the acceleration of the body (change in velocity u per unit time t) relative to the bottom or to fluid far away from the body. The coefficient α depends on the shape of the body; its value is 0.5 for a sphere, 1.0 for a cylinder whose long axis is oriented perpendicular to the direction of motion, 0 for a flat plate moving edgewise, and infinity for the same flat plate moving broadside. A high α means that the mass of fluid to be moved in addition to that displaced by the volume of the body is large. Unlike drag, which is roughly proportional to the surface area or cross-sectional area of the body, the acceleration reaction is proportional to the body's volume. It acts backward and in the same direction as drag when the body accelerates but forward and against drag when the body decelerates. The acceleration reaction does not, of course, affect bodies or parts moving at constant velocities, but almost all movement in animals is effected by organs that move to and fro in steps. As they do so, the organs change in velocity and are therefore affected by the acceleration reaction (Daniel, 1984).

When a symmetrical body moves in a direction parallel to its axis of symmetry and at constant velocity, the only force imposed by the fluid on the body is drag. If, however, the axis of symmetry is inclined at an angle (the angle of attack) to the direction of motion, the force acting on the body can be resolved into a backward-acting drag and a lift force acting perpendicular to the drag. Lift also acts on asymmetrical bodies. Flying animals (including those which "fly" under water) use planar wings or other extensions of the body to create an upward-acting force, whereas vertebrates that swim by lateral or up-and-down movements of the tail create lift that acts as a forward-driving thrust. A body or structure that creates lift is called an airfoil or hydrofoil, depending on whether the surrounding medium is air or water. Lift is created because fluid flows more rapidly over one surface than over the other surface of the airfoil or hydrofoil. Because the pressure of fluid is higher on the surface where fluid flow is lower, equalization of pressure results in a flow of fluid behind the body toward the region of lower pressure. Lift depends on RN, angle of attack, and aspect ratio (span of the airfoil or hydrofoil from tip to tip divided by its chord, the anteroposterior dimension). Its general form is given by

$$L = \rho u^2 S_P C_l / 2, \qquad (9.9)$$

where C_l is the coefficient of lift. Lift increases with an increasing angle of attack up to a critical angle (the stalling angle), which is typically about 20°. The stalling angle can be increased by the presence of slots in the airfoil or hydrofoil. Foils of a high aspect ratio (long narrow wings or high tail fins, for example) have a greater lift capacity than do foils with a lower aspect ratio. That lift is more effective for large bodies than for smaller and slower bodies is suggested by the observation that the coefficient of lift can be as high as 1.5 for bodies whose RN is 10^6 but only 1.0 for those whose RN is 10^3. Lift coefficients at low RNs may be high, but so are drag coefficients, so that lift at low velocities and at small body size is difficult to achieve (R. M. Alexander, 1968, 1977b, c).

SPECIALIZATIONS TO LOCOMOTION

Improvements in locomotor capacity can be achieved either by reducing counteracting forces—drag and acceleration reaction—or by increasing lift, thrust, and endurance. At the same time, because thrust is almost always produced by rhythmic or discontinuous movement or contraction, locomotion is enhanced when resistance to slippage in the direction opposite to that of locomotion is increased.

Drag may be reduced by streamlining the body, that is, by inducing laminar flow of particles around the contours of the body. A streamlined body is elongated, rounded in front and tapering behind, and has a smooth surface. Any projections on the surface are so small that they do not extend above the boundary layer—the thin film of fluid adhering to, and traveling at approximately the same velocity as, the moving body. If the projections extend beyond the boundary layer, laminar flow is interrupted in favor of turbulent flow, which is usually associated with a large increase in drag. The degree of streamlining that is required for the maintenance of laminar flow at a given velocity depends on the RN, and therefore on the size, of the body. For bodies whose RNs are approximately 10^5, projections of height $10^{-3}l$ create roughness that is sufficient to induce turbulence, whereas bodies with an RN of about 10^6 induce turbulent flow when projections are only $10^{-4}l$ in height (R. M. Alexander, 1968). In effect, this means that greater specialization toward streamlining is needed for a large body than for a small body at a given velocity. This is one of many interesting examples of the subtle effect of scaling on mechanical properties. The geometrical similarity of bodies differing only in size does not imply mechanical similarity.

Aptations that overcome the backward-acting acceleration reaction during forward motion have been discussed by Daniel (1984). They include a reduction in the weight of the skin or of external armor, a reduction in the mass of fluid that travels with the body, and streamlining. Maneuverability is enhanced by lateral compression, as in some disc-like ammonoids and in living butterfly fishes (Chaetodontidae), and by a body that is short relative to its height. Compression insures that a body can attain a high angular acceleration, because the resistance of the broad lateral face of the body against the fluid is high. Shortening of the long axis of the body enables the body to turn without a large volume of water near the ends of the body being put into motion.

Lift is created by bodies or parts of bodies that are shaped such that the flow of fluid is faster over the upper surface than over the undersurface. A high aspect ratio and, at high RNs, a streamlined profile of the airfoil or hydrofoil enhance lift.

LOCOMOTION IN GASTROPODS

Although gastropods are usually thought of as being slow, many snails have evolved the capacity to move rapidly from place to place. This locomotion—crawling, jumping, swimming, and burrowing—is often rapid enough to provide the chief means of defense against predators. Large *Strombus gigas* conchs, for example, attain speeds of 0.0250 m/sec, whereas the smaller *S. mutabilis* and *S. maculatus* escape at speeds of 0.0120 and 0.0160 m/sec, respectively, by jumping away from predaceous gastropods (S. L. Miller, 1974; C. J. Berg, 1978).

These speeds are evidently sufficient for escaping from gastropods and sea stars (Gonor, 1966; Kohn and Waters, 1966; Ansell, 1969). Escape of the South African limpet *Patella oculus* (0.0037 m/sec) and *P. granatina* (0.0036 m/sec) from predaceous sea stars and snails requires only modest speeds (Branch and Marsh, 1978; Branch, 1979). The columbellid *Amphissa columbiana* in Oregon escapes from gastropods with a velocity of 0.0058 m/sec (B. W. Kent, 1981a). Such escapes appear to be effective and indeed may be the primary method of defense for these gastropods. The attained velocities are so low that the RNs remain low despite the large size of some snails. For a large (0.30 m long) *S. gigas*, I estimate the RN at top speed to be 7.5×10^3, whereas that for a 0.02 m long *A. columbiana* is only 120. At such RNs, there are few morphological specializations to locomotion. It is therefore difficult to identify morphological features that are associated with escape at low velocities even when, as in gastropods fleeing from slow predators, escape is highly effective (A. R. Palmer, 1980).

Linsley (1977, 1978b), however, has identified one geometrical feature that is often associated with the crawling habit in gastropods. This is the so-called tangential aperture. The plane of the tangential aperture is coincident with the base of the body whorl and lies on or very close to the surface on which the snail crawls. In snails in which the shell is dragged behind the foot instead of balanced on the posterior portion of the foot, the ventral surface of the spire whorls also lies in the plane of the aperture. This condition contrasts with the radial aperture, whose plane does not rest on the crawling surface. The radial aperture occurs both in sedentary gastropods, such as the sundial shells (family Architectonicidae) which are associated with cnidarians, and in some gastropods which are capable of locomotion. It is especially common in small liotiids, vitrinellids, and omalogyrids, which often crawl in an upside-down position on plants. An interesting group of gastropods with a radial aperture is represented today by the carrier shells of the family Xenophoridae. The apertural plane of these shells is situated well above the sediment surface, with the body hanging down from the opening. The shell is propped up by a ventral flange around the periphery of the body whorl. Although this morphology would seem to be ill-suited for locomotion, xenophorids are capable of lunging movements and are therefore not sedentary (Linsley et al., 1978; Ponder, 1983). A tangential aperture usually is associated with the capacity for locomotion in gastropods, but its absence is not necessarily a sign of a sedentary habit.

Among closely related gastropods, higher speeds are often associated with a larger aperture and foot and therefore with greater power. At a temperature of 20 to 25°C, the maximum crawling speed of the melongenid whelk *Busycotypus spiratus*, which moves by means of ciliary action on the sole of the foot, is 0.0065 m/sec, about twice as fast as that of the related *Sinistrofulgur contrarium* (0.0033 m/sec). These speeds were measured in animals of equal shell length and weight, but the aperture (and therefore the foot) of *B. spiratus* is about twice as long as that of *S. contrarium* (B. W. Kent, 1983b).

The rate of burrowing in snails is also enhanced by a larger foot. Signor (1982c) found that slender high-spired terebrids with a tiny aperture and foot take longer to burrow than more compact species with a relatively larger foot and aperture. He correlated apical angle of the shell with a burrowing rate index (BRI), defined as the cube root of the animal's volume divided by the time between onset of burrowing movements at the surface of the sediment and complete burial of the animal. For terebrids with an apical half-angle less than 10°, such as *Terebra babylonia*, the burrowing rate index is less than 0.30, whereas for more compact species, such as *T. affinis* (13°) and *T. maculata* (20°), the index lies above 0.50. These relationships suggest that some incompatibility exists between the development of effective armor (especially the evolution of a small aperture) and the capacity for rapid movement, although the compact terebrids that I have just discussed still have extremely resistant shells. The incompatibility has evidently been overcome in members of the Olividae (olive shells), which despite the longitudinally elongate narrow aperture have a large powerful foot capable of extending over the outer surface of the shell. Narrow-apertured mitrids, costellariids, conids, and columbellids are extremely slow burrowers (BRI from 0.11 to 0.54 in species of these families from Guam), whereas olivids are quite fast (BRI greater than 3.00 in *Oliva minacea* and *O. annulata* from Guam) (Vermeij and Zipser, 1987).

Several shell features may indicate the ability of gastropods to burrow in sand or mud (Signor, 1982b, c, 1983; Signor and Kat, 1984). A shell that reduces drag, prevents back slippage, and reduces the mass of sediment that adheres to the outside of the animal as the gastropod burrows should be favored in species for which burrowing is important for defense or for food capture. Most burrowing gastropods have relatively streamlined shells whose surface is either smooth in the direction of movement (as in naticids, olivids, marginellids, and many conids) or adorned with what Signor has called ratchet sculpture: spirally arranged ribs whose anterior face is less steep than the posterior face (as in some terebrids, cerithiids, and mitrids). Ratchet sculpture prevents back slippage of the shell as the foot is extended forward during the burrowing cycle. If the ratchet sculpture is abraded artificially, complete burial requires a longer time and involves a greater number of burrowing cycles or "steps." In our study of burrowing gastropods from sandy habitats in Guam, we found that smooth or ratchet-sculptured species had significantly higher burrowing-rate indices than species with strong axial or spiral non-ratchet sculpture (Vermeij and Zipser, 1987).

In many gastropods, the ability to burrow is associated with the presence of folds (plicae) on the inner or columellar side of the aperture. These folds prevent slippage of the columellar muscle along the columella during the forceful contractions of this very long muscle. Contraction of this muscle pulls the shell forward after the foot has extended forward and downward into the sediment (Signor and Kat, 1984). Many gastropods that are incapable of burrowing (epifaunal vasids, mitrids, and costellariids, for example) also have columellar folds, but the greatest development of these folds occurs in burrowing mitrids, costellariids, olivids, volutids, marginellids, and cancellariids.

The amount of sediment entrained with the burrowing animal can be kept small by a reduction in external sculpture as well as by a flattened whorl profile. If whorls were strongly convex and separated by a deep suture, sediment caught between the whorl faces in the suture would be dragged along with the burrowing snail. An umbilical cavity would also entrain sediment. It is probably significant that the only burrowing gastropods with an umbilicus—some Olividae and most Naticidae—have shells which during burrowing are covered by extensions of the foot, so that sediment is not trapped in the umbilicus.

Finally, burrowing gastropods tend to possess an anterior notch or other expansion for the anterior and downward extension of the foot. Together with the other features I discussed in the preceding paragraphs, this attribute serves to identify the burrowing habit in most externally shelled gastropods, especially in those with a tall spire (Signor, 1982c). It must be remembered, however, that most burrowing gastropods are slow in spite of these specializations to burrowing.

Not all of the features associated with burrowing are found with the same expression in all unconsolidated sediments. In particular, ratchet sculpture is found only in species from fairly coarse, well-sorted sands, and not in species from soft mud. This statement also holds for pelecypods (Vermeij and Dudley, 1985).

Several shell-bearing gastropods have evolved the habit of swimming, which usually involves extensions of the foot (Farmer, 1970). Included are the Atlantidae in the prosobranch order Heteropoda; the trochid *Solariella nectonica* (Okutani, 1961); the thin-shelled, streamlined strombid *Terebellum*; and some members of the Naticidae and Olividae, as well as a number of opisthobranchs (*Akera, Gasteropteron*, and the thecosomate pteropods). All of these gastropods have light, thin shells with little or no sculpture, and they are often covered by extensions of the foot. As a group, they are morphologically difficult to characterize, so that the capacity to swim cannot be inferred on grounds of shell form alone. Gastropods that swim the fastest and with the greatest endurance have shown evolutionary trends toward shell loss and are therefore not recoverable at all from the fossil record.

Some of the swimming gastropods probably evolved progenetically from larval forms. Bandel and his colleagues (1984), for example, have argued that pteropods are a composite group whose members evolved from tonnacean and architectonicacean larvae. In other gastropods, however, swimming is intimately associated with, and may have evolved from, burrowers. This is likely in the Strombidae, Naticidae, and Olividae. The transition from burrower to swimmer has also occurred in various clades of pelecypods (Chapter 11) and brachyuran crabs.

THE HISTORY OF GASTROPOD BURROWING

Although gastropods have emphasized armor perhaps more than have other groups of animals, they have probably also increased in locomotor capacity over the course of time. Early in the Paleozoic, most gastropods were either slow-moving or sedentary animals crawling on the surface of rocks and soft sediments. The opercula of members of the Ordovician superfamily Macluritacea and of at least some Euomphalacea (Ordovician to Cretaceous) were of a shape ill-suited for balancing the shell during crawling (Linsley, 1978b; Yochelson, 1979, 1984a; J. H. McLean, 1981). Moreover, the aperture of the shell was typically radial (Linsley, 1977). Like most open-coiled gastropods of the Paleozoic, these gastropods were probably filter-feeders incapable of locomotion after larval settlement (Yochelson, 1971; Peel, 1975; Linsley, 1978b). Even the planispiral bellerophonts, some of which had a tangential aperture, are thought to have been very slow or sedentary (Harper and Rollins, 1982).

The burrowing habit may not have become established in gastropods until the Silurian and was probably never common during the Paleozoic. Harper and Rollins (1985) have interpreted several planispirally coiled, spirally ribbed, compact bellerophonts as infaunal or semi-infaunal burrowers. The earliest of these was *Beyrichidiscus* from the Late Silurian of Gotland, Sweden. Later representatives include *Praematurotropis* from the Middle Devonian of New York, *Euphemites* from the Carboniferous of North America, and *Euphemitopsis* from the Permian. The shell was covered by a secondarily deposited layer that obscures the original growth lines and the selenizone (the trace of the apertural slit). Harper and Rollins believe that the shell was covered by a greatly enlarged foot. Because these bellerophonts are found exclusively in rocks representing fine muddy sediment, they are believed to have been capable of complete or partial burial in the sediment, and to have had a mode of life (if not a mode of feeding) comparable to that of living naticid moon snails. Another possible burrower in the Paleozoic was the Early Devonian (Siegenian) murchisoniid *Ptychocaulus*. Signor (1982c) cites its flattened whorl profile, plicate columella, tangential aperture, and smooth shell as supporting evidence for this interpretation. Some Late Paleozoic subulitids with high-spired, smooth, flat-sided shells may also have been burrowers, as was perhaps *Girtyspira*, the oldest (Tournaisian stage of the Early Carboniferous) opisthobranch (Kollmann and Yochelson, 1976). Ratchet sculpture, though present in some Paleozoic trilobites, is unknown in Paleozoic gastropods, although it must be remembered that the type of environment in which ratchet sculpture is found most commonly today (coarse, well-sorted sand) is not well represented in the Paleozoic fossil record.

Even if all these interpretations are correct, burrowing species still constituted only a small fraction of Paleozoic gastropods. My analysis of Linsley's (1979) compilation of all Devonian gastropod genera shows that burrowing gastropods never accounted for more than 3% of the genera during any stage in the Devonian. Between 5 and 10% of Late Carboniferous and Late Triassic species may have been burrowers. These low percentages contrast sharply with those in the Late Mesozoic and Cenozoic. Analysis of the Maastrichtian (Late Cretaceous) Ripley Formation shows that about 37% of the gastropod species may have burrowed. Five diverse assemblages of gastropods from Recent shallow-water species in the Eastern and Western tropical Pacific are composed chiefly of actively burrowing species, which account for from 62 to 75% of the total species. In short, the temporal increase in the percentage of burrowers is in marked contrast to the decrease in the percentage of sedentary species (Figure 9.1).

Rapidly burrowing gastropods are probably restricted to the Mesozoic and Cenozoic. The burrowing Naticidae may have appeared as early as the Late Triassic or more likely during the Early Cretaceous, but the most rapidly burrowing members of this family—the large-footed Sininae and Polinicinae—originated during the Late Cretaceous (Marincovich, 1977). The Olividae, another family of fast burrowers, have a Late Cretaceous origin as well (J. D. Taylor et al., 1980). Species of the nassariid genus *Bullia* bury themselves completely in sand in less than a minute (A. C. Brown, 1982). The subfamily to which they belong (Dorsaninae) is known first from the Paleocene (Cernohorsky, 1984).

SUMMARY

Locomotion enables animals to escape from their enemies and to find resources. Components of movement that may enhance an individual's locomotor ability include high velocity, rapid acceleration, maneuverability, great endurance, and unobtrusiveness. Speed, acceleration, and efficiency increase when drag and body mass are re-

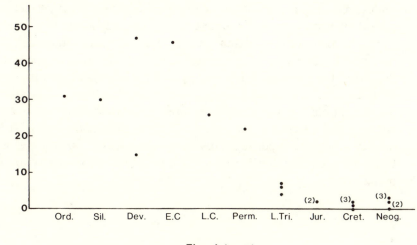

FIGURE 9.1 The occurrence of sessile or sedentary uncemented gastropods through time. For each assemblage, the percentage of sedentary species is plotted against temporal occurrence. (Numbers in parentheses refer to the number of samples when there is more than one per time period.) Sessile and sedentary forms include most Macluritacea, Euomphalacea, Bellerophontacea, open-coiled gastropods, and certain other gastropods with radial apertures. Data were compiled from references in Vermeij (1983c).

duced, and when muscles operating the propulsive organs increase in mass.

Gastropods exemplify these characteristics even though they are better known for their armor than for their powers of locomotion. The history of gastropods shows that sedentary epifaunal types predominated during the Paleozoic, whereas burrowing infaunal forms have been common since the Late Cretaceous (Figure 9.1). Rapid burrowers did not evolve until the Late Mesozoic and Cenozoic. This historical pattern is consistent with the general increase in the metabolic rate of animals through time (Chapter 4).

Difficulties in inferring the locomotor powers of gastropods from shell form have thus far prevented the construction of a more detailed historical account of locomotion in this group. To improve our understanding of movement in fossil gastropods, we must refine studies of living gastropods so that the correspondence between shell form and locomotor performance can be more firmly established. It may be especially important to study locomotion during escape from predators, for it is during the escape phase that the locomotor powers of individuals are tested most rigorously. In performing these studies, we should pay particular attention to the relationship between effectiveness of escape and external morphology of the shell. This may best be done experimentally by systematically modifying shell characteristics in various groups of individuals all exposed to similar predators.

CHAPTER TEN

Armor and Locomotion

in Cephalopods

ADAPTATIONAL DILEMMAS

The contrast between the evolutionary history of cephalopods and that of gastropods is fascinating. Whereas shell-bearing gastropods are perhaps more diverse today than they ever were in the past, cephalopods with external shells are represented in the Recent fauna by only a single family (the Nautilidae) of about five species (Saunders, 1981), and even the diversity of cephalopods with internal shells (Sepiidae and Spirulidae) is low. During the Paleozoic and Mesozoic eras, however, shell-bearing cephalopods constituted an important and highly diverse element of the marine fauna. In the Cenozoic, the shell-bearing forms have very largely been replaced by squids and octopuses in which the shell has been either lost entirely or reduced to an internal flexible rod (gladius).

From this well-known historical pattern, we can infer that passive shell armor has not proven to be successful for cephalopods in the long run. Following on important work on the biology of living cephalopods, especially that of Peter Ward, I shall try to show in this chapter that two factors—rapid locomotion and the necessity of compensating for changes in pressure as the animal moves vertically through the water—impose severe limitations on the elaboration of passive external armor in cephalopods. Increasing expression of armor characterized many cephalopod lineages during the Paleozoic and Mesozoic, but during the Late Cretaceous predation and competition may have become so intense that the adaptational conflicts resulting from the demands of armor, locomotion, and pressure compensation could no longer be reconciled in a body plan that retained the primitive external shell.

Like the shell of gastropods, that of cephalopods is a single logarithmically growing structure that is primitively external. The most profound difference between the shells of gastropods and cephalopods is that the cephalopod shell is divided by perforated septa into chambers, whereas the gastropod shell either lacks septa or has imperforate partitions that completely seal off the apical part from the occupied part of the shell. The bulk of the cephalopod body is housed in the most recently formed chamber (body chamber), which extends from the last septum to the apertural edge.

Cephalopods are predaceous animals that are, and probably for the most part always have been, capable of swimming by jet propulsion (Packard, 1972). In connection with this habit, cephalopods often move vertically as well as horizontally. Vertical movement is always associated with a change in ambient pressure. Animals capable of vertical movement must therefore have the capacity to compensate for changes in pressure; otherwise, the tissues will explode or implode. In primitive externally shelled cephalopods, the shell functions in part as an organ of buoyancy. Gas that is contained in the apical portions buoys up the shell, enabling the animal to maintain an overall density equal to that of sea water (1025 kg/m^3). The volume of this gas, which changes as the animal ascends and descends, is regulated by the siphuncle, a posterior tubular extension of the mantle that passes through the perforations (necks) in the septa, and that pumps liquid into and out of the chambers. Liquid is pumped out of the chambers when the animal ascends and into the chambers during descent. Between the septa, the siphuncle is encased in an organic connecting ring, whose circumference and wall thickness regulate the rate of liquid flux (Westermann, 1971; Ward, 1979). Movement of liquid is impeded by a thick siphuncular wall, and it is slower in cephalopods with a siphuncle of small diameter than in those with a large tube (Ward, 1982). At a given depth, the salt gradient between the liquid outside the siphuncle and that inside the siphuncle must correspond to the difference in pressure between the chamber and that of the water surrounding the animal.

Gas pressures inside the chambers of cephalopod shells are always about 0.9 atm (0.9 kg/cm^2) and do not vary with water depth (Denton,

1974). The shell and connecting rings must therefore be able to resist the difference between pressures inside and outside the chambers. Recent species of *Nautilus* implode at a depth of between 750 and 900 m, which corresponds to a pressure of 77 to 93 atm (Kanie et al., 1980; Westermann and Ward, 1980) and to the rupture strength of the connecting rings (Chamberlain and Moore, 1982). *Spirula*, whose evolute shell is internal, can tolerate pressures of at least 107 atm (Denton, 1974).

An important consequence of the independence of internal gas pressure and water depth is that shells of deep-water species must be more resistant to implosion than are the shells of shallow-water cephalopods (Westermann, 1971, 1975, 1977). The characteristics that enhance resistance to implosion, however, also influence the effectiveness of the shell as passive armor against predators, as well as the animal's locomotor capacity and the rate at which changes in pressure compensation can take place during the cephalopod's ascent or descent. Pressure compensation is potentially crucial both for effective vertical escape from predators and for coping with sudden traumatic changes in the animal's weight, as when the shell is damaged (Ward and Signor, 1983; Ward, 1985, 1986).

Let us consider, as an example of adaptational conflict, the effects of increasing the thickness of the shell. An obvious way of increasing resistance against implosion and of preventing breakage due to predators is to thicken the septa, connecting rings, and outer shell wall. Because the density of the shell material (calcium carbonate) is approximately 2.5 times that of sea water, a thicker shell is also a heavier shell, requiring a larger volume of gas to prevent the animal from sinking to the bottom. Moreover, a massive shell is incompatible with rapid acceleration during locomotion. A thickened connecting ring may prevent implosion, but it also reduces the flux of liquid into and out of the chambers and therefore is probably associated with a poor ability on the part of the animal to compensate for rapid changes in ambient pressure. Although the slow transfer of liquid across a thick connecting ring may be offset by a siphuncle with large surface area, this solution leaves less room for gas and may result in a weakening of the shell, because a large surface is exposed to the pressure differential between the chamber and the siphuncle (Ward, 1982). Given the pattern of adaptational incompatibility between shell, septal, and connecting-ring thickness, resistance to implosion, locomotion, and rate of compensation for changes in pressure, thickening would appear to have only a limited value as a means of preventing implosion or of protecting the shell against predator-induced breakage.

The interrelationship between septal geometry, septal strength, and shell shape in cephalopods also has important implications for the evolution of passive armor in this group. Pressure P which is exerted on a spherical membrane is directly proportional to membrane thickness T and inversely proportional to the membrane's radius of curvature R; that is,

$$P = 2ST/R, \tag{10.1}$$

where S is the strength of the membrane's material (Westermann, 1975). The cephalopod septum can be treated as a membrane that must resist the pressure differential between the inside of the chambers and the water outside the shell. If the calcium carbonate of which the septum is composed is regarded as having a constant strength, then a strength index T/R can be calculated to assess the relative effectiveness of variously shaped septa. For adult *Spirula*, the septum is about 0.17 mm thick and has a radius of curvature of about 3 mm; its strength index, therefore, is 0.057. Although the central part of the septum of *Nautilus* is much thicker (0.75 mm) than that of *Spirula*, its much larger radius of curvature (28 mm) gives it a much lower strength index (0.027) (Westermann, 1975). The difference in strength index between the septa of *Nautilus* and *Spirula* corresponds to the difference in maximum depth (and therefore pressure differential) that the animal can tolerate. Westermann (1975, 1977) has found that strongly curved septa (those with a small radius of curvature) are characteristic of cephalopods whose shells are longiconic (long, slender cones with a small apical angle); in such shells, the septa are spaced relatively far apart. Breviconic cephalopods, whose shells are broad and short, have relatively closely spaced septa with a large radius of curvature. If breviconic shells were to be buoyant, therefore, they would have had very thin septa to compensate for the close spacing. This argument has led Westermann to propose that most breviconic cephalopods were found in relatively shallow waters, whereas longiconic cephalopods were more apt to be found in deep waters where greater resistance to implosion is more important.

It has been suggested from time to time that septa confer increased strength to cephalopod shells by buttressing the outer shell wall from within (Westermann, 1971, 1975). This kind of architectural strengthening has the advantage of permitting the outer shell wall to be relatively thin, so that the shell as a whole is light in weight. Buttressing would seem to be especially effective when the junction (suture) between the septum and the outer shell wall is greatly elaborated and extended. In cephalopods with straight or curved shells, the suture is typically a simple curve, but in many coiled cephalopods the suture has become wavy. Extreme complication of the suture, in which the individual lobes and saddles are themselves frilled, are seen in many Mesozoic ammonoids. This frilling increases the length of the suture and decreases the unsupported area of the outer wall. Complex sutures also probably increase the strength of the septum itself, just as folds on the valve increase the strength of scallop shells (Pennington and Currey, 1984).

Several objections have been raised to the hypothesis that complex sutures are associated with a sturdy shell. Henderson (1984) points out that complexly lobed parts of the suture in Mesozoic ammonoids occur on the area of the septum that adjoins the preceding whorl—which presumably needs no buttressing—rather than on the area adjoining the outer shell wall. The idea that the septum served as the attachment area of the retractor muscles is more attractive to Henderson than is the strengthening function of a complexly lobed septum. It must be pointed out, however, that rejection of the hypothesis that complex sutures enhance strength is based on the premise that strength is chiefly needed to avoid implosion of the shell. No analysis has been made of the role of sutures in preventing attempts to crush the shell at only a few points on the exterior, as would be expected if the ammonoid were attacked by a predator. Moreover, the two hypotheses are not mutually exclusive.

Ward (1986) has argued that the buttressing function of the complexly sutured septum may be much less important than the role played by the septum in controlling the animal's weight. The angle between the septum and the outer wall is often rather low, so that buttressing is much less effective than it would be if the septum were set at right angles to the shell wall. Ward believes, however, that the region where the septum meets the shell constitutes an important site for the retention of small amounts of liquid. In species with complex sutures, the contact area between septum and shell is large, and substantial amounts of water can be stored there. Complication of the suture may therefore enhance liquid reserves and increase the animal's capacity to compensate for pressure changes.

The extent to which breakage-resistant armor can evolve depends on the consequences of attack by a shell-molesting enemy. These consequences may be especially grave for externally shelled cephalopods. When a fragment of shell is broken off by an attacker, the cephalopod suffers a catastrophic and immediate loss of weight. If this weight loss is not quickly compensated for by the pumping of liquid out of the chambers, the animal will ascend to the surface and suffer debilitation and possibly death. Features of the siphuncle, connecting rings, and septa that enhance the liquid reserve and the rate of liquid removal are therefore essential if cephalopods are to survive attacks to the shell (Ward and Signor, 1983; Ward, 1986). Complex sutures and a large, thin-walled siphuncle would thus be expected in well-armored cephalopod shells. Because resistance to implosion at greater depths requires a small, thick-walled siphuncle, deep-water cephalopods are expected to be subject to more stringent limitations on the evolution of breakage-resistant armor than are shallow-water forms.

It is conceivable that some injuries of the shell would be in the form of holes in the early chambers rather than damage to the apertural margin. Injuries to the phragmocone (chambered portion of the shell) would cause catastrophic flooding of one or more chambers and therefore would suddenly make the animal heavier. Again, rapid compensation for pressure changes would be required in order for the cephalopod to survive the rapid descent that would result from injuries of this type.

Although the limitations of pressure compensation set cephalopod shells apart from those of gastropods, the geometrical characteristics that enhance resistance to breakage in gastropods should confer similar benefits in cephalopods. A tightly coiled shell should be more effective than a loosely coiled, curved, or straight one. External sculpture and a thickened apertural lip should add strength which, in the case of sculpture, can be achieved with only a modest increase in the animal's weight. Apertural constriction and the presence of a cover should prevent enemies that attack their cephalopod victims by way of the aperture from being successful. Most of these expectations cannot be verified directly in cephalopods, because the single living family of externally shelled cephalopods is morphologically conservative and has only one of the features (tight coiling) that could be interpreted as enhancing armor. Nevertheless, it might be possible to show that the deployment of these features is correlated with a high incidence of repaired shell injuries in fossils.

THE HISTORY OF CEPHALOPOD ARMOR

The ecological limitations on cephalopod armor notwithstanding, much of the history of cephalopods is characterized by a general trend toward the elaboration of armor. This can already be seen in the Early Paleozoic, when cephalopods were perhaps the top predators in the sea.

The earliest cephalopods were generally small animals, sometimes not more than 20 mm long, with straight to gently curved shells whose tiny chambers were separated by simple septa spaced only 1 mm apart. The oldest known cephalopods are species of *Plectronoceras* from the Yenshou Member of the Fengshan Formation and equivalent horizons of probable Franconian (Middle Late Cambrian) age in China (Chen and Teichert, 1983). Size increase was a common evolutionary trend in the Late Cambrian descendants of these earliest cephalopods. Only 2 or 3 of the 34 Late Cambrian genera that have been described by Chen and Teichert had calcareous deposits in the siphuncle, which weighted the apical region and enabled the animal to assume a horizontal position. Calcareous deposits in the chambers (cameral deposits), which were to become so common during the Ordovician, were still unknown in the Late Cambrian.

Cephalopods diversified extensively during the succeeding Ordovician period (Flower, 1976; Crick, 1981). Most of these cephalopods, or "nautiloids," had straight or curved shells with simple septa, and many developed large siphuncles and massive deposits of calcium carbonate either in the siphuncle or in the chambers themselves. These features suggest relatively slow locomotion, especially in longiconic forms, as well as mechanical weakness and low rates of compensation for changes in pressure.

Despite the prevalence of simple-shelled cephalopods during the Ordovician, most of the important modifications that enhanced antipredatory resistance appeared first during this period. To trace the history of these modifications in shell-bearing cephalopods, I calculated the number of genera with the given modification divided by the total number of genera for each time interval of the Paleozoic (Figure 10.1). This analysis was based on Dzik's (1984) compilation of nautiloid genera and therefore excludes the Devonian to Cretaceous ammonoids. Dzik's treatment of nautiloids is highly conservative in that he has synonymized or placed in doubt a large number of genera that had been described previously. I am unable to evaluate Dzik's decisions on the systematic status of cephalopods, but his argument that many generic-level taxa are based on differences in preservation and especially on variation among individuals within a population seems highly plausible to me. Moreover, he presents a unified and consistent outline of the group. It therefore seems appropriate to use a single investigator's treatment as the data base for a generic-level analysis of form through time, even though experts may disagree with aspects of this treatment.

One of the earliest antipredatory modifications to appear in cephalopods was the constricted aperture—an aperture that in the adult stage is smaller than the diameter of the rest of the body chamber. The first cephalopods with a constricted aperture were the ellesmeroceratidans *Buehlersoceras* and *Burenoceras*, both of Tremadocian (earliest Ordovician) age. During the Ordovician, cephalopods with this apertural modification were minor components of the fauna, but by the Silurian many Oncoceratida and Discosorida evolved constricted apertures, which were sometimes T-shaped or even divided into separate small openings (Dzik, 1981a; Stridsberg, 1981). As can be seen in Figure 10.1, genera with constricted apertures contributed substantially to Silurian, Devonian, and Early Carboniferous cephalopod faunas, but thereafter they are no longer represented. Constricted apertures did, however, reappear in Mesozoic Ammonitida.

Another architectural type that appeared early in the Ordovician is the coiled shell. Most Ordovician coiled nautiloids were either evolute (with successive whorls disjunct) or partially coiled with a straight adult body chamber. Partially coiled forms account for 8 to 15% of nautiloid genera during the interval from the Arenigian (Late Early Ordovician) to the Wenlockian (Middle Silurian), but after the Wenlockian they dwindled to 5% or less of the fauna. Tightly coiled forms in which successive whorls are in contact arose during the Ar-

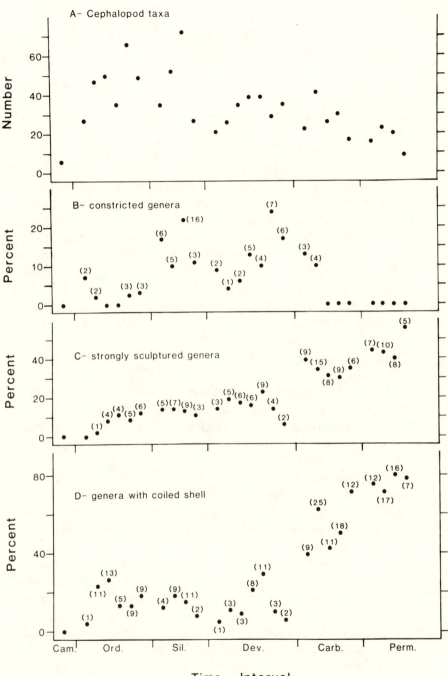

FIGURE 10.1 The history of predation-related shell traits in Paleozoic nautiloid cephalopods. For each Paleozoic stage, I have plotted the proportion of genera with the trait in question. The absolute number of such genera is given in parentheses. Data are based on my calculations and on an analysis of Dzik's (1984) monographic treatment of Paleozoic nautiloids.

enigian (*Hardmanoceras, Trocholitoceras*), and conispirally coiled nautiloids appeared in various Silurian and Devonian lineages (Flower, 1955). Coiled nautiloids show an irregularly increasing incidence during the Early and Middle Paleozoic with peaks during the Llanvirnian (Early Middle Ordovician), Wenlockian and Ludlovian (Middle and Late Silurian), and Eifelian and Givetian (Middle Devonian). Beginning in the Tournaisian stage of the Early Carboniferous, the incidence of coiled genera rose more or less steadily, reaching 75% by the last stage of the Late Carboniferous. With the possible exception of one orthoceratidan genus in the Triassic, all post-Permian nautiloids are coiled. The ammonoids, which arose from straight-shelled bactritid orthoceratidans during the Early Devonian, are a primitively coiled group of cephalopods that achieved a high diversity from the Devonian to the Cretaceous. Although many stocks of ammonoids evolved secondarily straight or partially uncoiled shells, especially during the Cretaceous, cephalopods with straight shells never achieved the numerical supremacy that they enjoyed during the pre-Devonian Paleozoic.

The coiled shells differ in several important ways from the primitive straight-shelled types. Cameral and siphuncular deposits are lacking, and the suture line, which was simple in most of the early cephalopods, is wavy. These differences probably mean that coiled shells were more maneuverable, mechanically stronger, and capable of faster and more effective compensation for pressure changes than were the earlier straight shells.

In the Paleozoic, most coiled shells had a lobate suture line in which the lobes and saddles were smoothly rounded (goniatitic type). Late in the Paleozoic, several lineages of ammonoids evolved a ceratitic suture line in which the lobes were dentate. Ammonoids with ceratitic sutures predominated during the Triassic. Ammonitic suture lines, in which both the lobes and the saddles were toothed or frilled, also evolved in the Late Paleozoic and again in the Triassic, but they became the dominant theme in the Jurassic and Cretaceous (W. J. Kennedy, 1977). Following Ward's (1986) analysis, I interpret this general complication of the suture line as a trend toward strengthening the shell and toward better compensation for pressure changes. Reversions from the ammonitic and ceratitic to the goniatitic type are known, and it would be most interesting to compare the morphology of secondarily simple ammonoids with that of co-occurring species that have retained the more complex sutures. Living nautiloids also have rather simple sutures, which compare in lobation to some of the simpler Paleozoic goniatites. In fact, although ammonoids often evolved complex sutures, coiled nautiloids evidently never did (Ward, 1980). Nautiloids generally may have had thicker shells which would have required less buttressing from within to maintain a given level of resistance against compression (Ward, 1980). Just how nautiloids differed from contemporaneous ammonoids is still unclear.

Calcareous coverings of the aperture evolved relatively late in cephalopod history. The first covering of any kind known from a cephalopod is a three-part structure (*Aptychopsis*) from an unidentified Late Wenlockian (Middle Silurian) orthoceratidan. According to Stridsberg's (1984) interpretation, this structure covered the aperture completely and therefore probably served the function of an operculum. Beginning in the Devonian, thin plate-like organic coverings (anaptychi) were found in association with various ammonoids. The first calcareous coverings are of Early Jurassic age. These aptychi are paired clam-like plates which often were thick and fit precisely into the aperture. Aptychi appeared in several stocks, including the Phylloceratida, Lytoceratida (Late Cretaceous Baculitidae and Scaphitidae), and most Ammonitida (Lehmann, 1975, 1981a, b; Kennedy and Cobban, 1976).

Strong external sculpture—another potential trait conferring resistance against shell-breaking predators—also became more common among cephalopods over the course of time. The first ribbed cephalopod is the Arenigian (Late Early Ordovician) *Hardmanoceras*. Other early cephalopods with strong sculpture include *Zitteloceras* (Middle to Late Ordovician), *Discoceras* (Late Ordovician), *Peismoceras* (Early to Late Silurian), and the knobbed *Tubiferoceras* (Middle Silurian). The incidence of strongly ribbed and tuberculate nautiloid genera rose through the Ordovician to a plateau in the interval from the Llanvirnian (Middle Ordovician) to the Gedinnian (Early Devonian). Thereafter, it rose again slightly during the Devonian to a peak in the Givetian and then sustained high values (30 to 56%) during the Late Paleozoic (Figure 10.1).

Among ammonoids, an increase in the incidence of strongly sculptured genera can be traced from the Late Paleozoic to the Cretaceous (Ward, 1981). Ward defined various categories of shell sculpture. A very strongly ribbed shell is one in which the ratio of rib width to whorl height is 0.15 or higher. Shells in which this ratio is 0.05 or less are defined as smooth or very weakly ribbed. According to Ward's (1981) findings, no Paleozoic ammonoid was very strongly ribbed, and only about 10% of Paleozoic ammonoid genera had spines, tubercles, or nodes. Some 73% of genera at that time may be described as smooth to very weakly ribbed. By the Cretaceous, the very strongly sculptured ammonoids greatly predominated over smooth forms in diversity. Ward (1985) has analyzed the Jurassic and especially the Cretaceous in greater detail (see Figure 10.2). He found that, although the incidence of strongly sculptured ammonoids rose during most of the later Mesozoic, it fell steadily after the Turonian stage of the Late Cretaceous. Moreover, there was a reduction not

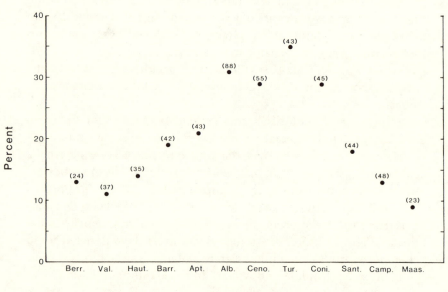

FIGURE 10.2 The incidence of strongly sculptured ammonoid genera during the Cretaceous. Numbers in parentheses denote the number of strongly sculptured genera living in each stage. Data were derived from Ward (1985).

only in the typical planispirally coiled ammonoids, but also in the so-called heteromorphs—ammonoids that depart markedly from the planispiral form by becoming secondarily uncoiled, conispirally coiled, or coiled in various other bizarre ways. The reduction in sculpture among post-Turonian ammonoids was caused by the differential extinction of highly sculptured forms, and it accompanied a sharp decline in the diversity of shallow-water genera. The only post-Turonian shallow-water ammonoids were highly streamlined forms. Most deep-water ammonoids of the post-Turonian Late Cretaceous were either delicate heteromorphs such as nostoceratids and diplomoceratids, which may have floated passively in the water as do modern cranchid squids (Ward and Westermann, 1977; Ward, 1979), or smooth planispiral tetragonitids, desmoceratids, and phylloceratids. It may be surmised that armor in ammonoids was no longer successful after the Turonian (Ward, 1985). I cannot help but notice the parallel with the ammo-knights in armor, who disappeared as a warrior class at the end of the Middle Ages in Europe.

By assessing the frequencies of repaired scars in local assemblages at various times, it is possible to test whether the temporal trends in cephalopod armor reflect increased shell-breaking predation through time. Unfortunately, most of the available data on the incidence of repair is anecdotal. Published data for Jurassic ammonoids indicate very low incidences of scars. Guex (1967), for example, found that only about 2.0% of the 8,000 ammonoids he examined from the Toarcian (Early Jurassic) of Aveyron, France, had scars, whereas N. Morton (1983) recorded an incidence of about 0.7% for the Middle Jurassic *Graphoceras* aff. *G. concavum*. This latter value is identical with that found by Bayer (1970) for Graphoceratinae of Middle Jurassic (Aalenian to Bajocian) age in Germany. Bayer explained this low incidence in *Graphoceras* by pointing out that ammonoids of this genus were probably good swimmers and were therefore capable of evading many potential enemies. For contemporaneous Sphaeroceratinae (incidence of scars 1.4%) and other Stephanoceratacea such as Ottoitidae and Normannitidae (incidence 9.7%), this explanation does not hold, because these ammonoids were either passive floaters or bottom-crawlers (Bayer, 1970). They possessed either a thickened adult lip or lappets and therefore show some indication of morphological protection against enemies, although lappets probably also had a reproduction-related function, because they are known only in males (Lehmann, 1981b). Even these higher incidences of scars are still low compared to those of most gastropods. Ward informs me that many Cretaceous ammonoids show very high incidences of repair. If documentation proves this impression to be correct, the Mesozoic ammonoids would closely parallel gastropods in showing a concomitant rise in scars and armor. Surveys of shell architecture and traces of predation throughout the Phanerozoic record of cephalopods are critically needed, for they would provide a powerful test of the hypothesis that shell breakage became more important through the Late Paleozoic and Mesozoic.

Several aspects of the foregoing account require qualification and comment. Most important, Figures 10.1 and 10.2 are based on the global appearances and disappearances of genera. This kind of global analysis treats cephalopods from all kinds of environments at all latitudes, and it provides, for each trait in question, only a single datum per geological stage. Spatial variation in architecture is therefore ignored, and temporal variation cannot be judged against it. This serious flaw can be corrected only by undertaking studies of local assemblages from each time interval. Moreover, the assertion that apertural restriction, strong sculpture, and shell coiling confer resistance against various predators is based entirely on analogies with gastropods and is therefore itself tenuous and subject to reinterpretation.

I began this chapter by noting that the history of armor in cephalopods contrasts markedly with that in gastropods. The latter group has, in general, emphasized armor, whereas in cephalopods the trend toward increased armor ended and then reversed during the Late Cretaceous. In addition to this contrast, the two groups differ in the kinds of armor that developed early in their history. In gastropods, calcareous opercula appeared in several lineages during the Early and Middle Paleozoic, whereas other apertural defenses (elongation, dentition, and thickened lips) were chiefly post-Paleozoic innovations. In cephalopods, the order of appearance of apertural characters was reversed. Constricted apertures appeared during the Early and Middle Paleozoic, whereas cephalopods with calcareous apertural coverings are all from the Mesozoic. I do not know what might account for these differences. Did the features that we now interpret as antipredatory aptations of cephalopods have other functions as well? Was there a greater representation of bottom-dwelling cephalopods during the Paleozoic than later? Put another way, were Paleozoic cephalopods ecologically more like living gastropods than were Mesozoic and Cenozoic shell-bearing cephalopods? We do not have the answers to these questions at present. Additional evidence bearing on cephalopod functional morphology and on the environments in which cephalopods lived is needed if we are to assess the architectural history of this group.

LOCOMOTION IN CEPHALOPODS

Swimming is the chief means of locomotion in the predaceous cephalopods. It is achieved by jet propulsion; that is, water is ejected through the funnel, an extension of the mantle. Cephalopods can attain very high speeds. The cuttlefish *Sepia officinalis* achieves a speed of 0.8 m/sec, and a large (100 g) *Loligo* squid can swim at 2.1 m/sec (Trueman and Packard, 1968). Oceanic squids are reported to have burst speeds of 15.3 m/sec, which is faster than that of most fishes and whales (Norris and Mohl, 1983). The shell-bearing *Nautilus*, with a top speed of 0.25 m/sec (Ward et al., 1977), is notably slower than the other cephalopods that have been studied.

Speed in cephalopods depends on the volume of the water that can be contained in and ejected from the mantle cavity and on the shape and mass of the animal. The amount of water in the mantle cavity is correlated with the size of the body chamber in shell-bearing cephalopods. A short, broad body chamber is probably a more feasible container for a large volume of water than is a long, narrow body chamber of similar volume, because the latter is associated with a shell in which the centers of gravity and buoyancy are close together, so that the animal rotates around the axis of the shell as water is forced out of the funnel. A shell with a short body chamber is relatively more stable in the water, because the center of buoyancy is located well above the center of gravity (Bayer, 1970; Lehmann, 1975, 1981b).

Acceleration can be increased by reducing the weight of nonmuscular tissue in favor of swimming musculature. For cephalopods, this means either a thin shell or no shell at all. Drag can be reduced and maneuverability can be increased by streamlining the body. In shell-bearing cephalopods, streamlining is associated with a compressed, smooth shell with a peripheral keel.

Although early straight-shelled cephalopods may have been well streamlined in some cases, the cameral and siphuncular deposits that characterized many of them must have made the animals heavy. Compared with later cephalopods, these early forms probably accelerated slowly and maneuvered clumsily. Short, squat brevicones, which apparently lived mainly in shallow water, were probably the Early Paleozoic cephalopods that were most capable of quick maneuvers (Westermann, 1977). Elimination of deposits in the phragmocone of coiled shells must have enabled cephalopods to achieve faster acceleration, but it is doubtful if any Paleozoic cephalopod attained high speeds. No Paleozoic ammonoid or coiled nautiloid possessed a streamlined shell. In fact, the typical Late Paleozoic goniatite ammonoid had a broad, squat shell with rounded venter (Raup, 1967; Ward, 1980; Saunders and Swan, 1984).

The proportion of ammonoid genera with streamlined shells increased considerably in the later Mesozoic (Figure 10.3). Ward (1985) defines a highly streamlined shell as one that is tightly coiled (shell-coiling parameter D ranging from 0 to 0.33; see Chapter 7 for a defi-

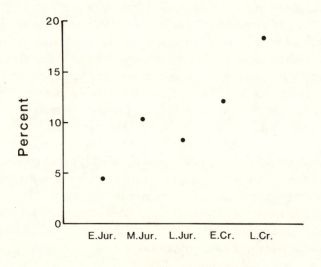

FIGURE 10.3 The incidence of highly streamlined ammonoid genera during the Jurassic and the Cretaceous. Data were derived from Ward (1985).

nition of D) and compressed (whorl breadth to whorl height ratio 0.50 or lower). Genera whose shells conform to these criteria make up 4.5% of the fauna in the Early Jurassic, 10.0% in the Middle Jurassic, 8.3% in the Late Jurassic, 12.0% in the Early Cretaceous, and 18.0% in the Late Cretaceous. After the Turonian stage of the Late Cretaceous, most shallow-water ammonoids were streamlined, the slow-swimming, highly sculptured forms having been restricted to deep water. In fact, about 29% of the ammonoid genera living in the Maastrichtian (the last stage of the Cretaceous) met the criteria for a highly streamlined shell (Ward, 1985).

Ultimately, the most successful way cephalopods increased their locomotor performance was by reducing and then losing the shell. Because calcium carbonate is about 2.5 times as heavy as sea water, a shell adds significantly to the acceleration drag of a swimming animal. Reduction of the shell therefore enabled cephalopods to achieve rapid acceleration. It also set the stage for the evolution of fins, which enable the animal to swim forwards as well as backwards (Reitner and Engeser, 1982).

A trend toward shell reduction was already under way in the Early Devonian. Three genera assigned by Bandel and colleagues (1983) to the belemnomorph order Aulacocerida are known from the Early Devonian Hunsrückschiefer of Germany. In these genera, the chambered part of the shell was much reduced relative to the simplified body chamber (proostracum). The guard (rostrum), which surrounded the apical part of the chambered phragmocone, augmented and later in ontogeny replaced the cameral liquid as the mechanism for maintaining the animal's horizontal orientation. Before becoming extinct during the Pliensbachian stage of the Early Jurassic, the Aulacocerida may have given rise to the Belemnitida in the Hettangian (earliest Jurassic). Like the Aulacocerida, the Belemnitida and the related Jurassic order Belemnoteuthida (Callovian to Tithonian) had an internal shell consisting of a delicate phragmocone and a solid, rather massive, cylindrical guard whose weight probably precluded rapid acceleration. The Belemnitida became extinct at the end of the Cretaceous (Jeletzky, 1966; Donovan, 1977; Reitner and Engeser, 1982). The only belemnomorph order that persists to the present day is the Spirulida, which originated in the Late Cretaceous and today is represented by a few deep-sea species with loosely coiled internal shells having simple septa.

The superorders Teuthomorpha (squids and octopuses) and Sepiomorpha (cuttlefishes) probably also arose from a belemnomorph stock. Most living cephalopods, including all the fast swimmers, belong to the order Teuthida, in which the internal shell has been reduced to an uncalcified rod-like remnant (gladius) of the body chamber. The earliest Teuthida belong to the suborder Loligosepiina and date from the Late Norian stage of the Late Triassic (Reitner, 1978). Primitive loligonaceans with small suckers on the arms had already evolved by Early Callovian (Middle Jurassic) time (Fischer and Riou, 1982).

In summary, cephalopods show many trends toward faster locomotion, manifested by shell coiling, shell streamlining, and shell reduction. Retention of a well-armored external chambered shell proved in the long run to be incompatible with rapid swimming. The history of cephalopods is thus similar to that of fishes (Packard, 1972): most armored lineages became extinct while more agile types persisted.

SUMMARY AND CONCLUSIONS

The history of cephalopods shows that in the long run external shell armor has been less successful in this class of molluscs than it has been in gastropods. Cephalopods are mobile predators whose chambered external shell functions to compensate for changes in water pressure as the animal ascends and descends in the water. The presence of gas in the chambers of the shell poses a limitation on the elaboration of armor, and the shell itself prevents the attainment of high speeds in swimming. The fossil record indicates that the incidence and expression of various forms of armor increased episodically during the Paleozoic and Mesozoic, especially during the Silurian, Devonian, and early and middle stages of the Cretaceous. After the Turonian stage of the Late Cretaceous, however, the incidence of armor among externally shelled cephalopods fell sharply. Extinction of the ammonoids at the end of the Cretaceous reduced externally shelled cephalopods to only one or two families. The post-Turonian decline in armor is interpreted as an indication that the incompatibility between resistance to shell-breaking agents and the requirements for rapid compensation for changes in pressure and for rapid locomotion could no longer be reconciled by improvements in the external chambered shell. Cephalopods that had either an internal shell or no shell and that excelled in locomotion became increasingly important during the Mesozoic, eclipsing the types with external shells in the Cenozoic.

This account is based entirely on a global analysis of genera and on inferences of shell functional morphology derived from analogies with gastropods. An understanding of cephalopod history on a finer scale will require the architectural and ecological study of well-preserved local faunas. Future work should include studies of unsuccessful shell-breaking predation, as well as the functional characterization of cephalopods from various water depths and different geographical areas within each of several time intervals.

CHAPTER ELEVEN

Armor and Locomotion in

Bivalved Animals

THE BIVALVE SHELL

Gastropods and cephalopods have shells consisting of a single unit that grows in size by the addition of skeletal material at one end. Calcified opercular structures that close the aperture when the foot is retracted into the shell may augment the effectiveness of such a one-piece skeleton, but these structures are evolved secondarily. Many animals have skeletons consisting of two units, or valves, which by the action of muscles can be moved with respect to each other. The bivalve form has evolved independently in many groups (Table 11.1), the largest of which are the molluscan class Pelecypoda (also known as the Bivalvia or Lamellibranchia) and the lophophorate phylum Brachiopoda.

The bivalve shell provides the capacity not only for effective protection of the vital organs from the external environment, but also for locomotion with the use of the skeleton. Like gastropods, pelecypods and other bivalved animals are generally thought of as sluggish, and indeed this is an apt description of the many pelecypods and brachiopods that live permanently fixed to a hard surface. Nevertheless, many bivalved animals have evolved various locomotor mechanisms. The purpose of this chapter is to review the history of armor and locomotion in bivalved animals.

The pelecypod shell consists of a right and a left valve connected dorsally by an elastic ligament. The mantle is attached to the inner surface of the valve along the pallial line, a series of muscle scars lying close to the free margins of the valves. Contraction of one or two adductor muscles, which insert more dorsally on the inner valve

TABLE 11.1   Survey of Bivalved Brachiopods and Molluscs

Phylum Brachiopoda

> Class Inarticulata: Early Cambrian to Recent (Williams and Hurst, 1977; Wright, 1979)

> Class Articulata: Early Cambrian to Recent (Williams and Hurst, 1977; Wright, 1979)

Phylum Mollusca

> Class Monoplacophora, Order Tuarangiida: Middle Cambrian (MacKinnon, 1982; Runnegar, 1983)

> Class Rostroconchia: Early Cambrian to Late Permian (Pojeta and Runnegar, 1976; Runnegar, 1978, 1983)

> Class Pelecypoda (= Bivalvia): Early Cambrian to Recent (Pojeta, 1978; Runnegar, 1983; Runnegar and Bentley, 1983)

> Class Gastropoda, Subclass Opisthobranchia, Order Sacoglossa, Family Juliidae: Eocene to Recent (E. A. Kay, 1968)

---

surface, causes the valves to shut and acts against the ligament. The valve edges near the ligament are usually provided with teeth and sockets which collectively make up the hinge.

The Paleozoic rostroconchs were bivalved molluscs whose valves were joined dorsally by a flexible region that is topographically a continuation of the valves and not a ligament (Pojeta and Runnegar, 1976). A similar type of hinge has evolved both in the pelecypod family Pinnidae (Yonge, 1953) and in ancestors of the bivalved juliid gastropods (E. A. Kay, 1968).

The brachiopod plan is basically at right angles to that of the pelecypod. The valves are denoted as brachial (or dorsal) and pedicle (or ventral). The hinge lies posteriorly rather than dorsally, and the plane of symmetry divides each valve into more or less equal right and left halves. If the animal is attached, it is either fixed by a pedicle—a muscular organ arising from a hole (foramen) between the valves or in the pedicle valve—or cemented by the lower valve. In the Inarticulata, there is no hinge, and the valves move vertically with respect to each other. The Articulata have a hinge, consisting of teeth and sockets, that either coincides with the axis of rotation of the valves (strophic condition) or is separate from it (nonstrophic condition) (Rudwick, 1959; R. M. Carter, 1967a, b; Jaanusson, 1971). In the inarticulates, valves open indirectly as a result of the contraction of longitudinal muscles, which cause the body to swell and the valves to gape (Gutmann et al., 1978). Articulate brachiopods have evolved diductor muscles which open the valves. In neither group of brachiopods is there an elastic ligament of the type seen in pelecypods. Several pelecypods have also evolved diductor muscles and have lost the ligament. In the rudists (superfamily Hippuritacea), the diductors arise from the adductors and insert on large flanges on the interior of the valves. The rock-boring and wood-boring Pholadacea (angel wings) have diductor muscles that are external to the main valves in the vicinity of the hinge, and that are covered by auxiliary shell plates (Purchon, 1955a; Skelton, 1978).

BIVALVE ARMOR

Perhaps the most important manifestation of armor in pelecypods and other bivalved animals is the tight seal. The capacity to seal the valves tightly has been shown to be highly effective in resisting attack in three ways. First, it prevents diagnostic cues from being released into the environment, so that many predators have difficulty detecting the bivalve (Carriker and Van Zandt, 1972). Second, tight closure enables a bivalve to spend many hours or even days in the digestive system of predators such as astropectinid sea stars, which swallow their prey whole. When it is finally voided by the predator, the bivalve is still alive (Christensen, 1970; Massé, 1975). Third, the tightly closed valves of many bivalves are difficult for snails and some crabs that attack their victims at the shell margin to open (Nielsen, 1975; D.A.S. Smith, 1975; B. W. Kent, 1983a; B. S. Morton, 1985).

A tight seal would be ineffective if the pelecypod were unable to keep the valves adducted for long periods. Comparative studies have indeed shown that species that are able to maintain a tight seal are extremely tolerant of anaerobic conditions and have low basal rates of metabolism (Christensen, 1970).

A feature that is often associated with the ability to seal the valves tightly is a denticulated or crenulated valve margin. It may also function to prevent the valves from shearing in the plane of the commissure when the valves are shut, but the geometry of the hinge and of the adductor muscles also plays an important role in this function (Coen, 1985), the ecological significance of which remains unstudied. Most predators that attack pelecypods by pulling the valves apart apparently produce forces that are perpendicular rather than parallel to the plane of the commissure. R.D.K. Thomas has suggested to me that another function of marginal crenulation might be to enable the valves to resist compression when the pelecypod is attacked by a shell-crushing predator, or when the adductor muscles attempt to shut the valves during an attack by a shell-opening sea star. The crenulations are V-shaped indentations on the inner valve surface at the margin. As a compressive force is applied to the valve, these inner buttresses prevent failure of the valves in the radial direction.

The radial ribs and folds of scallops (Pectinidae) and cockles (Cardiidae) probably do not function in effecting a tight seal or in buttressing the valves against compression, because the valves cannot be shut for long periods in these highly aerobic and active pelecypods, and because the troughs between the ribs in the radial direction (perpendicular to the valve margin) are parallel-sided rather than V-shaped, as are the troughs between the crenulations. In scallops, the ribs impart strength to the relatively thin valves, whereas in cockles strength as well as anchorage in the sediment is enhanced by strong radial folds (Waller, 1972; S. M. Stanley, 1981; Pennington and Currey, 1984).

In sharp contrast to tightly closing bivalved animals, many pelecypods have one or more permanent gapes between the adducted valves. In deeply burrowing pelecypods, there is often a permanent gape where the inhalant and exhalant siphons pass between the posterior ends of the valves (S. M. Stanley, 1970; Runnegar, 1974). An anterior gape is found in many rock-boring forms and in a surprisingly large number of freshwater unionaceans (S. M. Stanley, 1970; Vermeij and Dudley, 1985). Fast-swimming scallops have gapes on either side of the hinge as well as on both the anterior and posterior margins of the main disc of the shell. Similar gapes occur in limids, some of which also have the ability to swim (B. S. Morton, 1979). Gapes provide easy access for many kinds of would-be predators. Cancrid crabs can kill *Mya, Saxidomus,* and other permanently gaping deeply buried pelecypods by inserting the claws through the posterior siphonal opening (Boulding, 1984), and the nemertean ribbon worm *Cerebratulus* penetrates through the siphon of *Mya* in order to consume that clam (Kalin, 1984). Moon snails of the naticid genus *Neverita,* which normally drill their prey, insert the proboscis through the permanent pedal or siphonal gape of solenacean razor clams (H. J. Turner, 1955). In the same way, the muricid *Ceratostoma foliatum* attacks gaping pelecypods by inserting the proboscis through the siphonal gape (B. W. Kent, 1981b).

The presence of permanent gapes indicates that complete protection by enclosing armor is relatively unimportant, and that marginal valve damage does not substantially increase the risk of detection or successful subjugation. This might be so for one of two contrasting reasons: either the animal lives in a relatively "safe" environment where predators are rare or absent, as in endolithic habitats or far below the surface of unconsolidated sediments, or it has come to rely on defenses other than armor. Gaping scallops and razor clams, for example, have evolved rapid means of escape (see below), whereas most limids (especially the thin-shelled species of *Limaria*) and many leptonaceans have pallial tentacles that are capable of autotomy and of producing irritating substances (Gilmour, 1967; Stasek, 1967; B. S. Morton, 1973, 1975, 1979). Autotomy is also reported for the siphons of razor clams (Stasek, 1967). No pelecypod has entirely lost the shell, but many have augmented or functionally replaced it as the chief means of defense. This is true even for many pelecypods capable of complete closure.

The potential for the evolution of armor against predators that break the shell by attacking the valves at the free margins is probably much lower in bivalved animals than in retractable coiled gastropods (Vermeij, 1983b). As I pointed out in Chapters 1 and 2, breakage-resistant armor cannot evolve unless an individual bearing an armor-conferring mutant survives and leaves descendants after having withstood an attempt to break the shell. In many instances, such attempts cause damage to the valve margins. Crabs, clawed homarid lobsters, and spiny palinurid lobsters attack the valve edges if all else fails, thereby causing surviving victims to be damaged at the margins (Ebling et al., 1964; Seed, 1969; D.A.S. Smith, 1975; Elner and Jamieson, 1979; Griffiths and Seiderer, 1980; Boulding, 1984; Cunningham and Hughes, 1984). Predaceous whelks of the melongenid genus *Busycon*, which apply the lip of the shell to the commissure of a tightly closed victim, also inflict superficial damage to clams that they are unable to overpower (B. W. Kent, 1983a). If the edge of one or both valves is damaged, the valves cannot form a tight seal when they are shut. Not only can enemies penetrate between valves at the breach, but cues to which potential enemies could respond cannot be prevented from diffusing out of the clam. In other words, superficial shell damage is likely to expose a bivalved animal to damage greater than that a coiled gastropod would be exposed to through lip damage. Unfortunately, no quantitative estimates of this risk have been made.

Pelecypods have evolved at least three types of modification that have enabled them to overcome the detrimental effects of damage to the valve margins. The first is a flexible margin. In members of the pteriomorph superfamilies Pinnacea, Isognomonacea, Pteriacea, Pectinacea, and Anomiacea, the mantle is highly retractable, so that when the free edges of the valves are broken, the mantle is not injured. The flexible nature of the most recently formed portions of the valves ensures that the valves can still close more or less tightly even if the margins are damaged. This pterioid modification, which is made possible by the prismatic nature of the valve edges, is found chiefly in tropical and warm-temperate pelecypods that are either epifaunal (living on the surface of the substrate) or semi-infaunal (living partly buried in sand or mud, with the posterior end extending above the surface). Solenacean razor clams, many mactraceans, and solemyaceans have a wide fringe of flexible periostracum that extends beyond the calcified part of the shell at the growing margin. If the valve is cracked, a seal can still be maintained by the periostracum. This type of flexibility seems to have a worldwide distribution in infaunal as well as epifaunal pelecypods.

The second major modification is valve overlap (Figure 11.1). Several pelecypod stocks have evolved a type of bilateral asymmetry in which one valve overlaps the other along part or all of the commissure (Table 11.2). If overlap occurs along the free (that is, the anterior, posterior, or ventral) margins, the valve that juts out beyond the other may be damaged at the edge without the seal between the valves being broken. Pelecypods with this kind of valve overlap are mostly infaunal warm-water forms. R.D.K. Thomas has pointed out

FIGURE 11.1   *Anadara inaequivalvis*, collected June 1979, at Wom Village, north coast of Papua New Guinea. The crenulated edges of the valves of this species overlap, especially along the postero-ventral margin.

TABLE 11.2 Survey of Pelecypods with Overlapping Valves

Subclass Pteriomorphia

Order Arcoida, Superfamily Arcacea: many Cenozoic genera of Arcidae, Cretaceous to Recent genera of Cucullaeidae (Thomas, 1978a, b; B. S. Morton, 1981c)

Order Ostreoida: many members of the superfamily Aviculopectinacea, Family Pseudomonotidae (Early Carboniferous to Late Permian); many members of the superfamily Pectinacea, families Pectinidae and Propeamussiidae during the Cenozoic (Newell and Boyd, 1970; Reif, 1978; Waller, 1978)

Subclass Heteroconchia

Order Veneroida

Superfamily Veneracea, Family Petricolidae (Eocene to Recent): *Claudiconcha* (B. S. Morton, 1978)

Superfamily Tellinacea, Family Donacidae (Late Cretaceous to Recent): group of *Donax assimilis* (Coan, 1983)

Superfamily Dreissenacea: some Dreissenidae of the Recent (Nicol, 1958)

Order Myoida: Corbulidae, Erodonidae (Middle Jurassic to Recent) (Yonge, 1946; Lewy and Samtleben, 1979)

Subclass Anomalodesmata, Order Pholadomyoida

Superfamily Pandoracea: Pandoridae (Eocene to Recent) (Nicol, 1958; B. S. Morton, 1981a)

Superfamily Thraciacea: Thraciidae, Periplomatidae (B. S. Morton, 1981a)

Superfamily Poromyacea: Poromyidae, Cuspidariidae (Late Cretaceous to Recent) (B. S. Morton, 1981a)

---

to me that the overlapping valve margins also may have a function similar to that of marginal crenulation. Overlap permits the valves to resist strong compressive forces during valve adduction and during attacks by shell-crushing predators. The frequent occurrence of valve overlap in the posterior portion of the valves (as in arcids and donacids, for example) is consistent with the expectation that most attempts at shell crushing occur at or near the sediment surface, that is, near the posterior (upward-pointing) end of the pelecypod.

A variation on this theme is the development of scales or spines that project beyond the commissure from the valve margins. Such projections are common in the Pteriidae, Spondylidae, Chamidae, and Tridacnidae, as well as in productide, atrypide, and spiriferide articulate brachiopods. The spines may be broken without the seal being affected, or they may prevent some kinds of enemies from reaching the commissure. The geometry of productide and some strophomenide articulates suggests strongly that spines in these animals also functioned to stabilize the individual on the sediment surface, and to attach it to crinoid stems and other objects (R. E. Grant, 1966, 1968). Most bivalved animals with projecting spines and scales are epifaunal or semi-infaunal, but some very shallowly infaunal cardiids, lucinids, and venerids also have such spines. All occur in warm, shallow water.

The third modification that permits the evolution of breakage-resistant armor in bivalved animals is thickening of the valve margins. As in gastropods with thickened outer lips, the pelecypod is protected against shell breakage by a margin that does not shatter when it is grasped by a predator. Thick margins are common in some warm-water, shallow-burrowing arcid, lucinid, venerid, donacid, and corbulid pelecypods.

Attempts by predators to crush bivalved shells do not necessarily affect the valve margins. If, for example, a porcupine fish or a ray were to attempt to crush a clam between the tooth plates in the jaw, an unsuccessful attack might leave little more than a few tooth marks on the shell's exterior, and the commissure would be left intact. In such cases, there are few risks associated with nonlethal shell-breaking attacks, and little should stand in the way of the evolution of features that strengthen the valves. Most prominent among such strength-enhancing characteristics are a thick valve wall, high valve convexity, and the presence of radially or concentrically oriented folds.

Studies of the incidence of predation and of repaired injuries of Paleozoic brachiopods have enabled R. R. Alexander (1981, 1986) to identify several features that confer resistance against breakage in brachiopods. The architectural group with the highest incidence of shell repair was that of concavo-convex shells, in which one valve is concave and the other convex. This shape characterized many orthides and atrypides, but it was especially typical of strophomenides. Various strophomenides, including the common Silurian genera *Rafinesquina* and *Leptaena*, developed a strong angular bend (geniculation) in the adult valve, so that the free anterior valve margins point upward out of the plane of the rest of the shell. Often, one of the valves is strongly thickened at the point of geniculation, so that if the valve margin is damaged, the injury is prevented from spreading to the main part of the shell. Of 11 species of concavo-convex brachiopods, ranging in age from Late Ordovician to Late Carboniferous, for which more than 10 specimens per sample were available, 6 showed frequencies of repair in excess of 0.10.

Many Paleozoic brachiopods had biconvex shells (both valves convex) that were ornamented with plications, that is, with ribs that crenulated or scalloped the valve margins. This architecture was typical of most rhynchonellides and of many spiriferides and some orthides. It was associated with extremely low incidences of repaired injury. Of the 10 plicated brachiopod species that Alexander (1986) examined from the Paleozoic, none had frequencies of repair exceeding 0.08. Alexander attributes these low frequencies to the relative invulnerability of the valve margins to attack by shell-breaking agents.

The shells of pelecypods appear to be generally weaker than those of gastropods. B. W. Kent (1981c) remarked, for example, that the thick-shelled venerid clam *Mercenaria* was much more likely to shatter when dropped by gulls on a hard surface than were large gastropods. Large mussels (*Mytilus edulis*) in Britain are weaker (1400 N) than are co-occurring snails such as *Littorina littorea* and *Nucella lapillus* (Elner, 1978; R.M.L. Kent, 1981; Currey and Hughes, 1982). In the Chesapeake Bay region, large specimens of the corbiculid freshwater clam *Corbicula* and the estuarine mactrid *Rangia* have less compressive strength (400 and 200 N, respectively) than estuarine gastropods of comparable weight (Blundon and Kennedy, 1982a; Blundon and Vermeij, 1983; Kennedy and Blundon, 1983). I suspect that many shallow-water tropical clams have much greater compressive strength than the temperate species for which maximum loads have been measured, but no data on their strength exist.

As in gastropods, resistance against drilling predation seems to be achieved largely by thickening the valves. Thickening has the effect of lengthening the time required by predators to drill through the shell wall. In many pelecypods of the families Lucinidae and Veneridae, the central part of each valve is distinctly thicker than either the anterior or posterior end. Because naticacean gastropods often drill in the middle of the valves, this thickening is effective in preventing successful penetration by these predators. Kitchell and her colleagues (1981) have provided convincing evidence of the effectiveness of central valve thickening in *Stewartia floridana*, a lucinid in which more than 50% of the drills in the central part of the valves were unsuccessful (that is, incomplete).

Firm attachment to a hard object is another important way bivalved animals can prevent subjugation by enemies. Attachment may be by a flexible muscular pedicle (as in many brachiopods), by a fibrous byssus secreted by glands near the foot (as in many pelecypods), by cementing one of the valves (as in many pelecypods, craniacean inarticulates, and strophomenide articulates), or by a system of fibrils, which penetrate through the lower valve (as in orthotetacean articulates; see R. E. Grant, 1980). Fixation prevents dislodgment by predators, by animals that scrape algae from surfaces, and by waves and currents. Relatively few measurements of the strength of attachment are available. Wave-exposed *Mytilus edulis* in the northeast Pacific require a maximum of 104 N of force to be dislodged, whereas mussels of the same species in more sheltered conditions in both the Pacific and the Atlantic require only 7 to 23 N (Norton-Griffiths, 1967; Price, 1982; Witman and Suchaneck, 1984). The larger *M. californianus*, which is typically found on wave-exposed shores in the northeast Pacific, can sustain a pull of up to 242 N before coming loose (Witman and Suchaneck, 1984). The pedicle strength of the brachiopod *Terebratalia transversa* in Puget Sound is at least 90 N and is therefore comparable to that of *Mytilus* (Thayer, 1975b). The presence of a byssus in a pelecypod can usually be inferred from a persistent ventral byssal gape between the valves. In scallops (Pectinidae), the byssus is accommodated in a notch in the anterior ear in front of the hinge in the right valve. This notch is adorned with teeth (the ctenolium) which apparently function to prevent the valves from twisting (Waller, 1984). In brachiopods, the presence of a pedicle is usually indicated by an open apical hole in the lower (pedicle) valve.

PATTERNS OF ARMOR IN SPACE AND TIME

Although the armor of bivalved animals is generally less well developed that that of gastropods, its pattern of distribution is similar. In the modern fauna, pelecypods with tightly closing valves are especially characteristic of warm, shallow marine waters, and they are rare in fresh water as well as in temperate and polar seas. Genera of pelecypods having crenulated inner valve margins, for example, make up 30 to 50% of the epifaunal and infaunal pelecypod faunas on tropical and warm-temperate shores, but they account for less than 25% of the genera on cold-temperate coasts. They are virtually unknown in fresh water. Permanently gaping pelecypods show a complementary distribution. They are most common on cold-temperate shores, where they constitute 31% or more of infaunal genera, and are less well represented on warm-temperate and tropical shores (less than 25% of infaunal genera). Freshwater pelecypods also show a high incidence of gaping. An interesting difference between marine and freshwater pelecypods is that, whereas in the sea gaping is usually characteristic of deep-burrowing clams (S. M. Stanley, 1970), in fresh water it is found frequently among shallow-burrowing genera as well (Vermeij, 1978; Vermeij and Veil, 1978; Vermeij and Dudley, 1985).

The poor preservation of most Paleozoic pelecypods has made it difficult to trace the history of valve closure in this group. Several lines of evidence, however, suggest that crenulated valve margins and radially oriented folds were less common during the Paleozoic than in later eras. In their studies of the evolution of the superfamily Trigoniacea, the first representatives of which are recorded from the Silurian, Newell and Boyd (1975) found that all pre-Permian forms were smooth, whereas most post-Triassic forms were sculptured with strong radial or oblique ribs. Even the latter, however, exhibited only weak valve crenulation, and in this respect they differed from the radially ribbed, marginally crenulated cockles (Cardiacea), which largely replaced the Trigoniacea ecologically after the Cretaceous (S. M. Stanley, 1977b). Few Paleozoic scallops had radial folds, whereas the majority of Cenozoic ones had them (Waller, 1972). Paleozoic members of the heteroconchian veneroid superfamily Crassatellacea lacked radial ribs and crenulated margins, but beginning in the Triassic several crassatellids and members of the related Carditacea evolved these characters (Morris, 1978). Radial ribbing also seems to be a relatively late invention in the Arcacea (Thomas, 1978a, b).

These trends within clades are also evident in comparisons of local assemblages through time (Table 11.3). The incidence of crenulated and overlapping valve margins was very low in the Paleozoic, intermediate in the Jurassic and Cretaceous, and high in the Late Cenozoic. Recent temperate assemblages have incidences similar to those of warm-water Mesozoic assemblages. The incidence of gaping declined somewhat in warm-water assemblages from the Jurassic and Cretaceous to the Recent.

Because the Paleozoic record of brachiopods is extremely good, it should be possible to trace the architectural history of brachiopods in some detail. Unfortunately, this has not been attempted on any large scale. R. R. Alexander's continuing work suggests that the incidence of plicate species rose markedly during the Devonian and Carboniferous, and that the incidence of smooth biconvex species correspondingly decreased. After having originated in the Ordovician, concentrically ridged or frilled geniculate forms remained common throughout the Paleozoic, although the frill is often broken up into spines in Late Paleozoic representatives.

The first spines in bivalved animals date from the Ashgillian stage of the Late Ordovician, at which time small, delicate spines are known in well-preserved specimens of the orthide brachiopod genus *Platystrophia* (Richards, 1972). In a survey of warm-water North American brachiopod assemblages, I find that spiny brachiopods compose 5% or less of the species in any given assemblage. Chonetacean strophomenides, which developed delicate spines in the hinge area, evolved these spines during the Wenlockian stage of the Middle Silurian. These spines are generally interpreted as aids in the attachment of brachiopods to small objects or as devices that enable these animals to rest on the surface of soft sediments (Bassett, 1984). It is equally probable that the delicate spines served to increase the effective size of these small animals, thereby making them less available to small-mouthed predators. Spiny brachiopods constituted 0 to 9% of local Silurian assemblages in North America. A marked increase in the incidence of spiny brachiopods occurred during the Devonian. Evidence of this increase emerges in compilations of genera on a global scale (Signor and Brett, 1984), as well as in analyses of local

TABLE 11.3 Incidence over Time of Pelecypod Species with Crenulated Valve Margins (C) and Gaping Margins (G)[a]

Age and Locality	N[b]	C[c]	G
Late Devonian, Chemung Limestone, New York	19	0	0
Jurassic, Europe			
Hettangian	15	20%	20%
Sinemurian	20	15%	25%
Pliensbachian	32	13%	29%
Toarcian	25	20%	12%
Aalenian	34	18%	32%
Bajocian	48	22%	23%
Bathonian	58	27%	26%
Callovian	50	26%	24%
Oxfordian	51	26%	20%
Kimmeridgian	45	16%	24%
Late Cretaceous (Maastrichtian), Fox Hills Formation, U.S.	36	19%	28%
Paleogene (Eocene), Moodys Branch Formation, Mississippi	55	24%	16%
Neogene			
Pliocene			
North Sea Basin	76	39%	18%
Mediterranean Sea	172	34%	11%
San Diego Formation, California	68	29%	16%
Recent			
San Juan Island, Washington	25	20%	31%
Venado Beach, Panama	28	43%	18%
Guam	30	33%	3%
Las Lajas, Panama	17	47%	18%
West Malaysia and Singapore	90	31%	16%
West Florida	29	45%	21%
North Sea	77	28%	25%
Mediterranean Sea	114	31%	18%

[a] Data were compiled from McAlester (1962), Speden (1970), Hallam (1976), Dockery (1977), Morris and Purchon (1981), Raffi et al. (1985), S. M. Stanley (1986), and my own collections.

[b] N Number of species.

[c] The crenulated category includes species with anterior, ventral, or posterior valve overlap, as well as those with toothed or ribbed inner valve edges.

North American faunas. My study of 7 North American assemblages shows that the incidence of spiny brachiopod species varied from 9 to 25% during the Devonian, with no apparent trend in incidence through the period. The succeeding Carboniferous period witnessed a rise in the incidence of spiny species. They made up 23 to 39% of Carboniferous faunas and 32 to 56% of Permian ones. As Signor and Brett have pointed out, the rise of spiny brachiopods is attributable not only to the predominantly spiny Productacea, Chonetacea, and Strophalosiacea, but also to many spiriferides and atrypides.

With the catastrophic decline of the brachiopods at the end of the Paleozoic came the extinction of all spiny brachiopods. Thereafter, only a few Jurassic rhynchonellides evolved spines (Rudwick, 1970).

Epifaunal pelecypods evolved spines during the Early Carboniferous at the latest. The Pseudomonotidae, an important ostreoid family of the Late Paleozoic, had its first spiny members in the Permian (Newell and Boyd, 1970). Although the Mesozoic descendants of this family lacked spines, many other pelecypod stocks evolved spines. The most spectacular development of spines occurred in the Spondylidae (Jurassic to Recent) and the Chamidae (Late Cretaceous to Recent).

Attachment by cementation probably evolved first in various groups of brachiopods during the Llanvirnian stage of the Early Middle Ordovician. The earliest cemented forms were members of the order Craniida and the possibly acrotretacean genus *Eoconulus* (Williams and Hurst, 1977; A. D. Wright, 1979). Although *Eoconulus* and several other minor stocks of cemented brachiopods disappeared soon after their origin, the family Craniidae has survived to the present day. Articulate brachiopods first evolved cementation during the Wenlockian stage of the Middle Silurian (the strophomenide strophomenacean *Leptaenisca*), but the major diversification of encrusting or umbonally cemented brachiopods did not begin until the Early Devonian with the differentiation of the strophomenide superfamilies Davidsoniacea and Productacea. A culmination of the cemented habit came during the Late Paleozoic with the evolution of oyster-like Lyttoniacea and coral-like Richthofeniacea. All these cemented strophomenides became extinct at the end of the Paleozoic. Cementation reappeared in the thecidiacean terebratulides (Triassic to Recent).

The Early Carboniferous saw the first appearance of the cemented

habit in pelecypods (Table 11.4). Although many of the early cemented clades did not stand the test of time, cementation re-evolved
in many Mesozoic and Cenozoic lineages.

TRACES OF PREDATION

BREAKAGE

Almost all the studies undertaken thus far of shell breakage in pelecypods have been anecdotal, so that we have as yet no clear picture
of how effective pelecypod shells are against breakage, or whether
breakage has become more important over the course of time, as it
has in gastropods. Boyd and Newell (1972), for example, inferred that
predation by helodontid sharks must have been important to various
Permian pelecypods in Wyoming because most pelecypods were encountered as fragments with sharp unweathered edges. Tooth marks
of fishes have been detected in Carboniferous brachiopods and were
inferred by R. R. Alexander (1981) as indicating successful and unsuccessful predation. Similar inferences have been drawn for Cretaceous inoceramid pelecypods (Kauffman and Kesling, 1960; Speden,
1971; Kauffman, 1972). Among Recent pelecypods, I know of only
one estimate of effectiveness of the shell against a shell-breaking
predator. In a study of predation by the ghost crab *Ocypode ceratophthalma* on *Donax faba* in Kenya, D.A.S. Smith (1975) found that
52% of the individuals that had been attacked were still intact, although they were frequently chipped around the valve margins.

With few exceptions, repaired injuries are rare in fossil and living
bivalved shells. I have examined all Recent pelecypod species in my
collection for which 10 or more individuals (20 or more valves) are
available. Frequencies of repair (number of scars per individual) were
0.10 or higher in only 2 of 36 samples of infaunal species (0.20 in 10
Quidnipagus palatum from Pago Bay, Guam, and 0.67 in 15 *Ano-
malocardia flexuosa* from Rio de Janeiro, Brazil); in none of the 22
samples of epifaunal mytilids and arcids was the frequency of repair
higher than 0.10. A sample of 10 *Panopea* sp. from Rice's Pit (Yorktown Formation, Pliocene of Virginia, frequency 0.13) was the only
one among 27 samples in museum collections from the Early Miocene to Pliocene of Atlantic America with a repair frequency of 0.10

TABLE 11.4   Survey of Pelecypods That Attach by Cementation

Subclass Pteriomorphia, Order Ostreoida

    Superfamily Aviculopectinacea: some Late Paleozoic genera (Newell and Boyd, 1970)

    Superfamily Buchiacea, Pseudomonotidae: several lineages in Late Paleozoic, first in *Pachypteria* of Visean stage of Early Carboniferous (Newell and Boyd, 1970; Waller, 1978)

    Superfamily Pectinacea: *Hinnites* and other *Chlamys*-like Recent genera; Family Spondylidae (Jurassic to Recent) (Yonge, 1951b; Newell and Boyd, 1970)

    Superfamily Plicatulacea: Family Plicatulidae (Triassic to Recent) (Waller, 1978)

    Superfamily Dimyacea: Family Dimyidae (Triassic to Recent) (Waller, 1978; Yonge, 1978b)

    Superfamily Ostreacea (Triassic to Recent) (Newell and Boyd, 1970; Waller, 1978)

Subclass Heteroconchia

    Order Trigonoida, Superfamily Unionacea: Etheriidae (Recent, freshwater) (Yonge, 1962, 1978a)

    Order Veneroida, Superfamily Chamacea (Senonian stage of Late Cretaceous to Recent) (Yonge, 1967; W. J. Kennedy et al., 1970)

    Order Hippuritoida (Late Oxfordian stage of Late Jurassic to Early Paleocene) (Yonge, 1967; Skelton, 1978)

Subclass Anomalodesmata, Order Pholadomyoida, Superfamily Pandoracea, Families Myochamidae and Cleidothaeridae (Eocene to Recent) (B. S. Morton, 1974, 1981a)

---

or higher. In their survey of pelecypods from the Pliocene of Albenga, Italy, Robba and Ostinelli (1975) found that only 1 of 4 species (*Corbula gibba*, frequency 0.35) had a frequency of repair of 0.10 or higher. R. R. Alexander's (1986, and manuscript) studies of 17 North American species of Paleozoic biconvex brachiopods revealed only 1 (*Brachyspirifer eurytines* from the Middle Devonian Silica Shale of

Ohio, frequency 0.13) with a frequency higher than 0.10. This low incidence of repair was also found in other Paleozoic brachiopods (Sparks et al., 1980; B. Jones, 1982). The only morphological group in which repair was relatively common was that of Paleozoic concavo-convex or plano-convex brachiopods. Among the 15 North American species surveyed by Alexander, 5 had frequencies of repair of 0.10 or higher.

Taken at face value, these data suggest that bivalved animals with convex shells did not show an increase in the frequency of repair through time despite the fact that crenulated margins and perhaps other strengthening features have increased in incidence through the Phanerozoic. This apparent constancy may reflect the very high risks associated with sublethal injury to the valve margins. If this is the correct explanation, then we should still expect a temporal increase in repair among bivalved animals with such modifications as overlapping or flexible valve margins. Unfortunately, because of the difficulty of recognizing repair in some forms (Pteriacea and Ostreacea) and the insufficient number of specimens in most museum collections, it has proven difficult to compare Alexander's data on concavo-convex and plano-convex Paleozoic brachiopods with frequencies of repair in morphologically similar post-Paleozoic pelecypods. It is obvious that much work remains to be done on the distribution in space and time of repaired injuries in bivalved animals, but as things stand no evidence exists for a temporal increase in the incidence of repaired injuries in bivalved animals. These animals therefore contrast with gastropods, in which a temporal increase has been demonstrated.

Drilling

Although bivalved animals have died from drilling predation since at least Late Cambrian time, drilling remained a generally minor cause of death until after the Early Cretaceous (Figure 11.2). Species in which more than 10% of the individuals were drilled constitute a minority in 5 out of 6 Paleozoic brachiopod assemblages that have been systematically studied with respect to drilling (all in North America), but they are the majority in 7 out of 9 warm-water Cenozoic pelecypod assemblages surveyed. Pelecypods in the Late Triassic *Palaeonucula strigillata–Dentalium undulatum* community of

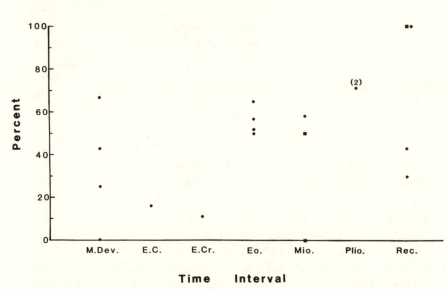

FIGURE 11.2 Drilling as a cause of death of bivalved animals during the Phanerozoic. For each local assemblage, denoted by either a circle (warm-water assemblage) or a square (cold-water assemblage), the percentage of species in which the frequency of drilling (number of drilled valves divided by half the total number of valves) exceeds 0.10 is plotted in relation to temporal occurrence. Paleozoic data refer to brachiopods, whereas all post-Paleozoic data pertain to pelecypods. Only species in which the total number of valves was 20 or more were considered. Calculations were based on data in P. H. Fischer (1966), J. D. Taylor (1970), Laws and Laws (1972), Adegoke and Tevesz (1974), Hoffman et al. (1974), Watkins (1974), Robba and Ostinelli (1975), Ausich and Gurrola (1979), Darragh and Kendrick (1980), Vermeij (1980a), Commito (1982), Sander and Lalli (1982), Taylor et al. (1983), Colbath (1985), and S. A. Smith et al. (1985).

the Italian Alps have an aggregate drilling frequency of only 0.016 (Fürsich and Wendt, 1977). Only 5 of 45 pelecypod species (11%) from the Albian (Early Cretaceous) Blackdown Greensand of England have drilling frequencies exceeding 0.10 (J. D. Taylor et al., 1983).

Evidence of effective resistance adaptation against drilling is observed in a slightly higher proportion of Cenozoic pelecypods than gastropods. Of the 34 pelecypod species that have been surveyed systematically in fossil assemblages, and in which the number of drill holes is 10 or more, 9 (26.4%) have an effectiveness of 0.30 or higher (Table 11.5). The comparable value for gastropods is 6.4% (3 of 47 species; see Table 7.3). Pelecypods with a high incidence of incomplete drill holes are generally thick-shelled or strongly sculptured

TABLE 11.5 Effectiveness of Bivalved Animals against Drilling by Gastropods

Site and Species	Effectiveness[a]	Reference
Recent, Dodinga Bay, Halmahera		Vermeij (1980b)
Anadara (Tegillarca) granosa	0.410	
Pliocene, Yorktown Formation,		B. W. Kent (1986),
Virginia		Vermeij (unpub-
Mercenaria corrugata	0.550	lished data)
Macrocallista sp.	0.140	
Eucrassatella undulata	0.330	
Lucinisca sp.	0.050	
Dosinia sp.	0	
Callocardia sp.	0.010	
Puberella sp.	0	
Angulus spp.	0	
Merisca spp.	0.008	
Caryocorbula spp.	0.530	
Glycymerella spp.	0	
Tucetona spp.	0	
Pliocene, Jackson Bluff Formation,		B. W. Kent (1986)
Florida		
Eucrassatella undulata	0.050	
Late Miocene, St. Mary's		B. W. Kent (1986)
Formation, Maryland		
Mercenaria sp.	0.330	
Middle Miocene, Choptank		Dudley and Dudley
Formation, Maryland		(1980), B. W. Kent
Anadara elevata	0.150	(1986)
Astarte thisphila	0	
Mercenaria plena	0.270	
Eucrassatella sp.	0.200	
Macrocallista sp.	0.140	
Middle Eocene, France		J. D. Taylor (1970)
Glycymeris pulvinata	0	
Ostrea plicata	0	
Venericardia imbricata	0	
Crassatella dilatata	0	
Calpitaria distincta	0	
Aphrodina nitidula	0.130	
Corbula rugosa	0.040	

Table 11.5 (*continued*)

Middle Eocene, France		P. H. Fischer (1966)
Glycymeris pulvinata	0.060	
Ostrea plicata	0.200	
Venericardia imbricata	0.080	
Sunetta semisulcata	0.710	
Meretrix laevigata	0.210	
Aphrodina nitidula	0.160	
Bicorbula gallica	0.540	
Corbula rugosa	0.310	
Eocene, Nigeria		Adegoke and Tevesz
Cardita costaeinodulosus	0.058	(1974)
Nuculana sp.	0.059	
Varicorbula amekiensis	0.120	
Viséan, Northern Ireland		Brunton (1966)
Productina margaritacea	0.043	
Wenlockian, Arctic Canada		Rohr (1976)
Dicaelosia spp.	0.033	
Epitomyonia sp.	0	
Early Pliocene, Victoria		Hingston (1985)
Glycymeris halli	0.032	
Notocorbula ephamilla	0.180	
Placamen subrostratum	0.350	
Sunetta gibberula	0.034	

[a] Effectiveness, defined as the number of incomplete drill holes divided by the total number of drill holes (complete as well as incomplete), was calculated for each sample in which the total number of drill holes was ten or more.

(Vermeij, 1980b; Kitchell et al., 1981; Hingston, 1985). Incomplete naticid drill holes appear to be much less common than incomplete drill holes made by muricaceans (Adegoke and Tevesz, 1974; Kojumdjieva, 1974). Almost 50% of the drill holes observed in Devonian brachiopods from New York by S. A. Smith and her colleagues (1985) were incomplete. Although the form of the holes suggests a naticid-like predator, the high incidence of predatory failure is inconsistent with a naticacean perpetrator, pointing instead to a predator that drilled its victims extremely slowly on or near the sediment surface.

B. W. Kent's (1986) work on the history of drilling by the muricacean *Ecphora* on thick-shelled pelecypods from the Neogene of the Atlantic Coastal Plain of the United States sheds some light on the dynamics of the escalation between drillers and their victims. Kent showed that the effectiveness of *Mercenaria* and *Eucrassatella* increased from the Middle Miocene Choptank Formation to the Late Miocene St. Mary's Formation, and then again from the Late Miocene to the Early Pliocene lower Yorktown Formation, even though *Ecphora* became larger over this time interval and drilled deeper holes (Table 11.5). The increased effectiveness was achieved mainly by an extraordinary increase in valve thickness. A specimen of *Mercenaria "corrugata"* 110 mm long, from the Yorktown Formation of Virginia, had a valve whose wall was 30 mm thick. Contemporaneous *Ecphora* were able to drill holes up to 15 mm in depth and 5 mm in diameter. Both *M. "corrugata"* and *Ecphora* became extinct during the Middle Pliocene.

LOCOMOTION IN PELECYPODS

In the discussion on armor, I pointed out that rapid locomotion is one of several alternative forms of defense that has been widely adopted by pelecypods. It is also an essential component of food gathering by deposit-feeding pelecypods such as protobranchs and tellinacean heterodonts.

Burrowing is the most widespread and ancient form of locomotion in pelecypods. It is typically accomplished by the foot, which extends ventrally or anteriorly from the opening between the valves. To understand how the shell affects burrowing, we must review briefly the burrowing sequence of the typical pelecypod. The burrowing sequence consists of six phases: (1) downward probe of the foot into the sediment, first accompanied by raising the shell into the probing position with the hinge axis parallel to the sediment surface, and later done with the shell held stationary in the sediment through the action of the ligament, which serves to push the valves laterally against the sediment; (2) dilation of the foot, permitting it to serve as an anchor in the sediment; (3) closure of the exhalant and inhalant si-

phons, so that water cannot be ejected or taken in through the siphons; (4) shutting (adduction) of the valves, accompanied by
ejection of water ventrally so that the sediment is loosened; (5)
downward movement of the shell, accomplished by retracting first
the anterior, then the posterior pedal retractor muscles, which are inserted on the valves on the dorsal side of the adductor muscles; and
(6) relaxation of the adductors, enabling the elastic ligament to force
the valves open again (Trueman et al., 1966; Trueman, 1968; S. M.
Stanley, 1970). The consecutive contractions of the anterior and posterior pedal retractors cause the shell to rotate in the plane of the
commissure during downward penetration into the sediment. While
rotation does not occur in elongate pelecypods such as *Ensis*, it involves an angular deviation of 10° in rounded or ovate clams such as
Mercenaria and as much as 45° in the tellinacean *Strigilla* (S. M.
Stanley, 1969, 1975b).

The features that are associated with rapid burrowing in pelecypods into unconsolidated sediments have been carefully studied by
S. M. Stanley (1969, 1970, 1975b, 1977b, 1981). He devised a method
of comparing the burrowing performance of pelecypods by establish-
ing a burrowing rate index (BRI) that is independent of the animal's
size. This index, as noted in Chapter 9, is defined as the cube root of
the animal's mass divided by the time required for complete burial in
the sediment. Very rapid burrowers have a BRI greater than 6, and
rapid burrowers an index of 2 to 5. Donacid beach clams are among
the fastest burrowers. Wade (1967), for example, reported on a 25 mm
long *Donax denticulatus* from Jamaica that buried completely in less
than 2 seconds.

Rapidly burrowing pelecypods are characterized by a streamlined
shell, usually of light weight (except in donacids). It is compressed,
normally in the form of a disc, wedge, blade, or elongate flattened cylinder, and has a smooth shiny surface and a strong elastic ligament
behind the umbo. The foot is usually large, a condition that is reflected in the shell by a large anterior portion and by the umbones'
placement behind the shell's middle. In several lucinids (*Divaricella*,
Divalinga) and tellinids (*Strigilla*, *Scissula*, *Scissulina*), rapid burrowing is enhanced by obliquely disposed delicate ridges whose anterior and ventral slopes are less steep than the posterior and dorsal
slopes. The cuesta form of each ridge permits the sculpture to resist
upward movement of the animal in the sediment as the shell rotates

during stage 5 of the burrowing sequence (S. M. Stanley, 1969). Asymmetrical sculpture is also found in less well streamlined cardiids (*Trachycardium, Americardia, Fragum*, among others); lucinids (*Fimbria*); trigoniaceans (many Jurassic and Cretaceous genera and the living *Neotrigonia*); venerids (*Anomalocaria, Ventricolaria, Periglypta, Nioche*, and so on); and tellinids (*Eurytellina, Tellidora*). This sculpture, which is concentrically disposed, probably also permits the sediment to be gripped during burrowing. An additional function may be to keep the animal anchored in the sediment once the final life position has been achieved. The lunule—a depression formed anterior to the umbones as a result of forward coiling of the beaks during growth—also serves, as Ansell (1961) suspected and S. M. Stanley (1975b) showed experimentally, to prevent the shell from being pushed upward and backward in the sediment as the posterior retractor contracts.

Pelecypods that fulfill the morphological criteria for rapid burrowing account for 22 to 42% of infaunal species in warm-water assemblages of burrowing pelecypods in sandy, shallow-water habitats, and for about 15% of similar cold-temperate assemblages. Except for a few solenaceans in Southeast Asia, rapid-burrowing pelecypods are unknown in fresh water (Vermeij and Dudley, 1985).

It is a remarkable fact, first pointed out by S. M. Stanley (1968), that pelecypods that are specialized for rapid or very rapid burrowing did not become common until the later Mesozoic. No Paleozoic pelecypod is known to have had burrowing-enhancing asymmetrical sculpture, and most of the groups that contain rapid burrowers in modern seas did not become established until the Early Cretaceous (Tellinidae, Solenidae) or even later (Mactridae, Solecurtidae, Cultellidae, Veneridae, Donacidae). Protobranchs of the superfamily Nuculanacea may have been the only rapid burrowers among Paleozoic pelecypods. Even in *Palaeostraba*, the earliest member of this group (Wenlockian stage of the Silurian of Gotland, Sweden), the beaks were located posterior to the middle of the shell, a position indicating the presence of a large foot (Liljedahl, 1984).

The diversification of rapid burrowers during the Mesozoic is part of the large-scale evolutionary deployment of pelecypods with various degrees of mantle fusion and eulamellibranch gills (see Chapter 4). Both these characters permit the gills (ctenidia) to create strong water currents that carry food and oxygen in and expel excrement and carbon dioxide from the animal. This capacity is necessary for the evolution of deeply buried and highly active pelecypods (S. M. Stanley, 1968, 1977a).

As is true for the history of burrowing in gastropods, it is necessary to inject a cautionary note about the interpretation of the history of burrowing in pelecypods. The sandy habitats in which rapidly burrowing pelecypods are most common and most specialized in form are poorly represented in the fossil record, especially in Paleozoic strata. Proper documentation of the history of burrowing must await analyses of the incidence of burrowing-enhancing morphology in local assemblages from various times in the Phanerozoic.

Leaping is a locomotor mechanism that in pelecypods is closely tied to burrowing. Cockles (Cardiidae), the trigoniacean *Neotrigonia*, and some solenids and mactrids are especially well known for their capacity to flee from predators by jumping (Ansell, 1967; S. M. Stanley, 1970, 1977b; Nielson, 1975; McMahon and McMahon, 1983). When contacted by the sea star *Asterias*, the cockle *Laevicardium crassum* can move at a maximum speed of 0.9 m/sec by taking up to five leaps, the first of which takes the animal 0.3 m from its enemy (Ansell, 1967). The razor clam *Ensis minor* moves backward a distance of 0.15 m per leap after contact with the naticid predator *Neverita duplicata* (McMahon and McMahon, 1983). A large foot and high metabolic rate are necessary prerequisites for leaping. In cockles, a large foot is indicated by a shell that is higher than it is long.

Another mode of locomotion that has evolved in diverse pelecypods is swimming (S. M. Stanley, 1970; Thayer, 1972). The most adept swimmers are scallops of the family Pectinidae. As the valves are closed by the rapid contraction of the centrally placed adductor muscle, water is forced out either from the ventral margin or from the ear-shaped auricles on either side of the hinge line. B. S. Morton (1980b) reports that *Amusium pleuronectes* in Hong Kong attains a maximum speed of 0.73 m/sec and swims for a distance of at least 10 m. The North Atlantic *Placopecten magellanicus* swims for a distance of 4 m at a maximum speed of 67 cm/sec (Caddy, 1967). The ability to swim is associated with a thin, flattened, smooth or very finely ribbed shell that gapes permanently at the auricles and anteriorly as well as posteriorly on the main disc. More sedentary species usually have more strongly sculptured shells that are often thick and strongly convex, with a strong byssal notch that is usually well developed. Other pelecypods known to swim include the cardiid *Laevicardium*, the solecurtid razor clam *Ensis*, the protobranch *Solemya*, and file shells of the family Limidae (S. M. Stanley, 1970; B. S. Morton, 1979). *Ensis minor* swims over 1 meter in 20 seconds as a maneuver to escape from predators before reburrowing in the sand (McMahon and McMahon, 1983). All swimming pelecypods have thin, smooth shells with smooth valve margins.

The history of swimming in pelecypods has not been considered comprehensively by other authors. Swimming may already have been possible in pelecypods as early as the Middle Ordovician, when the limoid stock was first differentiated (Waller, 1978). *Pernopecten*, a probable swimmer as inferred from the structure of its ligament and from its thin smooth shell, and related entoliids evolved in the Early Carboniferous (N. D. Newell, 1937). True Pectinidae and Propeamussiidae date with certainty from the Early Triassic (Waller, 1984). Swimming in cockles, razor clams, and solemyids evolved secondarily among burrowers and is probably a rather recent development.

Summary

The study of bivalve armor and its history is in its infancy. In general, the potential for evolution of breakage-resistant armor is expected to be lower than in retractable coiled gastropods because marginal injuries that result from unsuccessful attacks by shell-breakers are apt to place a bivalved animal at a high risk to other dangers. Several shell features that increase resistance against breakage or that prevent a tight seal from being broken when the valves are attacked at the margins have nevertheless evolved. These include crenulations on the inner valve surface at the margins, a flexible margin, overlap of one valve by the other along part or all of the margin, and marginal spines. The incidence of crenulated valves has increased through time, whereas the incidence of permanent gapes between valves has declined somewhat from the Mesozoic to the Late Cenozoic. Preliminary data from Paleozoic brachiopods and Cenozoic pelecypods suggest that the incidence of repaired shell injuries has remained roughly constant from the Ordovician to the Recent. Drilling has become a more important agency of death since the Early Cretaceous.

Many pelecypods have emphasized locomotion instead of armor as the principal mode of antipredatory defense. Characteristics associated with rapid burrowing are known chiefly from the Late Cretaceous onward, but rapid swimming may be a more ancient form of locomotion.

A great deal remains to be learned about how pelecypods and brachiopods cope with enemies. Surveys are needed to evaluate the compressive strength and antipredatory effectiveness of shells, to trace the history of the various armor-related and locomotion-related shell features, and to assess the risks associated with marginal valve injury during unsuccessful attacks by shell-breakers.

CHAPTER TWELVE

Armor and Locomotion in

Articulated Animals

FUNCTIONAL POSSIBILITIES AND LIMITATIONS

A skeleton consisting of a single rigid unit can serve as highly effective passive armor, as beautifully exemplified in many gastropods. In addition, it can be used, together with muscles in the foot, in locomotion and retaliation. For example, shell sculpture in some species sometimes enhances burrowing in sand, while in other species it occasionally inflicts injury to would-be attackers. Limpets are able to push away competitors and attack predaceous snails with the outer edge of the shell. Some degree of active retaliation using hard parts has been achieved in strombids, in which the operculum has been modified into a pushing or stabbing device, but this path to retaliation has been taken only rarely. Formidable methods of retaliation do occur in univalved molluscs—the harpooning by the radular teeth of *Conus*, for example—but they involve non-skeletal structures.

The locomotor and retaliatory capacities of bivalved animals are only a little less constrained than those in univalved animals. Clams are capable of swimming and burrowing by alternating the adduction of the valves with the opening action of the ligament. Rapid adduction for trapping prey is most closely approached in tridacnid giant clams, but it probably has never played more than a subsidiary role in the biology of bivalved animals. Entrapment of prey does occur in the so-called septibranch anomalodesmatan pelecypods (Allen and Morgan, 1981; B. S. Morton, 1981a, b), but it does not involve the shell valves directly and is effective only against small copepods.

Animals with an articulated skeleton—one consisting of many

rigid elements joined together at flexible hinges—have carried powered locomotion and active defense much further. When rigid elements can change position relative to one another by the action of muscles, forces can be created that enable the animal both to overcome the resistance of the surrounding medium during locomotion and to acquire and defend resources. Walking, flying, swimming, burrowing, biting, pounding, and stinging all lie within the evolutionary reach of animals with multipartite articulated skeletons, without diminishing the role of passive armor necessarily.

Another constraint on the one-piece skeleton of gastropods and the two-piece skeleton of pelecypods and brachiopods is the logarithmic growth pattern. Increase in size must occur at one end, and shape can change only by systematic alteration in the logarithmic parameters through the course of ontogeny, or by secondarily resorbing previously formed parts of the skeleton. The elements of a jointed skeleton, on the other hand, must each grow, either by accretion or by periodic molting. Because growth takes place at so many sites, however, the capacity to adapt the skeleton as the animal grows is very considerable (Vermeij, 1970).

The various groups of animals with articulated skeletons—chitons, arthropods, echinoderms, and vertebrates—face an adaptational dilemma: although a flexible skeleton makes locomotion and retaliation with the use of hard parts possible, a rigid mosaic of plates provides better protection as armor than does a skeleton consisting of plates connected by flexible hinges. Potential attackers often concentrate their attention on the articulations, which therefore constitute the weak areas of the skeleton. A dominant theme in the development of articulated armor should therefore be rigidification, which can be accomplished either by buttressing existing joints or by reducing the number and size of joints. The former can be accomplished by having a substantial zone of overlap between plates, so that no part of the body will be covered by less than one thickness of rigid armor, even when the animal changes shape. The alternative method, that of rigidly suturing the plates together, in effect produces the univalved or bivalved condition; it therefore eliminates or greatly reduces the locomotor and retaliatory role of the skeleton.

Little is known about the ecological risks associated with the evolution of skeletons of articulated armor. Probably the risks vary from group to group. Joint research by ecologists and functional morphologists is needed in order to evaluate the following hypotheses about the risks that are engendered by an evolutionary emphasis on armor in articulated skeletons.

ARTHROPOD MOLTING

The mode of growth of arthropods engenders several potentially severe limitations on the expression of passive external armor in this superphylum, which is by far the most diversified, and ecologically most widespread, group of animals today. Growth in post-larval arthropods occurs episodically by molting, a process in which the skeleton (instar) is periodically shed in favor of a larger one. The chief problem with molting comes immediately after the hard exoskeleton has been shed and while the new skeleton is still soft. During this critical period, the skeleton is of little value in protection, feeding, or locomotion. Only after the protein component has become tanned and the mineral component is in place can the skeleton resume its full range of functions. The duration of the vulnerable phase is therefore potentially critical. Skeletal hardening probably takes longer for thick, heavily mineralized skeletons—that is, in heavily armored arthropods—than for delicate ones. Data on this important point have not been systematically gathered. Aiken (1980) reports that the American lobster (*Homarus americanus*) spends about 20 minutes in the vulnerable phase, whereas the duration of this phase in the spiny lobster *Panulirus* may be as short as 3 minutes. Shedding is often done while the animal is in a shelter. If the animal is large, however, adequate shelters may be difficult to find. Consequently, the duration of the soft phase exposes arthropods to great risks. That this is indeed so is indicated by Neville (1975), who cites earlier work showing that dragonfly nymphs suffer a 15% mortality during molting. Although some of this mortality results from the inability of the nymph to free itself from the old exoskeleton, some is due to birds and other predators which find the just-molted insect easy prey.

An indication that molting is incompatible with strong armor and with many other functions comes from the observation that many arthropods do not molt in the adult stage. Many crustaceans (including crabs) and almost all flying insects have a terminal adult molt whose exoskeleton is often much sturdier than that in the growing phase (Neville, 1975; Maiorana, 1979; Hartnoll, 1983). Because there has been no systematic survey of the distribution of terminal molting in Crustacea, it is at present too early to determine if cessation of molting is functionally related to the deployment of heavy adult armor.

Another indirect indication of the possible risks that molting presents to the elaboration of armor is provided by the pattern of distribution of large terrestrial crustaceans. Finding a safe place to molt becomes increasingly difficult for a larger crustacean. A large, heavily armored crustacean would therefore be evolutionarily feasible only in situations where enemies that could take advantage of the recently molted crustacean are rare or absent. The largest living terrestrial Crustacea are members of the hermit crab family Coenobitidae. Most species of this family molt while they are inside the shell of a gastropod, but the largest species, the heavily armored and pugnacious coconut crab (*Birgus latro*), has secondarily lost the shell-dwelling habit. The coconut crab is found on small islands throughout the tropical Western Pacific and Indian Oceans, but it is apparently absent from the continental shores of Africa and Asia, as well as from the large continental islands of Indonesia, New Guinea, and Australia (Reyne, 1938, 1939; Holthuis, 1959, 1963). Predators that are common on the large land masses are perhaps absent or of reduced importance on smaller islands. Research on the coconut crab with this hypothesis in mind would be very interesting, for it would go a long way toward answering the larger question of whether molting and its attendant risks limit the size of crustaceans and other arthropods on land.

CONGLOBATION IN ARTHROPODS

Despite the limitations of molting, many lineages of arthropods have evolved substantial armor. Unfortunately, our understanding of the functional morphology and distribution of arthropod armor is still rudimentary. Aside from a few studies of exoskeletal architecture in trilobites, isopods, and barnacles, most statements in the literature about arthropod armor are anecdotal and rather generalized. Nothing seems to be known about the effectiveness of arthropod armor relative to that of other groups. Therefore, while a vast amount of fascinating work awaits the ambitious investigator who is willing to tackle the awesomely diverse arthropods, at this point I can do little more than discuss some common themes of armor in arthropods and suggest patterns in their adaptive histories.

A common theme in animals with jointed exoskeletons is conglobation, the capacity to roll up into a ball or to fold up, so that the anterior end of the animal touches and often intermeshes with the posterior end. Conglobation was extremely widespread in Paleozoic trilobites (Figure 12.1) and xiphosurans, in which it has been reasonably well studied, and it is still found in many diplopod myriapods (millipedes), some insects, and many isopod Crustacea (Table 12.1). Outside the Arthropoda, chitons (polyplacophoran molluscs) and some mammals have evolved the capacity to roll up.

Several benefits of rolling the body up into a ball suggest themselves. The most obvious is that the exoskeleton, which is least fortified ventrally, can protect the entire body when the latter is attacked in the rolled-up state. Eisner and Davis (1967), for example, showed that ants, jays, and grasshopper mice were unable to break open or to gain a purchase on the smooth cuticle of rolled-up millipedes of the genus *Sphaerotherium*. The banded mongoose (*Mungos mungo*) was able to overwhelm the millipede only by hurling the creature against rocks. If, as is so often the case in conglobating animals, the anterior and posterior parts of the body fit snugly against one another or are held in position in the rolled-up state by interlocking furrows and ridges (coaptive structures), an attacker would be unable to unroll the victim. Coaptive structures are well known and of diverse origins in Ordovician trilobites (Bergström, 1973; Clarkson and Henry, 1973; Henry and Clarkson, 1975; Fortey and Owens, 1979); living diplopods (Manton, 1954); and living oniscoid isopods (Vandel, 1943, 1948, 1959). Comparative studies of the effectiveness of the different styles of coaptation and conglobation have not been carried out, but they would be quite feasible in isopods, in which two fundamental types of conglobation occur. These are exoantennal conglobation, in which the antennae are protected in a groove on the outside of the ball, and endoantennal conglobation, in which the antennae are concealed within the ball. The degree of protection and the intricacy of the furrows and ridges that hold the ball together vary greatly from species to species.

Another potential benefit of enrollment is that the animal's small-

FIGURE 12.1   The Middle Devonian trilobite *Phacops rana* in a rolled-up state.

TABLE 12.1   Survey of Conglobating Animals

Phylum Mollusca, Class Polyplacophora (chitons): marine, Late Cambrian to Recent; early forms with valves not overlapping and therefore perhaps unable to roll up (van Belle, 1975; Runnegar et al., 1979; Runnegar, 1983)

Phylum Trilobita: marine, Early Cambrian to Late Permian; conglobation absent in primitive (Early Cambrian) Olenellacea and Redlichiacea, secondarily lost in Ordovician Remopleurididae, Cyclopygidae, and Olenidae (Bergström, 1973; Clarkson and Henry, 1973; Henry and Clarkson, 1975; K.S.W. Campbell, 1977; Fortey and Owens, 1979)

Phylum Crustacea, Class Malacostraca, Order Isopoda (mainly Cenozoic)

Suborder Oniscoidea: terrestrial; antennae outside ball in Spelaeoniscidae and Tendoniscidae; antennae inside ball in Tylidae, Buddelundiellidae, Porcellionidae (*Cylisticus, Porcellium*), Armadillidiidae, Eubelidae, Actoeciidae, and Armadillidae (Vandel, 1943, 1948, 1959)

Suborder Asellota: freshwater or marine; Asellidae (*Haploniscus*), Cirolanidae (*Faucheria*), Sphaeromidae (Vandel, 1948)

Phylum Chelicerata, Class Xiphosura (horseshoe crabs): marine, some Paleozoic forms perhaps aquatic to semiterrestrial (Middle Cambrian to Recent); many Paleozoic forms able to roll up, as in *Pseudoniscus* (Late Silurian) and *Euproops* (Late Carboniferous) (Bergström, 1975; D. C. Fisher, 1977)

Phylum Uniramia

Subphylum Myriapoda, Class Diplopoda (millipedes): terrestrial, Late Silurian to Recent

Order Oniscomorpha: Late Carboniferous to Recent (Manton, 1954; Hannibal and Feldmann, 1981)

Order Polydesmoidea: *Oniscodesmus* (Manton, 1954)

Order Limacomorpha: *Glomeridesmus* (Manton, 1954)

Subphylum Insecta, Class Hexapoda

Order Blattaria (= Dictyoptera) (cockroaches): terrestrial, Late Carboniferous to Recent; *Pseudoglomeris, Derocalymma, Perisphaerides* (Vandel, 1943)

Order Coleoptera (beetles): terrestrial, Permian to Recent; Silphidae (*Agathidium*), Clambidae, Acanthoceridae, Anobiidae (*Dorcatoma, La-*

Table 12.1 (*continued*)

---

sioderma), Nitidulidae (*Cybocephalus*), Scydmaenidae (larva of *Cephennium*) (Vandel, 1948; Crowson, 1981)

Phylum Chordata, Class Mammalia: terrestrial, Late Triassic to Recent

Order Xenarthra: Dasypodidae (armadillos) (*Tolypeutes*)

Order Pholidota (pangolins): *Manis*

Order Insectivora: Tenrecidae (*Setifer*), Erinaceidae (hedgehogs)

Order Monotremata: Tachyglossidae (spiny anteaters) (Eisenberg, 1981)

---

est dimension, which in the unrolled position is the dorsoventral dimension, is increased, so that a would-be predator would have difficulty swallowing the victim. This benefit would accrue only to animals that are able to maintain the enrolled state for long periods, and it would be ineffective against predators with a large gape. Because so little is known about the predators of conglobating animals, it is difficult at present to assess the importance of this function of enrollment in the course of evolution.

D. C. Fisher (1977) has elegantly approached the functional morphology of conglobating animals by performing experiments both with models of actual fossils and with models in which features have been artificially deleted or enhanced. He has shown that rolling up in the Late Carboniferous xiphosuran genus *Euproops* and in morphologically similar trilobites is intimately associated with the presence of spines, which stabilize the descent through the water of a potential rolled-up victim. For a swimming animal, immobility or the appearance of immobility is an effective defense against visual hunters (Kerfoot, 1977, 1978; Kerfoot et al., 1980). When a swimming *Euproops* or another conglobating animal rolled up, it was unable to use its appendages for locomotion. If the resulting descent to the bottom had been unsteady, however, the victim would have appeared to be in motion, and it might therefore have elicited further attack by its pursuer. By removing or by lengthening the ophthalmic spines on models of *Euproops*, Fisher was able to show that descent would have been uneven unless the spines were of the length actually observed in fossils. Other spines on the body of *Euproops* probably had a resistance function, or they served to make the rolled-up victim effectively large and difficult to grasp.

F. R. Schram has suggested to me that conglobating trilobites might have benefited from enrollment not so much in defense as in feeding. He points out that the gnathobases of the walking legs, which were used as feeding organs, would all be placed in a small area when the trilobite was in a rolled-up position (Bergström and Brassel, 1984). Food could be transplanted directly from the nest of gnathobases to the mouth without being passed forward between the legs, as is the pattern seen in primitive Crustacea (Manton, 1977). I believe there is merit in this interpretation, but its validity neither detracts from nor contradicts the importance of conglobation as a defensive aptation. Instead, the benefits for feeding can be construed as a happy consequence of the ability to roll up, which was evolved originally as an effective way of employing articulated armor. As is so often true of aptations, rolling up could thus be said to have a joint benefit.

Conglobation is not free of faults or limitations. It is effective only when the animal is actually rolled up. While moving or feeding, the animal (especially its ventral side) is vulnerable to attack by predators that would otherwise be incapable of coping with the victim's armor (Schmalfuss, 1978). In the rolled-up position, most other functions must also cease, including respiration. If the precise lock-and-key type of enrollment that is observed in many Ordovician trilobites had to be maintained for long periods, the animal would effectively be sealed up and inactive, and ultimately it would be incapable of maintaining itself. K.S.W. Campbell (1977) believes that this potential difficulty was partially overcome in some Silurian and Devonian Dalmanitacea, in which the seal between the anterior end (cephalon) and the fused posterior segments (pygidium) was incomplete. In the rolled-up position, two or more holes, often associated with projecting spines, were left between cephalon and pygidium, so that water currents could still pass, thereby permitting respiration. These trilobites also had special furrows for the antennae, which could be used to sense danger or to fulfill other functions, even when the trilobite was nearly completely enrolled.

Inspection of Table 12.1 reveals the striking fact that conglobation in living animals is chiefly a terrestrial phenomenon. In diplopods and some isopods, it is often (though by no means always) associated with burrowing in soil or leaf litter. The only large marine group of conglobating animals is the Polyplacophora. Nothing is known about the effectiveness of enrollment in these animals, but they never exhibit coaptive devices of the type seen in trilobites, diplopods, and isopods. In the Xiphosura, conglobation seems to be characteristic only of Paleozoic forms. The Recent genus *Limulus* can enter into a flexed position, but it does not roll up (D. C. Fisher, 1977). The largest of all the marine conglobating groups, the trilobites, declined sharply after the Devonian and was extinct by the end of the Permian. Attractive as conglobation would seem to be as a passive defense, it evidently has not been the method of choice in post-Paleozoic animals. Even on land, where conglobation is still fairly widespread, it is augmented in its defense capacity by the presence of retaliatory defenses such as spines (in tenrecs, hedgehogs, and some isopods and diplopods) and chemicals (in most diplopods and some isopods) (Sutton, 1972; Eisenberg, 1981; Hannibal and Feldmann, 1981).

THE CARAPACE FORM IN ARTHROPODS

The joints between parts of the exoskeleton are places where the animal is especially vulnerable to attack by predators. The octopod *Eledone* injects toxin into the thin cuticle in the joints of the legs and beneath the carapace of crabs (Boyle and Knobloch, 1981). The spider *Nephila maculata* bites and injects venom at the joints of insects (Robinson and Robinson, 1973). Passive resistance to this kind of predation can be enhanced by a reduction in the number or accessibility of the vulnerable joints. This requirement is at odds with conglobation, because enrollment requires many flexible joints which in the rolled-up state are exposed to the outside. The development of a carapace in crabs and of a hard skeleton of elytra in beetles and some other insects is therefore potentially effective against skeleton-piercing enemies.

One form of carapace that has evolved independently several times in the Crustacea is the bivalved carapace, which encloses the appendages when the valves are adducted. Bivalved carapaces are characteristic of the Notostraca (or Branchiopoda, Early Devonian to Recent); Diplostraca (including the Conchostraca, Devonian to Recent, and the Cladocera, Permian or Oligocene to Recent); Phyllocarida (Middle Cambrian to Recent, including many peculiar forms in the Middle Cambrian Burgess Shale); Ostracoda (Cambrian to Recent); Phosphatocopina (Early Cambrian); some Cirripedia (barnacles, Silurian to Recent); and some Middle Cambrian groups of uncertain affinity, including the genera *Odaraia* and *Perspicaris* (Schram, 1982; Briggs, 1983). As in conglobation, withdrawal of the appendages into the bivalved carapace leads to a cessation of movement. It is therefore not surprising that passive sinking is a common form of defense in the Cladocera (Kerfoot, 1977, 1978; Kerfoot et al., 1980). The problems with this kind of defense have already been noted. Evolutionarily speaking, therefore, bivalved carapaces are expected to be less successful today than they were in the past. The large number of bivalved crustacean groups in the Cambrian is consistent with this hypothesis. Moreover, most living Crustacea with a bivalved carapace are small-bodied, and many live in freshwater environments where other skeletonized groups also show a markedly reduced expression of armor.

The crab form has evolved repeatedly in bottom-dwelling crustaceans of the order Decapoda. It is, of course, characteristic of the true crabs (infraorder Brachyura), but in addition it has evolved independently several times in the infraorder Anomala, which includes such familiar groups as the mole crabs (Hippidae), porcelain crabs (Porcellanidae), and several secondarily crab-like derivatives of the hermit crabs (Lithodidae and the deep-sea genus *Probeebei*) (Wolff, 1961; Stevčić, 1971; McLaughlin, 1983). The essential features of the crab form are a dorsal carapace covering the fused cephalothorax, with the walking legs inserting laterally beneath the carapace and the abdomen tucked beneath the carapace posteriorly. The ventral surface is covered with sternites, and in some forms (Calappidae, some Xanthidae) the claws—modified first pair of walking legs used for fighting and feeding—cover the anterior face of the animal.

Some very heavily armored animals have the crab form. Many xanthid, carpiliid, menippid, and parthenopid crabs have thick, highly mineralized carapaces and claws. On land, conspicuously heavy armor has evolved in the Coleoptera, Hemiptera, certain ant-associated dipteran flies (E. O. Wilson, 1971), and soil-dwelling harvestmen of the arachnid order Opiliones. True crabs evolved in the Early Jurassic, but the history of other crab-like crustaceans and of heavily armored land arthropods remains largely unknown.

SIZE IN LAND ARTHROPODS

An interesting historical trend among land arthropods is the decline through time in the maximum body size of adults. Many Paleozoic myriapods, insects, and arachnids greatly exceeded the size of living land arthropods. The aquatic (or possibly terrestrial) Late Silurian diplopod myriapod *Necrogammarus* is estimated by Rolfe (1980) to have had a total body length of 27.5 cm. *Gigantoscorpio*, an amphibious scorpion from the Tournaisian (Early Carboniferous) of Scotland, attained a length of 50 cm. The spiny Carboniferous diplopod *Acantherpestes* was 30 cm long (Kraus, 1974), and the contemporaneous *Arthropleura*, a member of the extinct myriapod order Arthropleurida, had the almost unbelievable length of 180 cm. Even the Early Devonian *Eoarthropleura* from the Alken beds of Germany was a 100-cm animal (Störmer, 1976). Some Late Carboniferous members of the insect order Protodonata had a wingspread of perhaps 75 cm (F. M. Carpenter, 1976).

The extraordinarily large size of many Paleozoic terrestrial arthropods is especially surprising in view of the generally poor fossil record of terrestrial arthropods. If animals of this size were rare or unusual, the spotty fossil record would almost certainly have failed to reveal them. That we find them at all indicates a great abundance of large species. As I have already suggested in the discussion of the coconut crab, problems associated with molting may be so great today that large size coupled with a heavily sclerotized skeleton may have become unfeasible in most post-Paleozoic terrestrial environments.

That risks for large arthropods were generally lower during the Paleozoic is further suggested by the replacement of paleopterous insects by neopterous ones during the Permian. In paleopterous forms, the wings could not be folded against the body while the animal was at rest (F. M. Carpenter, 1976). Not only were these insects susceptible to enemies by virtue of the fact that the wings could not protect the rest of the body, but they also could not have fit easily into shelters. With the ability to fold their wings, the various neopterous groups overcame these limitations and evidently were able to exist in a terrestrial world in which the dominant large competitors and predators were vertebrates.

BARNACLES

Despite the widespread use of the chitinous exoskeleton as passive armor, many arthropods have evolved alternative forms of armor that are more reminiscent of molluscs than of typical arthropods. The barnacles of the crustacean class Cirripedia are one group in which this alternative form of armor has been developed very far. Unlike most other crustaceans, barnacles are permanently attached as adults to a hard object or surface. Their external skeleton consists of chitinous or calcareous plates that surround the "mantle" and the filter-feeding appendages (cirri).

In the discussion below, I shall consider only the so-called thoracican barnacles, which live on either hard, inanimate objects or the outside of various other animals. The acrothoracicans are endolithic animals (see Chapter 5), whereas the Rhizocephala and Ascothoracica are internal or gall-forming parasites that are practically unknown as fossils.

Fortunately, because some excellent biomechanical and ecological research has been carried out on thoracican barnacles, several aspects of the functional skeletal morphology of this ecologically important group are now well documented. Studies of the compressive force that a barnacle test can withstand when a flat metal plate is lowered onto the aperture from above show that species with intricately interlocking parietal (lateral) plates and a calcareous base are considerably stronger than species with less tightly articulated plates and a membranous base. British species of *Balanus*, representing the first type, can withstand compressive forces of 57 to 145 N when the test volume is standardized at 1 cm^3, whereas species of the second type, belonging to *Chthamalus*, *Elminius*, *Semibalanus*, and *Verruca*, can withstand loads of only 23 to 50 N (Gubbay, 1983). Large *Balanus balanus* break at loads exceeding 1000 N, whereas large *Semibalanus balanoides* fail at 390 N (Murdock and Currey, 1978). Species of the tropical and subtropical genus *Tetraclita* have not been measured, but they are probably even stronger than *Balanus*.

Another trait that enhances the compressive strength of *Balanus*,

Tetraclita, and other well-armored thoracican barnacles is the presence of tubules in the walls. Not only do these tubes stop propagation of cracks, but they also permit the construction of a thick lateral wall without large quantities of calcium carbonate, so that rapid outward growth is not compromised as it would be if the walls were solid (Murdock and Currey, 1978; Stanley and Newman, 1980).

Menge and Lubchenco (1981) point out that many tropical barnacles, including species of *Catophragmus*, *Chthamalus*, and *Tetraclita*, have limpet-like, flattened, thick tests that are difficult for fishes to graze and for crabs to grasp. This shape is also associated with a very small aperture. Tall, cylindrical barnacles with relatively large apertures, such as species of *Megabalanus*, are more characteristic of temperate shores, or are found in the tropics on steep faces where grazing by fishes may be less intense than on more gently sloping or horizontal faces.

In his studies of drilling of barnacles by thaidid gastropods in the northeastern Pacific, A. R. Palmer (1982, 1983b) has found that attacks at plate junctions are two to three times more likely to be successful than attacks through the parietal plates. This observation strongly suggests that barnacles with relatively few plates (and thus with a reduced number of vulnerable junctions) would have an antipredatory advantage over forms with a large number of lateral plates. The ecological distribution of barnacles is consistent with this expectation. Species with six or eight lateral plates are found either on large marine vertebrates (turtles, manatees, sea snakes) or on the upper reaches of rocky seashores. Drilling predation in these environments is either lacking or much reduced compared with that on rocky surfaces in the lower intertidal and shallow subtidal zones. The species of *Chelonibia* living on large vertebrates show several other characteristics that suggest the absence of drilling predation. They have a large aperture and smooth lateral plates. By contrast, species of *Tetraclita* and *Balanus* tend to have a small opening and ribbed lateral plates. Palmer has speculated that thaidids may be unable to distinguish between ribs on parietal plates and the junctions between plates. The drillers are therefore apt to make many mistakes in choosing a drilling site.

Although the fossil record of barnacles is rather sparse, it portrays a history that is consistent with the hypothesis that barnacles have emphasized armor as a defense over the course of time. The account given here is taken chiefly from the work of W. A. Newman and his collaborators (Newman et al., 1969; Newman and Ross, 1976, 1977; Newman, 1979, 1982; Schram and Newman, 1980; Stanley and Newman, 1980).

Paleozoic free-living barnacles were uncalcified, stalked lepadomorphs or goose-necked barnacles. The date of the earliest barnacle is still a matter of controversy. Collins and Rudkin (1981) regarded *Priscansermarinus*, a genus based on a single specimen from the Middle Cambrian Burgess Shale of British Columbia, as the earliest barnacle because of its general resemblance to living lepadomorphs. Because neither cirri nor an undoubted capitulum (the main part of the body and its enclosing plates) was preserved, however, Briggs (1983) has seriously questioned the inclusion of *Priscansermarinus* (the "first marine goose") among the Cirripedia, and he even doubts that the object is a fossil. The next oldest barnacle, *Cyprilepas*, is known from the exoskeleton of a Late Silurian eurypterid. The only other described genus of Paleozoic lepadomorphs is *Praelepas* from the Carboniferous. Like *Cyprilepas*, its capitulum consisted of uncalcified plates.

The Triassic genus *Eolepas* is the first lepadomorph with a partially calcified test. It lies at the base of a lineage of Mesozoic and Cenozoic lepadomorphs in which plate number had a tendency to increase. From this group arose several groups of so-called acorn barnacles, which are distinguished from lepadomorphs by the absence of a stalk and by closure of the aperture by two or four opercular plates. Rather than being attached by a flexible stalk, the adult acorn barnacle adheres by a membranous or calcareous base. At least three lines of sessile barnacles, all arising from scalpellid-like lepadomorphs, have evolved since the mid-Mesozoic. The first group was the Late Jurassic to Miocene Brachylepadomorpha, about which very little seems to be known. The Verrucomorpha, comprising three genera, arose in the Albian or Cenomanian (middle Cretaceous). They differ from the Balanomorpha by having two opercular plates instead of four, and by having six rather than eight lateral plates in the primitive members.

Most Cenozoic acorn barnacles belong to the suborder Balanomorpha. The oldest genus is the Senonian (Late Cretaceous) *Pachydiadema* which, like other members of its family (the Pachylasmidae), had a solid-walled, loosely articulated test of eight main lateral plates. Although some Early Cenozoic pachylasmids lived in shallow water, all post-Oligocene species are deep-water forms (Buckeridge, 1983). Loose articulation and solid-walled plates have been retained in members of the superfamily Chthamaloidea (Early Miocene to Recent), which are found chiefly in the upper intertidal zone on rocky shores and mangrove trees. Calcareous bases and tubiferous walls were evolved independently in several lineages, including the Tetraclitidae and Balanidae (both Oligocene to Recent), two families that are prominent worldwide in shallow waters today. Most balanomorph lineages also have undergone a reduction in the number of lateral plates (A. R. Palmer, 1982). This number was primitively eight, but it has been reduced to six in five lineages, to four in ten lineages, and to a single fused test in three lineages. The reduction to one plate occurs in coral-associated pyrgomatids, a family with Early Miocene origins (Ross and Newman, 1973; Newman and Ladd, 1974). Because these barnacles live partially enclosed by coral polyps, however, the reduction in plate number is unlikely to be the result of selection by predators, because the latter typically avoid nematocyst-bearing corals.

In short, the history of barnacles closely parallels that of gastropods. Both groups are capable of a tight seal of soft tissues within the skeleton, and both show elaboration of armor, especially in the Cretaceous and Cenozoic.

ECHINODERMS

One of the most distinctive and peculiar groups of animals is the Echinodermata, whose five-part radial symmetry, water-vascular system, and mesodermal skeleton composed of calcium carbonate plates with a mesh-like structure have been unique features since the group became differentiated in the Atdabanian stage of the Early Cambrian. The main part of the body is enclosed in the theca, but many echinoderms (especially the Crinozoa and the Stelleroidea) have arms that contain extensions of the water-vascular system, and the theca may be reduced or even absent. Many echinoderms, especially in the Paleozoic, attached to the sea bottom by means of a stem (pelma) or branches (cirri) of the stem (C. E. Brett, 1981). Others, including stelleroids (sea stars and brittle stars), echinoids (sea urchins), and holothuroids (sea cucumbers), move from place to place by means of tube feet. A few holothurians, which have lost external armor, even swim, as do some comatulid crinoids.

In contrast to many other skeletonized animals, echinoderms fundamentally have a porous skeleton. Respiratory structures such as tube feet and epispires (see below) extend through or between the plates. Tube feet are also essential for feeding in many groups; therefore they must be capable of extending to the animal's exterior. A tight seal of the type seen in molluscs, rolled-up arthropods, and barnacles is thus impossible in echinoderms.

Nevertheless, echinoderms as a group are slow-moving animals that more often than not depend on armor and associated attributes for resistance and defense. Major themes in the evolution of the echinoderm skeleton have been the strengthening, especially the rigidification, of the theca and the protection of exposed structures and tissues. Paul (1977) has given a detailed inferential account of the history of echinoderms, and much of what follows is drawn from his paper.

The earliest echinoderms had a flexible theca, which was composed either of imbricating plates that overlapped one another like shingles on a roof (seven Early Cambrian genera) or of tessellated plates that were not tightly joined by collagenous sutures (two Early Cambrian genera). Most Paleozoic sea urchins, which originated during the Late Ordovician, also had flexible tests, as did all members of the order Echinocystitoida and the six Late Paleozoic (Early Carboniferous to Permian) genera of the order Archaeocidaroida (Kier, 1965; A. B. Smith, 1984). Among living thecate echinoderms, a flexible theca is known only in the echinoid order Echinothuroida, a primitive group of deep-sea urchins whose fossil record extends back to the Oxfordian stage of the Late Jurassic (Kier, 1974). Several sea stars such as the bun-shaped, coral-eating *Culcita* (family Oreasteridae) also have a flexible theca-like skeleton, but the skeleton is extremely sturdy (D. B. Blake, 1983).

Various mechanisms for strengthening the theca and for rendering it incapable of being deformed evolved first in the latest Early Cambrian (Elankian stage) and became common during the Ordovician (Paul, 1977; Paul and Smith, 1984). The genera *Gogia* and *Lichenoides*, which are primitive representatives of the subphylum Pelmatozoa (stemmed echinoderms), are among the earliest echinoderms with a rigid skeleton. Several echinoderms independently evolved a cup-like theca whose rim was reinforced marginally by well-developed plates. Such a structure is known from the Middle Cambrian onward. Early examples from the Middle Cambrian include the pelmatozoans *Ctenocystis* and *Cyclocystis*, the primitive eleutherozoan *Cambraster*, and the cinctan carpoids *Gyrocystis* and *Trochocystites*. A rigid theca also characterized the Late Middle Ordovician genus *Bothriocidaris*—formerly regarded as the earliest echinoid but now considered by A. B. Smith (1984) to be a primitive holothuroid—as well as members of the echinoid family Palaechinidae (Late Silurian to Late Permian) and many Ordovician to Devonian cystoids and blastoids.

Plate thickening and rigidification also characterize Middle Paleozoic and later crinoids (Signor and Brett, 1984). Compact bowl-shaped thecae, in which the plate thickness equaled or exceeded one third the thecal radius, appeared in all three subclasses of crinoids (Camerata, Flexibilia, and Inadunata) during the Devonian and characterized more than 80% of Carboniferous and Permian species. Mesozoic articulate crinoids were also dominantly thick-plated and rigid. Only with the appearance of the articulate order Comatulida in the Pliensbachian stage of the Early Jurassic did the rigid, highly fortified crinoid types begin to make way for mobile, more lightly built forms.

Echinoderms in which the main body is enclosed in a theca show a general trend toward increasing protection of respiratory structures through time (Paul, 1977). Respiration in early echinoderms seems to have been accomplished by passive gas exchange at the surface of the theca, either in epispires (external openings of the coelomic cavities between thecal plates, seen in all Early Cambrian pelmatozoan and eleutherozoan echinoderms) or in podia that penetrated between plates (as in all three Early Cambrian helicoplacoids). Passive gas exchange requires that a large surface area of thin-walled tissue be exposed to an oxygen-rich medium; it is therefore basically incompatible with the development of tightly enclosing armor. Active gas exchange evolved first during the Tremadocian stage of the Early Ordovician in diplopore cystoids. Deoxygenated fluid rises up one canal to the peripore, where gas exchange takes place; water then re-enters the body cavity at the peripore through a second canal. The peripore, located on the external surface of the theca, probably was covered by tissue during life and generally was not protected by spines or plates. The Late Ordovician to Middle Silurian diploporan family Holocystitidae and the Ordovician fistuliporite rhombiferans did evolve calcified pore structures, but even these protective devices had a geologically short duration. Bothriocidarids (ranging from the Llanvirnian stage of the Middle Ordovician to the Ludlovian stage of the Late Silurian) and echinoids effectively protect respiratory and other soft tissues on the external surface of the theca (test) with spines. The earliest (Late Ordovician) echinoids had spines that were located in pits rather than on tubercles as in most later forms, and the tube feet were not surrounded or enclosed by ambulacral plates. Tubercles and complete enclosure first evolved during the Late Silurian, and long spines appeared for the first time with the origin of the archaeocidarids in the Tournaisian stage of the Early Carboniferous (Kier, 1965; A. B. Smith, 1984). The Early Carboniferous also marked the appearance of urchins which, by virtue of the presence of suckered aboral tube feet and uniformly short aboral spines, are able to cover themselves with debris, presumably in order to camouflage themselves. This habit, which is widespread in echinoids today, is known first in the Early Carboniferous genus *Lepidesthes* (A. B. Smith, 1984). The fact that sea urchins persisted to the Recent whereas well-armored cystoids lasted only until the end of the Middle Devonian is perhaps attributable to the better protection of surficial tissues in some Late Paleozoic and most Mesozoic and Cenozoic echinoids (Paul, 1977).

A potentially more effective mode of protecting tissues involved in active gas exchange evolved in rhombiferan cystoids, which are known first in the Tremadocian stage of the Early Ordovician, and independently in the blastoids (Middle Silurian to Late Permian). These animals had hydrospires, structures within the theca that communicated to the outside via narrow slits. Why this form of protection of respiratory structures died out with the extinction of the blastoids remains an intriguing mystery.

Sea urchins are the only thecate echinoderms to have persisted beyond the Paleozoic. Their history, which has been beautifully documented by Kier (1965, 1974, 1982) and A. B. Smith (1984), follows along two divergent lines: rigidification and increased armor among urchins living on open, hard surfaces and reduction of armor in urchins that burrowed in sand and mud. Each line of development will be discussed in turn.

Most pre-Jurassic sea urchins had thecae composed of numerous simple (elemental) plates, whereas most later forms had thecae in which two or more elemental plates coalesced around a tubercle to form a compound plate. This new arrangement, which is first seen in the Carnian (Late Triassic) *"Cidaris" dilleri*, permitted the evolution of sturdier spines and allowed for a higher density of tube feet. Kier (1974) also believes that the evolution of compound plates led to the strengthening of the test, but Strathmann (1981) disputes this assertion on the basis of his observations of the destruction of tests when a weight is dropped on them from above. Strathmann found that breaks tended to occur within rather than between plates, especially in such relatively fragile-shelled genera as *Lytechinus* and *Diadema*, and that the spines reduced and spread the impact of the force on the test. Only in the thick-walled *Eucidaris*, a stout-spined echinometrid, did the test disintegrate along plate boundaries. Strathmann proposes that a flexible test is better able to withstand impacts than a rigid one is, and that the spines confer the flexibility that in most Paleozoic and some Recent deep-water forms is conferred by imbrication of the plates. Although this interpretation is likely to be correct, the type of impact Strathmann studied may not be the kind of breakage to which sea urchins are exposed in life. Destruction of the tests by biting or crushing fishes, crabs, hermit crabs, and sea otters is perhaps resisted by very different skeletal properties than those that resist impact. A flexible or yielding surface is highly effective against a blow (Currey et al., 1982), whereas an unyielding rigid surface is perhaps better against concerted crushing or biting. Future research must resolve the differences between Strathmann's and Kier's interpretations by concentrating on the ways sea urchin tests disintegrate when loads of various magnitudes and types are applied with and without the presence of spines on the test.

The number of plates making up an echinoid compound ambulacral plate increased from two in the Late Triassic *"Cidaris" dilleri* to three in the Hettangian (Early Jurassic) *Diademopsis*, *Palaeopadina*, and *Jeannetia* to four in the later (Toarcian) *Pseudodiadema*. Whereas pores were lined up in a vertical row in pre-Jurassic echinoids, a triserial arrangement became possible as the number of elemental plates constituting a compound plate exceeded two (Kier, 1974).

The junction between plates also underwent modification. Temnopleurids (Late Cretaceous to Recent) were the first urchins whose adjacent plates were joined by peg-and-socket articulations, which spread loads throughout the test. This form of articulation is characteristic of most warm-water Cenozoic epifaunal echinoids (A. B. Smith, 1984).

The second theme in post-Paleozoic sea urchin history is the reduction of armor as echinoids invaded unconsolidated bottoms. This path culminated in the evolution of irregular sea urchins, which depart from other echinoids in being able to burrow and feed beneath the surface of sands and muds. Reduction in armor can already be seen in Early Jurassic (Late Hettangian or Early Sinemurian) species of *Plesiechinus*. Although these urchins were not yet burrowers, they were capable of covering themselves with camouflaging debris that adhered to the dorsal (aboral) part of the test through the action of suckered tube feet. Modern urchins that characteristically cover themselves, such as *Lytechinus* and *Tripneustes*, have thinner tests than urchins that do not cover themselves. The path toward reduced armor may thus have been initiated in lineages that relied more on camouflage than on resistance as their chief antipredatory defense. *Plesiechinus*, or an urchin much like it, gave rise in the Toarcian stage of the Early Jurassic to *Eogaleropygus*, the first urchin capable of burrowing in coarse unconsolidated sediment. The tests of most burrowing urchins are delicate, and the spines are usually short and thin. The sediment rather than the test serves to protect the urchins from their enemies (Kier, 1982; A. B. Smith, 1984).

Despite their burrowing habits, sand dollars (order Clypeasteroida re-evolved sturdy, highly resistant tests. The first known member of the order is *Togocyamus*, a Paleocene genus from West Africa. By the Middle Eocene, the tests of *Periarchus* and *Protoscutella* possessed wedges and thorns that served to interlock adjacent plates; they also had internal struts that enabled the strongly flattened, cookie-shaped tests to resist predation (Seilacher, 1979; Kier, 1982). Burial itself, however, continued to be as important for the protection of sand dollars as it was for the thin-walled irregular urchins that had evolved earlier. Ghiold (1984) has pointed out that most members of the clypeasteroid suborder Scutellina feed on particles from the surface of the sediment by collecting the particles from below rather than from above, as most surface-dwelling animals do. The typical sand dollars belonging to the Scutellina have developed food-gathering structures on the upper (aboral) side, whereas the mouth on the underside has become less important in feeding.

The functional morphology and history of the other living classes of echinoderms have not been as well studied, but these classes seem to have emphasized three themes: armor, autotomy (voluntary loss of parts when the animal is threatened), and toxicity. Each of these methods of defense is particularly well developed in species from shallow tropical waters.

A heavily calcified sturdy skeleton capable of resisting both large and small predators has evolved in many tropical sea stars (class Stelleroidea), especially in members of the valvatidan families Oreasteridae, Ophidiasteridae, and Goniasteridae and in the Late Jurassic genus *Sphaeraster* (D. B. Blake, 1983, 1984). An important feature of these sea stars is that a very firm body wall, composed of numerous well-connected overlapping ossicles and plates and a thick dermal layer, still permits a degree of flexibility of the body and arms. This flexibility enables the sea star to fit snugly into shelters and to cling to irregular surfaces as it feeds on immobile epifaunal animals such as corals, sponges, and bryozoans. *Sphaeraster* had a more rigid shell-like calcified armor and represents a type of construction unknown in living sea stars. Another important feature of the modern heavily armored families is that the ambulacral furrows and mouth are small and well protected by ossicles and spines. These protective devices severely limit the ability of the animal to adhere to the substrate and to open shell-bearing prey by the use of force. D. B. Blake (1983) believes that this limitation may be responsible for the near absence of molluscivorous sea stars in the shallow-water tropics, where predators and other enemies posing risks to relatively unarmored sea stars would be especially common. He points out that the only molluscivorous tropical sea stars are astropectinids and luidiids, whose habit of burrowing shallowly into sand and mud may compensate for the lack of armor as a means of protection.

Little is known about the effectiveness of armor in sea stars. Wickler and Seibt (1970) showed that the armor of *Linckia*, an Indo-West-Pacific ophidiasterid, limits predation by the shrimp *Hymenocera*, but they gave no quantitative information on the effectiveness of *Linckia*, and no data are available on the fate of less heavily armored sea stars.

The second theme, which is very common in brittle stars (ophiuroids), crinoids, and to a lesser extent sea stars, is the development of a highly flexible skeleton that promotes more or less rapid locomotion and permits some species to autotomize part of the body. This trend has been documented best, by Meyer and Macurda (1977), in living stalkless comatulid crinoids. In comatulids, the theca has been reduced to insignificance, and the oral disc is protected only by the proximal pinnules (branchlets) of the highly flexible arms. The pinnules may be stiffened or spine-bearing, especially in Western Pacific species, but most other skeletal features of comatulids suggest an emphasis on flexibility, locomotion and autotomy. A large majority of ophiuroid and crinoid individuals have arms that either have regenerated or are in the process of regenerating after autotomy or partial removal by enemies (Bowmer and Keegan, 1983; Mladenov, 1983).

The third theme in living echinoderms is exemplified by the sea cucumbers (Holothuroidea), in most of which the skeleton has been reduced to small spicules in the body wall. These animals rely on chemical deterrents, retaliation (including the deployment of Cuvierian tubules, which form a sticky mass of threads when thrown out into the water), and various forms of locomotion as defenses against enemies (Bakus, 1973, 1974; Bakus and Green, 1974; Margolin, 1976; Kropp, 1982). Holothuroids were already established by the Middle Ordovician (Pawson, 1980), but the history of skeletal reduction in this group remains completely unknown.

In summary, most Paleozoic echinoderm groups showed a trend toward better armor. This trend began in the Middle Cambrian, with important additions coming in the Early and Middle Ordovician, Early Silurian, and especially the Devonian and Early Carboniferous. Echinoids and some sea stars continued this trend during the Mesozoic and especially during the Cenozoic. Other groups (especially crinoids), however, came to rely either on locomotor deterrence or on forms of resistance other than armor. Future work should concentrate on measuring the effectiveness of echinoderm armor, documenting historical trends at the level of the local assemblage, and assessing with greater temporal accuracy the dynamics of adaptive change.

ARMOR IN VERTEBRATES

The last group of articulated animals whose armor I shall consider is the phylum Chordata, which contains the vertebrates. Although vertebrates with a rigid outer covering of scales, bony plates, or scutes constitute a minority of living species, they have evolved in many groups (Table 12.2) that have been, and remain, highly successful.

The basic element of vertebrate armor is the odontode, a scale-like structure composed of a cap of enamel, a middle layer of dentine, and a basal layer of dermal bone. These odontodes either coalesce to form larger units or form the superficial layer of dermal bone plates. During the course of evolution, the enamel cap and dentine layer have often been lost, so that only the bone layer remains (Reif, 1982).

I agree with Reif (1982) that the primary function of vertebrate armor is protection against enemies, and that predation was the agency of selection chiefly responsible for the evolution of armor in primitive fishes. Romer (1933) was among the first to draw attention to the antipredatory function of armor in Late Silurian and Early Devonian fishes. He pointed out that armored members of the Agnatha often occur in fossil deposits with large eurypterid arthropods. Kjellesvig-Waering (1961) expanded on this theme, showing that some armored fishes co-occurred with large Early Devonian pterygotid eurypterids in which the big pointed teeth of the grasping organs (chelicerae) were well worn. That this armor was effective is indicated by Reif's (1982) observation that the surface of heterostracan armor was often abraded and healed, presumably as the result of attacks by predators. In his studies of the anterior plate armor of the stickleback *Gasterosteus*, Reimchen (1983) has shown that the plates provide support for the dorsal and pelvic spines, which are sometimes broken and displaced laterally as predators such as trout attempt to subdue the stickleback. The armor also protects the internal organs from being compressed and damaged by would-be attackers. That the carapace of turtles is effective against some predators is suggested by Erickson's (1984) findings of healed punctures in the dorsoposterior parts of the carapace of Paleocene turtles that were apparently attacked by *Leidyosuchus*, a large crocodile with pointed teeth. In none of these studies have estimates been made of the number of individuals that survived attacks, nor is anything known about the forces used by predators or the resistance offered by the external armor of potential vertebrate prey.

An alternative interpretation of the function of vertebrate armor is expressed by Gutmann (1967), who holds that the primary function of armor in primitive fishes is to stiffen the cross-sectional shape of

TABLE 12.2   Survey of Armored Vertebrates

Class Agnatha: marine and some freshwater, Late Cambrian to Recent; armor in Heterostraci (Late Cambrian to Late Devonian) and Osteostraci (Late Silurian to Late Devonian), but loss of armor in both groups and in Cyclostomata (lampreys and hagfishes), Late Carboniferous to Recent (Romer, 1966; Repetski, 1978; Reif, 1982)

Class Placodermi: marine and freshwater, Early to Late Devonian; most groups armored, but loss or reduction of armor in some Athrodira (Early to Late Devonian), Rhenanida (Early Devonian), and freshwater Antiarcha (Middle to Late Devonian) (Denison, 1978)

Class Acanthodii: marine and freshwater, Early Silurian to Late Permian; less heavily armored than other Middle Paleozoic vertebrates, and some late freshwater forms without armor (Denison, 1979)

Class Osteichthyes (bony fishes): Early Devonian to Recent

Subclass Dipnoi (lungfishes): early ones marine, most later forms freshwater; loss of armor beginning in Early Carboniferous (Westoll, 1949)

Subclass Actinopterygii (ray-finned fishes) (Lauder and Liem, 1983)

Infraclass Cladistia, Order Polypteriformes, including bichirs of the family Polypteridae: mainly freshwater, Late Paleozoic to Recent (Greenwood, 1984)

Infraclass Chondrostei, Order Acipenseriformes (sturgeons): mainly freshwater but some marine, Late Cretaceous to Recent; loss of armor in Polyodontidae (paddlefishes) (Gardiner, 1984)

Infraclass Ginglymodi, Lepisosteidae (gars): mainly freshwater, Cretaceous to Recent (Wiley, 1976)

Infraclass Halecostomi

Teleosti: Middle Triassic to Recent

Superorder Ostariophysi, Order Siluriformes (catfishes), Callichthyidae, Loricariidae: freshwater, Late Cretaceous to Recent (Reif, 1982)

Superorder Acanthopterygii (spiny-rayed fishes)
Order Gasterosteiformes: includes the mainly marine Syngnathidae (pipefishes), Early Eocene to Recent

Table 12.2 (*continued*)

Order Scorpaeniformes, Agonidae: temperate marine poachers

Order Tetraodontiformes, Ostraciidae (trunkfishes): marine, Eocene to Recent (Tyler, 1980)

Class Amphibia: Late Devonian to Recent
Order Labyrinthodontia: terrestrial, Late Carboniferous to Permian (Dissorophidae); terrestrial, Late Triassic (Plagiosauridae, *Gerrothorax*) (DeMar, 1966, 1968; Romer, 1966)

Class "Reptilia": Early Carboniferous to Recent

Subclass Anapsida

Order Procolophonia, Infraorder Pareiasauria: terrestrial plates on back, Middle to Late Permian (Romer, 1966)

Order Placodontia: marine, Early to Late Triassic (Westphal, 1975)

Order Chelonia (turtles): marine, freshwater, Middle Triassic to Recent; terrestrial since Late Cretaceous; some loss of armor in marine forms (Bramble, 1974; Pritchard, 1979; Bramble et al., 1984)

Order Squamata, Suborder Lacertilia (lizards): Late Permian or Triassic to Recent; flexible armor in Scincidae, some Anguidae (Early Cenozoic *Placosaurus*) (Estes, 1983)

Subclass Diapsida: Late Permian to Recent

Order Thecodontia, Suborder Aetosauria: terrestrial, Late Triassic (Romer, 1966)

Order Crocodilia: marine, freshwater, and some terrestrial, Middle Triassic to Recent; reduction of armor in some marine forms (Steel, 1973)

Order Ornithischia, Suborder Ankylosauria: terrestrial, Middle Jurassic to Late Cretaceous (Coombs, 1978b; Galton, 1983)

Class Mammalia
Order Xenarthra, Infraorder Loricata: terrestrial, Paleocene to Recent (armadillos) (Romer, 1966; Edmund, 1985)

---

the body in all directions except the horizontal plane in which undulatory movements for swimming are made. At the time external armor evolved, vertebrates lacked a stiff internal skeleton, which in many modern fishes provides the stiffening and muscle support regarded by Gutmann as advantageous. Although I believe that stiffening of the body was an important consequence of the development of external armor, I doubt that it can be construed as the chief benefit of armor.

It is a pity as well as a matter of some surprise that armored vertebrates have been so little studied from the point of view of protection against enemies. We have essentially no understanding at all of the ecological risks associated with damage to the external armor. Something is known, however, about how the possession of heavy armor affects other functions. The incompatibility between armor and other requirements suggests that armor in vertebrates has only a limited potential as an effective defense.

Like echinoderms, vertebrates cannot seal the inner organs tightly within the external armor, but the reasons for this inability are quite different. In echinoderms, the skeleton is fundamentally porous, whereas in the vertebrates permanent openings remain between the skeletal elements to allow for the protrusion of the head and appendages. The closest approximation to a tight seal occurs in certain freshwater turtles (Pelomedusidae, Trionychidae, Kinosternidae, and Emydidae—all of Paleocene to Recent age), in which the ventral plate (plastron) contains a hinge so that one or both ends of the shell can be closed off (Bramble, 1974; Bramble et al., 1984). Even in these turtles, however, the hindlimbs protrude through permanent openings. In many vertebrates, the armor is concentrated dorsally. This is true for the South American catfish families Callichthyidae and Loricariidae, in which the ventral side is often naked, and for most Cenozoic crocodiles (Steel, 1973).

As in other animals, the development of heavy armor carries with it limitations on speed. Not only does armor typically make fishes heavier than sea water, but it prevents animals from accelerating rapidly, although maneuverability is apparently not impaired (R. W. Blake, 1976, 1977, 1981). Just how profoundly armor affects speed is illustrated by the trunkfishes of the family Ostraciidae, whose body is encased in a box-like armor of bony plates. Their maximum speed, as recorded by R. W. Blake (1981), is only about 12 cm/sec. Sea horses of the syngnathid genus *Hippocampus* typically travel at a speed of 5 cm/sec (R. W. Blake, 1976). These speeds are strikingly low compared with those of fishes with more flexible, less armored bodies, such as the anchovy *Engraulus* (162 cm/sec), the mullet *Mugil* (442 cm/sec), or even the gobies of the genus *Gobius* (27 cm/sec) (Beamish, 1978).

Turtles show the same pattern. Young marine green turtles (*Chelonia mydas*), which have a flexible shell 11 cm long and 8 cm wide in which the head and appendages cannot retract, can travel at speeds as high as 143 cm/sec. Freshwater turtles of comparable size (*Mauremys caspica, Chrysemys scripta,* and *Kinosternon subrufum*) have rigid shells into which they can retract the head and appendages completely, but they can attain speeds of no more than 23 cm/sec (Davenport et al., 1984).

Some armored fishes have partially overcome the adaptational conflict between armor and speed by developing a flexible armor of overlapping plates or scales. This type of armor, which permits lateral displacement of parts of the body during swimming, is seen in the sticklebacks and in gars of the family Lepisosteidae (Wiley, 1976; Reimchen, 1983).

Heavy armor also seems to be associated in vertebrates with a low metabolic rate. This is perhaps most compellingly demonstrated in mammals, few of which have external plate armor. Armadillos of the family Dasypodidae and pangolins of the order Pholidota have relatively low body temperatures (below 35° C) and low metabolic rates (McNab, 1980). There are very few armored forms among endothermic vertebrates, whose metabolic rates exceed those of ectotherms by an average factor of about six (Pough, 1980). Among turtles, too, basal metabolic rates are higher in fast-swimming species with reduced armor (Trionychidae and Cheloniidae) than in the more sedentary and more heavily armored groups such as the Emydidae and Testudinidae (Davenport et al., 1984).

My impression from reviewing the history of armor in vertebrates is that armor has become ecologically more restricted in this group through the course of time, especially in aquatic environments. That armor has become reduced or lost in many marine and freshwater groups has been well known for more than one hundred years, but the conditions that favored such loss remain poorly understood. Jawless (agnathan) fishes were well-armored during the Silurian and Devonian, but the two surviving groups (hagfishes of the order Myxinoida and lampreys of the order Petromyzonida), which already existed in the Carboniferous, are naked. Some early amphibians, notably the Permian Dissorophidae and some Triassic Plagiosauridae, were heavily armored dorsally (DeMar, 1966, 1968), but among living amphibians only the caecilians of the order Caecilia or Gymnophiona have remnants of armor in the skin (Reif, 1982). Lungfishes of the osteichthyan subclass Dipnoi appeared in the Early Devonian with thick body scales, but by Carboniferous time the scales were already thin. The three living genera have the thin scales immersed in the skin and therefore effectively lack armor (Westoll, 1949). Many osteichthyan bony fishes of the Paleozoic were more or less armored, but through the course of time most lineages have undergone a reduction in the extent of armor. Good examples of this reduction include the gars (family Lepisosteidae) and sturgeons and paddlefishes (order Acipenseriformes). The paddlefishes have lost all traces of external armor (Wiley, 1976; Gardiner, 1984).

Reduction and loss of armor have been especially prominent in marine lineages. Modern marine turtles of the families Dermochelyidae and Cheloniidae (Late Cretaceous to Recent), for example, differ from their probable ancestors by having nonretractile head and limbs and relatively flexible shells (Romer, 1966; Pritchard, 1979). The Thalattosuchia, a suborder of fish-eating marine crocodiles of Late Jurassic to Early Cretaceous age, had strongly reduced armor compared with their freshwater ancestors and with contemporaneous freshwater Bernissartiidae and later Crocodylidae (Steel, 1973). Relict armored osteichthyan fishes, such as sturgeons, gars, and bichirs (Polypteridae), are found either exclusively or chiefly in fresh water, and even then they typically show less armor than did their Paleozoic ancestors.

Living armored vertebrates fall into three ecological groups: (1) small slow-moving fishes in cluttered marine habitats such as reefs and areas of dense vegetation, especially in the tropics and warm-temperate zones; (2) freshwater fishes, crocodiles, and turtles, often of large size; and (3) slow-moving terrestrial turtles, lizards, and mammals, often of large size. Large marine armored forms are still represented by a few turtles, but this ecological group was apparently much more prominent during the Paleozoic. At that time, there were marine armored Agnatha, Placodermi, and some Dipnoi, Acanthodii, and Osteichthyes. In the Triassic, placodont reptiles were large armored marine forms. Modern armored marine fishes are mostly small animals such as trunkfishes (Ostraciidae), pipefishes (Syngnathidae), and the cold-water poachers (Agonidae), all of which are of Cenozoic origin (Tyler, 1980; Lauder and Liem, 1983). In fresh water, the relict armored vertebrates tend to be large (1 m or longer), whereas the large number of South American armored catfishes, which originated during the Late Cretaceous or Cenozoic, are mainly small species less than 30 cm long. Geographically, the most widespread group of freshwater armored vertebrates is the turtles. Mechanisms to close the shell anteriorly and posteriorly have evolved independently in several families during the Cenozoic. In some groups, notably some American species of *Kinosternon* which make occasional excursions on land, considerable protection against predators has been achieved. In *K. scorpioides*, for example, the margin of the ventral part of the shell (plastron) is recessed within the dorsal part (carapace), and the plastron prevented from being forced against the carapace during a predatory attack by the presence of buttressing ridges on the pygal and peripheral bones that underlie the carapace (Bramble et al., 1984).

On land, heavily armored vertebrates with ponderous gaits evolved in the Permian (dissorophid amphibians and pareiasaur reptiles), Triassic (aetosaur reptiles and plagiosaurid amphibians), Middle Jurassic (the first ankylosaur dinosaurs), and Early Cenozoic (testudinid tortoises and loricate xenarthrans, or anteaters). Whereas in the Late Paleozoic and Mesozoic these armored forms were found worldwide, during the Late Cenozoic and today the very large terrestrial armored vertebrates are known chiefly from islands and from the small island-like continents of South America and Australia.

These historical patterns in vertebrate armor suggest that armor has become ecologically restricted, especially in large animals weighing more than 10 kg. I interpret this trend to mean that armor has not proved to be an effective defense for most large marine and terrestrial vertebrates, save for those with low metabolic rates. Only in freshwater habitats and on small land masses, where competition may be generally less intense than elsewhere, are very large armored vertebrates still reasonably successful. In the sea, armored fishes are successful if they are small and are capable of hiding in cluttered environments. Although the timing and extent of this ecological restriction are still far from being well established, restriction seems to be concentrated during the Mesozoic. Vertebrates therefore seem to parallel cephalopods, crinoids, and many clades of pelecypods in that elaboration of armor occurs in only the most sedentary groups.

SWIMMING IN VERTEBRATES

Swimming has been a characteristic form of locomotion in vertebrates ever since the fish-like vertebrates evolved during the Late
Cambrian or Ordovician. Although the habit is primitive in the fish-
like classes, it has re-evolved in amphibians, reptiles, and mammals
as well. Swimming has been well described in some living vertebrates, but work on the precise relationships between form and performance has only just begun, so that statements about the capacities of fossil vertebrates are by necessity still imprecise and
provisional.

Water-adapted vertebrates have evolved a great diversity of swimming styles. These have been discussed at length by R. M. Alexander
(1977b) and P. W. Webb (1982, 1984) but the most comprehensive
classification of swimming modes is that of Braun and Reif (1985).
Propulsive force is generated by the whole body, the tail (caudal fin),
the unpaired fins (anal and dorsal fins), or the paired fins (pectoral and
pelvic fins of fishes, forelimbs and hindlimbs of tetrapods). The most
primitive mode of swimming in vertebrates is referred to as the axial
undulatory mode, in which several propulsive waves travel backward along the long slender body. This mode characterizes not only
many eel-like fishes, but also sea snakes and the salamander *Siren*.
The body is flexed either laterally or dorsoventrally. The eel *Anguilla*
can achieve a speed of 1.7 m/sec with this mode of swimming. In
more advanced forms of axial swimming, the waves are confined to
the posterior part of the body while the anterior part is stiffened. The
fastest swimming vertebrates (scombrid mackerels, thunnid tunas,
lamnid sharks, and cetacean whales) have a highly streamlined lu-
nate tail attached to the body by a narrow flexible peduncle, so that
drag is low. Among these vertebrates, the fastest species are those
that maintain high body temperatures. The thunnids *Acanthocy-
bium* (wahoo) and *Thunnus* (tuna) attain the astonishingly high
speed of 21 m/sec (Beamish, 1978). The spotted porpoise (*Stenella at-
tenuata*) reaches a speed of 11 m/sec after only 2 sec of acceleration
(Lang, 1974), and the sperm whale (genus *Physeter*) may reach similar
speeds during short bursts (Norris and Mohl, 1983). Some penguins,
such as Adelie's penguin (genus *Pygoscelis*), which swim by means
of the modified wings, attain a top speed of 10 m/sec (R. W. Blake,
1981), but other species are slower (Clark and Bemis, 1979).

Axial oscillatory swimming may have evolved as early as the Famennian stage of the Late Devonian, when the shark *Cladoselache* may have exemplified this mode of locomotion, but Reif believes this shark was not a rapid swimmer. It was not until the Mesozoic that rapid axial oscillatory swimming was evolved. Lamnid sharks arose during the mid-Cretaceous, whereas scombroid teleosts and cetacean mammals are of Cenozoic derivation. It is unlikely that fossil reptilian groups attained the speeds of these high-energy fishes and mammals. Mesozoic plesiosaurs probably used both pairs of limbs for swimming (a form of underwater flying) and may have been faster than contemporaneous ichthyosaurs, mosasaurid lizards, and metriorhynchid crocodiles (Braun and Reif, 1985).

Most swimming animals have not emphasized high speed but instead have increased such other components of locomotion as rapid acceleration and maneuverability. Unsteady swimming of this kind is inferred in many fossil vertebrates, including predaceous Paleozoic palaeoniscoid bony fishes. Early unsteady swimmers, however, had a heavy fin-ray skeleton, diffuse and weak supports of the anal and dorsal fins, an unrestricted notochord, weak neural and haemal arches, and a heavily armored body, so that acceleration drag was probably high and thrust slight. Greater thrust force became possible in Mesozoic halecostome and teleost bony fishes through reduction of the fin-ray skeleton, greater flexibility of the fins, abbreviation of the notochord, and reduction of external armor (P. W. Webb 1982). Unsteady swimming is especially common in the acanthopterygian (spiny-rayed) teleosts of the Late Cretaceous and Cenozoic. It is typically accompanied by retaliatory defenses such as spines, secondary external armor, and toxicity (Patterson, 1964; Kaufman and Liem, 1982; P. W. Webb, 1982, 1984). Just how these modern slow swimmers differed from ecological antecedents in locomotor performance and mechanical defenses remains unknown.

RUNNING IN LAND ANIMALS

Both the arthropods and the vertebrates have evolved the ability to run rapidly on land. In the arthropods, running is associated with long appendages, short pace durations, and a lightweight body. The highest recorded speeds for various arthropod groups are as follows: 0.28 m/sec for millipedes (genus *Polydesmus*); 0.42 m/sec for centipedes (genus *Scutigera*); 0.59 m/sec for spiders of the Araneae (genus *Tegenaria*); 0.62 m/sec and 1.15 m/sec for coleopteran beetles (the carabid *Cicindela* and the tenebrionid *Onymacris*, respectively); 1.30 m/sec for dictyopteran cockroaches (genus *Periplaneta* at 35° C); and 4.00 m/sec for the arachnid order Solpugida (genus *Galeodes*) (Manton, 1952, 1954; Evans and Forsythe, 1984; Nicolson et al., 1984). Little is known about the evolution of rapid locomotion in running arthropods, except that carabid beetles closely resembling living fast-running Nebriinae already existed during the Mesozoic (Evans and Forsythe, 1984).

Vertebrates have achieved far higher speeds than most arthropods. Small ectothermic scincid and agamid lizards in the Kalahari Desert and in Israel run at speeds of 1.6 to 2.5 m/sec (Hertz et al., 1982, 1983; Huey, 1982). Large inertial homeotherms and endotherms which are morphologically unspecialized for running can still achieve impressive speeds: 5.0 m/sec in the Komodo dragon lizard (*Varanus komodoensis*), 11.0 m/sec in elephants, and 12.0 m/sec in humans (R. M. Alexander, 1977a; Auffenberg, 1981; Garland, 1983). Specialized runners attain remarkable speeds. Examples include the ostrich *Struthio camelus* (17.0 m/sec), hares of the genus *Lepus* (20.0 m/sec), the spotted hyena *Crocuta crocuta* (17.0 m/sec), and the cheetah *Acinonyx jubatus* (31.0 m/sec) (Kruuk, 1972; Schaller, 1972; R. M. Alexander et al., 1979; Garland, 1983). Many smaller bovid ungulates, such as gazelles, exceed 25.0 m/sec in running speed, and even mouse-sized rodents typically run at speeds of 3.5 m/sec or faster. Such speeds exceed those of similarly sized lizards.

Two kinds of data can yield evidence about the speeds of fossil vertebrates. These are measurements from tracks left in soft sediments by running animals and anatomical information. R. M. Alexander (1976) pioneered the quantitative study of vertebrate trackways. If a trackway of a single individual can be followed for several strides, an estimate of speed can be obtained by applying the following relation, which is based on mammals ranging in size from horse to man:

$$\lambda/h = 2.3(u^2/gh)^{0.3}, \tag{12.1}$$

where λ is stride length (distance between corresponding points on successive prints of the same foot), h is the height of the hip (pivot of the leg) above the ground, u is the animal's speed, and g is the gravitational constant. The interpretation of the calculated speed depends not only on the applicability of this equation to vertebrates other than large mammals, but also on the conditions under which the trackway was produced. If the animal was merely ambling along, its calculated speed is obviously well below top running speed. Moreover, the soft terrain that is most apt to preserve trackways is not conducive to fast running.

Nevertheless, several speeds of dinosaurs have been estimated using Alexander's method. Two Late Triassic dinosaur trackways, perhaps made by bipedal *Anchisaurus*, indicate a strolling pace of 1.25 and 2.20 m/sec, respectively (Tucker and Burchette, 1977). A giant Campanian (Late Cretaceous) ornithischian from Colorado, whose track was discovered and described by Russell and Beland (1976), walked at a speed of 2.40 m/sec, according to Thulborn's (1981) interpretation. The fastest speed calculated from trackways is 12 m/sec for an Aptian or Albian bipedal carnivorous theropod saurischian in Texas (Farlow, 1981). Only one estimate of dinosaur speed can reasonably be interpreted as running speed. In the Winton Formation of Early Cenomanian (Early Late Cretaceous) age in Queensland, hundreds of tracks of coelurosaurs and ornithopods are associated with the track of a single *Tyrannosaurus*-like carnosaur. It had an estimated speed of 2.26 m/sec. This predator apparently provoked a stampede of the other dinosaurs, which fled from the approaching carnosaur in a northeasterly direction at speeds of 3.62 to 4.31 m/sec (Thulborn and Wade, 1979).

Bakker (1971, 1980, 1983) and Coombs (1978a) have summarized the anatomical and physiological features that are associated with, and therefore serve as indicators of, the capacity for rapid running in land vertebrates. These features include (1) long limbs relative to other body dimensions; (2) an erect posture, in which the long axis of the limbs is more or less perpendicular to the long axis of the rest of the body, so that the belly is held off the ground; (3) hinge-like joints in the limbs, so that only fore-aft movement is permitted and movement in other directions is impossible; (4) limbs with a short, massive proximal bone (femur in the hindlimb, humerus in the forelimb) and long, slender distal elements; (5) fusion of the bones of the lower hindlimb (tibia and fibula) and forelimb (radius and ulna) into single units; and (6) symmetrical form of the foot (pes) and hand (manus), with the lateral fingers and toes reduced or absent. High speed and endurance are also associated with high metabolic rates and high activity temperatures. Bakker (1971), for example, notes that short-limbed lizards such as geckoes have low activity temperatures and are active at night, whereas fast lizards such as agamids have long limbs, maintain high activity temperatures (34 to 40° C) by sunning themselves, and are active during the day. Large reptiles maintain high temperatures by virtue of the slow loss of body heat. Birds and mammals are strictly endothermic, producing heat in various tissues in the body.

Some vertebrates rely more on modifications of the vertebral column than on aptations of the limbs to attain high running speeds. In the cheetah and other fast carnivores, for example, the vertebral column is very flexible, so that stride length is increased even though the limbs are not as highly modified as are the limbs of fast ungulates (Bakker, 1983). Morphological criteria, therefore, are not always reliable indicators of an animal's capacity.

In an attempt to estimate top running speeds of dinosaurs from body proportions, Coombs (1978a) used a relation between speed and the hindlimb dimensions that was proposed by Bakker (1975) for large animals weighing 10 kg or more. This relation is

$$V = 4.132RHL - 14, \tag{12.2}$$

where V is velocity and RHL (relative hindlimb length) is the sum of the lengths of the femur, tibia, tarsus, and longest metatarsal, divided by the cube root of body mass. Coombs's calculations are consistant with other anatomical criteria indicating high speed; they show that the fastest dinosaurs (ornithomimid coelurosaurian ornithischians) ran at a speed of 27 m/sec, which is comparable to the speed of the fastest living mammals. Even the huge Late Jurassic brachiosaurid sauropods, weighing some 78,000 kg, could travel at a speed of 13 m/sec, and the renowned predaceous Late Cretaceous theropod *Tyrannosaurus* could attain a speed of 18 m/sec, according to Coombs's calculations.

Bakker (1972, 1975, 1980, 1983) has traced the history of terrestrial locomotion in vertebrates. According to his interpretation, the earliest land vertebrates (amphibians of the Late Devonian and Early Carboniferous) were faster than their dipnoan ancestors. The first land reptiles—cotylosaurs and pelycosaurs of the Late Carboniferous

and Early Permian—had short limbs and might have had performance levels similar to those of the living Komodo dragon. Mammal-
like reptiles of the order Therapsida replaced pelycosaurs and other
reptiles as the common large herbivores and predators in the Late
Permian. Despite their probable capacity to maintain constant high
body temperatures (McNab, 1978, 1983), therapsids may not have
been significantly faster than pelycosaurs, because they maintained
the sprawling gait of their antecedents. No Permian or Triassic therapsids had long limbs or other features that in modern vertebrates
would indicate the capacity to run rapidly. With the rise of thecodonts in the Early Triassic, running performance probably increased.
At least three vertebrate groups—rauisuchiid thecodonts, terrestri-
suchid crocodiles, and saurischian dinosaurs—evolved erect gaits,
and therefore the capacity for more sustained running, during the
Triassic (Bakker, 1971; Bonaparte, 1984; Crush, 1984). Although two
of these groups became extinct during the Triassic, the dinosaurs became the dominant large land vertebrates of the Jurassic and Cretaceous. Within the ranks of the dinosaurs are several forms that can
be described as subcursorial (large ornithopods, theropods, and ceratopsians) or cursorial (small theropods and ornithomimid coelurosaurs). Hopson (1977) and Coombs (1978a) believe that, with the exception of coelurosaurs, most dinosaurs could not achieve the fast
speeds of many living mammals such as lagomorphs (hares), ungulates, and carnivores. Coombs points out, for example, that the tibia
and fibula of dinosaurs were always separate elements of more or less
equal size in the hindlimb, although in *Albertosaurus* of the Late
Cretaceous the fibula was embraced by the tibia so that the lower leg
functionally became a single unit. Moreover, most dinosaurs (again
with the exception of some theropods and all coelurosaurs) had relatively small brains, a condition which, in living vertebrates, Hopson
(1977) associates with poor development of sensory and motor skills
and with low speeds. Bakker (1980), on the other hand, believes that
the performance of dinosaurs was generally similar to that of Cenozoic and living mammals. Although I agree with Bakker that dinosaurs had high levels of activity, I am inclined to accept Hopson's and
Coombs's interpretation that most dinosaurs did not run as quickly
as do most living mammals.

There is general agreement that cursorial dinosaurs became extinct at the end of the Cretaceous, and that the first cursorial mam-

mals did not evolve until the Late Paleocene or Early Eocene. Mesonychid carnivores of the Late Paleocene and Eocene increased in running speeds in a stepwise fashion as genera replaced one another through time, but after their extinction at the end of the Eocene, they were replaced as large carnivores by hyaenodontids, which even by the Middle Oligocene had not regained the level of cursorial ability that had characterized Middle Eocene mesonychids (Bakker, 1983). Throughout the Cenozoic, ungulates increased in running speed. Although early mesonychid predators of ungulates in the Late Paleocene probably outran their prey, the ungulates soon surpassed their predators in speed and continued to outperform their predators until the Late Miocene, when the fast dogs and cats (canids and felids) evolved in the continents of the northern hemisphere. In South America and Australia, the level of cursorial ability never transcended that of Eocene mesonychids (Bakker, 1983).

The salient points of this interpretative history are that significant advances in running performance in terrestrial vertebrates took place shortly after the first colonization of land in the Late Paleozoic, during the Triassic with the rise of thecodonts, and in a stepwise fashion during the Cenozoic. Living vertebrates having performance levels similar to those of earlier runners are found only on islands and small continents, or they inhabit environments such as deserts where long periods of metabolic quiescence are typical. Bakker's (1983) analysis further indicates that the rate of increase in running performance was generally faster in prey species than in predators, and that temporary improvement in locomotion is therefore possible in prey vertebrates that depend on speed for escape from predators.

Replacements of taxa may be associated with increases in speed in some instances, but in other replacements there is either no increase in speed or actually a decrease. According to Bakker's interpretations, replacements are not accompanied by improved performance when one group becomes extinct before subsequent ones take its place. These so-called noncompetitive replacements are thought to have occurred when therapsids replaced pelycosaurs in the Early Late Permian (Bakker, 1975, 1977), when dinosaurs eclipsed thecodonts in the latest Triassic (Tucker and Benton, 1982; Benton, 1983), and when mammals filled the void left by the extinction of the dinosaurs at the end of the Cretaceous (Bakker, 1980).

POWERED FLIGHT

Potentially the fastest form of locomotion is powered flight, which is achieved by the muscular flapping of wings. Even unremarkable fliers attain impressive speeds: the herring gull (*Larus argentatus*), 10.0 m/sec; the gadwall (*Anas strepera*), 20.7 m/sec; and the frigate bird (*Fregata magnificens*), 10.0 m/sec (Tucker and Schmidt-Koenig, 1971; Schnell, 1974; Schnell and Hellack, 1979). The fastest known insects are deer botflies (11.0 m/sec), but other fast forms include the dragonfly *Anax* (8.5 m/sec) and the honeybee *Apis mellifera* (6.5 m/sec) (Nachtigall, 1968).

Padian (1983) has carefully studied the features that distinguish powered fliers from gliders and other vertebrates. These features include internally stiffened wings (stiffened by feathers in birds, by fingerbones in bats); a large robust breastbone (sternum) from which the pectoralis muscles powering the downstoke of the wings originate; a keel or other expansion of the sternum indicating enlarged flight muscles; bracing of the shoulder against the sternum by coracoid bones (in birds) or by the clavicles (in bats); and a great expansion and lengthening of the distal forelimb elements relative to the proximal elements. Moreover, the main wing vein is located behind the middle of the wing in active fliers, whereas it divides the wing into equal anterior and posterior halves in birds that have become flightless (Feduccia and Tordoff, 1979). Except for the development of wing-like expansions, gliders show none of these skeletal specializations, and the "wings" typically are flexible rather than stiffened.

Although these features identify powered fliers among vertebrates, no reliable methods have yet been developed to estimate flight speeds from skeletal morphology. This is due in part to the small number of reliable measurements of flight speed. However, it has been possible to reconstruct the probable pathway by which flight evolved. Caple and his colleagues (1983) have shown by aerodynamic analysis that flapping wings in vertebrates probably originated in an animal that was already capable of jumping and was not in a gliding form. Elaboration of the wings and the forelimb movements required for powered flight enabled the animal to improve stability during landings and to stabilize its attitude and directional movement during the jump. The wings thus initially increased the length and directional precision of the jump.

The available evidence suggests that, although gliding evolved many times in amphibians, reptiles, and mammals and was a mode of locomotion for some vertebrates as early as the Late Permian (Emmons and Gentry, 1983; Benton, 1984b), powered flight evolved only three times among tetrapod vertebrates. The first group to evolve flapping wings was the reptilian order Pterosauria (Late Triassic to Late Cretaceous). Birds (class Aves) evolved powered flight by the Tithonian stage of the latest Jurassic, and bats of the mammalian order Chiroptera are recorded first in the Eocene (Padian, 1983). Gliding vertebrates may be either ectothermic or endothermic, but all vertebrates capable of powered flight were and are endotherms.

Powered flight among insects may have evolved only once (Kukalova-Peck, 1978). The earliest insects capable of powered flight are known from the Namurian stage of the Middle Carboniferous, but the possibility that flight evolved earlier cannot be discounted (F. M. Carpenter, 1976). As in vertebrates, the ability to maintain high body temperatures relative to air temperature may have been a prerequisite for the evolution of flapping wings in insects. Kingsolver and Koehl (1985) have shown with models of fossil insects that an increase in wing length confers aerodynamic advantages only when wings are already long. When wings are short, an increase in length has a negligible effect on aerodynamic properties, but it has a large positive effect on the ability of the insect to elevate body temperature. Kingsolver and Koehl therefore argue that selective agencies favoring higher body temperatures were initially responsible for the development of the movable insect wing, and that agencies favoring flight elicited elaboration of the wings once they had achieved a certain critical size.

SUMMARY

Articulated animals (arthropods, echinoderms, vertebrates, and a few other forms), in which skeletal elements move relative to each other by the action of muscles, have a high potential for evolving armor, locomotion, and retaliatory defenses. A common early theme in arthropod armor was conglobation, the ability to roll or fold the body into a ball so that appendages and other delicate structures are encased. Conglobation, which today is found chiefly in small terrestrial arthropods and in a few marine isopods, seems generally to be associated with low activity levels, and it carries with it the potential disadvantage of many joints in the armor being vulnerable to attack in the rolled-up state. Dorsal protection by a one-piece carapace, as seen in living crabs, and a greater emphasis on locomotion have been common themes in marine post-Paleozoic crustacean arthropods. Sessile barnacles, however, have evolved effective armor, especially during the Late Mesozoic and Cenozoic. Heavy armor has evolved in many terrestrial insects, but large armored terrestrial arthropods have been rare since the Late Paleozoic, possibly because the risk of predation during molting is particularly high in the vertebrate-dominated terrestrial environments that have characterized land since the post-Carboniferous.

Many echinoderm groups, especially Paleozoic stemmed echinoderms, epifaunal echinoids (sea urchins), and Cenozoic sand dollars (clypeasteroid echinoids), show a pattern of increasing armor through time. Locomotion has been emphasized by brittle stars and post-Jurassic crinoids.

External armor in vertebrates, though secondarily evolved in small marine fishes of cluttered habitats and in some slow turtles and mammals, is a primitive feature of marine and freshwater fishes. In most vertebrate stocks, increases in locomotor performance (speed, acceleration, maneuverability, and powered flight) have been common, especially during the Mesozoic and Cenozoic.

In no group of articulated animals, with the possible exception of barnacles and sea urchins, are the trends in armor and locomotion documented adequately, even on the global scale. Not only must we learn a great deal more about the performance and capacity of living animals—about the magnitudes of forces they can resist or muster, the speeds they can attain, and their effectiveness in coping with enemies—but we must also undertake studies to assess how morphology is related to function. Only in this way can we make inferences about the capacities of fossil forms. For arthropods, the risks attending molting must be measured if we are to understand the possibilities and limitations of design in this group.

PART FOUR

Escalation, Diversification,

and Extinction

CHAPTER THIRTEEN

The Dynamics of Escalation

A REVIEW OF THE EVIDENCE PERTAINING TO ESCALATION

In Chapter 3, I predicted that if selection among individuals predominates over other processes of evolutionary change, and if enemies are the most important agencies of this selection, the incidence and expression of traits that enable individual organisms to cope with their enemies (competitors and predators) should be found to increase within specified habitats over the course of time. Species unable to cope with enemies should either become extinct or be restricted to low-energy environments where enemies are rare and have limited capability. Living species in high-energy environments are not necessarily better able to cope with their enemies than ancient ones were with theirs, but the biological surroundings have in an absolute sense become more challenging.

Though often frustrating in its incompleteness and inadequacy, the fossil record is the only source of evidence pertaining to escalation between organisms and their enemies on a time scale of more than a few centuries. On the whole, the evidence from fossils is in accord with the hypothesis of escalation. The best-documented trends that indicate escalation, as discussed in detail in Parts Two and Three, are: (1) an increasing metabolic rate in suspension-feeding animals, echinoderms, and vertebrates; (2) an increase in the representation and power of shell-breaking predators in the sea; (3) an increase in the per-capita rate of sediment reworking and in the depth of burial of infaunal animals; (4) an increase in the depth of penetration into rock by excavating grazing animals and by endolithic species; (5) an increase in the dental specialization of herbivorous vertebrates on land; (6) an increase in the incidence and expression of armor in gastropods, cephalopods (until the Turonian stage of the

Late Cretaceous), pelecypods, barnacles, epifaunal echinoids, encrusting calcareous algae, and stemmed Paleozoic echinoderms; (7) an increase in the incidence of repaired breakage-related shell damage in gastropods; (8) an increase in the role of drilling predators as agents of mortality in gastropods, pelecypods, and barnacles; (9) an increased emphasis on locomotion in mobile animals, including gastropods, cephalopods, pelecypods, echinoderms, fishes, and land vertebrates; and (10) the ecological restriction of formerly shallow-water marine groups to environments in deep, cold, and dark waters. Many other trends indicative of escalation may have occurred, but the available evidence is insufficient for an adequate appraisal.

Phenomena and trends that are in apparent conflict with the hypothesis of escalation include (1) the temporal constancy and low values of the incidence of repaired breakage-induced damage in the shells of bivalved animals; (2) the reduction in the incidence of constricted apertures, which presumably limit access to predators, among shell-bearing cephalopods after the Early Carboniferous; (3) the elimination at the end of the Permian of echinoderms with well-protected internal respiratory structures, and their failure to reappear subsequently; (4) the diversification late in the Phanerozoic of predators that use drilling, a slow and therefore potentially risky form of predation; and (5) the invasion, at various times in the Phanerozoic, of the more highly escalated communities in the sea by species from freshwater habitats, where the level of escalation has generally remained low. Further work is needed to determine if these points of conflict are valid and, if they are, to explain them.

Potentially more damaging to the hypothesis of escalation is the observation that the evolution of enemies and of aptations against them has proceeded at a highly variable rate. Some intervals of time—the Early Cambrian, Middle Paleozoic, and later Mesozoic, for example—were characterized by particularly rapid and profound escalation, whereas other intervals were marked by much slower change or sometimes even by reversals. These variations in rate and direction suggest that, although selection among individuals may be of primary importance, it is not the only agency that has shaped the course of adaptive evolution.

The purpose of the remainder of this chapter is to summarize the chronology of escalation during the Phanerozoic, and to probe the geographical and climatic context in which the most dramatic episodes of adaptive escalation took place. The role of extinction in modifying escalation will be taken up in the next chapter.

A CHRONOLOGY OF PHANEROZOIC ESCALATION

THE PALEOZOIC

The record of escalation extends back to the earliest phases of the Phanerozoic. The first indication of escalation during this eon came during the Riphean, the period immediately preceding the Paleozoic era. Coniform stromatolites declined in abundance and diversity perhaps as a consequence of the evolution of metazoans. During the succeeding Vendian and Cambrian periods, there were further reductions in the diversity and ecological range of stromatolites as grazing and bioturbating animals diversified (see Chapter 5).

The "Cambrian explosion" was perhaps the most far-reaching of any of the great Phanerozoic episodes of escalation. It was heralded by the appearance of *Cloudina*, a small mineralized tubular skeleton, in the latest Vendian Nama Group of Namibia. At about the same time or perhaps somewhat later, the infaunal environment was invaded substantially for the first time by burrowing animals. A major diversification of skeletonized animals soon followed during the Tommotian stage of the earliest Cambrian (Crimes and Anderson, 1985; Nowlan et al., 1985). These animals included molluscs (monoplacophorans and perhaps pelecypods), articulate and inarticulate brachiopods, hyolithans, archaeocyathans, and lapworthellids, as well as tubular fossils with possible affinities with cnidarian corals (Brasier, 1979; A. D. Wright, 1979; Runnegar, 1981, 1983; Cloud and Glaessner, 1982; Tynan, 1983; Jell, 1984; MacKinnon, 1985). Trilobites and ostracode crustaceans are recorded first from the succeeding Atdabanian stage, and the first echinoderms appeared either during the Atdabanian or the succeeding Botomian stage of the Early Cambrian (Paul, 1977; Whittington, 1979; Briggs, 1983). With the exception of the reef-building archaeocyathans, these early skeletonized organisms were small, the greatest linear dimension not exceeding a few millimeters in most cases. Larger soft-bodied animals belonging to various extinct groups were also common in the Early Cambrian.

Fossils of the Early Cambrian yield the earliest evidence of endolithic animals, competition among modules in colonial organisms (archaeocyathans), and shell-breaking and shell-drilling predation, but animals with obvious morphological specializations for these modes of predation are unknown until the Late Silurian or Devonian. S. M. Stanley (1976) and Brasier (1979) estimated that the Early Cambrian explosion lasted perhaps ten million years, a remarkably short time in view of the many biological innovations that evolved. This Cambrian event was accompanied by an exponential rise in the diversity of marine invertebrates, an increase that was evident at all levels of the taxonomic hierarchy from the species to the phylum (Sepkoski, 1979; Missarzhevskiy, 1983).

The next probable phase of adaptive escalation began in the Late Cambrian and accelerated in the Early Ordovician. This episode marks the appearance and rapid diversification of predaceous cephalopods, and also a great diversification of skeletonized marine invertebrates both globally and in local assemblages (Bambach, 1977, 1983; Sepkoski, 1979; Sepkoski et al., 1981). During this phase, many antipredatory features evolved or became common for the first time: constricted apertures, strong sculpture, and coiling in shell-bearing cephalopods; radial plications and geniculation in articulate brachiopods; spines in echinoderms; calcareous opercula and tight coiling in gastropods; and conglobation (enrollment into a ball) in trilobites.

The Middle and Late Ordovician witnessed several additional important events. Colonial reef-building invertebrates became established during the Middle Ordovician, and colonial integration among corals reached a plateau by the Late Ordovician. The Llanvirnian stage of the Early Middle Ordovician marks the appearance of predaceous eurypterids and of cemented brachiopods. Two groups of bioeroders—ctenostome bryozoans and *Corallidomus*, a facultatively boring pelecypod—appeared during the Ashgillian stage of the Late Ordovician. The incidence of strong shell sculpture in cephalopods reached a plateau in the Llanvirnian, whereas shell coiling in this group rose irregularly through the Middle and Late Ordovician. The presence of spore tetrads suggests that plants may have occupied the land as early as the Caradocian stage of the Early Late Ordovician. Bioturbation during the Early Middle Ordovician was clearly more intense than it had been earlier in the Paleozoic, and remained at this level until the latest Silurian (Bambach, personal communication).

The mid-Paleozoic event, spanning the Silurian and Devonian periods (Signor and Brett, 1984), was marked by important increases in armor and bioturbation and by large-scale colonization of the land. The first phase of this event was already well underway during the Llandoverian stage of the Early Silurian. Externally armored fishes, jaw-bearing sea urchins (echinoids), jawed fishes (acanthodians), and cephalopods with constricted apertures were well represented during the Llandoverian. Pterygotoid eurypterids, with massive chelicerae potentially capable of piercing vertebrate armor, arose during the Llandoverian and diversified through the Late Silurian into the Devonian. Spine-bearing and cemented articulate brachiopods as well as spiny stemmed (pelmatozoan) echinoderms appeared during the Wenlockian (Middle Silurian) and diversified markedly during the Early Devonian. Undoubted vascular land plants appeared during the Ludlovian (Late Silurian), although some evidence points to a substantial land flora as early as the Llandoverian. Land plants diversified extensively during the Early Devonian. Leaves had evolved by the Siegenian stage of the Early Devonian, and forest trees by the Givetian stage of the Late Middle Devonian. Several groups of fishes whose jaws were morphologically specialized for shell-crushing evolved during the Early Devonian and diversified during the Givetian and Frasnian (Late Devonian) stages. Trends toward enhanced locomotor capacities are indicated both by the appearance of internally shelled cephalopods during the Early Devonian and by the reduction of external armor among many Devonian lineages of fishes. A major increase in bioturbation had occurred by the Early Devonian. The plateau in the incidence of strongly sculptured cephalopod genera that had been maintained from the Llanvirnian stage of the Early Middle Ordovician to the Gedinnian stage of the earliest Devonian was followed by an increase in these animals during the succeeding Siegenian stage. Two new groups of rock-destroying animals originated during the Devonian, the Phoronidea (Siegenian stage) and the acrocephalan barnacles (Givetian stage).

Some characteristics related to the competitive and defensive performance of individuals did not change appreciably during the Middle Paleozoic. There was apparently no increase in colonial integration among coral-like invertebrates, no decline or ecological restriction of immobile animals living unattached on the surface of unconsolidated sediments, and no striking increase in either the incidence or the depth of boring of endolithic animals.

The number of marine animal families, which previously had peaked during the Ashgillian stage of the latest Ordovician, rose to new maxima during the Devonian. From the Siegenian stage of the Early Devonian to the Frasnian stage of the Late Devonian, the number of marine families attained a level not exceeded until the Late Cretaceous (Van Valen and Maiorana, 1985). Peaks in the probability of origination of marine families came during the Wenlockian stage of the Middle Silurian and the Emsian and Givetian stages of the Middle Devonian.

The Early and Middle Paleozoic episodes of escalation are punctuated by several intervals of reversal. This can be seen most clearly from data on the presumably antipredatory features of cephalopod shells (Figure 10.1). The Llandeilian stage of the Middle Ordovician and the interval from the Pridolian stage of the latest Silurian to the Emsian stage of the Early Middle Devonian were periods in which there was a low incidence of constricted apertures and coiled shells. Signor and Brett (1984) thought that the low incidence of spiny crinoids during the Late Devonian relative to earlier and later intervals might be an artifact of poor collecting in Upper Devonian rocks, but a similar reduction is evident in predation-related features of cephalopod shells and in colonial integration of coral-like invertebrates.

Further escalation took place in the Late Paleozoic. Spiny brachiopods, cemented pelecypods and brachiopods, armored crinoids, strongly sculptured and coiled cephalopods, and shell-crushing fishes diversified rapidly during the Early Carboniferous and expanded in numbers as morphologically less specialized forms declined. Immobile hermatypic animals—those with algal cells in their tissues—evolved during the Late Carboniferous and became important contributors to reef communities in the Permian. Mechanically weak umbilicate and open-coiled gastropods were not as well represented in the Late Paleozoic as they were during the Devonian. On land, the evolution of the seed during the Famennian stage of the latest Devonian was followed by a major diversification of land plants during the Carboniferous. Insect flight evolved not later than the Namurian stage of the Middle Carboniferous, and the first wood-boring insects and large herbivorous vertebrates appeared by the Stephanian stage of the latest Carboniferous. Endothermic vertebrates probably appeared during the Kazanian stage of the Late Early Permian. The number of marine families of animals was somewhat lower in the Late Paleozoic than in the Devonian, but the early part (Tournaisian stage of the Early Carboniferous) of the Late Paleozoic escalation coincided with a high rate of origination of families (Figure 13.1).

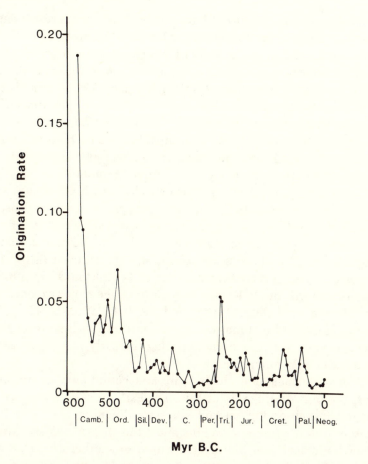

FIGURE 13.1 Rate of origination of marine animal families (number of newly appearing families divided by number of families surviving from preceding stage) through time. High rates of origination are seen early in the Phanerozoic, as well as after major episodes of extinction. The high rates during the Cenomanian (Early Late Cretaceous) and Early Cenozoic are associated with the attainment of hitherto unsurpassed levels of global familial diversity. (Figure redrawn from Van Valen and Maiorana, 1985, with permission.)

THE MESOZOIC AND CENOZOIC

The Triassic marks a time of important changes on land but only minor innovation in the sea. Major advances in vertebrate locomotion, including the evolution of erect posture and later of powered flight, occurred during the Middle and Late Triassic. Herbivores capable of grinding and shearing resistant plant tissues appeared, as did vertebrates capable of browsing high off the ground. Enclosed ovules evolved in various groups of land plants. That the marine Triassic was not very different from the Late Paleozoic is indicated by the architectural and taxonomic similarity between Triassic and Permian gastropods, fishes, echinoids, ammonoids, and pelecypods (Batten, 1973; Kier, 1973; Schaeffer, 1973; W. J. Kennedy, 1977; Vermeij et al., 1982; Flügel and Stanley, 1984; Yin, 1985). Those symptoms of escalation that did crop up during the Triassic occurred mainly late in the period. They include the attainment of intense bioturbating capacity in arenicolid annelids and glypheoid decapod crustaceans; an increased incidence of strongly sculptured ammonoid cephalopods; the appearance (or perhaps reappearance) of drilling predators; and the evolution of potentially shell-crushing palinuran spiny lobsters, reef-forming scleractinian corals, deeply excavating rock-boring pelecypods, and (as inferred from trace fossils) bioeroding sea urchins. The single-celled skeletonized plankton was heralded in the Late Triassic by the first coccolithophores and planktonic foraminifers, but, as with the other Triassic innovations, the major expansion of this element came later (Tappan and Loeblich, 1973a, b). The probability of origination of marine animal families was high throughout most of the Triassic, especially during the early part of the period after the end-Permian crisis.

Changes in adaptive architecture and community composition were profound during the Jurassic. The Early Jurassic witnessed the origin of shell-prying asteriid sea stars (Pliensbachian stage), intensely bioturbating sea urchins and axiid thalassinidean shrimps (Toarcian stage), and the diversification of shell-crushing elasmobranch fishes. By the Bathonian stage of the Middle Jurassic, the incidence of both crenulated margins among infaunal pelecypods and various antipredatory apertural features among gastropods exceeded earlier values in local warm-water assemblages, whereas the incidence of crushing-vulnerable umbilicate gastropod species was reduced to levels below those of earlier times. Other important Jurassic events include the evolution of conchicolous hermit crabs, bioeroding pelecypods, grazing-resistant encrusting coralline algae, and calcareous operculate structures in cephalopods; the diversification of mineralized plankton (foraminifers, radiolarians, coccolithophores, and the first diatoms); the assumption of the hermatypic habit in scleractinian corals; and the rise of herbivorous land vertebrates with grinding dentition.

Perhaps most important among Cretaceous innovations was the origin of flowering plants (angiosperms) during the Barremian stage. Not only did these plants greatly increase the diversity of life on land, including that of insects, but they also invaded the sea. As a result, unconsolidated marine bottoms came to be occupied by a highly productive community whose primary producers were angiosperms. Predaceous gastropods diversified remarkably from Aptian (Early Cretaceous) time onward. In their ranks were drillers, which after a brief appearance in the Late Triassic were unknown until the Albian stage of the Late Early Cretaceous. Sessile barnacles with calcareous lateral and opercular plates arose during the Late Jurassic, but they diversified throughout the Cretaceous. The incidence of streamlining and of presumably antipredatory strong sculpture increased through much of the Cretaceous in shell-bearing cephalopods, although strong armor declined after the Turonian stage of the Late Cretaceous. Other manifestations of escalation during the Cretaceous include the diversification of operculate cheilostome bryozoans, the multiple origins and diversification of bryozoans with highly integrated colonies, the achievement of the hermatypic condition in rudist (hippuritacean) pelecypods and in several groups of foraminifers, the disappearance of stalked crinoids and most brachiopods from shallow-water communities, the evolution of many families of shell-breaking predators, and the rise of deeply excavating bioturbators and bioeroders. The incidence of repaired shell damage in gastropods rose to modern levels by the Early Maastrichtian stage of the Late Cretaceous.

There is insufficient evidence at hand to provide much insight into the dynamics of the Mesozoic revolution. It is possible, for example, that escalation proceeded in several phases that were separated in time by intervals of relative quiescence. Data on gastropod shell architecture, the incidence of mortal drilling in gastropods, and colonial integration in corals suggest temporal stability through much of the Cretaceous, but this apparent constancy may reflect poor sampling rather than reality. Ward's (1986a) evidence from ammonoids points to important architectural changes in these animals throughout the Cretaceous.

The unprecedented changes that mark the Mesozoic revolution coincided with a large-scale increase in both local and global organic diversity. During the post-Ordovician Paleozoic, the number of marine animal families oscillated around a plateau. This level was not transcended until the Late Cretaceous, when diversification of "modern" groups—gastropods, pelecypods, foraminifers, fishes, and decapod crustaceans, for example—was well under way (Valentine, 1969; Bambach, 1977, 1983; Sepkoski, 1981, 1984; Sepkoski et al., 1981; Van Valen and Maiorana, 1985).

Further escalation took place during the Cenozoic, following the end-Cretaceous crisis. Although stomatopods and brachyuran crabs were already well differentiated in the Jurassic and Cretaceous, they did not evolve shell-breaking limbs until the Paleocene and Eocene. At the same time, the shell-crushing habit evolved in many acanthopterygian and elasmobranch fishes. Modern frequencies of lethal drilling predation were established by the Middle Eocene. Scleractinian corals and cheilostome bryozoans achieved very high levels of colonial integration during the Eocene and especially during the Miocene. Balanomorph barnacles evolved calcareous bases and thick tubiferous walls during the Oligocene. In the Mesozoic, burrowing echinoids were all delicate thin-shelled animals, but in the Early Cenozoic flat, internally buttressed sand dollar types evolved and diversified. Many animals capable of deep excavation in rock evolved throughout the Cenozoic. Animal-mediated mating (pollination) and dispersal of plants, though known since Late Carboniferous time, reached unprecedented levels of specialization during the Cenozoic and characterized the majority of plant species in most tropical and warm-temperate Cenozoic floras.

Renewed escalation may have taken place during the Miocene and Pliocene. Gastropods and infaunal pelecypods showed an increase in the incidence and expression of armor, and the frequency of repair in at least some gastropod groups was higher in the Miocene than in the earlier Cenozoic. Whales and other marine mammals capable of reworking large amounts of sediment per day evolved during the Miocene. Beginning in the Oligocene, but accelerating in the Miocene, terrestrial herbivorous mammals evolved specialized grinding dentition as well as horns, antlers, and other organs of combat and display. The relative size of the brain increased greatly in mammals and birds from the Late Eocene to the Recent (Jerison, 1973; Radinsky, 1978), and the locomotor performance (especially the velocity) of mammals increased markedly during the Late Miocene. Perhaps last, but certainly not least, the Pliocene witnessed the evolution of humans, the most potent competitors and predators of all time.

Rapid diversification characterized the Early Cenozoic, especially the Early Eocene, and the number of marine animal families passed the previous Maastrichtian peak by the Middle Eocene. By the Pleistocene, the number of marine families was almost double the Paleozoic maximum reached during the Devonian.

THE GEOGRAPHICAL COMPONENT OF ESCALATION

There is an important geographical dimension of escalation during the Cenozoic. Not only are there conspicuous variations in the nature, incidence, and expression of aptations against competitors and predators along gradients of latitude, altitude, and water depth, but there are in addition surprising variations among oceans and climatically comparable land masses. In the Recent shallow-water marine tropics, the incidence and expression of defensive and competition-related traits are highest in the Western Pacific and Indian Oceans (Indo-West-Pacific region), intermediate in the Eastern Pacific, and lowest in the Atlantic (Vermeij, 1978, 1983a). These patterns are illustrated by armor in hard-bottom gastropods, the frequency of shell repair in neritid gastropods, shell-crushing features in the claws of crabs, and the diversity and host specificity of coral-associated decapod crustaceans (Vermeij, 1974a, b, 1976, 1977a; Bruce, 1976; Castro, 1976; Abele et al., 1981). In the Western Pacific, comatulid crinoids, especially those species that perch on the tops of reefs by day, show a higher incidence of spine-like basal pinnules that protect the calyx and visceral mass than do comatulids in the southern Caribbean part of the Atlantic (Meyer and Macurda, 1977; Meyer, 1985). Associations between shell-dwelling sipunculans and solitary scleractinian corals are confined to the Indo-West-Pacific and Eastern Pacific regions (Gill and Coates, 1977), and hermatypic tridacnid clams are known only from the Indo-West-Pacific region (Rosewater, 1965). Didemnid ascidians of the Western Pacific are more specialized to house the alga *Prochloron* than are Atlantic members of this group. Glynn (1982b) found that the coral-eating sea star *Acanthaster planci* showed higher frequencies of nonlethal damage to the arms and disc in Guam in the Western Pacific than in Pacific Panama. Many coral-eating predators, including *Acanthaster* and other sea stars, tetraodontid pufferfishes, the muricacean gastropods *Drupella* and *Quoyula*, the nudibranch *Phestilla*, and the hermit crabs *Aniculus* and *Trizopagurus*, are confined to the Pacific and Indian Oceans. Scarid parrotfishes, though present in the Atlantic, graze corals mainly or exclusively in the Pacific and Indian Oceans (Randall, 1967, 1974; Glynn et al., 1972; Rudman, 1981, 1982; Glynn, 1982a, 1983; Glynn and Wellington, 1983). In a survey of lithophagine mussels that bore into living corals, Highsmith (1980b) found that all species with this habit occur in the Pacific and Indian Oceans.

The Eastern Pacific ranks behind the other tropical oceans with respect to several other interactions among species, but the Western Pacific and Indian Oceans invariably harbor the most highly escalated species. Terebrid gastropods living in sand have a lower frequency of repaired injuries in the Eastern Pacific than in the Western Atlantic, but those from the Indo-West-Pacific region are morphologically more specialized against lip-peeling by crabs and are more apt to have scars than are their counterparts from either the Pacific or Atlantic coasts of tropical America (Vermeij et al., 1980). The diversity and host specificity of carapid pearlfishes in sea cucumbers are lowest in the Eastern Pacific, intermediate in the Western Atlantic, and greatest in the Indo-West-Pacific (Trott, 1970). The same pattern is seen in the number and specificity of fishes living in sea anemones, snails on diadematid sea urchins, and pyrgomatid barnacles in scleractinians (Ross and Newman, 1973; Ponder and Gooding, 1978; Dunn, 1981; Vermeij, 1983a).

On land, mammals from North America, Eurasia, and Africa tend to have greater locomotor capacities than those from the smaller continents of South America, Australia, and Madagascar (Bakker, 1980, 1983). The development of horns or antlers as weapons and organs of display is common in ungulates that evolved in Africa and Eurasia, rare in mammals originating in North America, and absent in those with Australian or South American origins (Janis, 1982). Africa has a larger representation of venomous snakes, mound-building termites, and large indigenous predaceous mammals than does South America (Janzen, 1976). Many other examples pertaining to species on land as well as in fresh water could be given. The important point is that aptations against enemies have evolved to different levels in physically similar environments in different parts of the world.

THE ROLE OF EXTRINSIC EVENTS

The finding that competitive and defensive capacities of individuals within specified habitats have increased over the course of the Phanerozoic is, I believe, consistent with the expectation that natural selection due to enemies is an important evolutionary process that has shaped not only the character of individuals, but also the communities comprising these individuals. Nevertheless, it is clear that escalation between species and their biological surroundings has varied in rate over time, and that factors other than enemy-induced selection among individuals have intervened.

Two important questions must now be posed. First, how have the geographical and climatic events of earth history influenced adaptive escalation, either by preventing or by stimulating it? Second, can escalation take place in the absence of such extrinsic events? The rest of this chapter and the next one suggest some possible answers to these questions.

In Chapters 1 and 2, I pointed out that, although the potential for improvement relative to a given hazard nearly always exists, this potential often is not realized. The traits that confer improvement may be inconsistent with already existing aptations and risks, or they may be unable to spread because the selective agency varies on a small scale relative to the dispersal distance of a typical individual. Moreover, individuals may locate and occupy situations to which they are already adequately adapted rather than adapt further. As a result, species often achieve a certain phenotypic stability, or stasis. This tendency toward adaptive conservatism suggests that population-wide changes in the nature and strength of selection may be needed to bring about adaptive change, and that improvement is either limited or impossible in the absence of a shift in the selective regime.

To evaluate this hypothesis, we must monitor populations and their surroundings over time. Although numerous studies of lineages exist, and many cases of changes in gene frequency have been described in Recent populations, most of these studies fail to reveal details about the dynamics of adaptation. Even when components of the environment have been monitored independent of the changes in the traits of individuals through time, adaptational interpretations of the changes typically are based on *a posteriori* arguments of plausibility and not on predictions. In short, evidence pertaining to the presence of, and possible aptations to, hazards is usually neither sought nor provided in studies of evolutionary change in populations and lineages. If an adaptational interpretation is rejected, it is usually done in the absence of adequate information on environmental hazards and the possible ways individuals in a lineage might cope with them.

The evolution of color patterns that mimic those of unpalatable or dangerous animals has been perhaps the most fruitful subject for the study of adaptation, in part because the selective agents (visual predators) and the effectiveness (degree of resemblance between the species being studied and the model) can be evaluated and inferred with some confidence. I find J.R.G. Turner's (1981, 1983) work on mimicry in *Heliconius* butterflies from tropical America particularly appealing because, although he lacks direct evidence from specimens of the past, he can reconstruct the course of evolution in several of the species. Moreover, he has worked out the genetic basis of many of the characters involved in mimicry. The pattern of mimicry evolved in a particular population of *Heliconius* depends on the species composition of potential models. If a particular model becomes extinct, the resemblance of *Heliconius* to that species is no longer useful in protecting it against predators. Instead, mutants that somewhat increase *Heliconius*'s resemblance to other potential models, which were either already present or have been newly introduced, will be favored. The initial resemblance, as Turner points out, need not be particularly close. Modifier genes control the improvement in resemblance until the *Heliconius* closely resembles the new model. The pattern of adaptation that Turner envisions therefore consists of two phases: (1) an initial phase of rapid adaptation, which sets the course of subsequent evolution and represents an improvement with respect to previous conditions, and (2) a slower phase of gradual modification, which increases the effectiveness of the aptation. The model is influenced by the evolution of the mimics and may itself undergo modification, for if the mimics become too common, the effectiveness of the pattern which initially served as a warning against potential attackers will decline. The extent to which the resemblance between mimic and model can increase depends on the population sizes of the potential prey species as well as on the species composition and abundance of predators. If a change takes place in any of these components, the strength and even the nature of selection in favor of resemblance between mimic and model may also change. Heliconiine butterflies may be especially sensitive to local variations in the populations of models and predators, because they are relatively sedentary as adults and are therefore capable of adapting to local conditions (J.R.G. Turner, 1981, 1983; L. E. Gilbert, 1983). If the biological environment of *Heliconius* remains constant, further adaptation eventually ceases, and the population becomes adaptively static.

I have studied the course of adaptation of two intertidal gastropods before and after the introduction of a shell-breaking predator, the green crab *Carcinus maenas*. The crab, first noticed north of Cape Cod, Massachusetts, in 1905, steadily moved northward and eastward until it reached Nova Scotia in 1954. Both *Nucella lapillus* and *Littorina littorea* in the affected area showed an increase in the incidence of repaired shell breaks. This indicated that an increasing proportion of the populations of these gastropods was affected but not killed by shell-breaking agencies. In previous experiments with shell-breaking predators, others had determined that a thick shell provided protection against breakage in *Littorina*, and that a thick outer lip, tall spire, and small narrow aperture served a similar function in *N. lapillus* (Kitching et al., 1966; Hughes and Elner, 1979; Currey and Hughes, 1982). I therefore expected an increase in these features after the introduction of the green crab. This expectation was fulfilled for *N. lapillus*, a species with little dispersal during the larval stage, but not for *L. littorea*, whose larvae disperse widely during their planktonic existence. Although both species were affected by the change in predator composition, only *N. lapillus* showed a change in shell characters that is consistent with an adaptational interpretation. The possibility that the morphological change was not evolutionary cannot be excluded, because the genetic basis of the characters has not been definitely established. The results of this work show that evolutionary change occurred only when there was a change in the biological surroundings, and that some species may be unable to evolve even then.

The idea that extrinsic change is necessary to stimulate evolution is also supported by analyses of the evolutionary history of particular lineages in the fossil record. Hallam (1978), for example, pointed out that sudden changes in morphology within two Jurassic lineages of the oyster genus *Gryphaea* coincided with falling sea levels (marine regressions). In his detailed studies of evolution in the scallop genus *Argopecten*, Waller (1969) concluded that speciation (splitting of the lineage into separate species) coincided with geographical changes that led to the genetic isolation of inshore (bay) scallops from populations of the open-ocean coast. Other examples of evolutionary change coinciding with climatic events have come from studies of Neogene mammals in Africa (Vrba, 1984). Still another interesting example comes from the Pliocene and Pleistocene of the Atlantic Coastal Plain of the United States (Cronin, 1985). Morphological changes in ostracode lineages coincided with long-term changes in climate and oceanic circulation, especially during the Middle Pliocene (3.1 to 3.6 million years ago) when the Atlantic and Pacific Oceans became separated by the uplift of the Central American isthmus. Morphological changes did not occur when climatic and oceanographical conditions remained constant, or when fluctuations were very rapid and frequent, as during the Pleistocene when glacial and interglacial episodes followed each other in rapid succession. It is not known in these fossil examples whether the morphological changes were adaptive and, if they were, to which selective agencies the lineages were responding evolutionarily, but the fact that morphological change was concentrated during times of extrinsic change seems to be well established.

This observation is consistent with the view that species, once established, are adaptationally static entities. When they do undergo change, it is rapid. The pattern of long-term stasis and short-term change has come to be known as punctuated equilibrium (Eldredge and Gould, 1972). In its strict form, the theory of punctuated equilibrium holds that morphological change is concentrated at times of speciation (lineage splitting), but as Hallam (1978), J.R.G. Turner (1981, 1983), and others have pointed out, the alternation between stasis and change is characteristic also of lineages in which no splitting takes place. The extent to which punctuated equilibrium in this broader sense is typical is still a matter of debate. Proponents argue that it is the dominant pattern in the fossil record (S. M. Stanley,

1975a, 1979; Hallam, 1978; Gould, 1982; S. M. Stanley et al., 1983). Skeptics believe that gradual change within established species is the rule, and that the pattern of stasis arises from artifacts of preservation, sampling, and taxonomic practice in the fossil record (Gingerich, 1977; Schopf, 1981; M. L. McKinney, 1985). They point to several case studies, mostly involving small planktonic species and some Eocene mammals, in which gradual change is documented during intervals of time that are judged to be represented by relatively gap-free strata. The scarcity of cases in benthic (bottom-dwelling) marine species is explained by the fact that most stratigraphic sections in which benthic species are preserved are riddled with gaps, and the the chronology of evolutionary events on the geologically small time scale of hundreds to thousands of years is blurred by bioturbation and other agencies of sedimentary disturbance. If fossils are scarce in a given sedimentary stratum, one must combine specimens from several adjacent strata in order to obtain a sample large enough for statistical analysis, but these samples would then span a long time interval. Preservation and sampling therefore would result in time-averaging, which could have the effect of producing discontinuities between temporally static morphologies, when in fact the undisturbed sediments would have yielded continuous change. Finally, Schopf (1981) has brought forth the argument that many characteristics that might change gradually remain undetected in the fossil record because these characteristics refer to either molecular structure or soft tissues.

I agree with some of the skeptics that the fossil record cannot provide unequivocal evidence with respect to the incidence of punctuated equilibrium and gradual change in evolution. It can yield cases of morphological stability as well as instances of gradual change if certain conditions of preservation are met. To demonstrate stasis, one must have two or more samples from several stratigraphically separated horizons, and these must be statistically indistinguishable with respect to the morphological characters being studied. A demonstration of continuous gradual change requires independent evidence from sedimentology and detailed stratigraphy showing that the sequence that contains abundant specimens of the chosen taxon throughout its thickness is temporally continuous, that is, without gaps and without bioturbation. Because these conditions are seldom met in even the best stratigraphic sections (Schindel, 1980, 1982), unequivocal cases of gradual change are apt to be extremely rare. Accordingly, it seems to me pointless to keep a scorecard of the number of cases of punctuated equilibrium and gradual change. Rather, the important question from the perspective of this book is whether adaptive changes in the competitive and defensive attributes of species come only when enemies appear and disappear, or whether they can also occur when the populations of species and their enemies are persistent without great fluctuation. The evidence from living species favors the former viewpoint, but in the fossil record the question remains an unresolved and important issue.

CONDITIONS FAVORABLE TO ESCALATION

In most of the cases of adaptive escalation dealt with in this book, escalation involves two components which in principle are independent of each other. These are adaptation, leading to an increase in the expression of an aptation that enables an individual to cope with enemies, and differential diversification, leading to the increased representation of highly escalated species in a clade within a specified environment. To understand which conditions are especially favorable for escalation, we must probe the factors that stimulate adaptation and diversification.

If adaptation is generally prevented by ecological limitations, as I argued in Chapter 2, a necessary condition for escalation is the removal of some of these limitations. Perhaps the best indication that a limitation has been relaxed is that a population is able to increase. Populations may increase when (1) climate changes, so that some populations spread while others decline; (2) geographical barriers to a population's dispersal are removed as the result of changes in the configuration of oceans and land masses and in patterns of wind and water circulation; and (3) primary productivity rises. Such population increases should be especially effective in high-energy situations, where the potential for adaptive improvement is high, and in conditions that can support very large populations. As I pointed out in Chapter 1, selection can proceed further (and is less confounded by random population fluctuations) in large populations than in small ones. Large biogeographical regions where climates are warm and productivity is high (such as the Indo-West-Pacific region today) should be particularly favorable for escalation.

An additional important consequence of shifts in climate and in the position of geographical barriers is that biotas that had developed in mutual isolation are brought together. This biological interchange places large numbers of species in contact with new potential enemies and may therefore stimulate adaptation in many species simultaneously.

In addition to reducing the usual ecological limitations on adaptation, conditions that enable populations to increase may also favor speciation by founder populations. As populations spread or shift, new parts of the range are colonized first by small numbers of individuals, which may give rise to new species provided they are isolated for a sufficiently long time. During the colonization and divergence phase, the stringent ecological limitations that affected the parent population are either relaxed or changed (Carson, 1975; Carson and Templeton, 1984). Climatic and geographical change therefore are favorable for speciation (Valentine, 1968; Vrba, 1984).

Many instances of speciation probably do not involve a reduction or even a change in ecological limitations in the affected populations (Valentine and Jablonski, 1983b; Barton and Charlesworth, 1984) and may therefore not accompany escalation. This is especially so for cases of vicariant speciation, in which the daughter species arise from the isolated remnants of a continuously distributed ancestral population that became subdivided by a geographical barrier or by the extinction of geographically intermediate populations.

In short, the conditions that favor adaptive improvements in many species simultaneously favor speciation by founder populations. The most favorable conditions for escalation therefore include climatic warming (or the latitudinal spread of warm climates), increasing continental areas that are covered by moist lowland forest or productive shallow marine waters, and the coming together of previously separated biotas.

If the increased expression of aptations to enemies is often accompanied by an increased incidence of such aptations among species in a given environment, lineages that undergo substantial escalation should be either more susceptible to speciation or less prone to extinction than are lineages in which adaptation against enemies is modest. Sexual competition may be an important mechanism linking speciation and escalation. West-Eberhard (1983) has observed that populations in which competition for mates and the ability to recognize members of the same species at a distance are important are especially susceptible to genetic isolation and divergence once founding populations arise. Even a minor change in species-recognition traits is sufficient to cause a founding population to achieve genetic independence from the parent population and from other founding populations with which members of the expanding isolate could come into contact. I suggested in Chapter 2 that the characteristics that make individuals conspicuous at a distance and that are therefore important in mate recognition also enable enemies to detect these individuals. Species with conspicuous traits are therefore expected to be well adapted against enemies. Bright coloration, ornate external features, distinctive sounds, and great differences between male and female are trademarks of species in which competition for mates is important, and which are conspicuous to friend and foe alike. Such traits are common in angiosperms, arthropods, vertebrates, and even many molluscs. These groups are rich in species, and they have highly developed competitive and defensive capacities.

In short, although adaptation against enemies is distinct from speciation by founders, the two processes are stimulated by the same kinds of extrinsic change. The characteristics that are enhanced as a result of selection by enemies stimulate escalation further by lowering genetic barriers to speciation. In the next sections I explore some aspects of climatic and geographical history to determine if times of profound extrinsic change do in fact coincide with episodes of large-scale escalation.

MARINE TRANSGRESSIONS AND PRODUCTIVITY

For marine organisms, an increase in sea level permits large-scale population expansion and adaptive escalation. A rise in sea level is linked with mild climates worldwide and with high productivity in the photic zone. As sea level rises, continental areas are transformed into vast expanses of shallow sea. Especially in the early stages of transgression, these shallow seas lie almost entirely within the photic zone, thus providing ideal opportunities for large-scale high primary productivity. If the epicontinental seas are located at low latitudes, where evaporation exceeds rainfall, most of the world's deep ocean water originates in these warm seas as a result of the sinking of dense high-salinity water. As it sinks, this water is quickly depleted of oxygen through the metabolic activities of organisms, so that the deep ocean water tends to be warm as well as oxygen-poor (Brass et al., 1982; Wilde and Berry, 1984). Geological evidence linking transgression with warm climates and high productivity comes from the widespread deposition during transgressive phases of carbon-rich black shales and of phosphorites (Fischer and Arthur, 1977; Jenkyns, 1980; Leggett, 1980; Hallam, 1981b, 1984). In the modern ocean, phosphate-rich sediments are usually associated with upwelling and with other oceanographical conditions favorable to high productivity (Sheldon, 1981; Riggs, 1984).

It has been known for some time that major episodes of organic diversification have coincided with times of marine transgression (Fischer and Arthur, 1977; Hallam, 1984), but the link between transgression and escalation has not hitherto been appreciated. Thanks to Hallam's (1984) summary of fluctuations in sea level through the Phanerozoic, it is now possible to examine the temporal correspondence between transgressions and episodes of escalation. Although there are points of contention over the precise timing and duration of some transgressions, Hallam's account is in broad agreement with those of Schopf (1980), Boucot (1983), and others who have considered the evidence.

A rise in sea level seems to have accompanied the Cambrian explosion (Brasier, 1982; Hallam, 1984). This transgression apparently lasted until the end of the Early Cambrian and resumed again later in the period. Cook and Shergold (1984) also have suggested that the Cambrian explosion was associated with the widespread deposition of phosphorites at low latitudes, and they have proposed that this deposition implied high primary productivity. Stratigraphic uncertainties make it difficult to ascertain if phosphorite deposition and adaptive escalation occurred simultaneously during the Early Cambrian, but approximate contemporaneity is indicated by the available evidence.

The Ordovician diversification, like that in the Cambrian, was accompanied by a general rise in sea level. Marine transgression over the continents began during the Tremadocian (earliest Ordovician), but an even more extensive rise in sea level took place during the Llandeilian stage of the Middle Ordovician and culminated in a very high stand of the sea during the Caradocian (Early Late Ordovician). The Middle Ordovician was also a time of widespread phosphorite deposition. The first likely traces of land-plant life are recorded from the Caradocian (Gray et al., 1982). If the land was being colonized by organisms at this time, the possibility of enhanced productivity in near-shore marine environments during the Late Ordovician cannot be dismissed.

There seems to be some relationship between marine transgression and escalation during the Middle Paleozoic. After a rapid fall in sea level at the end of the Ordovician, there was a rapid rise during the Llandoverian (Early Silurian). Sea level fell through most of the rest of the Silurian, except toward the end of the period (Ludlovian stage). A prolonged period of more or less consistent rise in sea level spanned the Middle and Late Devonian, culminating in a peak in sea level during the Frasnian stage of the Early Late Devonian. Like the Ordovician, the Silurian and Devonian were characterized by the widespread occurrence of black shales, especially at times of transgression (Berry and Wilde, 1978; Leggett, 1980; House, 1985). Near-shore marine productivity may have been stimulated further by the colonization of the land by plants and arthropods, a process that began during the Late Silurian and proceeded especially rapidly during the Siegenian stage of the Early Devonian.

The Late Paleozoic was a time when sea levels generally fell and escalation was for the most part slow. Only during the Early Carboniferous (especially the Tournaisian stage) was there substantial marine transgression. This transgression coincided with the diversification of spinose brachiopods, shell-breaking fishes, and other products of the earlier Devonian escalation. Despite the widespread occurrence of vascular plants in lowland forests, marine productivity during the Carboniferous and Permian was for the most part low (Tappan, 1968, 1970; Berner and Raiswell, 1983), a situation that probably favored the evolution of the hermatypic habit in several groups of immobile marine animals. One observation that conflicts with this interpretation is that the Permian witnesses the deposition of one of the largest known phosphorite accumulations, the Phosphoria Formation of the western United States (Sheldon, 1981).

In the Early Triassic, sea levels rose from the low stand in the Late

Permian. Further transgression took place during the Anisian stage (Early Middle Triassic), with a maximum stand of the sea being attained during the Ladinian and Carnian stages of the Late Triassic. These transgressions, and the biological escalation associated with them, were modest in comparison with those in the later Mesozoic.

Sea level generally rose during the Jurassic and Cretaceous, peaking in the Late Campanian and Early Maastrichtian stages of the Late Cretaceous, although reversals of various magnitudes frequently interrupted this trend. Several episodes of widespread phosphorite deposition also occurred in the Middle and Late Mesozoic, especially during the Middle and Late Jurassic and in the Late Cretaceous. They usually followed or accompanied the deposition of black shales (Jenkyns, 1980; Cook and Shergold, 1984; de Graciansky et al., 1984). The rise of the angiosperms during the Cretaceous may also have enhanced near-shore marine productivity.

Escalation during the Cenozoic is also broadly coincident with marine transgression. Although sea level generally fell during the Cenozoic, it rose during the Eocene, again from the Late Oligocene to the Middle Miocene, and yet again briefly during the Pliocene. These episodes correspond with both an increase in the antipredatory characteristics of molluscs and the diversification of marine mammals (Lipps and Mitchell, 1976). The Eocene, Miocene, and to a lesser extent the Pliocene and Pleistocene were periods of widespread phosphorite deposition in many parts of the world (Sheldon, 1981; Riggs, 1984).

The link between marine transgression and adaptive escalation may be an artifact of sampling and preservation. Fossils accumulating at times of transgression are more accessible and better sampled than are those interred during times of falling sea level. Properties such as diversity, which depend on sampling effort, will therefore appear to increase as sea level rises. For other characteristics, such as the percentage of species in a local assemblage with a certain adaptive trait or the degree of expression of that trait, however, these sampling-dependent biases may not be important.

If rise in sea level stimulated escalation in the sea, did a drop in sea level bring on escalation on land? Although the Devonian was generally a time of marine transgression, low stands of sea level and episodes of regression marked the Early Devonian, when some of the most far-reaching ecological changes on land were taking place. The

Late Carboniferous and Middle Permian were not important for marine escalation, but they were significant for events on land. Sea levels generally fell during this time. The Triassic, especially the late part of the period, witnessed important steps in herbivory and locomotion in vertebrates. Sea level had reached a peak during the Carnian stage of the Late Triassic. Cretaceous events on land were roughly contemporaneous with marine escalation and occurred during a phase of generally rising sea level. During the Cenozoic, when sea level generally fell, there were important events in escalation in the Late Paleocene, Eocene, and Miocene. With the data that is presently available, therefore, we cannot establish a clear connection between escalation on land and an increase in habitable area (as inferred from falling sea level).

CLIMATIC CHANGE AND BIOTIC INTERCHANGE

Times of diversification and of adaptive innovation were often associated with climatic and geographical change. Cooling at the end of the Cretaceous allowed communities of forest mammals from Asia to spread southward to North America and to diversify there as the dinosaurs waned in the marshlands and savannas (Van Valen and Sloan, 1977). Climatic warming during the Paleocene and Early Miocene coincided with the northward migration and subsequent diversification of many groups of mammals in North America (Gingerich, 1977). In Africa, lineage splitting in bovid artiodactyls (antelopes and their relatives) came after population contractions that resulted from climatically induced changes in vegetation (Vrba, 1984). When glaciers retreated northward some 11,000 years ago in the northern hemisphere, populations of mammals living south of the glaciers spread northward and underwent rapid evolution in areas with a lush plant cover and high productivity (Geist, 1983). North American mammals walking across the Central American isthmus during the Late Miocene, Pliocene, and Pleistocene not only invaded South America, but also diversified there (S. D. Webb, 1977; Marshall, 1981; Marshall et al., 1982). Similar evolutionary trends have been documented for some North Pacific molluscs that invaded the North Atlantic during the Pliocene (Strauch, 1972), and for Asian brachiopods that invaded North America during the Llandoverian stage of the Early Silurian (Sheehan, 1975).

It is both interesting and puzzling that few if any of the thousands of species that humans have introduced to areas not previously occupied by those species have diverged from ancestral populations or have diversified. Some possibly adaptive changes in color and in body size have been documented for both European house sparrows in North America (Johnston and Selander, 1964, 1971, 1973) and Asian myna birds in New Zealand (Baker and Moeed, 1979; Baker, 1980), but no detectable morphological changes have occurred in European face flies, land snails (Cepaea nemoralis), and green crabs in North America (Almaça, 1963; Brussard, 1975; E. H. Bryant et al., 1981), or even in European rabbits in Australia (Richardson et al., 1980).

These examples of morphological constancy demonstrate that the colonization of a new area is not sufficient to cause change and speciation. This point was made forcefully by Ehrlich and Raven (1969), who emphasized that changes in the nature and intensity of selection were required for a daughter population to diverge from its ancestor. It is possible that the biological environment of many of the species transported by humans was much the same in the new habitat as it was in the old, for most of the species that have been studied are found in close association with man. They are therefore less apt to interact with native species than are individuals of populations that spread into new biotas without human agency (Vermeij, 1986b). All these points represent potentially fruitful areas for further investigation.

Little is known about continental or biotic movements during the Paleozoic. Generally, the pattern seems to have been one of unification of continents by means of the closure of seaways. By the Permian, a single continent (Pangaea) had become established (Valentine and Moores, 1970; Boucot, 1975; Schopf, 1980). Several episodes of biotic exchange of marine faunas are known during the Early Silurian and the Late Eifelian stage of the Middle Devonian, but although other migrations doubtless occurred, little can be said of the consequences of these exchanges (Boucot, 1983).

The Mesozoic and Cenozoic were characterized by continental fragmentation. During the Triassic, Pangaea consisted of a northern portion (Laurasia) and a southern one (Gondwanaland), which were separated by a broad and deep Tethyan embayment. As the two portions drifted apart, this broad embayment became a continuous seaway (Tethys) in the tropical belt. Further continental break-up during the Middle Late Jurassic resulted in the opening of the North Atlantic Ocean. Gondwanaland began to break up during the Early Cretaceous, with India and Madagascar separating from Africa. Later, New Zealand drifted away from Antarctica, and South America broke away from Africa. With the northward drift of South America away from Antarctica, a continuous southern seaway (circum-Antarctic ocean) was established by the Late Eocene or Early Oligocene. The Tethys seaway was gradually narrowed and broken up by the northward movement of Africa, so that by Early Miocene time a land barrier existed between the Indian Ocean and the Mediterranean. North and South America became connected by a definitive land bridge during the Pliocene, 3.1 to 3.6 million years ago. The northward drift of Australia and New Guinea resulted in a collision, during the Pleistocene, with the plates carrying the islands of Indonesia and the Philippines (Berggren and Hollister, 1977; Keigwin, 1978; van Andel, 1979; Hallam, 1981a, b). In the northern hemisphere, a connection between North America and Asia was periodically established and broken during the Miocene, Pliocene, and Pleistocene; and an oceanic connection between the Arctic Ocean and the North Atlantic was formed during the Pliocene as the result of the flooding of a high-Arctic land bridge (Strauch, 1970, 1972).

Although continental movements created some important barriers for marine organisms—barriers such as the Central American isthmus between the Pacific and Atlantic Oceans and the Isthmus of Suez between the Red Sea and the Mediterranean—they eliminated other barriers. Migration of reef-associated species from the Central and Western Pacific to the Eastern Pacific was made possible by the northward movement of the Line Islands, which in post-Pliocene time have acted as a "steppingstone" for planktonically dispersing larvae (Dana, 1975). Submergence of the high-Arctic Thule land bridge permitted large-scale interchange between the biotas of the North Atlantic and North Pacific (Strauch, 1970, 1972). In the case of the trans-Pacific migration, there is some evidence that the immigrant species are architecturally better armored than the native Eastern Pacific species (Vermeij, 1978). If this migration persists, it could result in an increased incidence of armor among Eastern Pacific molluscs.

Other instances of marine biotic exchange could have resulted from more vigorous oceanic circulation. Because many marine species disperse by means of planktonic larval stages, a more vigorous circulation could transport larvae over greater distances and therefore result in the geographical expansion of many species. In post-Pliocene time, for example, some molluscs from the Indian Ocean and the west coast of Africa have dispersed to the Atlantic coast of the Americas by way of the westward-flowing North and South Equatorial Currents. The velocity of these currents probably increased after the uplift of the Central American isthmus during the Pliocene, when circulation in the Atlantic generally became faster (Petuch, 1982; Vermeij and Petuch, 1986). This increase in circulation, according to some interpretations (Berggren and Hollister, 1977; Fischer and Arthur, 1977), has been generally characteristic of the world ocean since the mid-Mesozoic. Slight changes in the position and speed of currents could therefore have had great consequences for the geographical ranges and patterns of speciation of planktonically dispersing species. That such changes in currents occur frequently is demonstrated by the oceanographic changes that accompany the so-called El Niño events. These climatically induced events, which occur every seven years or so, cause warm-water conditions to extend abnormally far south along the west coast of South America, interrupting the normal pattern of upwelling. The worldwide impact of El Niño events on the distribution of planktonic larvae is an important phenomenon that deserves to be investigated.

Much remains to be learned about the long-term consequences for evolution of biotic interchange, but it is clear from present evidence that conditions during the Mesozoic and Cenozoic provided many opportunities for biotic interchange. Geographical barriers appeared and disappeared in many parts of the world, oceanic circulation became more vigorous, and climatic fluctuations were frequent.

The various events and processes that stimulate adaptation and speciation on a grand scale are probably causally interrelated. Continents move as a consequence of sea-floor spreading, which takes place at long ridges standing high above the ocean floor. As these ridges increase in volume and the continents are pushed apart, sea water is displaced over the continents. Rapid continental movement is therefore associated with marine transgression (Hallam, 1984; A. G. Fischer, 1985). Higher sea levels, in turn, ameliorate world climates. Volcanic activity, which is the ultimate source of most of the world's carbon, is also greatest during times of rapid continental drift and marine transgression (A. G. Fischer, 1985). By interrupting circumglobal circulation in the tropical oceans, the unification of continents that has characterized the last 15 million years may be linked to cooling and eventual glaciation at high latitudes. Weyl (1968) and Keigwin (1982), for example, proposed that the increase in north-to-south circulation in the Atlantic following the uplift of the Central American isthmus during the Pliocene brought on or accelerated glaciation in the northern hemisphere. The creation of the circum-Antarctic seaway near the Eocene-Oligocene boundary also led to a general cooling of the deep ocean (Keigwin, 1980). Climate, sea level, primary productivity, and the configuration of oceans and land masses are therefore interrelated.

INTRINSIC FACTORS: THE IMPORTANCE OF NUTRIENTS

In addition to the extrinsic climatic and geographical events that stimulate adaptation and diversification, events that are set in motion by the process of escalation itself have propelled that process further. The importance of such intrinsic mechanisms was already emphasized in the discussions of the evolution of land plants (Chapter 4), the evolution of bioeroding organisms (Chapter 5), and the development of the conchicolous habit (Chapter 8). In the case of the colonization of the land, which I have interpreted as a response to increasing risks faced by organisms in aquatic environments, greater production of oxygen and the establishment of soils from which nutrients (nitrates and phosphates) could flow into near-shore waters enabled individual organisms to maintain higher metabolic rates and therefore to "explore" pathways of adaptation that were closed to earlier species. These effects may have been enhanced further during the Cretaceous, when angiosperms arose and spread worldwide. Moreover, Thayer has suggested to me that nitrogen fixation, a partnership between bacteria and plants that may have evolved for the first time with the origin of the Leguminosae during the Late Cretaceous, may have increased the quantity of nutrients available to both terrestrial and marine species.

The endolithic habit and the evolution of conchicoly also may have been stimulated by increasing risks posed by other organisms. The rock cavities and tunnels made by deeply excavating endolithic animals provided shelters for many other kinds of animals, including predators, which in turn could have stimulated further adaptation in potential prey species. Conchicoles had the effect of increasing the abundance and life span of shells, thereby making more available a particular resource on which predators could specialize.

These three effects are not the only examples of intrinsic factors in escalation. Rhoads and Morse (1971) suggested that the attainment of a minimal level of oxygen in the atmosphere enabled some animals to precipitate mineralized skeletons for defense and other animals to burrow into unconsolidated sediments beginning in the Early Cambrian. The evolution of larger animals brought with it a more rapid cycling of nutrients than was possible when only fungi and other decomposers exploited the tissues of primary producers (A. G. Fischer, 1984). These ideas are speculative, but they point to the likelihood that escalation has consequences which in turn may stimulate further escalation.

SUMMARY AND CONCLUSIONS

The theory of adaptation by natural selection among individuals predicts escalation between species and their biological surroundings. As species within particular environments improve their capacities to acquire and defend resources, they pose increasingly high risks to each other and engender further adaptive improvement. Evidence from the Phanerozoic fossil record is generally consistent with this interpretation. Selection is not sufficient, however, to propel escalation. Studies of adaptive change in living populations point to the conclusion that extrinsic change may be necessary to stimulate adaptation. In the absence of such change, ecological limitations imposed by incompatibilities between new variants and already existing risks and traits prevent adaptation even though individuals fall short of being 100% effective in coping with most kinds of hazard. Escalation becomes possible when ecological limitations are relaxed or changed, especially in high-energy environments. Conditions favoring escalation include climatic warming (or the latitudinal spread of warm climates), increasing continental areas that are covered by moist lowland forest or productive shallow marine waters, and the coming together of previously separated biotas.

Differential diversification of highly escalated groups is a second fundamental component of escalation. Sexual selection may provide the link between adaptation against enemies on the one hand and the propensity for speciation by founders on the other.

The historical record of escalation suggests that the greatest episodes (Early Cambrian, Ordovician, Middle Paleozoic, later Mesozoic, and Cenozoic) coincided both with periods of increased primary productivity and with biotic interchanges resulting from shifts in the configuration of oceans and land masses. At least some of the increase in primary productivity is a consequence of the colonization of the dry land by plants during the Middle Paleozoic and of the origin and diversification of angiosperms in the Cretaceous. This is one of several examples of a process that is both stimulated by and favorable to escalation.

Much remains to be learned about the episodic nature of escalation. The precise timing and duration of most of the currently recognized events are not well established. We are equally uncertain about the chronology of marine transgressions and continental movements, especially during the Paleozoic.

Important questions about the dynamics of adaptation in living populations remain unanswered. How much extrinsic change is necessary to reduce ecological limitations sufficiently to allow adaptive improvement to take place, and what effect does the frequency of extrinsic change have on adaptation? What has prevented many species that humans have transported to new surroundings from diverging and diversifying rapidly in these new habitats, where selection is presumably different from that in the original range of the species? Do the populations of successfully introduced species differ in some general way from "natural" colonists whose arrival in new surroundings is often followed by rapid divergence and speciation? Alternatively, is adaptive evolution always unlikely, even in populations that spread without the intervention of humans? Has insufficient time elapsed for species-level evolution to have become evident in human-introduced populations?

If we are to answer such questions, it is important that we monitor introduced populations and their environments. Carefully tracking the hundreds of species that have invaded the Mediterranean from the Red Sea through the Suez Canal may enable us to detect evolutionary divergence, because this kind of biotic interchange, involving a large number of species, is probably a better model for the great episodes of interchange of the past than are the instances of introduction of single species. Moreover, data of this kind will help us to understand the consequences of constructing a sea-level canal across the Isthmus of Panama. This plan would effectively reunite the tropical Western Atlantic and Eastern Pacific biotas, which have been separated for at least three million years (Vermeij, 1978; Glynn, 1982a).

The emphasis in this chapter has been on the extrinsic and intrinsic factors that stimulate adaptation and diversification. Like the events that stimulate escalation, extinction may result in the relaxation of ecological limitations, yet it tends to interfere with escalation, at least in the short run. The role of extinction is the subject of the next chapter.

CHAPTER FOURTEEN

Extinction

A THEORY OF EXTINCTION

The Phanerozoic record of life consists almost entirely of extinct species. This simple observation forcefully reminds us that extinction—the disappearance of populations and species—has been the fate of most lineages, and that population decline must be a common phenomenon even for the survivors. Accordingly, no account of adaptive escalation is complete without an understanding of extinction. The timing, frequency, magnitude, consistency, selectivity, and consequences of extinction all potentially influence the rate of adaptive escalation. My aim in this chapter is to develop a predictive theory of extinction, and to review the empirical evidence bearing on extinction.

Extinction is an extraordinary event in the sense that the normal causes of mortality of individuals in a population are transcended by hazards of unusual severity, frequency, or kind. Agencies of extinction include circumstances with which individuals are unable to cope and to which they are accordingly not adapted, or under which reproduction is prevented. The effects of extinction on escalation depend both on the nature of the agencies that cause populations to decline and become extinct and on the environments in which these agencies are most prevalent. I shall try to show that most extinctions recorded in the fossil record are caused by weather-related catastrophes, and that they tend to interfere with adaptive escalation.

Consider first the possibility that predators or superior competitors are chiefly responsible for extinction. If this were so, escalation would be accelerated, because species with inferior defensive and competitive abilities would be eliminated selectively. Enemies are important agents of death, of course, and they may reduce their victims' populations to the point where the latter become vulnerable to extinction, but they are unlikely to bring about extinction directly in most situations. This assertion rests on the following argument. For an agency of death to eliminate a population, it must be potent enough to kill all the individuals that encounter it, and "safe" places where the agency does not operate must be wholly absent. I pointed out in Chapter 1 that most predators and competitors are much less than 100% effective in locating or encountering potential victims or in causing death during an encounter. As those involved in pest management know only too well (Debach, 1974), biological enemies are therefore rarely able to extinguish an entire population except under unusual circumstances, such as during the invasion of insular biotas by continental species or following the arrival of technologically advanced humans. Even then, complete eradication of a population occurs only if safe places are unavailable.

Knoll (1984) has argued that plants may constitute an important exception to the general unimportance of biologically caused extinction. He points out that plants cannot move away from changes in the environment, and that all plants fundamentally "do the same thing," that is, photosynthesize and tap nutrients. If foreign plant species are introduced into a community, Knoll argues, extinction often results. As supporting evidence, Knoll cites the replacements of ancient plants by mechanically stronger forms with a more effective water-conducting xylem at times when extinctions among plants were frequent. These replacements may well have taken place, but whether the coincident extinctions were caused by superior competitors is in my opinion doubtful. Except on oceanic islands, introduced plants have not generally caused the extinction of native species, even in North America, Europe, and Australia, where hundreds of foreign species have become well established. I therefore view with skepticism Knoll's suggestion that competition played an important role in plant extinctions during the Phanerozoic.

If, on the other hand, extinctions were caused chiefly by the action of weather-related agencies that inhibited primary production, populations of biologically superior species would be at highest risk. This would be so because many of the characteristics that enable individuals to cope effectively with enemies require high metabolic rates and therefore the uninterrupted availability of food. If biological activity were interrupted for a substantial duration, species with high metabolic requirements throughout the life cycle would be susceptible to catastrophic mortality. Sheehan and Hansen (1986) and Arthur and associates (1986) have further pointed out that animals that depend directly or indirectly on living plants for food would be the first to suffer during a crisis in primary production, whereas animals that feed on decomposing organic matter, the availability of which is less tightly bound over the short term to variations in primary productivity, should be less prone to extinction, at least during brief crises. Animals and plants that are able to store food or pass part of the life cycle in a dormant state may be relatively resistant to extinction, even if they are biologically superior during the active part of their life history.

The so-called "safe places" of Chapter 5—environments and situations in which species with poorly developed competitive and defensive capacities live—should be less affected by crises in productivity. These safe places fall into two broad categories: those in which primary production does not take place, and in which fluctuations in physical conditions are small, and those in which interruptions in primary production from day to day or over the seasons are already frequent. Environments at high latitudes and altitudes are examples of safe places of the second type, whereas the deep sea and caves exemplify the first type.

Some weather-related changes may stimulate rather than interfere with primary production. This might occur, for example, when warm climates spread to higher latitudes. Extinction resulting from such changes would fall most heavily on cold-adapted species with relatively modest competitive and antipredatory abilities. The pattern of selectivity of extinction would thus be indistinguishable from that arising from extinction due to enemies.

Two factors—small population size and small geographical range (endemism)—should be associated with high probabilities of extinction. Compared with abundant species, those whose populations are already small and at low density before a crisis may be less able to recoup their losses because the population remnant may be too small to permit successful breeding (Boucot, 1975; Diamond, 1984). Because of their generally smaller population sizes, large-bodied species should be more prone to extinction than small-bodied forms. Similarly, endotherms should be more susceptible than ectotherms, and predators are expected to be more prone than herbivores, omnivores, and decomposers. A small geographical range places a species at risk to catastrophes of relatively small areal extent. Widely distributed populations would decline locally during such crises, but they could survive in unaffected areas (Jackson, 1974). These effects of population size and geographical range are likely to be important, because many (if not most) modern episodes of extinctions and those in the past have been linked to great reductions in habitat area (Schopf, 1974; Hallam, 1976, 1986; Diamond, 1984). No matter which agencies—physical or biological—bring about the final demise of species, reductions in suitable habitats should place high-energy species, which have well-developed methods of coping with enemies but also relatively small population sizes, at greatest risk.

An apparent contradiction to this expectation arises from Schopf's (1974) observation that habitat reduction results in more intense competition and predation among survivors as they are squeezed into ever smaller spaces. High-energy species, according to this argument, should do especially well as the costs of losing to enemies increase in the diminishing habitats. This argument may not, however, be valid. As deterioration continues, the habitats that remain become increasingly island-like and are therefore expected to lose species whose individuals have the highest food requirements (and therefore the smallest population sizes) first. Accordingly, there may be no contradiction between the claim that competition intensifies during episodes of habitat reduction and my assertion that high-energy species are especially prone to extinction at those times.

In short, habitat reduction and the imposition of production-inhibiting crises should halt or reverse escalation, whereas extinctions caused by global warming or by biological agents should generally stimulate it. High-energy animals that migrate over long distances, as well as intensely bioturbating animals, may be the only exceptions to the generalization that high-energy species are most susceptible to extinction.

There are good reasons to believe that the negative effects of extinction on escalation should predominate over the stimulative effects. I have already argued that enemies are unlikely to bring about extinction except under a limited number of circumstances. Similarly, global warming (which stimulates primary production) is expected to be associated with a low rate of extinction. The chief reason for this is that a steep latitudinal gradient in organic diversity has apparently always existed during the Phanerozoic. At high latitudes, the diversity of life is generally low, whereas in the tropics it is very high (A. G. Fischer, 1960; Stehli et al., 1969). There are some exceptions to this pattern—oaks, pines, brown algae, chitons, limpets, forcipulate sea stars, and articulate brachiopods are today more diverse in the temperate zones than in the tropics, for example—but they are insufficient to invalidate the generalization that the number of species increases toward the equator. During times of global warming, relatively few species will become extinct because the number of vulnerable high-latitude and deep-sea species is small compared to the number of low-latitude species whose populations are able to expand and spread. Conversely, cooling adversely affects a very large number of low-latitude species and benefits only a relative handful of cold-adapted species. This argument thus leads to the expectation that cooling and other production-inhibiting changes precipitate the extinction events of greatest magnitude.

If interruption in primary production is the most important cause of extinction, the species that are well adapted to tolerate low food supplies for long periods should have an advantage during times of crisis. These species fall into two contrasting groups. The first group is characterized by low metabolic demands, slow growth, and low fertility. Species of this kind are found in areas where there is little or no primary production, or in environments where activity is strongly constrained by enemies (see Chapter 2). These species thus occupy habitats that are little affected by the agency of extinction, or they are themselves resistant to the agencies of extinction.

The second group of species consists of opportunists. In the active phase of the life cycle, they have high growth rates, high fertility, and typically a short life span. They are able to take advantage of temporarily plentiful resources, and they pass through unfavorable conditions in an inert or resting phase (see Chapter 4). These attributes make opportunists highly successful in environments that have been devastated by fires, storms, floods, earthquakes, and human-caused habitat destruction. In populations that are able to expand exponentially, mutants that further increase fertility are apt to be favored over those that promote individual survival (Lewontin, 1965). With their high fertilities and growth rates, opportunists are best described as short-lived high-energy organisms with a great evolutionary potential. Moreover, most opportunists are of small body size, an attribute usually associated with a lack of adaptive specialization in any particular direction (S. M. Stanley, 1973; Gould, 1977). The combination of small size and high energy flux in opportunistic species therefore provides an evolutionary point of departure for any of several adaptive pathways, most of which would be unavailable in larger species.

Survival during a crisis is no guarantee of subsequent participation in evolutionary events. This is particularly true for species with low metabolic demands. As environments that are affected by a crisis are occupied by organisms, low-energy species with low rates of population increase will be at a disadvantage relative to opportunists, which are adapted to exploit suddenly favorable situations. Even if low-energy species did become established, they would remain at risk for a long time, because their population size would increase only slowly, and because small populations are less likely than large ones to persist should another catastrophe strike. Accordingly, I expect opportunists to be the most important members of post-crisis communities in high-energy environments, and surviving low-energy species to remain confined to low-energy situations (Vermeij, 1986b).

Recolonization of environments by opportunists after crises has important implications for escalation. Because of their high energy flux, opportunists are expected to have a large adaptive scope (Chapter 2) and to provide excellent "raw material" for the evolution of enhanced competitive and antipredatory characteristics. Because many of the usual enemies will have been exterminated during the crisis, ecological barriers in early post-crisis communities are expected to be low, and many otherwise unavailable evolutionary pathways may be entered by descendant lineages. Thus, although the short-term consequence of a crisis that is brought on by an interruption in primary production is that escalation ceases, the long-term effects may allow escalation to proceed further than was possible before the crisis.

It must be emphasized that, even if a population possesses all the characteristics that have been linked to a low probability of extinction, the population may still disappear during a crisis. If the population is very small, for example, there may be so much inbreeding that many deleterious or lethal recessive mutants will be expressed in homozygous form, and many individuals will die. We should therefore expect patterns of selectivity during episodes of extinction to reflect statistical biases rather than absolute rules.

How does the empirical record of extinction accord with expectation? Despite a huge burst of important recent work on extinction, we still know shockingly little about this process. Nevertheless, in the remainder of this chapter I shall try to show that the available data are on the whole consistent with the general framework outlined above.

TIMING, FREQUENCY, AND MAGNITUDE OF EXTINCTION

If extinction generally interfered with escalation, the timing of major extinction events should correspond to periods when escalation ceased or reversed. To evaluate this hypothesis, we must briefly review the chronology of extinction during the Phanerozoic.

Thanks to Sepkoski's (1982b) worldwide compilation of the stratigraphic ranges of families of marine animals, it has become possible to specify the timing, frequency, and magnitude of the crises. Sepkoski's data are accurate to the level of the stage. In the Mesozoic and Cenozoic, the average stage has a duration of about six million years, but Paleozoic stages are usually longer than ten million years in duration. Determining the frequency and magnitude of extinction is therefore substantially less precise for the Paleozoic. More taxa will appear to have become extinct during a Paleozoic stage than during a post-Paleozoic one, because several episodes of extinction may have occurred in a single stage and have been counted as a single end-of-stage event.

The choice of the family as the level at which to measure the impact of extinction is in many ways not ideal. The family is an artificial unit in most instances; its limits are determined by a subjective appraisal of the degree of similarity among its members. Moreover, specialists whose work is confined to one period or era may be unaware that a given family extends to intervals of time whose fossils are unfamiliar to them. The magnitudes of extinction given below and by Sepkoski and his colleagues must therefore be regarded as rough estimates that are subject to all the biases and artifacts inherent in a global analysis of data at a high taxonomic rank (Chapter 3). One can do little more than hope that the preservational and taxonomic problems cancel out, and that the observed pattern retains the essential properties of the species-level history of extinction.

Despite these problems, several time intervals in Phanerozoic history seem to have been associated with unusually high rates of extinction. These episodes, which usually can be recognized in local stratigraphic sequences, have come to be known as *mass extinctions*, implying that the extinctions were sudden and simultaneous for many species in many places. Although only the Cretaceous-Tertiary boundary crisis has been sampled sufficiently to warrant the conclusion that extinctions were concentrated during a very short period of time, I shall use the term "mass extinction" for convenience.

Two major, three or four somewhat smaller, and nine or ten minor extinction events have been documented thus far from Sepkoski's compilation of marine animal families (Raup and Sepkoski, 1982, 1984; Sepkoski, 1982a, 1984, 1986; Rampino and Stothers, 1984; Van Valen, 1984, 1985a). The events that in terms of magnitude of extinction departed most dramatically from those of adjacent time intervals are the crises at the end of the Permian and Cretaceous periods. Slightly less dramatic increases in the magnitude of extinction relative to those of surrounding stages occurred at the end of the Ordovician (Ashgillian event), during the Late Devonian, at the end of the Triassic, and perhaps during the Vendian.

Little is known about extinction early in the Phanerozoic. The earliest episode of extinction documented thus far came during the Early to Middle Vendian (or Ediacarian), when about 70% of the species of acritarchs and other organic-walled organisms, which are thought to have been phytoplankters, disappeared (Vidal and Knoll, 1982). Trilobites and perhaps other invertebrates were decimated at least five times during the Cambrian. These crises, which particularly affected tropical species, came at the end of the Early Cambrian (top of the "olenellid" biomere), in the Middle Cambrian (top of the corynexochid biomere), and three times in the Late Cambrian (tops of the marjumiid, pterocephaliid, and ptychaspid biomeres). Careful work by A. Palmer (1984) has shown that the extinctions of the Late Cambrian occurred over narrow stratigraphic intervals (within a rock thickness of 1 to 10 cm).

The last stage of the Ordovician (Ashgillian stage) marked one of the great extinction events of the Phanerozoic. About 27% of existing families of marine animals became extinct during this stage, which lasted about five million years. Sheehan (1986) has shown that, at least among brachiopods, taxa inhabiting tropical epicontinental seas were more affected than those from oceanic and high-latitude sites.

The Late Devonian was another time of biological impoverishment. The extinction was especially marked at the Frasnian-Famennian boundary (McLaren, 1982), but the rate of extinction was high throughout the Late Devonian. About 15% of the standing diversity of families disappeared during the Givetian stage, as compared to 20% during the succeeding Frasnian and 24% during the Famennian (Van Valen, 1984). Ammonoids, brachiopods, trilobites, echinoderms, stromatoporoid sponges, and corals (Rugosa and Tabulata) were particularly affected. Extinction was especially profound in warm, shallow waters (Copper, 1977). Only 4% of shallow-water rugosans of Frasnian age persisted into the Famennian, as compared with 40% of deep-water species (Pedder, 1982). The only group that diversified across the Frasnian-Famennian boundary was the hexactinellid sponges, many of which may have lived in cold, deep water (McGhee, 1982). For most warm-water species, then, conditions during the Late Devonian went from feast to Famenne.

The greatest crisis of the Phanerozoic occurred during the Late Permian. High rates of extinction began during the Leonardian stage, when about 13% of the marine families then living became extinct. Losses were far greater during the succeeding Guadelupian (44%) and Dzhulfian (36%) stages of the Late Permian (Van Valen, 1984). Groups that were completely eradicated include blastoid echinoderms, trilobites, eurypterids, rugosan corals, fusulinacean foraminifers, acanthodian fishes, rostroconch molluscs, and all crinoid subclasses except the ancestors of the Articulata. Echinoids, ammonoids, ostracodes, and articulate brachiopods were reduced to a handful of surviving species.

Hard on the heels of this end-Paleozoic crisis came the extinctions of the Late Triassic. About 15% of Carnian and 21% of Norian marine families disappeared. More than half of the genera of Triassic pelecypods died out by the end of the Triassic (Hallam, 1981c). Conodonts were eliminated, as were most ammonoids and many articulate brachiopods.

Extinction during the Jurassic and Cretaceous was generally regional and of small magnitude. Only about 7% of families disappeared during the Pliensbachian or early Toarcian stage of the Early Jurassic, and about 8% were eliminated during the Tithonian stage of the latest Jurassic. Neither of these extinctions was worldwide in extent (Hallam, 1986). Additional minor episodes occurred at the ends of the Aptian (Early Cretaceous) and Cenomanian (Early Late Cretaceous).

Although the crisis at the end of the Cretaceous eliminated only 15% of marine families, the magnitude of this extinction exceeded that of any other post-Triassic event. This crisis marked the final appearance of ammonoid and belemnoid cephalopods, most marine reptiles, and rudist pelecypods, and it decimated planktonic coccolithophores, dinoflagellates, and foraminifers.

Because family-level extinctions during the Cenozoic occurred at a very low rate, it is difficult to discern times of mass extinction. The late Eocene has been suggested as a time of substantial extinction in planktonic foraminifers (Thunell, 1981), and the latest Miocene and perhaps the Pliocene witnessed the extinction of many molluscan genera (Stanley and Campbell, 1981; Vermeij and Petuch, 1986; Raffi et al., 1986). Corliss and his colleagues (1984), however, have disputed the claim that the end of the Eocene marked an important time of extinction; they suggest instead that the Late Eocene and Oligocene were punctuated by several episodes of impoverishment.

Episodes of extinction on land have not been as well documented as those in the sea chiefly because the terrestrial fossil record is much spottier. The most important episodes at the familial level among tetrapods occurred during the Early and Middle Permian (Sakmarian to Artinskian), the Late Permian to Early Triassic (Tatarian to Scythian), the Late Triassic to Early Jurassic (Norian to Hettangian), the Late Cretaceous (Maastrichtian), the Middle Oligocene, the Late Miocene, and perhaps the Late Jurassic (Bakker, 1977; Tucker and Benton, 1982; Benton, 1985; Prothero, 1985). According to Benton's (1985) analysis, the magnitudes of these episodes have decreased over time. Several episodes of extinction punctuate the Neogene record, the most recent being at the end of the Pleistocene (Marshall, 1981; Marshall et al., 1982; Diamond, 1984; Martin, 1984; S. D. Webb, 1984).

Catastrophic extinction apparently has not occurred among land plants. Knoll (1984), who has provided the most thorough overview of extinction in plants, finds broad extinction peaks during the Late Devonian and Early Carboniferous, the Early Permian to Middle

Triassic, and the Late Cretaceous. None of the great crises indicated by the record of marine animals and terrestrial vertebrates is visible in the plant record. As Knoll points out, the persistence of species does not mean that productivity continued uninterrupted through these crises. Plants may lose all their leaves and still survive to re-initiate activity when conditions become favorable once again, provided that they have mechanisms for storing food and for tolerating extreme conditions.

Inspection of the family-level data suggested to Raup and Sepkoski (1984) that the great crises of the Phanerozoic, especially those of the Mesozoic and Cenozoic eras, have occurred periodically at intervals of every 26 million years. The length of this period has been estimated by others to be 30 to 32 million years (Thomson, 1976, 1977; Fischer and Arthur, 1977; Kitchell and Pena, 1984; Rampino and Stothers, 1984). I believe the family-level data are as yet too crude to warrant unequivocal acceptance of this hypothesis, even if statistical tests of the pattern of extinction show a highly significant departure from random. If the crises do in fact come approximately every 30 million years, we should expect to see periodic halts or reversals in adaptive escalation.

THE DECREASE IN EXTINCTION RATE

Analysis of Sepkoski's (1982b) compilation of the stratigraphic ranges of marine animal families reveals that the rate of extinction of families has declined through the Phanerozoic. This decrease, which proceeded at a faster rate during the Paleozoic than during the Mesozoic and Cenozoic, applies to the crises as well as to the times of "normal" or "background" extinction (Raup and Sepkoski, 1982; Kitchell and Pena, 1984; Sepkoski, 1984; Van Valen, 1984, 1985a). The reasons for this decrease are still uncertain. Raup and Sepkoski originally suggested that the decline in the rate of extinction implies that the fitness of species, or at least that of families, generally increased through time. Sepkoski (1984) and Flessa and Jablonski (1985) have since modified this interpretation. They suggest that ancient families with few members, high rates of diversification in the early history of the higher categories to which they belong, and high rates of extinction were replaced in time by larger families whose rates of diversification and extinction were lower. In support of this claim, Sepkoski has pointed out that the dominant groups of the Mesozoic and Cenozoic (gastropods, pelecypods, echinoids, malacostracan crustaceans, and osteichthyan fishes, for example) had low rates of extinction even during the Paleozoic, whereas groups that reached their maximum diversity during the Paleozoic (trilobites, crinoids, conodonts, and brachiopods, for example) have always had higher rates of extinction. Moreover, post-Paleozoic extinction rates of anthozoans, malacostracan and ostracode crustaceans, and bryozoans were significantly lower than Paleozoic rates in these same groups (Van Valen, 1985a). On the other hand, no temporal decline in the rate of extinction is detectable in echinoids, gastropods, pelecypods, brachiopods, cephalopods, foraminifers, and poriferans (Van Valen, 1985a). Evidently, the Mesozoic and Cenozoic families of these groups are not more resistant to extinction than were their Paleozoic antecedents, so that the explanation put forward by Flessa and Jablonski does not apply in all cases (Van Valen, 1985a).

I suspect that the decrease in the rate of extinction at the family level is partly an artifact of classification and preservation, and partly a consequence of the progressive occupation of marginal environments where extinction is rare. Families persist even if only one of their component species survives a crisis. As sampling in the fossil record improves toward the Recent, the expected size of families increases, as does the probability that at least one of the species will be found to persist during a crisis. Moreover, if habitats such as the deep sea, cryptic surfaces, and deep sediment layers are invaded by even a few species in a family, the probability that the family will survive a catastrophe is rather high.

If the reduction in the rate of extinction is real, it might have contributed to escalation during the Mesozoic and Cenozoic. We can speculate, for example, that the more recent mass extinctions slowed or reversed escalation less than did the earlier extinctions. At present it is impossible to evaluate this possibility.

SELECTIVITY

INTERRUPTIONS IN PRIMARY PRODUCTION

With so many episodes of mass extinction, it is perhaps unreasonable to expect that a single underlying cause is responsible. Nevertheless, S. M. Stanley (1984) has amassed data from various lines of evidence (sedimentology, oceanography, geology, and geochemistry) to argue persuasively that most of the mass extinctions were associated with unusually cold (though not necessarily glacial) conditions. His generalization applies even to the renowned mass extinction at the end of the Cretaceous, which is marked in the sedimentological record by unusually high concentrations of iridium and other elements that have been linked to both the impacts of extraterrestrial bodies and intense volcanism (Alvarez et al., 1984a, b; Officer and Drake, 1985). Most mass extinctions coincided with low stands of sea level, which could have resulted from the incorporation of sea water into glaciers. High-latitude glaciation has been documented for the crises at the end of the Ordovician and during the Late Devonian and the Late Cenozoic (Crowell, 1982). Cooling during the Eocene and Oligocene was associated with the extinction of many planktonic organisms (Corliss, 1979; Keigwin, 1980; Corliss et al., 1984; Prothero, 1985). The Late Cretaceous is not known to have been a time of ice formation, but high levels of dust and smoke resulting from extraterrestrial impacts, volcanism, and associated large-scale forest fires could have shielded the earth's surface from sunlight and thus brought on cooling (Hsu et al., 1982; Wolbach et al., 1985). The only episode of glaciation for which no mass extinctions are recorded occurred in the Late Paleozoic in the southern hemisphere.

Additional evidence that warm-water organisms have historically been more prone to extinction during crises than those in cold or deep waters comes from many groups. This pattern has been found in Late Cambrian trilobites (A. Palmer, 1984); Late Ordovician and Early Silurian brachiopods (Sheehan, 1975, 1982); Late Devonian brachiopods and corals (Copper, 1977; McLaren, 1982; Pedder, 1982); Late Cretaceous gastropods and cephalopods (Kollmann, 1979; Birkelund and Hakansson, 1982); and Cretaceous and Cenozoic planktonic foraminifers (Lipps, 1970; Cifelli, 1976; Fisher and Arthur, 1977; Hsu et al., 1982) and pelecypods (Raffi et al., 1985).

Cold is not the only agency that has been cited as a cause of the great crises, but the other agencies that have been suggested would similarly have brought about an interruption in primary production. Geochemical evidence for such an interruption has been found near the base of the Cambrian (Hsu et al., 1985) and in the boundary clay separating Maastrichtian (Late Cretaceous) from Danian (Paleocene) strata (Arthur et al., 1986). Some paleoceanographers have held rapid changes in salinity responsible for the end-Permian and end-Cretaceous crises (Holser, 1977; Thierstein and Berger, 1978). These changes were brought about when deep basins in which various salts (notably gypsum and halite) were being precipitated were re-invaded by the ocean as sea level rose (Degens and Stoffers, 1976). A sudden increase or decrease in salinity might have serious consequences for photosynthesis by phytoplankton even if it did not extinguish whole populations.

A more plausible suggestion is that widespread extinction was brought on by large-scale upwelling of high-salinity water that was rich in nutrients but poor in oxygen (Wilde and Berry, 1984). The high concentration of sulfur compounds and other toxins in this water would be lethal to most marine organisms. Such widespread upwelling is, of course, rare. During times of warm worldwide climate, deep waters in the ocean are formed by the sinking of warm high-salinity waters at low latitudes, where the rate of evaporation is higher than the rate of precipitation (Brass et al., 1982). Because the oxygen content of warm water is lower than that of cold water, this high-salinity water rapidly loses oxygen as a result of respiration by organisms. Therefore, deep water during time of warm worldwide climate tends to be poor in oxygen. Usually, the deep water stays below the photic zone, but when climates at high latitudes cool, the chief source of deep ocean water becomes cold, oxygen-rich, low-salinity water from high latitudes. At the time of this switch, various oceanographic conditions can bring about a massive upwelling of the anoxic deep water that originated in the tropics. Small-scale modern episodes of mass mortality caused by this switch from oxygen-rich productive to oxygen-poor unproductive waters accompany the well-known El Niño events of the coasts of Peru and Chile.

That cooling had a disproportionate effect on species with well-developed antipredatory traits is suggested by data on extinction in tropical America after the Pliocene uplift (3.1 to 3.5 million years ago) of the Central American isthmus (Vermeij and Petuch, 1986). The subgeneric diversity of gastropods in the so-called Atlantic Gatunian region (corresponding to the Atlantic coasts of Central and South America and the West Indies) was reduced by about 36%, whereas in the Pacific Gatunian area (corresponding to the Pacific coast of tropical America) subgeneric diversity declined by only 15%. Hard-bottom gastropods with a narrow or thick-lipped aperture (cypraeaceans, conids, mitrids, and columbellids, for example) had an extinction probability of 32% in the Atlantic Gatunian region, whereas forms with a broad or thin-lipped aperture (many buccinids, fasciolariids, and muricaceans, for example) had only an 8% probability of extinction there. Gastropods with morphological specializations for resisting crushing or extraction by predators were therefore substantially more prone to extinction than were other gastropods in the Western Atlantic. In the Pacific Gatunian, however, the well-armored forms actually had a somewhat lower probability of extinction (3%) than that of the morphologically less-specialized open-apertured group (15%). The reasons for this interoceanic difference in selectivity of extinction are unclear, but it is notable that the pattern of selectivity in the Western Atlantic, where Neogene temperatures fluctuated considerably and productivity declined, is consistent with what would be expected under such conditions. In the Eastern Pacific, there was no reduction in productivity (Keigwin, 1982), and there was less extinction.

Further support for the hypothesis that large-scale extinction selectively eliminates well-armored molluscs comes from the observation that there exists a negative correlation between the degree of development of antipredatory characteristics and the post-Pliocene magnitude of extinction in gastropods. The tropical Indo-West-Pacific region has suffered the least biological impoverishment and retains the most specialized array of armored gastropods, whereas the tropical Atlantic has sustained the greatest loss of species and now boasts the least expression of armor. The tropical Eastern Pacific holds an intermediate position in both respects (Vermeij and Petuch, 1986; Vermeij, 1986b).

Other interesting examples come from the Neogene history of infaunal pelecypods. Both the North Sea and Mediterranean faunas suffered substantial losses (41% and 51% of infaunal burrowing pelecypod species, respectively) as the result of cooling during the later Pliocene and Pleistocene. My calculations from data presented by Raffi and colleagues (1985) show that infaunal burrowing species with crenulated valve margins (which promote shell strength and tight valve closure) were somewhat more prone to extinction in both the North Sea (18 of 28 species, 64%) and the Mediterranean (18 of 53 species, 34%) than were permanently gaping species (in the North Sea 6 of 14 species, 43%; in the Mediterranean 4 of 19 species, 21%). Extinction was less marked in California (S. M. Stanley, 1982b, 1986). Only 12 of the 68 infaunal burrowers (18%) in the Late Pliocene San Diego Formation are extinct today. By my calculations based on Stanley's data, crenulated species suffered somewhat more extinction (7 of 20 species, 35%) than did gaping forms (2 of 11 species, 18%). None of the differences between crenulated and gaping species with respect to extinction is statistically significant, but all differences are in the expected direction.

Data in Table 14.1 reveal that, for most of the Jurassic stages, the probability of extinction of European infaunal burrowing pelecypods with crenulated valve margins approximated that of species with gaping valves. The only possible exception occurred during the Pliensbachian (Early Jurassic), when overall extinction was high, but the number of species is too small for adequate statistical analysis. Again, the difference during the Pliensbachian, an interval that Raup and Sepkoski (1984) identify as one of the minor mass extinctions, is in the expected direction.

TABLE 14.1 **Extinction Probabilities of Jurassic Pelecypod Species with Crenulated Valve Margins or Gaping Margins[a]**

Stage	Crenulated Species		Gaping Species		Other Species	
	N[b]	%[c]	N	%	N	%
Hettangian	6	0	8	0	15	7
Sinemurian	3	33	5	20	12	17
Pliensbachian	4	100	9	67	19	74
Toarcian	5	20	3	33	17	35
Aalenian	13	0	11	0	34	5
Bajocian	11	18	11	18	26	31
Bathonian	16	25	15	13	27	33
Callovian	13	15	12	25	25	20
Oxfordian	13	31	10	20	28	29
Kimmeridgian	7	14	11	18	27	44

[a] Calculations are from Hallam's (1976) compilation of European infaunal burrowing pelecypod species.
[b] N   Number of species.
[c] %   Percentage of species that became extinct.

A similar pattern emerges from my analysis of Dzik's (1984) compilation of the stratigraphic ranges of Paleozoic nautiloid cephalopod genera (Table 14.2). Taxa with antipredatory armor (strong sculpture, shell coiling, or a constricted aperture) were similar to morphologically unspecialized genera in susceptibility to extinction in most Paleozoic stages, but they were significantly more prone to extinction than were other nautiloids ($p < 0.05$, Chi-square test) during the Ludlovian (Late Silurian) and Visean (Early Carboniferous). The pattern during the Givetian and Frasnian stages of the later Devonian was similar but not statistically significant. Only during the Caradocian stage of the Late Middle Ordovician was the susceptibility of armored nautiloids to extinction less than that of other nautiloids ($p < 0.05$). It is noteworthy that the Ashgillian stage (latest Ordovician), one of the great crises, was a time when there was no clear distinction between armored and unarmored nautiloids in susceptibility to extinction.

The tendency for large-bodied animals to have been especially prone to extinction is consistent with expectation. Large size is usu-

TABLE 14.2 Extinction Probabilities of Paleozoic Nautiloids with and without Specializations for Armor[a]

Stage	Armored Genera N[b]	%[c]	Other Genera N	%	Stage	Armored Genera N[b]	%[c]	Other Genera N	%
Arenigian	13	60	34	59	Emsian	8	38	27	42
Llanvirnian	15	47	35	57	Eifelian	14	50	25	52
Llandeilian	9	44	26	35	Givetian	19	74	20	50
Caradocian	16	31	50	60	Frasnian	13	54	16	19
Ashgillian	17	71	32	75	Tournaisian	12	17	11	18
Llandoverian	12	33	23	35	Viséan	31	71	13	15
Wenlockian	17	35	35	17	Namurian	13	46	13	46
Ludlovian	24	83	44	52	Westphalian	19	31	11	36
Pridolian	7	43	44	52	Stephanian	19	31	5	60
Gedinnian	6	33	15	27	Sakmarian	12	8	4	25
Siegenian	7	14	19	21	Artinskian	17	35	6	33

[a] Calculations are based on Dzik's (1984) compilation of Paleozoic nautiloid cephalopod genera. Armored cephalopods include those with strong ribs or tubercles, shell coiling, or a constricted aperture.

[b] N Number of species.

[c] % Percentage of species that became extinct.

ally associated with competitive dominance and with effective protection against predators (Morse, 1974; Vermeij, 1982d). Moreover, large animals usually have small population sizes. Documentation of the greater susceptibility of large species to extinction is available for mammals and other terrestrial vertebrates (Van Valen, 1975; Bakker, 1977; Diamond, 1984; Martin, 1984); solitary corals (Coates and Jackson, 1985); Jurassic pelecypods and ammonoid cephalopods (Hallam, 1975, 1978); Neogene pelecypods (S. M. Stanley, 1986); and Pleistocene North Sea gastropods (Vermeij, 1986a).

Another indication that competitively superior animals are especially prone to extinction comes from the history of hermatypic animals (see Chapter 5). Hermatypic foraminifers, brachiopods, and pelecypods were completely eliminated during the end-Permian crisis. Rudist pelecypods were wiped out at the end of the Cretaceous. The high susceptibility of hermatypic animals to extinction during the crises may help to explain the observation that reefs, which owe their construction in large part to hermatypes, were virtually eliminated during some of the crises, notably those of the Devonian and Permian (N. D. Newell, 1971; Copper, 1974).

There are some troublesome exceptions to these generalizations. Despite their large size and aggressive behavior toward enemies, freshwater crocodiles have persisted with little change since the Late Cretaceous (E. R. Meyer, 1984). Similar immunity may have been enjoyed by large freshwater fishes.

My final example of selective extinction of animals with developed antipredatory traits comes from the Devonian gastropods tabulated by Linsley (1979). I divided the gastropod genera into two groups: immobile forms (macluritaceans, euomphalaceans, some platycerataceans, and several other gastropods with a radial aperture, loose coiling, and other characteristics indicating a sedentary mode of life) and mobile forms. In most Devonian stages, the two groups did not differ significantly in magnitude of extinction, but in the Givetian stage, when the magnitude of extinction reached a peak, immobile gastropods were far more prone to extinction than were mobile forms ($p < 0.01$; see Table 14.3). I have no ready explanation for this pattern.

The foregoing examples suggest that species or genera with well-developed capacities to deal with competitors and predators have a greater than average susceptibility to extinction at times when the magnitude of extinction is high. A great deal remains to be learned

TABLE 14.3   Extinction Probabilities of Devonian
Immobile and Mobile Gastropods[a]

Stage	Immobile Genera		Mobile Genera	
	N[b]	%[c]	N	%
Gedinnian	12	8	58	7
Siegenian	18	39	83	28
Emsian	11	36	85	26
Eifelian	15	13	77	11
Givetian	17	65	82	39
Frasnian	7	14	62	21
Famennian	6	17	51	18

[a] Calculations are based on Linsley's (1979) compilation of stratigraphic ranges of Devonian gastropod genera. Immobile gastropods include most Bellerophontacea, Macluritacea, Euomphalacea, Platyceratacea, and open-coiled forms.
[b] N   Number of species.
[c] %   Percentage of species that became extinct.

about the selectivity of extinction, however, before we can accept with confidence the hypothesis that interruptions or reductions in primary productivity differentially affect highly escalated species. In determining patterns of selectivity, it is especially important that we compare times of "background" (or minor) extinction with times of mass extinction.

One way in which mass extinctions may have accentuated long-term trends in escalation is by favoring the persistence of intense bioturbators at the expense of species that either live on the surface of sediments or disturb the sediment only slightly (Thayer, 1983). This pattern of selectivity accords with the hypothesis that agencies of extinction are most effective at the earth's surface and increasingly less potent at greater depths in the sediment, where most intense bioturbators live (see Chapter 5).

BIOLOGICAL AGENTS

I suggested earlier in this chapter that enemies generally have not been important in bringing about extinction and that, when they do cause the demise of populations, adaptive escalation is promoted. To be sure, there are many documented cases of extinction in which biological agents played a crucial role. Most of these cases involved man or species that man has introduced onto small islands (Diamond, 1984; Vermeij, 1986a). Rats, malarial protists, and the clearing of forests by man have all been implicated in the disappearance of over half the endemic bird fauna of the Hawaiian Islands since the arrival of Polynesians there less than a thousand years ago (Olson and James, 1982). Grazing by sheep and competition from introduced African grasses eliminated about 4 of the 25 native plant species on Ascension Island in the South Atlantic (Cronk, 1980). Introduction of the predaceous snail *Euglandina rosea* from Florida for the purpose of controlling the herbivorous African land snail *Achatina fulica* on Moorea in Polynesia is causing the elimination of the endemic partulid land snails within a matter of a few years (B. Clarke et al., 1984). Similar stories can be related for most of the world's insular biotas as they are exposed to animals and plants brought there from the continents, and as man himself is altering the environment.

A strong (and in my opinion compelling) case has been made for the hypothesis that humans have brought about the extinction of large land mammals and reptiles on the continents (Diamond, 1984; Martin, 1984), especially on those continents that man has invaded within the last 30,000 years (Australia and the Americas). Deforestation, pollution, drainage of marshes, damming of rivers, and hunting are man-induced changes that have caused the extinction of continental birds, butterflies, flowering plants, freshwater animals, and a few large, coastal marine birds and mammals. It is difficult to attribute any extinction during the Late Pleistocene or the succeeding Holocene to agencies other than man. The disappearance of large scavenging birds in North America may have been caused by the extinction of land mammals on whose carcasses they depended (Steadman and Martin, 1984), but the mammals themselves were most likely the victims of man.

The pattern of selectivity of human-caused extinctions is complicated by the fact that some of the extinctions resulted from competition and predation by man and the species that accompanied him, whereas others were caused by the destruction of habitat. Extinctions of the first type should have been most frequent among species with poor competitive and defensive abilities, whereas biologically superior species would have been most susceptible in cases of habitat destruction.

The most dramatic extinctions that have been linked to human intervention have occurred in regions colonized by man within the last 30,000 years. Until the great technological advances of the last two centuries, extinction in Africa and Asia was less dramatic than in Australia or the Americas, continents occupied by man only from the late Pleistocene onward. This fact accords well with the long history of the genus *Homo* in Africa and Asia (Klein, 1984; Martin, 1984). Martin has pointed out that the only large mammals that survived in North America after the arrival of man were Pleistocene immigrants from Asia (deer, bison, sheep, musk ox, and bear, for example). The most devastating effects of man have been felt on oceanic islands, whose biotas are characterized by low fecundities, slow individual growth rates, and generally slight development of competitive and defensive traits. Flightless birds, slow-moving (and often armored) mammals and reptiles, diurnal ground-dwelling birds, and colonial nesters have been especially prone to extinction on islands and on the continents of Australia, South America, and Madagascar (Cassels, 1984; Diamond, 1984). Animals on islands that were already occupied by rats and land crabs before man's arrival were apparently less affected by man and associated domesticated animals than were islands in which rats and crabs were not indigenous (Diamond, 1984). Now that habitat destruction is becoming the most important cause of extinction, the selective loss of biologically inferior species may well be reversed (Vermeij, 1986b).

The arrival of foreign species into a biota is not a phenomenon unique to man's migration. Exchanges between biotas with different evolutionary histories must have occurred frequently throughout the Phanerozoic and could have precipitated extinction at times when continents came together or seaways were opened between previously isolated oceans. The possibility that some extinctions resulted from biotic interchange on land cannot be excluded, but in the sea the evidence suggests that extinctions due to biological agents have been rare.

Faunal interchange between North and South America began about 9 million years ago during the Late Miocene and greatly accelerated after the Panama land bridge was definitely established during the Middle Pliocene, about 3.1 to 3.5 million years ago (Marshall,

1981; Marshall et al., 1982). In his thorough analysis of the interchange, Marshall (1981) has pointed to several possible examples of competitive exclusion by predators. Predaceous borhyaenid marsupials disappeared at about the same time during the Late Miocene that North American Procyonidae (raccoons and their relatives) first appeared in South America. The last thylacosmilids coincided with the first North American machaerodontine saber tooths in the early Pleistocene. Dogs and cats (Canidae and Felidae) may also have excluded South American predaceous birds of the family Phororacidae. As Marshall points out, however, these inferences are based on broad morphological and ecological similarities between native and immigrant groups, not on estimates of effectiveness, and they are therefore weak.

S. D. Webb (1984) has suggested that migrants from Asia were possibly responsible for the extinctions of native North American mammals during the Pliocene and Pleistocene. Migration began during the Pliocene (Blancan land-mammal age) and continued intermittently throughout the Pleistocene. It brought deer, cricetid rodents, bears, and the flying squirrel *Cryptopterus* during the Pliocene, and large deer, bovids, mammoths, and small microtine rodents during the Middle and Late Pleistocene. No studies of the correspondence between the timing of extinction and the timing of immigration have been undertaken; neither have details about the possible mechanisms by which Asian mammals might have overwhelmed their North American counterparts been worked out.

Earlier episodes of interchange between continental biotas may also have led to extinction. In their detailed studies of the extinction of dinosaurs in western North America, R. E. Sloan and associates (1986) point out that the disappearance of dinosaurs is coincident with the arrival from Asia of primitive ungulate and multituberculate mammals, which possibly outcompeted the dinosaurs. There was also significant biotic interchange between North and South America during the Late Cretaceous.

Interchanges between marine biotas have also occurred frequently in the course of earth history, but evidence that they brought about extinction is weak at best. The most thoroughly studied interchange is that between the Red Sea and the Mediterranean via the Suez Canal, which was opened for navigation in 1869. Nearly all the movement of species has been one way from the Red Sea to the Mediterranean. Many changes in population size and distribution of native Mediterranean species have taken place at a result of interactions with immigrants from the Red Sea (Ben-Yami and Glaser, 1974; Vermeij, 1978), but only one case of extinction has been proposed, and it is only weakly supported. It involves the complete replacement, after 1955, of the Mediterranean sea star *Asterina wega* by *A. gibbosa* from the Red Sea (Achituv, 1973). Because the sea stars are rare, nothing is known about the attributes that enabled *A. gibbosa* to overwhelm *A. wega*. The Suez case is particularly interesting, however, because the interchange involves two biotas whose members differ strikingly in the development of aptations to predation (Vermeij, 1978). Molluscs of the Red Sea show a higher incidence and greater expression of crushing-resistant armor than do their Mediterranean counterparts, and there is a corresponding difference in claw morphology between mollusc-crushing crabs in the two seas. If the interchange between such very different biotas as those of the Red Sea and Mediterranean did not result in widespread extinction, interchanges involving less divergent biotas are even more unlikely to have done so.

Two other cases of biotic interchange in the sea yield similar results. The wholesale invasion of the North Sea basin by species from the North Pacific during Pliocene time seems not to have led to the extinction of endemic North Sea molluscs, yet extinction devastated both the immigrant and the endemic component of the North Sea fauna during the Pleistocene, presumably because of glacial cold (Strauch, 1970, 1972; Vermeij, 1978; Raffi et al., 1985). Immigration of species from the tropical Central Pacific to the Eastern Pacific after the Pliocene has not resulted in any obvious depletion of Eastern Pacific endemics despite the fact that the reef-associated immigrants came from a biota with a higher incidence and greater expression of molluscan shell armor (Vermeij, 1978).

Numerous other interchanges between biotas result from the transport of species by humans, but Simberloff (1981) has pointed out that new colonists rarely bring about the extinction of native species except on small oceanic islands. Even the agents of biological control that have been brought in to eliminate troublesome insect pests are rarely, if ever, so effective that they cause the extinction of the pests (Debach, 1974). Introduced bacterial and viral diseases may eliminate most of a population of native species (the chestnut blight in eastern North America, for example), but to my knowledge no instance of extinction can be attributed to disease, at least in mainland populations.

I conclude from this evidence that biologically induced extinction is generally rare, especially in the sea. Present information suggests that it is confined to small oceanic islands and to some continental species that have been exterminated by humans. The possibility that immigrant species caused the extinction of native North American dinosaurs at the end of the Cretaceous, and of North and South American mammals during the Pliocene and Pleistocene, cannot be eliminated and deserves careful study.

POPULATION SIZE AND GEOGRAPHICAL RANGE

Several patterns of selectivity of extinction are related to such population properties as geographical range and population size. The best known of these is the tendency for species with a small geographical range to be especially prone to extinction. This tendency has been well documented for gastropods in the Late Cretaceous and Cenozoic (Hansen, 1978, 1980, 1982; Scheltema, 1978; Jablonski, 1980, 1982; Valentine and Jablonski, 1982, 1983b). Gastropod subgenera from the Pliocene of marine tropical America illustrate this point particularly well. Some subgenera are known as Pliocene fossils in only one of the three biogeographical regions of tropical America—the Atlantic Gatunian, Pacific Gatunian, and Caloosahatchian (or Floridian) regions. These so-called endemics had extinction probabilities of greater than 0.50, no matter which of the three regions they inhabited. By contrast, Pan-American taxa (those found as fossils in all three regions, had extinction probabilities of 0.25 or less. Because the proportion of endemic subgenera during the Pliocene was higher in the Atlantic Gatunian (15%) and Caloosahatchian (28%) than in the Pacific Gatunian (5%), part of the difference in extinction between the Atlantic and Pacific molluscs is attributable to the effect of geographical range (Vermeij and Petuch, 1986).

An important question from the perspective of this book is whether there are statistical associations between the size of a species's geographical range and the presence and expression of aptations against enemies. The greater susceptibility to extinction of Pliocene tropical American gastropods with narrow or thick-lipped apertures is not due to a difference in geographical range between armored and unarmored subgenera (Vermeij and Petuch, 1986), but continuing studies of the distribution and predation-related architecture of tropical Western Pacific gastropods do reveal differences between narrowly distributed and wide-ranging species (Vermeij et al., 1984).

In his studies of extinction of birds on Barro Colorado Island, a hill that was isolated as an island in Gatun Lake when the Panama Canal was dug in 1914, Karr (1982a, b) has pointed to the importance of population stability in preventing extinction. He found that species whose populations are subject to large year-to-year variations in size were apt to be sensitive to changes in the abundance of food, which in turn depends on the duration and amount of rainfall in the rainy season. In fact, species with greatly fluctuating populations have proven to be more susceptible to local extinction on Barro Colorado than species that have maintained fairly stable populations. How these population characteristics are related to the traits of individuals has not been studied.

POST-CRISIS CONDITIONS

There is some evidence from the fossil record that the communities that immediately followed the great crises were composed chiefly of opportunistic forms, or at least of morphologically simple species. Specializations related to survival and to adaptation against enemies were evidently not as well developed in the post-crisis assemblages as they were in the communities before the extinctions took place. In the Cambrian, for example, trilobites occurring just above biomere boundaries were less ornately sculptured than trilobites in earlier and later assemblages (A. Palmer, 1984). Ammonoids after each of the eight Devonian episodes of extinction documented by House (1985) had simpler sutures and less external sculpture than did the larger ammonoids that evolved from them. Planktonic foraminifers that survived the crises at the end of the Cretaceous and during the Eocene belonged to a few smooth or simply keeled species that were not ornamented with spines, whereas antecedents and descendants often had well developed surface sculpture (Lipps, 1970; Cifelli,

1976; Fischer and Arthur, 1977). I know of no counter-examples to this. Communities of bottom-dwelling invertebrates and planktonic foraminifers changed rapidly in composition before achieving stability again some 35 thousand to 350 thousand years after the end-Cretaceous crisis (Birkelund and Hakansson, 1982; Hsu et al., 1982; Smit, 1982). Cyst-forming "diaster" species of phytoplankton were found by Percival and Fischer (1977) to be common only during and immediately following the end-Cretaceous event, and to be rare during the rest of the Late Cretaceous and Paleocene. Mammals of the Early Paleocene were generally small and unspecialized for running, grinding plant tissues, or taking large prey (Bakker, 1977). In these ways they differed both from earlier vertebrates of the Late Cretaceous and from later mammals of the Late Paleocene and Eocene.

If post-crisis opportunists provided ideal "raw material" for subsequent adaptive escalation, we should expect episodes of escalation consistently to follow the great crises. This expectation is realized for some crises but not for others. The extinctions of the Late Cambrian were followed by the great diversification of skeletonized animals in the Ordovician, and the Ashgillian crisis at the end of the Ordovician may have set the stage for the mid-Paleozoic escalation. The early phases of the mid-Mesozoic escalation may be linked to the end-Triassic event, and escalation during the Early Cenozoic came soon after the end-Cretaceous crisis. The Late Devonian extinction events may have helped to propel the mid-Paleozoic escalation into the Early Carboniferous. The greatest of all crises, that at the end of the Permian, was followed by only a very modest escalation event in the Triassic, even though the rate of diversification of animal families was high immediately after the crisis. Triassic molluscs, echinoids, bryozoans, and sponges had a distinctly Paleozoic cast, both architecturally and taxonomically; and the chief difference between Middle Triassic reefs and Permian ones was the absence in the Triassic reefs of several Permian reef-building groups (Batten, 1973; Kier, 1973; Hartman et al., 1980; Boardman, 1984; Flügel and Stanley, 1984; Yin, 1985).

In short, mass extinctions provided opportunities for population expansion, especially of weedy species characterized by a high energy flux. Some crises may have been so widespread and so destructive of highly escalated species that a considerable lag occurred between the time of extinction and the onset of significant escalation. If some regions in the photic zone were spared, and if some superior predators and competitors survived, escalation could ensue shortly after a crisis.

SUMMARY

Extinction—the disappearance of populations and species—is the fate of most evolutionary lineages. Weather-related agencies of mortality that inhibit or interrupt primary production are expected to cause the extinction of a larger number of species than are biological agents or those physical events (such as warming) that enhance primary productivity. Species with highly expressed enemy-related aptations are expected to be more susceptible to extinction during production-inhibiting events than are opportunistic species or those inhabiting "safe" environments where rates of metabolism and interactions with enemies are reduced. This is so not only because individuals of highly escalated species are less tolerant of interruptions in primary production, but also because the environments occupied by these species are generally more affected by weather-related calamities. Episodes of extinction involving many species therefore interfere with adaptive escalation, whereas the effect of minor episodes of extinction on escalation may be either neutral or stimulative.

Although careful studies of the selectivity of extinction are still scarce, especially for the so-called minor episodes of extinction, they are generally in accord with the above expectations. Thus, the major biological crises (mass extinctions) of the Late Ordovician, Late Devonian, Late Permian, Late Triassic, and Late Cretaceous, as well as the regional extinctions in the Atlantic Ocean during the Neogene, were especially calamitous for warm-water species, among which there is a particularly high incidence of antipredatory and competition-related traits. Extinctions due to predators and competitors are documented only for the Recent biotas of small oceanic islands on which continental species have been introduced and for large continental animals exterminated by humans. Interchange between biotas with contrasting degrees of escalation probably has not led to the extinction of marine species, but it may have caused the extinction of terrestrial vertebrates.

Despite the negative effect of the major episodes of extinction on adaptive escalation over the short term, the crises have not prevented adaptive escalation from proceeding further. It is possible that metabolically active opportunistic species lay at the base of most of the important groups that diversified after the great crises. In any case, species in "safe" environments such as the deep sea, caves, and deep layers of the sediment did not re-invade high-energy environments and were not important contributors to post-crisis evolution there. In the long run, therefore, the crises may have stimulated escalation because they favored the survival and rapid expansion of high-energy opportunists in which the potential for escalation is high.

CHAPTER FIFTEEN

Implications, Difficulties, and

Future Directions

THE DOMAIN OF INDIVIDUAL ADAPTATION

In his theory of evolution, Darwin (1872) assigned a central role to organisms as agents of selection. Indeed, he coined the phrase "struggle for life" to emphasize that individuals must cope with their biological surroundings in order to survive and to leave offspring. Many traits are interpretable as aptations enabling individuals to acquire and defend resources that are under the control of other organisms. Van Valen (1973, 1976, 1983, 1985b) expanded on Darwin's theme by pointing out that improvement of one species causes the environment of co-occurring species to deteriorate. If these other species fail to adapt, they become ecologically restricted or even extinct. I have introduced the term escalation to describe the process in which the capacities of enemies as well as the competitive and defensive performances of potential victims increase through time in a given environment.

Evaluation of the role of escalation in evolution requires the identification and characterization of aptations and biological agents of selection. Despite many inadequacies in the data, the available evidence from the fossil record is consistent with the assertion that escalation between species and their biological surroundings has been a pervasive theme in the history of plants and animals during the Phanerozoic. Natural selection at the level of the individual organism has therefore played a key role in determining which kinds of organisms persist and which do not, just as Darwin had surmised from the even less adequate evidence available to him.

This conclusion does not mean, however, that processes acting at levels above that of the individual have been unimportant. Indeed, the ability of species to adapt as conditions change depends on such population characteristics as geographical range and the number of reproducing individuals. Perhaps even more important, as Vrba (1983, 1984) points out, susceptibility to speciation and hence escalation is influenced by the traits of individuals. The ability to disperse, for instance, enables populations to spread rapidly and often may be associated with speciation by founders. Sexual competition among individuals may promote rapid divergence of isolated populations and is itself often associated with well-developed survival-related aptations. In short, escalation may have occurred not only because there is usually room for adaptive improvement in individuals relative to their enemies, but also because species with the greatest expression of survival-related aptations are especially susceptible to adaptive change and to speciation.

There is thus a curious paradox. Adaptive escalation arising from natural selection at the level of the individual organism is arguably the dominant pattern in the Phanerozoic history of life, yet it could not be sustained without the imposition of agencies that create selection among populations and among species. In the absence of extrinsic change, adaptive improvement ceases and evolution stagnates. When geographical and climatic conditions do change, they cause some populations to decline precipitously and others to expand and spread. It is therefore no exaggeration to say that extinction is necessary for further evolution. The fate of individuals is perhaps dictated by aptations for coping with hazards, but the fate of populations and species to which these individuals belong depends on the imposition of agencies to which individuals are not adapted. Escalation occurs because selection among populations and among species generally favors the differential diversification of highly escalated species, and because the effects of diversification generally overshadow those of extinction.

EVOLUTIONARY PROGRESS

Does the claim that escalation characterizes the Phanerozoic history of life imply evolutionary progress? The idea that natural selection leads to a kind of progress is old. As Gould (1985) points out, it formed part of the world view of most of the great evolutionary synthesists, including Darwin. The principal difficulty with the concept of biological progress is that no operational criteria for recognizing and studying it has ever been outlined. Like most who have written on this subject, Huxley (1953) thought of progress as an increase in efficiency of mechanical design or as the differential deployment of species with such advanced design. As obvious examples of progress, he pointed to the trend toward increasingly high-crowned, grinding-adapted teeth and to the trend toward higher running speeds in horses over time. Another obvious example, elaborated at some length by Huxley, is the evolution of the highly integrated vertebrate nervous system, which reached an unsurpassed intricacy of design in the human species. Stebbins (1969) inferred evolutionary progress from structural complexity—the number of cell types or the degree of functional integration of the body—and made the point that the most complex groups are also the most recently evolved.

It seems to me that these interpretations of progress rest on a strongly one-sided analysis in which the traits of individuals are considered without reference to the ecological context in which they function. The benefits are somehow intrinsic or absolute and are therefore independent of the individual's surroundings. Yet, the evolution of horses' teeth cannot be considered without reference to the evolution of the plants ground up by the teeth. The vertebrate nervous system did not evolve in a black box. In fact, Jerison (1973) makes an intriguing case for the view that many of the important features of the mammalian brain arose when an emphasis on vision in reptiles was replaced or augmented by an emphasis on smell. If we think of progress in the human condition, we tend to measure our ability to prevent famine, fight disease, reduce crime, and ensure that all members of the population have economic opportunity. Progress is thus a measure of how well we cope with problems; it implies improvement *relative to* hazards in the environment.

A reasonable definition of biological progress might be that the ability of individuals to cope with hazards in their environment increases over the course of geological time. I argued in Chapter 3 that unilateral improvement with respect to an individual's biological surrounding is at best temporary, for the adaptive improvements attained by one species are typically compensated for by adaptation in other species. Long-term directional changes in the incidence and expression of aptations have occurred, and the average individual living today may be competitively and defensively superior to individuals of the past (or to those in modern environments that biologically resemble environments of the past), but there is no compelling evidence that Recent species are better able to cope with their biological surroundings than ancient species were with theirs.

It is possible, however, that species have improved in their capacity to survive in the *physical* environment. Many of the characteristics associated with competitive and defensive superiority—large body size, high body temperature, parental care of the young, and a tightly sealing exoskeleton, for example—also buffer individuals against short-term fluctuations in temperature and other physical factors. Consequently, individuals are able to carry on normal activity, or at least to survive, when physical conditions are temporarily unfavorable. Without such characteristics, individuals would be able to persist in a much smaller range of physical conditions. If progress with respect to the physical environment did occur, we should expect to see an increase in the mean age of individuals at the time of death. This proposition remains untested but can, I think, be investigated empirically in species whose skeletons faithfully record daily growth increments.

OBJECTIONS AND ALTERNATIVE INTERPRETATIONS

In addition to the inadequacy of the evidence bearing on escalation, there are some more basic objections that could be raised to the interpretation that the competitive and defensive capacities of individuals have increased within specified habitats over time as risks from enemies have increased. I shall try to deal with some of these objections and alternative interpretations here.

The first objection might be that I emphasize individual survival rather than reproduction. Survival to maturity is necessary for an individual to reproduce, but individuals that leave no offspring play no role in evolution no matter how well they are able to cope with potential agencies of death. Given that high fertility is one way of persisting in a high-risk environment, it is legitimate to ask why there was not a trend toward higher fertility through time. Why should we expect a trend in any direction at all if individual survival and high fertility are favored as alternative responses to a high risk of death?

My justification for dwelling on survival-related aptations rests both on practical grounds and on arguments about population dynamics. In practice, it is far easier to recognize and characterize survival-related traits in fossil species than it is to estimate such components of reproductive performance as lifetime female fecundity, age of first reproduction, number and size of eggs or seeds per brood, number of broods, and proportion of successfully breeding or reproducing adults. The problems inherent in characterizing these traits should not prevent us from looking intensively for situations in which these components are measurable in fossils, but it does mean that sufficient data for the evaluation of temporal patterns in reproduction will be difficult to assemble.

There are at least two theoretical arguments in favor of the assertion that trends in survival-related aptations should be more evident than long-term trends in fertility. The first is that fertility-enhancing variants are clearly favored only when population size is increasing. Because the supply of resources is finite, however, populations inevitably either reach saturation or are maintained at low levels by predators. In fact, most individuals are members of populations whose numbers are stable or falling. Periods of outbreak, or population increase, are typically short in duration compared with periods of population stability or decline. Accordingly, survival-related variants should be favored more often, and in a greater number of individuals, than fertility-enhancing variants. The second argument is that successful reproduction in many sexual species involves some form of competition for or by mates. This competition, which must be especially keen in populations of low density, often favors attributes such as the ability to communicate at a distance and the ability to combat rivals. Even if the aptations that enhance mating success are not themselves beneficial for survival, they are often evolutionarily derived from traits that are beneficial. This point is well illustrated by the evolution of horns in herbivorous land vertebrates (Chapter 4). In short, sexual competition may often promote the evolution of survival-enhancing variants.

A second major objection to the hypothesis of escalation is that the underlying causes of evolutionary change are so numerous and so complexly interrelated that no long-term trends are to be expected.

Any single evolutionary change may have a fairly simple causal explanation, but subsequent changes may have quite different causes. The resulting pattern of change in a lineage is a "random walk" rather than a consistent trend (Raup and Gould, 1974; Raup and Crick, 1981; Gould, 1985; D. C. Fisher, 1986).

Part of the reason for the erratic course of evolution in many lineages is that the physical environment varies over time, and lineage splitting often involves a change in environment. When conditions cool, for example, primary productivity is reduced, and selection due to enemies generally may become less intense. With climatic amelioration, primary productivity increases, and enemies may impose more intense selection. The history of particular lineages, then, confounds escalation with change in the physical environment and is therefore not likely to reveal consistent trends in the directions expected when physical conditions remain constant. This is why the hypothesis of escalation can be evaluated properly only in temporal comparisons of similar nonbiological environments.

A related objection is that biological agents (especially competitors) may not play the central role in selection that I claimed in Chapter 1. If competitors and predators were not usually the most important selective agents, escalation would not be a pervasive trend if physical conditions did remain the same. The idea that competition among individuals elicits aptations and controls the local distribution of species was fundamental to Darwin's view of evolution and remained axiomatic for most later evolutionists and ecologists (R. A. Fisher, 1958; Kohn, 1959; MacArthur, 1965, 1972; G. C. Williams, 1966). Perhaps the most significant challenge to this view has come from critics (Simberloff, 1978; Strong et al., 1979, for example) who charge that the inference of competition was often erroneously drawn from the observation that species living in similar habitats and eating similar foods are not usually found living together. A great deal of careful experimental work, partially summarized by T. W. Schoener (1983), was spawned by these well-founded critiques and shows that well over 70% of the species that have now been investigated are limited in their numbers and distribution by competition at some time during the life cycle. More important from the evolutionary viewpoint is the finding that loss during competitive encounters, although often not fatal, results in greater risk of predation (Jeffries and Lawton, 1984), less access to food, and reduced reproductive potential (Morse, 1980). Competition is important not only for large animals that engage in conspicuous displays of fighting, but also for plants and even for suspension-feeding and deposit-feeding invertebrates. Competition need not be only for food or living space, but may be also over mates and over sites that offer protection from predators. It is not confined to populations at saturation; in fact, competition for mates may be very keen for rare species.

The importance of predators as agents of selection has not been seriously questioned. Some ecologists, however, have concluded that predation is a relatively minor agency of population regulation for species in which competition is important, and that the reverse is also true (Connell, 1961a, b, 1972; Hay, 1981a, b, among others). Although predation and competition may be alternative mechanisms of population regulation when competition is over food or living space, no such complementarity is evident when the resources over which competition occurs are sanctuaries from predators. In fact, species in which predation is intense may often be exposed to severe competition as well (Bertness, 1981a; Jeffries and Lawton, 1984). The claim that competition is an important (and perhaps a universal) agency of selection or population regulation therefore does not contradict the assertion that predation is a process of comparable evolutionary significance.

The claim that escalation results in temporal trends in the expression of aptations may be challenged by pointing to cases of adaptation and counter-adaptation involving chemical or immunological defenses. This kind of adaptation is especially common when individuals are consumed slowly or only partially by organisms that may broadly be labeled as parasites (P. W. Price, 1980). Adaptive improvement in both the host and the parasite is achieved not by the greater production of a given defensive substance, but by the introduction of novel substances (Ehrlich and Raven, 1964; McKey, 1979; Berenbaum, 1983). The result is a greater diversity of substances rather than an increased expression of any single defense.

I pointed out in Chapter 2 that low-energy plants and animals frequently have well-developed chemical defenses. Slow-growing plants, for example, often contain high concentrations of tannins (Janzen, 1974), and Coley's (1980, 1983) surveys of herbivore-induced leaf damage in saplings of the Panamanian rain forest show clearly that fast-growing species are chemically less effectively defended than slow-growing species with more persistent leaves. Observations such as these might be interpreted as arguments against the generalization that escalation between species and their enemies proceeds furthest in high-energy situations. It is known, however, that the proportion of alkaloid-bearing plants and of saponin-bearing holothurians (sea cucumbers) increases toward the equator (that is, along a gradient of increasing mean temperature), and that slow growth at high altitudes, where temperatures are typically low, is not usually accompanied by high concentrations of defensive substances in plants (Bakus, 1974, 1981; Levin, 1975, 1976; Levin and York, 1978). Moreover, although passive chemical defense is not hindered in low-energy situations, it occurs in high-energy species as well (in mature leaves of fast-growing plants, for example). Forms of defense that rely on rapid mobilization of substances to parts of the body under attack are also best developed in high-energy situations. Plants and animals with such defenses have evolved internal systems of transport for the rapid translocation of substances. These are well known in many terrestrial and marine plants, but the most spectacular manifestation of such defenses is seen in the complex immune systems of warm-blooded vertebrates. Further research should concentrate on the relationship between metabolic rate and the deployment of chemical and immunological defenses, and on the establishment of criteria for the recognition of such defenses in fossil species. Present evidence, however, suggests that high-energy situations permit a greater diversity of chemical deterrents than is possible under conditions of lower energy flux.

A final important objection to the hypothesis of escalation is that, even if consistent trends through time were detected within lineages or in cases of replacement of one lineage by another, these trends would not necessarily be the result of consistent selection. As discussed at length in Chapter 2, evolutionary change is constrained by many genetic, developmental, and ecological limitations. These limitations will be especially evident as a lineage approaches some extreme form. If lineages often arise close to one extreme, most change is apt to be away from that extreme and will therefore result in an apparently consistent evolutionary trend. If, for example, the ancestor of a lineage is small and opportunistic, descendants are apt to be bigger and less opportunistic simply because it may be difficult for individuals to be even smaller and weedier. According to this view, consistent trends may be statistical regularities arising from the fact that ancestors of lineages consistently portray certain characteristics such as small body size and simple construction (D. C. Fisher, 1986).

The fact that evolutionary changes are far more likely in some directions than in others also means that many trends that are inconsistent with escalation could occur. Strathmann (1978a, b), for example, points out the evolutionary *loss* of a feeding planktonic larval stage in marine invertebrates has occurred far more often (and may therefore be less constrained by genetic, developmental, and ecological factors) than the evolutionary *acquisition* of such a stage, even though a feeding planktonic phase might be highly adaptive in many species lacking it. If "easy" evolutionary pathways often led to dead ends or were inconsistent with adaptation to enemies, many lineages would violate the predictions of escalation, and would eventually face ecological restriction or extinction.

This argument forcefully underscores the importance of seeking independent evidence of the action of selective agencies when a long-term trend is being investigated. A trend is not in itself sufficient evidence that consistent selection has occurred, although it makes a selection-based explanation at least plausible.

In short, although the empirical evidence concerning escalation is still inadequate, I believe the concept of escalation withstands several points of criticism. The hypothesis that organisms are the primary agencies of selection and that escalation with respect to survival-related traits between species and their biological surroundings has been a pervasive pattern in the Phanerozoic history of plants and animals is supported both by theoretical arguments and by data from the fossil record.

FUTURE RESEARCH

I am only too keenly aware of the fact that nearly every aspect of the history of life that I have treated in this book requires further documentation. In the course of the various chapters, I pointed to particularly fruitful directions for future research, but it is appropriate now to emphasize the areas that require the most urgent attention.

The first area falls under the general heading of adaptational biology. We need systematically collected data on the effectiveness of individuals with and without ostensible aptations to various hazards. Closely related to this need is the necessity of obtaining information on absolute capacities of organisms—growth rates, locomotor performance, resistance to predation, time required for dealing with prey, depth of burial and rate of reworking of sediment by infaunal animals, and so on. We need to develop a systematic understanding of the nature and spatial variation of ecological risks. Only when data on risks become available on a large scale can we begin to understand why one evolutionary pathway is taken instead of another. Is armor in arthropods really limited by risks during molting? Is marginal damage to pelecypod valves really so dangerous that the chances of an individual's surviving and giving rise to more heavily armored descendants are small?

With respect to the fossil record, perhaps the most urgent need is for the systematic collection of data on species composition, incidence of interactions with organisms, and incidence and expression of enemy-related aptations among species in local assemblages. Most of the historical patterns I have discussed in this book have been discerned at the global level, but analysis at this level is beset with time-dependent biases and with artifacts of preservation and taxonomy. Evaluation of temporal trends requires that spatial variation at the various scales be distinguishable from changes through time.

Several important paths of adaptation that I have had to gloss over in this book deserve careful scrutiny in the fossil record. I have already mentioned chemical defense as one subject that can be studied profitably in plants. Another is sensory capacity in animals. Trends in visual performance can be studied profitably in trilobites and some other arthropods in which the eyes are well preserved. Some inferences about vision and hearing can be drawn from studies of the skull in vertebrates. Knowledge of the sensory capacities of animals is crucial to an understanding of the evolution of mechanisms of predation, competition for mates, detection of food, and detection of enemies.

Besides studies on the incidence and expression of traits in local assemblages through time, it is important to study pathways of evolution by inferring ancestor-descendant relationships within lineages. Without such phylogenetic studies, it will be impossible to reconstruct how and when species become ecologically restricted, to assess the relative frequency of adaptive change and ecological restriction, and to disentangle evolutionary convergence (the attainment of a particular adaptive form from separate origins by two or more lineages) from similarity by virtue of common descent.

The dynamics of adaptive change are perhaps best studied in living species. At least two avenues of research are available for the study of adaptive evolution in living species. First, individuals and the environments they live in may be monitored in order to detect adaptive change. Although this may be done with any species, it seems particularly important and informative to carry out such monitoring in species that have been brought into new surroundings. The second avenue is to compare living populations with those of the recent past. This may be done when large samples of well-dated and well-documented specimens are available in museum collections. The fact that historical changes may be traced by studying material of living species in museums underscores the importance of continuing to make ecologically well-documented collections of common species.

Although extinction has come to be an important topic for study, we will know too little about its selectivity and its immediate and long-term effects on surviving species. How consistent is the pattern of selectivity through the Phanerozoic? Are there important differences between extinction episodes of small magnitude and the great crises? Did the rate of extinction of species decline through the Phanerozoic and, if so, is the greater resistance of more recently evolved species a consequence of the greater effectiveness of individuals in coping with the physical environment or an effect of population characteristics?

A central issue in the dynamics of escalation is the role played by population- or species-level characteristics in determining the susceptibility of a lineage to speciation. How do population size and the extent of geographical range influence speciation? Is it really true that species with well-developed defenses and competitive mechanisms are more susceptible to isolation and subsequent divergence from the parent stock? These questions are best approached in settings where extinction has been minor. The Neogene record in the tropical Western Pacific would seem to be especially rewarding; so would continuous sequences of sediment laid down after the great crises.

Finally, what are the circumstances surrounding the evolution and success of man? Was there something unique about the Pliocene, when the genus *Homo* first differentiated, or were the origin and ascendancy of man possible at many other times in earth history? Will the evolutionary trends that have characterized the whole of the Phanerozoic continue in the presence of man, or will the most potent of all competitors and predators put an end to the magnificent history of life on earth?

Appendix

Presented below is a geological timetable for the Phanerozoic eon. The time of beginning for each unit of geological time is given in millions of years before present. Estimate 1 is taken from Harland et al. (1982); estimate 2 is derived from Odin et al. (1982, 1983).

Interval	Time of Beginning	
	1	2
Paleozoic era	670	670
Vendian (Ediacarian) period	670	670
Cambrian period	590	530
Early Cambrian (Manykay, Tommotian, Atdabanian, Botomian, Lenian)	590	530
Middle Cambrian	540	
Late Cambrian (Dresbachian, Franconian, Trempeauleauian)	523	
Ordovician period	505	495
Tremadocian stage	505	495
Arenigian stage	488	
Llanvirnian stage	478	
Llandeilian stage	468	
Caradocian stage	458	
Ashgillian stage	448	
Silurian period	438	418
Llandoverian stage	438	418
Wenlockian stage	428	
Ludlovian stage	421	
Pridolian stage	414	
Devonian period	408	400
Gedinnian stage	408	400

Siegenian stage	401	
Emsian stage	394	
Eifelian stage	387	385
Givetian stage	380	
Frasnian stage	374	375
Famennian stage	367	
Early Carboniferous period	360	360
Tournaisian stage	360	360
Viséan stage	352	
Late Carboniferous period	333	320
Namurian stage	333	320
Westphalian stage	315	
Stephanian stage	296	
Permian period	286	290
Asselian and Sakmarian stages	286	290
Artinskian stage	268	
Kungurian stage	263	
Kazanian and Ufimian stages	258	
Tatarian stage	253	
(Late Permian may also be divided into Guadelupian and Dzhulfian)		
Mesozoic era	248	245
Triassic period	248	245
Scythian stage	248	245
Anisian stage	243	239
Ladinian stage	238	
Carnian stage	231	229
Norian stage	225	
Rhaetian or Sevatian stage	219	
Jurassic period	213	204
Hettangian stage	213	204
Sinemurian stage	206	
Pliensbachian stage	200	
Toarcian stage	194	
Aalenian stage	188	
Bajocian stage	181	178
Bathonian stage	175	
Callovian stage	169	

Oxfordian stage	163	150
Kimmeridgian stage	156	
Tithonian stage	150	
Cretaceous period	144	130
Berriasian stage	144	130
Valanginian stage	138	
Hauterivian stage	131	
Barremian stage	125	
Aptian stage	119	
Albian stage	113	
Cenomanian stage	97	95
Turonian stage	91	
Coniacian stage	88	
Santonian stage	87	
Campanian stage	83	
Maastrichtian stage	73	
Cenozoic era	65	65
Paleogene period	65	65
Paleocene stage	65	65
Eocene stage	55	53
Oligocene stage	38	34
Neogene period	25	23
Miocene stage	25	23
Pliocene stage	5.10	5.50
Pleistocene stage	2.00	2.00
Recent stage	0.91	0.01

References

Abbott, I., L. K. Abbott, and P. R. Grant. 1977. Comparative ecology of Galápagos ground finches (*Geospiza* Gould): Evaluation of the importance of
floristic diversity and interspecific competition. *Ecol. Monogrs.* 47:151–
184.

Abele, L. G. (ed.). 1982. *The Biology of Crustacea*. Vol. 1: *Systematics, the
Fossil Record, and Biogeography*. Academic Press, New York.

Abele, L. G., K. L. Heck, Jr., D. S. Simberloff, and G. J. Vermeij. 1981. Biogeography of crab claw size: Assumptions and a null hypothesis. *Syst.
Zool.* 30:406–424.

Achituv, Y. 1973. On the distribution and variability of the Indo-Pacific sea
star *Asterina wega* (Echinodermata: Asteroidea) in the Mediterranean Sea.
Mar. Biol. 18:333–336.

Adams, A. E. 1984. Development of algal-foraminiferal-coral reefs in the
Lower Carboniferous of Furness, northwest England. *Lethaia* 17:233–249.

Adegoke, O. S., and M.J.S. Tevesz. 1974. Gastropod predation patterns in the
Eocene of Nigeria. *Lethaia* 7:17–24.

Aiken, D. E. 1980. Molting and growth. In J. S. Cobb and B. F. Phillips (eds.),
The Biology and Management of Lobsters. Vol. 1: *Physiology and Behavior*, pp. 91–163. Academic Press, New York.

Akpan, E. B., G. E. Farrow, and N. Morris. 1982. Limpet grazing on Cretaceous algal-bored ammonites. *Palaeontology* 25:361–367.

Alberch, P. 1983. Morphological variation in the neotropical salamander genus *Bolitoglossa*. *Evolution* 37:906–919.

———. 1985. Developmental constraints: Why St. Bernards often have an
extra digit and poodles never do. *Amer. Nat.* 126:430–433.

Alberch, P., and E. A. Gale. 1985. A developmental analysis of an evolutionary trend: Digital reduction in amphibians. *Evolution* 39:8–23.

Alexander, R. M. 1967. *Functional Design in Fishes*. Hutchinson, London.

———. 1968. *Animal Mechanics*. University of Washington Press, Seattle.

———. 1976. Estimates of speeds of dinosaurs. *Nature* 261:129–130.

———. 1977a. Terrestrial locomotion. In Alexander and Goldspink (eds.),
1977, pp. 168–203.

———. 1977b. Swimming. In Alexander and Goldspink (eds.), 1977, pp. 222–
248.

———. 1977c. Flight. In Alexander and Goldspink (eds.), 1977, pp. 249–278.

———. 1981. Factors of safety in the structure of animals. *Sci. Progr. Oxford*
67:109–130.

———. 1984. Optimum strengths of bones liable to fatigue and accidental
fracture. *J. Theoret. Biol.* 109:621–636.

Alexander, R. M., and G. Goldspink (eds.). 1977. *Mechanisms and Energetics of Animal Locomotion*. Chapman and Hall, London.

Alexander, R. M., G.M.O. Maloiy, R. Njau, and A. S. Jayes. 1979. Mechanics of running of the ostrich (*Struthio camelus*). *J. Zool. London* 187:169–178.

Alexander, R. R. 1981. Predation scars preserved in Chesterian brachiopods: Probable culprits and evolutionary consequences for the articulates. *J. Paleont.* 55:192–203.

———. 1986. Resistance to and repair of shell breakage induced by durophages in Late Ordovician brachiopods. *J. Paleont.* 60:273–285.

Allen, J. A. 1978. Evolution of the deep sea protobranch bivalves. *Phil. Trans. Roy. Soc. London* (B) 284:387–401.

Allen, J. A., and R. E. Morgan. 1981. The functional morphology of Atlantic deep water species of the families Cuspidariidae and Poromyidae (Bivalvia): An analysis of the evolution of the septibranch condition. *Phil. Trans. Roy. Soc. London* (B) 294:413–546.

Aller, R. C. 1982. Carbonate dissolution in nearshore terrigenous muds: The role of physical and biological reworking. *J. Geol.* 90:79–95.

Almaça, C. 1963. Sur le problème de l'origine de *Carcinus maenas* (L.) du littoral américain. *Rev. Fac. Cienc. Univ. Lisboa* (2, C) 11:121–136.

Alpert, S. P., and J. N. Moore. 1975. Lower Cambrian trace fossil evidence for predation by trilobites. *Lethaia* 8:223–230.

Alvarez, W., E. G. Kauffman, F. Surlyk, L. W. Alvarez, F. Asaro, and H. V. Michel. 1984a. Impact theory of mass extinctions and the invertebrate fossil record. *Science* 223:1135–1141.

Alvarez, W., L. W. Alvarez, F. Asaro, and H. V. Michel. 1984b. The end of the Cretaceous: Sharp boundary or gradual transition? *Science* 223:1183–1186.

Ambros, R. F., and B. V. Nelson. 1983. Predation by *Octopus vulgaris* in the Mediterranean. *Mar. Ecol.* 4:251–261.

Andel, T. H. van. 1979. An eclectic overview of plate tectonics, paleogeography, and paleoceanography. In Gray and Boucot (eds.), 1979, pp. 9–25.

Anderson, P. K., and A. Birtles. 1978. Behaviour and ecology of the dugong, *Dugong dugon* (Sirenia): Observations in Shoalwater and Cleveland Bays, Queensland. *Austral. Wildl. Res.* 5:1–23.

André, M. 1932. Coquilles utilisées par les araignées. *J. Conchyliol.* 76:213–215.

———. 1937. Coquilles vides de bivalves habitées par des crustacés. *J. Conchyliol.* 81:72–81.

Ansell, A. D. 1961. The functional morphology of the British species of Veneracea (Eulamellibranchia). *J. Mar. Biol. Assoc. U.K.* 41:489–517.

———. 1967. Leaping and other movements in some cardiid bivalves. *Anim. Behav.* 15:421–426.

———. 1969. Defensive adaptations to predation in the Mollusca. *Proc. Symp. Mollusca* 2:487–512.

Ansell, A. D., and N. B. Nair. 1969. A comparative study of bivalves which bore mainly by mechanical means. *Amer. Zool.* 9:857–868.

Ansell, A. D., and A. Trevallion. 1969. Behavioural adaptations of intertidal molluscs from a tropical sandy beach. *J. Exp. Mar. Biol. Ecol.* 4:9–35.

Ar, A., H. Rahn, and C. V. Paganelli. 1979. The avian egg: Mass and strength. *Condor* 81:331–337.

Arnold, A. J., and K. Fristrup. 1982. The theory of evolution by natural selection: A hierarchical expansion. *Paleobiology* 8:113–129.

Arnold, E. N. 1984. Evolutionary aspects of tail shedding in lizards and their relatives. *J. Nat. Hist.* 18:127–169.

Arnold, J. M., and K. O. Arnold. 1969. Some aspects of hole-boring predation by *Octopus vulgaris*. *Amer. Zool.* 9:991–996.

Arnold, S. J. 1983. Morphology, performance and fitness. *Amer. Zool.* 23:347–361.

Arntz, W. E. 1978. The food of adult cod (*Gadus morhua* L.) in the western Baltic. *Meerforsch.* 26:60–69.

Aronson, R. B., and C. A. Harms. 1985. Ophiuroids in a Bahamian saltwater lake: The ecology of a Paleozoic-like community. *Ecology* 66:1472–1483.

Arthur, M. A., J. C. Zakos, and D. Jones. 1986. Primary productivity and the Cretaceous/Tertiary boundary event in the oceans. *Cretaceous Research*, in press.

Auffenberg, W. 1981. *The Behavioral Ecology of the Komodo Monitor*. University Presses of Florida, Gainesville.

Augspurger, C. K., and C. K. Kelly. 1984. Pathogen mortality of tropical tree seedlings: Experimental studies of the effects of dispersal distance, seedling density, and light conditions. *Oecologia* (Berlin) 61:211–217.

Ausich, W. I., and R. A. Gurrola. 1979. Two boring organisms in a Lower Mississippian community of southern Indiana. *J. Paleont.* 53:335–344.

Avery, R. A., J. D. Bedord, and C. P. Newcombe. 1982. The role of thermoregulation in lizard biology: Predatory efficiency in a temperate diurnal basker. *Behav. Ecol. Sociobiol.* 11:261–267.

Awramik, S. M. 1971. Precambrian columnar stromatolite diversity: Reflection of metazoan appearance. *Science* 174:825–827.

Bak, R.P.M., J.J.W.M. Brouns, and F.M.L. Heys. 1977. Regeneration and aspects of spatial competition in the scleractinian corals *Agaricia agaricites* and *Montastrea annularis*. *Proc. Third Intern. Coral Reef Symp.* 1:144–148.

Baker, A. J. 1974. Prey-specific feeding methods of New Zealand oystercatchers. *Notornis* 21:219–233.

———. 1980. Morphometric differentiation in New Zealand populations of the house sparrow (*Passer domesticus*). *Evolution* 34:638–653.

Baker, A. J., and A. Moeed. 1979. Evolution in the introduced New Zealand populations of the common myna, *Acridotheres tristis* (Aves: Sturnidae). *Canad. J. Zool.* 57:570–584.

Bakker, R. T. 1971. Dinosaur physiology and the origin of mammals. *Evolution* 25:636–658.

———. 1972. Anatomical and ecological evidence of endothermy in dinosaurs. *Nature* 238:81–85.

Bakker, R. T. 1975. Experimental and fossil evidence for the evolution of tetrapod energetics. In D. Gates and R. Schmerl (eds.), *Perspectives of Biophysical Ecology*, pp. 365–390. Springer, New York.

———. 1977. Tetrapod mass extinctions—a model of the regulation of speciation rates and immigration by cycles of topographic diversity. In Hallam (ed.), 1977, pp. 439–468.

———. 1978. Dinosaur feeding behaviour and the origin of flowering plants. *Nature* 274:661–663.

———. 1980. Dinosaur heresy–dinosaur renaissance: Why we need endothermic archosaurs and a comprehensive theory of bioenergetic evolution. In R.D.K. Thomas and E. C. Olson (eds.), *A Cold Look at the Warm-Blooded Dinosaurs*, pp. 351–462. Westview Press, Boulder.

———. 1983. The deer flees, the wolf pursues: Incongruencies in predator-prey coevolution. In Futuyma and Slatkin (eds.), 1983, pp. 350–382.

Bakus, G. J. 1967. The feeding habits of fishes and primary productivity of Eniwetok, Marshall Islands. *Micronesica* 3:135–149.

———. 1969. Energetics and feeding in shallow marine waters. *Intern. Rev. Gen. Exp. Zool.* 4:275–369.

———. 1973. The biology and ecology of tropical holothurians. In O. A. Jones and R. Endean (eds.), *Biology and Geology of Coral Reefs*. Vol. 2: *Biology* I, pp. 326–367. Academic Press, New York.

———. 1974. Toxicity in holothurians: A geographical pattern. *Biotropica* 6:229–236.

———. 1981. Chemical defense mechanisms on the Great Barrier Reef, Australia. *Science* 211:496–499.

Bakus, G. J., and G. Green. 1974. Toxicity in sponges and holothurians. *Science* 185:951–953.

Ballinger, R. E. 1979. Intraspecific variation in demography and life history of the lizard, *Sceloporus jarrovi*, along an altitudinal gradient in southeastern Arizona. *Ecology* 60:901–909.

Baluk, W., and A. Radwański. 1984. New data on the Korytnica Basin, its organic communities and ecological relationships between species (Middle Miocene, Holy Cross Mountains, central Poland). *Acta Geol. Polon.* 34:179–194.

Bambach, R. K. 1977. Species richness in marine benthic habitats through the Phanerozoic. *Paleobiology* 3:152–167.

———. 1983. Ecospace utilization and guilds in marine communities through the Phanerozoic. In Tevesz and McCall (eds.), 1983, pp. 719–746.

Bandel, K. 1976. Observations on spawning, embryonic development and ecology of some Caribbean lower Mesogastropoda (Mollusca). *Veliger* 18:249–271.

Bandel, K., J. Reitner, and W. Stürmer. 1983. Coleoids from the Lower Devonian black slate (Hunsrückschiefer) of the Hunsrück (West Germany). *Neues Jb. Geol. Paläont. Abh.* 165:397–417.

Bandel, K., A. Almogi-Labin, C. Hemleben, and W. G. Duser. 1984. The conch of *Limacina* and *Peraclis* (Pteropoda) and a model for the evolution of planktonic gastropods. *Neues Jb. Geol. Paläont. Abh.* 168:87–107.

Bar, Z. 1978. Nest-building by the solitary bee *Osmia pinguis* in aestivating *Trochoidea seetzenii* in Israel. *Levantina* 14:152–153.

Barghusen, H. R. 1975. A review of fighting adaptations in dinocephalians (Reptilia, Therapsida). *Paleobiology* 1:295–311.

Barraclough, D. A. 1983. The biology and immature stages of *Sepedon* snail-eating flies in Natal (Diptera: Sciomyzidae). *Ann. Natal Mus.* 25:293–317.

Barrette, C. 1977. Fighting behavior of the muntjac and the evolution of antlers. *Evolution* 31:169–176.

Barton, N. H., and B. Charlesworth. 1984. Genetic revolutions, founder effects, and speciation. *Ann. Rev. Ecol. Syst.* 15:133–164.

Bartonek, J. C., and J. J. Hickey. 1969. Selective feeding by juvenile diving ducks in summer. *Auk* 86:443–457.

Bassett, M. G. 1984. Life strategies of Silurian brachiopods. *Special Pap. Palaeont.* 32:237–263.

Batenburg, L. H. 1981. Vegetative anatomy and ecology of *Sphenophyllum zwickaviense*, *S. emarginatum*, and other "compression species" of *Sphenophyllum*. *Rev. Palaeobot. Palynol.* 32:275–313.

Batten, R. L. 1973. The vicissitudes of the gastropods during the interval of Guadelupian-Ladinian time. *Canad. Soc. Petroleum Geol. Mem.* 2:596–607.

Baumfalk, Y. A., R. A. Fortuin, and R. P. Mok. 1982. *Talpinella cunicularia* n.gen., n.sp., a possible foraminiferal parasite of Late Cretaceous *Orbitoides*. *J. Foram. Res.* 12:185–196.

Bayer, U. 1970. Anomalien bei Ammoniten des Aaleniums und Bajociums und ihre Bedeutung zur Lebensweise. *Neues Jb. Geol. Paläont. Abh.* 135:19–41.

———. 1982. Ammonite maneuverability—a new look at the function of shell geometry. *Neues Jb. Geol. Paläont. Abh.* 164:154–156.

Bayne, B. L., and C. Scullard. 1978. Rates of feeding by *Thais (Nucella) lapillus* (L.). *J. Exp. Mar. Biol. Ecol.* 32:113–129.

Beal, B. F. 1983. Predation of juveniles of the hard clam *Mercenaria mercenaria* (Linné) by the snapping shrimp *Alpheus heterochaelis* Say and *Alpheus normanni* Kingsley. *J. Shellfish Res.* 3:1–9.

Beamish, F.W.H. 1978. Swimming capacity. In W. S. Hoar and D. J. Randall (eds.), *Fish Physiology*. Vol. 7: *Locomotion*, pp. 101–187. Academic Press, New York.

Bell, J. D., J. J. Burchmore, and B. A. Pollard. 1978. Feeding ecology of three sympatric species of leatherjackets (Pisces: Monacanthidae) from a *Posidonia* seagrass habitat in New South Wales. *Austral. J. Mar. Fresh-W. Res.* 29:633–643.

Belle, R. A. van. 1975. Sur la classification des Polyplacophora: I. Introduction et classification des Paleoloricata, avec la description de *Kindbladochiton* nom. nov. (pour *Eochiton* Smith, 1964). *Inf. Soc. Belg. Malacol.* 4:121–131.

Belle, R. A. van. 1977. Sur la classification des Polyplacophora: III. Classification systématique des Subterenochitonidae et des Ischnochitonidae (Neoloricata: Chitonina). *Inf. Soc. Belg. Malacol.* 5:15–40.

———. 1978. Sur la classification des Polyplacophora: V. Classification systématique des Chitonidae (Neoloricata: Chitonina). *Inf. Soc. Belg. Malacol.* 6:19–28.

Bengtson, S. 1968. The problematic genus *Mobergella* from the Lower Cambrian of the Baltic area. *Lethaia* 1:325–351.

Bennett, A. F., and B. Dalzell. 1973. Dinosaur physiology: A critique. *Evolution* 27:170–174.

Bennett, B., C. L. Griffiths, and M.-L. Penrith. 1983. The diets of littoral fish from the Cape Peninsula. *South Afr. J. Zool.* 18:343–352.

Benton, M. J. 1979a. Increase in total global biomass over time. *Evol. Theory* 4:123–128.

———. 1979b. Ecological succession among Palaeozoic and Mesozoic tetrapods. *Palaeogeogr., Palaeoclimatol., Palaeoecol.* 26:127–150.

———. 1979c. Ectothermy and the success of dinosaurs. *Evolution* 33:983–997.

———. 1983. Dinosaur success in the Triassic: A noncompetitive ecological model. *Quart. Rev. Biol.* 58:29–51.

———. 1984a. Tooth form, growth, and function in Triassic rhynchosaurs (Reptilia, Diapsida). *Palaeontology* 27:737–776.

———. 1984b. The relationships and early evolution of the Diapsida. *Symp. Zool. Soc. London* 52:575–596.

———. 1985. Mass extinction among non-marine tetrapods. *Nature* 316:811–814.

Ben-Yami, M., and T. Glaser. 1974. The invasion of *Saurida undosquamis* (Richardson) into the Levant Basin—an example of biologic effect of interoceanic canals. *Fish. Bull.* 72:359–373.

Bequaert, J. 1925. The arthropod enemies of mollusks, with description of a new dipterous parasite from Brazil. *J. Parasitol.* 11:201–212.

Berenbaum, M. 1983. Coumarins and caterpillars: A case for coevolution. *Evolution* 37:163–179.

Berg, C. J., Jr. 1978. Development and evolution of behavior in mollusks, with emphasis on changes in stereotypy. In G. M. Burghardt and M. Bekoff (eds.), *The Development of Behavior: Comparative and Evolutionary Aspects*, pp. 3–17. Garland STPM Press, New York.

Berg, C. O. 1964. Snail-killing sciomyzid flies: Biology of the aquatic species. *Verh. Intern. Verein. Limnol.* 15:926–932.

Berger, W. H., and J. C. Crowell (panel co-chairmen). 1982. *Studies in Geophysics: Climate in Earth History*. National Academy Press, Washington.

Berggren, W. A., and C. D. Hollister. 1977. Plate tectonics and paleocirculation—commotion in the ocean. *Tectonophysics* 38:11–48.

Berg-Madsen, V., and J. S. Peel. 1978. Middle Cambrian monoplacophorans from Bornholm and Australia, and the systematic position of the bellerophontiform molluscs. *Lethaia* 11:113–125.

Bergman, K. M., R. W. Elner, and M. J. Risk. 1982. The influence of *Polydora websteri* borings on the strength of the shell of the sea scallop, *Placopecten magellanicus*. *Canad. J. Zool.* 60:2551–2556.

Bergström, J. 1973. Organization, life, and systematics of trilobites. In A. Martinsson (ed.), *Fossils and Strata*. Vol. 2, pp. 1–69. Universitetsforlaget, Oslo.

———. 1975. Functional morphology and evolution of xiphosurids. In A. Martinsson (ed.), *Fossils and Strata*. Vol. 4: *Evolution and Morphology of the Trilobita, Trilobitoidea and Merostomata*, pp. 201–307. Universitetsforlaget, Oslo.

Bergström, J., and G. Brassel. 1984. Legs in the trilobite *Rhenops* from the Lower Devonian Hunsrück Slate. *Lethaia* 17:67–72.

Berner, R. A., and R. Raiswell. 1983. Burial of organic carbon and pyrite sulfur in sediments over Phanerozoic time: A new theory. *Geochimica et Cosmochimica Acta* 47:855–862.

Bernstein, B. B., and N. Jung. 1979. Selective pressures and coevolution in a kelp canopy community in southern California. *Ecol. Monogrs.* 49:335–355.

Berry, W.B.N., and P. Wilde. 1978. Progressive ventilation of the oceans—an explanation of the distribution of the Lower Paleozoic black shales. *Amer. J. Sci.* 278:257–275.

Bertness, M. D. 1981a. Conflicting advantages in resource utilization: The hermit crab housing dilemma. *Amer. Nat.* 118:432–437.

———. 1981b. Pattern of plasticity in tropical hermit crab growth and reproduction. *Amer. Nat.* 117:754–773.

———. 1981c. Predation, physical stress, and the organization of a tropical rocky intertidal hermit crab community. *Ecology* 62:411–425.

———. 1982. Shell utilization, predation pressure, and thermal stress in Panamanian hermit crabs: An interoceanic comparison. *J. Exp. Mar. Ecol. Biol.* 64:159–187.

———. 1985. Fiddler crab regulation of *Spartina alterniflora* production on a New England salt marsh. *Ecology* 66:1042–1055.

Bertness, M. D., and C. Cunningham. 1981. Crab shell-crushing predation and gastropod architectural defense. *J. Exp. Mar. Biol. Ecol.* 50:213–230.

Bertness, M. D., S. D. Garrity, and S. C. Levings. 1981. Predation pressure and gastropod foraging: A tropical-temperate comparison. *Evolution* 35:995–1007.

Berzins, I. K., and R. K. Caldwell. 1983. The effect of injury on the agonistic behavior of the stomatopod *Gonodactylus Bredini* (Manning). *Mar. Behav. Physiol.* 10:83–96.

Bigelow, H. B., and W. C. Schroeder. 1953. Fishes of the western North Atlantic. Part II: Sawfishes, guitarfishes, skates and rays. *Mem. Sears Found. Mar. Res.* 1:585 P.

Birkeland, C. 1977. The importance of rate of biomass accumulation in early successional stages of benthic communities to the survival of coral recruits. *Proc. Third Intern. Coral Reef Symp.* 1:15–21.

Birkeland, C. 1982. Terrestrial runoff as a cause of outbreaks of *Acanthaster planci* (Echinodermata: Asteroidea). *Mar. Biol.* 69:175–185.

Birkeland, C., and R. H. Randall. 1982. Facilitation of coral recruitment by echinoid excavations. *Proc. Fourth Intern. Coral Reef Symp.* 1:695–698.

Birkeland, C., D. Rowley, and R. H. Randall. 1982. Coral recruitment patterns at Guam. *Proc. Fourth Intern. Coral Reef Symp.* 2:339–344.

Birkelund, T., and E. Hakansson. 1982. The terminal Cretaceous extinction in Boreal shelf seas—a multicausal event. *Geol. Soc. Amer. Special Paper* 190:373–384.

Bishop, M. J. 1980. Helicinid land snails with apertural barriers. *J. Molluscan Stud.* 46:241–246.

Black, R. 1976. The effects of grazing by the limpet, *Acmaea insessa*, on the kelp, *Egregia laevigata*, in the intertidal zone. *Ecology* 557:265–277.

———. 1978. Tactics of whelks preying on limpets. *Mar. Biol.* 46:157–162.

Blackstone, N. W. 1985. The effects of shell size and shape on growth form in the hermit crab *Pagurus longicarpus*. *Biol. Bull.* 168:75–90.

Blair, G. M., and R. R. Seapy. 1972. Selective predation and prey location in the sea slug *Navanax inermis*. *Veliger* 15:119–124.

Blake, D. B. 1981. The new Jurassic sea star *Eokainaster* and comments on life habits and the origins of the modern Asteroidea. *J. Paleont.* 55:33–46.

———. 1983. Some biological controls on the distribution of shallow water sea stars (Asteroidea; Echinodermata). *Bull. Mar. Sci.* 33:703–712.

———. 1984. Constructional morphology and life habits of the Jurassic sea star *Sphaeraster* Quenstedt. *Neues Jb. Geol. Paläont. Abh.* 169:74–101.

Blake, J. A., and J. W. Evans. 1973. *Polydora* and related genera as burrowers in mollusk shells and other calcareous substrates. *Veliger* 15:235–249.

Blake, R. W. 1976. On seahorse locomotion. *J. Mar. Biol. Assoc. U.K.* 56:939–949.

———. 1977. On ostraciiform locomotion. *J. Mar. Biol. Assoc. U.K.* 57:1047–1055.

———. 1981. Mechanics of ostraciiform propulsion. *Canad. J. Zool.* 59:1067–1071.

———. 1983. Energetics of leaping in dolphins and other aquatic animals. *J. Mar. Biol. Assoc. U.K.* 63:61–70.

Bloom, S. A. 1975. The motile escape response of a sessile prey: A sponge-scallop mutualism. *J. Exp. Mar. Biol. Ecol.* 17:311–321.

Blundon, J. A., and V. S. Kennedy. 1982a. Mechanical and behavioral aspects of blue crab, *Callinectes sapidus* (Rathbun), predation on Chesapeake Bay bivalves. *J. Exp. Mar. Biol. Ecol.* 65:47–65.

———. 1982b. Refuges for infaunal bivalves from blue crab, *Callinectes sapidus* (Rathbun), predation in Chesapeake Bay. *J. Exp. Mar. Biol. Ecol.* 65:67–81.

———. 1986. Manuscript in preparation.

Blundon, J. A., and G. J. Vermeij. 1983. Effect of shell repair on shell strength in the gastropod *Littorina irrorata*. *Mar. Biol.* 76:41–45.

Boag, P. T., and P. R. Grant. 1981. Intense natural selection in a population of Darwin's finches (Geospizinae) in the Galapagos. *Science* 214:82–85.

Boardman, R. S. 1984. Origin of the post-Triassic Stenolaemata (Bryozoa): A taxonomic oversight. *J. Paleont.* 58:19–39.

Boekschoten, G. J. 1966. Shell borings of sessile epibiontic organisms as palaeoecological guides (with examples from the Dutch coast). *Palaeogeogr., Palaeoclimatol., Palaeoecol.* 2:333–379.

Boesch, C., and H. Boesch. 1983. Optimisation of nut-cracking with natural hammers by wild chimpanzees. *Behaviour* 83:265–286.

Bonaparte, J. F. 1982. Faunal replacement in the Triassic of South America. *J. Vert. Paleont.* 2:362–371.

———. 1984. Locomotion in rauisuchid thecodonts. *J. Vert. Paleont.* 3:210–218.

Bonem, R. M. 1977. Comparison of cavities and cryptic biota in modern reefs with those developed in Lower Pennsylvanian (Morrowan) bioherms. *Proc. Third Intern. Coral Reef Symp.* 1:76–80.

Bosence, B.W.J. 1983. Coralline algal reef frameworks. *J. Geol. Soc. London* 140:365–376.

Böss, H.-P. 1982. Locomotion and feeding in Mesozoic durophagous fishes. *Neues Jb. Geol. Paläont. Abh.* 164:167–171.

Böttger, C. R. 1935. Die Abwehr der Landschnecken gegen Käfer. *Sitzungsber. Ges. Naturforsch. Freunde Frankfort am Main,* pp. 93–102.

Botton, M. L. 1984. Diet and food preferences of the adult horseshoe crab *Limulus polyphemus* in Delaware Bay, New Jersey, USA. *Mar. Biol.* 81:199–207.

Boucot, A. J. 1975. *Evolution and Extinction Rate Controls.* Elsevier, Amsterdam.

———. 1983. Does evolution take place in an ecological vacuum? *J. Paleont.* 57:1–30.

Boucot, A. J., and C. Janis. 1983. Environment of the Early Paleozoic vertebrates. *Palaeogeogr., Palaeoclimatol., Palaeoecol.* 41:251–287.

Boulding, E. G. 1984. Crab-resistant features of shells of burrowing bivalves: Decreasing vulnerability by increasing handling time. *J. Exp. Mar. Biol. Ecol.* 76:201–223.

Bousfield, E. L. 1983. An updated phyletic classification and palaeohistory of the Amphipoda. In Schram (ed.), 1983, pp. 257–277.

Bovbjerg, R. V. 1956. Mammalian predation on mussels. *Proc. Iowa Acad. Sci.* 63:737–740.

Bowmer, T., and B. F. Keegan. 1983. Field survey of the occurrence of and significance of regeneration in *Amphiura filiformis* (Echinodermata: Ophiuroidea) from Galway Bay, west coast of Ireland. *Mar. Biol.* 74:65–71.

Bowsher, A. L. 1955. Origin and adaptation of platyceratid gastropods. *Kansas Univ. Paleont. Contrib.* 17 (*Mollusca,* Article) 5:1–11.

Boyd, D. W., and N. D. Newell. 1972. Taphonomy and diagenesis of a Permian fossil assemblage from Wyoming. *J. Paleont.* 46:1–14.

Boyle, P. R., and D. Knobloch. 1981. Hole boring of crustacean prey by the octopus *Eledone cirrhosa* (Mollusca, Cephalopoda). *J. Zool. London* 193:1–10.

Braber, L., and S. J. de Groot. 1973. The food of five flatfish species (Pleuronectiformes) in the southern North Sea. *Netherlands J. Sea Res.* 6:163–172.

Brace, R. C. 1977. The functional anatomy of the mantle complex and columellar muscle of tectibranch molluscs (Gastropoda: Opisthobranchia), and its bearing on the evolution of opisthobranch organization. *Phil. Trans. Roy. Soc. London* (B) 277:1–56.

Bradbury, R. H., and P. C. Young. 1982. The race and the swift revisited, or is aggression between corals important? *Proc. Fourth Intern. Coral Reef Symp.* 2:351–356.

Bramble, D. M. 1974. Emydid shell kinesis: Biomechanics and evolution. *Copeia* 1974:707–727.

———. 1978. Origin of the mammalian feeding complex: Models and mechanisms. *Paleobiology* 4:271–301.

Bramble, D. M., J. H. Hutchison, and J. M. Legler. 1984. Kinosternid shell kinesis: Structure, function and evolution. *Copeia* 1984:456–475.

Branch, G. M. 1976. Interspecific competition experienced by South African *Patella* species. *J. Anim. Ecol.* 45:507–529.

———. 1979. Aggression by limpets against invertebrate predators. *Anim. Behav.* 27:408–410.

Branch, G. M., and A. C. Marsh. 1978. Tenacity and shell shape in six *Patella* species: Adaptive features. *J. Exp. Mar. Biol. Ecol.* 34:111–130.

Branson, B. A. 1963. Notes on snail distribution and leech feeding habits in Oklahoma. *Nautilus* 76:148–149.

Brasier, M. D. 1976. Early Cambrian intergrowths of archaeocyathids, *Renalcis*, and pseudostromatolites from South Australia. *Palaeontology* 19:223–245.

———. 1979. The Cambrian radiation event. In House (ed.), 1979, pp. 103–159.

———. 1982. Sea-level changes, facies changes and the Late Precambrian-Early Cambrian evolutionary explosion. *Precambrian Res.* 17:105–123.

Brass, G. W., E. Saltzman, J. L. Sloan, II, J. R. Southam, W. W. Hay, W. T. Holser, and W. H. Peterson. 1982. Ocean circulation, plate tectonics, and climate. In Berger and Crowell (eds.), 1982, pp. 77–82.

Braun, J., and W.-E. Reif. 1985. A survey of aquatic locomotion in fishes and tetrapods. *Neues Jb. Geol. Paläont. Abh.* 169:307–332.

Breder, C. M. 1950. Factors influencing the establishment of residence in shells by tropical shore fishes. *Zoologica* 135:153–158.

Breen, P. A. 1972. Seasonal migration and population regulation in the limpet *Acmaea digitalis*. *Veliger* 15:133–141.

Brenchley, G. A. 1982. Mechanisms of spatial competition in marine soft-bottom communities. *J. Exp. Mar. Biol. Ecol.* 60:17–33.

Bretsky, S. S. 1976. Evolution and classification of the Lucinacea (Mollusca; Bivalvia). *Palaeontographica Americana* 8:219–337.

Brett, C. E. 1981. Terminology and functional morphology of attachment structures in pelmatozoan echinoderms. *Lethaia* 14:343–370.

Brett, C. E., and J. F. Cottrell. 1982. Substrate specificity in the Devonian tabulate coral *Pleurodictyum*. *Lethaia* 15:247–262.

Brett, C. E., W. D. Liddell, and K. L. Derstler. 1983. Late Cambrian hard substrate communities from Montana/Wyoming: The oldest known hardground encrusters. *Lethaia* 16:281–289.

Brett, J. R. 1979. Some morphological and behavioural adaptations of pile perch (*Rhacochilus bacca*) feeding on mussels (*Mytilus edulis*). *Canad. J. Zool.* 57:658–664.

Brewer, B. A. 1975. Epizoic limpets on the black turban snail, *Tegula funebralis* (A. Adams, 1855). *Veliger* 17:307–310.

Briggs, D.E.G. 1983. Affinities and early evolution of the Crustacea: The evidence of the Cambrian fossils. In Schram (ed.), 1983, pp. 1–22.

Briggs, D.E.G., and J. D. Mount. 1982. The occurrence of the giant arthropod *Anomalocaris* in the Lower Cambrian of southern California, and the overall distribution of the genus. *J. Paleont.* 56:1112–1118.

Bromley, R. G. 1975. Comparative analysis of fossil and Recent echinoid bioerosion. *Palaeontology* 18:725–739.

Bromley, R. G., and N.-M. Hanken. 1981. Shallow marine bioerosion at Vardö, Arctic Norway. *Bull. Geol. Soc. Denmark* 29:103–109.

Brown, A. C. 1982. The biology of sandy-beach whelks of the genus *Bullia* (Nassariidae). *Oceanogr. Mar. Biol. Ann. Rev.* 20:309–361.

Brown, S. C., S. R. Cassuto, and R. W. Loos. 1979. Biomechanics of chelipeds in some crustaceans. *J. Zool. London* 188:153–169.

Bruce, A. J. 1976. Coral reef Caridea and "commensalism." *Micronesica* 12:83–98.

Brun, E. 1971. Predation of *Chlamys islandica* (O. F. Müller) by eiders *Somateria* spp. *Astarte* 4:23–29.

Brunton, H. 1966. Predation and shell damage in a Visean brachiopod fauna. *Palaeontology* 9:355–359.

Brussard, P. F. 1975. Geographic variation in North American colonies of *Cepaea nemoralis*. *Evolution* 29:402–410.

Bruton, D. L. 1981. The arthropod *Sidneyia inexpectans*, Middle Cambrian, Burgess Shale, British Columbia. *Phil. Trans. Roy. Soc. London* (B) 295:619–656.

Bryant, E. H., H. van Dijk, and W. van Delden. 1981. Genetic variability of the face fly, *Musca autumnalis* de Geer, in relation to a population bottleneck. *Evolution* 35:872–881.

Bryant, J. P. 1981. Phytochemical deterrents of snowshoe hare browsing by adventitious shoots of four Alaskan trees. *Science* 213:889–890.

Buckeridge, J. S. 1983. Fossil barnacles (Cirripedia: Thoracica) of New Zealand and Australia. *New Zealand Geol. Surv. Paleont. Bull.* 50:151 P.

Buehler, E. J. 1969. Cylindrical borings in Devonian shells. *J. Paleont.* 43:1291.

Buffetaut, E. 1983. Wounds on the jaw of an Eocene mesosuchian crocodilian as possible evidence for the antiquity of crocodilian intraspecific fighting behaviour. *Paläont. Z.* 57:143–145.

Buit, M.-H. du. 1978. Rémarques sur la denture des raies et sur leur alimentation. *Vie et Milieu* 28-29:165–174.

Bulkley, P. S. 1968. Shell damage and repair in five members of the genus *Acmaea. Veliger* 11 (Suppl.): 64–66.

Bullock, T. H. 1955. Compensation for temperature in the metabolism and activity of poikilotherms. *Biol. Revs.* 30:311–342.

Bumpus, H. C. 1899. The elimination of the unfit as illustrated by the introduced sparrow, *Passer domesticus.* (A fourth contribution to the study of variation.) *Biol. Lectures, Mar. Biol. Labs. Woods Hole,* pp. 209–226.

Burchette, T. P., and R. Riding. 1977. Attached vermiform gastropods in Carboniferous marginal marine stromatolites and biostromes. *Lethaia* 10:17–28.

Burton, P.J.K. 1974. *Feeding and Feeding Apparatus in Waders: A Study of Anatomy and Adaptations in the Charadrii.* British Museum of Natural History, London.

Buss, L. W. 1979. Bryozoan overgrowth interactions—the interdependence of competition for space and food. *Nature* 281:475–477.

———. 1983a. Somatic variation and evolution. *Paleobiology* 9:12–16.

———. 1983b. Evolution, development, and the units of selection. *Proc. Nat. Acad. U.S.A.* 80:1387–1391.

Buss, L. W., and J.B.C. Jackson. 1981. Planktonic food availability and suspension-feeder abundance: Evidence of *in situ* depletion. *J. Exp. Mar. Biol. Ecol.* 49:151–161.

Butler, R. W., and J. W. Kirbyson. 1979. Oyster predation by the black oystercatcher in British Columbia. *Condor* 81:433–435.

Buzas, M. A., and S. J. Culver. 1984. Species duration and evolution: Benthic Foraminifera on the Atlantic continental margin of North America. *Science* 225:829–830.

Buzas, M. A., C. F. Koch, S. J. Culver, and N. F. Sohl. 1982. On the distribution of species occurrence. *Paleobiology* 8:143–150.

Caddy, J. F. 1967. Underwater observations on scallop (*Placopecten magellanicus*) behaviour and drag efficiency. *J. Fish. Res. Bd. Canada* 25:2123–2141.

Cain, A. J. 1977. Variation in the spire index of some coiled gastropod shells and its evolutionary significance. *Phil. Trans. Roy. Soc. London* (B) 277:377–428.

———. 1978a. Variation of terrestrial gastropods in the Philippines in relation to shell shape and size. *J. Conchol.* 29:239–245.

———. 1978b. The deployment of operculate land snails in relation to shape and size of shell. *Malacologia* 17:207–221.

Cairns, S. D., and G. D. Stanley, Jr. 1982. Ahermatypic coral banks: Living and fossil counterparts. *Proc. Fourth Intern. Coral Reef Symp.* 1:611–618.

Caldwell, R. L., and H. Dingle. 1975. Ecology and evolution of agonistic behavior in stomatopods. *Naturwissenschaften* 62:214–222.

Calkins, D. G. 1978. Feeding behavior and major prey species of the sea otter, *Enhydra lutris*, in Montague Strait, Prince William Sound, Alaska. *Fish. Bull.* 76:125–131.

Cameron, B. 1967. Oldest carnivorous gastropod borings, found in Trentonian (Middle Ordovician) brachiopods. *J. Paleont.* 41:147–150.

Campbell, H. W. 1969. The effects of temperature on the auditory sensitivity of lizards. *Physiol. Zool.* 42:183–210.

Campbell, K.S.W. 1977. Trilobites of the Haragan, Bois d'Arc and Frisco Formations (Early Devonian), Arbuckle Mountains region, Oklahoma. *Oklahoma Geol. Surv. Bull.* 123:1–227.

Campbell, S. E. 1982. Precambrian endoliths discovered. *Nature* 299:429–431.

Capapé, C. 1976. Etude du régime alimentaire de l'aigle de mer, *Myliobatis aquila* (L., 1758) des côtes tunésiens. *J. Cons. Int. Explor. R. Mer* 37:29–35.

Caple, G., R. P. Balda, and W. R. Willis. 1983. The physics of leaping animals and the evolution of preflight. *Amer. Nat.* 121:455–476.

Carlquist, S. 1975. *Ecological Strategies of Xylem Evolution*. University of California Press, Berkeley.

Carpenter, F. M. 1976. Geological history and evolution of the insects. *Proc. 15th Intern. Congr. Entomol.* (Washington) 1:63–70.

Carpenter, F. M., and L. Burnham. 1985. The geological record of insects. *Ann. Rev. Earth Planet. Sci.* 13:297–314.

Carpenter, K., and D. Lindsey. 1980. The dentary of *Brachychampsa montana* Gilmore (Alligatorinae; Crocodylidae), a Late Cretaceous turtle-eating alligator. *J. Paleont.* 54:1213–1217.

Carriker, M. R. 1951. Observations on the penetration of tightly closing bivalves by *Busycon* and other predators. *Ecology* 32:73–83.

———. 1972. Observations on removal of spines by muricid gastropods during shell growth. *Veliger* 15:69–74.

Carriker, M. R., and D. Van Zandt. 1972. Predatory behavior of a shell-boring muricid gastropod. In H. E. Winn and B. L. Olla (eds.), *Behavior of Marine Animals.* I: *Invertebrates*, pp. 157–244. Plenum, New York.

Carriker, M. R., and E. L. Yochelson. 1968. Recent gastropod boreholes and Ordovician cylindrical borings. *U.S. Geol. Surv. Prof. Pap.* 593-B:B1-B23.

Carson, H. L. 1975. The genetics of speciation at the diploid level. *Amer. Nat.* 109:83–92.

Carson, H. L., and A. R. Templeton. 1984. Genetic revolutions in relation to speciation phenomena: The founding of new populations. *Ann. Rev. Ecol. Syst.* 15:97–131.

Carter, J. G. 1978. Ecology and evolution of the Gastrochaenacea (Mollusca, Bivalvia) with notes on the evolution of the endolithic habitat. *Yale Univ. Peabody Mus. Nat. Hist. Bull.* 41:92 P.

———. 1980. Environmental and biological controls of bivalve shell mineralogy and microstructure. In Rhoads and Lutz (eds.), 1980, pp. 69–113.

Carter, J. W. 1982. Natural history observations on the gastropod shell-using amphipod *Photis conchicola* Alderman, 1936. *J. Crustacean Biol.* 2:328–341.

Carter, R. M. 1967a. On the nature and definition of the lunule, escutcheon and corcelet in the Bivalvia. *Proc. Malacol. Soc. London* 37:243–263.

———. 1967b. On Lison's model of bivalve shell form, and its biological interpretation. *Proc. Malacol. Soc. London* 37:265–278.

———. 1968. The biology and palaeontology of some predators of bivalved Mollusca. *Palaeogeogr., Palaeoclimatol., Palaeoecol.* 4:29–65.

Case, T. J., and M. E. Gilpin. 1974. Interference competition and niche theory. *Proc. Nat. Acad. Sci. U.S.A.* 71:3073–3077.

Cassels, R. 1984. The role of prehistoric man in the faunal extinctions of New Zealand and other Pacific islands. In Martin and Klein (eds.), 1984, pp. 741–767.

Castilla, J. C., and I. Bahamondes. 1979. Observaciones conductuales y ecológicas sobre *Lutra felina* (Molina) 1782 (Carnivora: Mustelidae) en las zonas central y centro-norte de Chile. *Arch. Biol. Med. Exper.* 12:119–132.

Castro, P. 1976. Brachyuran crabs symbiotic with scleractinian corals: A review of their biology. *Micronesica* 12:99–110.

Cernohorsky, W. O. 1984. Systematics of the family Nassariidae (Mollusca: Gastropoda). *Bull. Auckland Inst. Mus.* 14:356 P.

Chaloner, W. G., and A. Sheerin. 1979. Devonian macrofloras. *Special Pap. Palaeontol.* 23:145–161.

———. 1981. The evolution of reproductive strategies in early land plants. In Scudder and Reveal (eds.), 1981, pp. 93–100.

Chamberlain, C. K. 1975. Recent Lebensspuren in nonmarine aquatic environments. In Frey (ed.), 1975, pp. 431–458.

Chamberlain, J. A., Jr., and W. A. Moore. 1982. Rupture strength and flow rate of *Nautilus* siphuncular tube. *Paleobiology* 8:408–425.

Chao, L. N. 1973. Digestive system and feeding habits of the cunner, *Tautogolabrus adspersus*, a stomachless fish. *Fish. Bull.* 71:565–586.

Chapin, D. 1968. Some observations of predation on *Acmaea* species by the crab *Pachygrapsus crassipes*. *Veliger* 11 (Suppl.): 67–68.

Charlesworth, B., R. Lande, and M. Slatkin. 1982. A neo-Darwinian commentary on macroevolution. *Evolution* 36:474–498.

Chelazzi, G., S. Focardi, J. L. Deneuborg, and R. Innocenti. 1983. Competition for the home and aggressive behaviour in the chiton *Acanthopleura gemmata* (Blainville) (Mollusca: Polyplacophora). *Behav. Ecol. Sociobiol.* 14:15–20.

Chen, Y.-Y., and C. Teichert. 1983. Cambrian Cephalopoda of China. *Palaeontographica Abt.* (A) 181:1–102.

Christensen, A. M. 1957. The feeding behavior of the sea-star *Evasterias troschelii* Stimpson. *Limnol. Oceanogr.* 2:180–197.

———. 1970. Feeding biology of the sea-star *Astropecten irregularis* Pennant. *Ophelia* 8:1–134.

Cifelli, R. 1976. Evolution of ocean climate and the record of planktonic Foraminifera. *Nature* 264:431–432.

Clark, B. D., and W. Bemis. 1979. Kinematics of swimming of penguins at the Detroit Zoo. *J. Zool. London* 188:411–428.

Clarke, A. 1983. Life in cold water: The physiological ecology of polar marine ectotherms. *Oceanogr. Mar. Biol. Ann. Rev.* 21:341–453.

Clarke, B., J. Murray, and M. S. Johnson. 1984. The extinction of endemic species by a program of biological control. *Pacific Sci.* 38:97–104.

Clarke, R. D. 1972. The effect of toe clipping on survival in Fowler's toad (*Bufo woodhousei fowleri*). *Copeia*, no. 1:182–185.

Clarkson, E.N.K., and J.-L. Henry. 1973. Structure coaptatif et enroulement chez quelques trilobites Ordoviciens et Siluriens. *Lethaia* 6:105–132.

Clegg, M. 1972. Carrion crows feeding on marine molluscs and taking fish. *Bird Study* 19:249–250.

Cloud, P., and M. F. Glaessner. 1982. The Ediacarian period and system: Metazoa inherit the earth. *Science* 217:783–792.

Clutton-Brock, T. H. 1982. The functions of antlers. *Behaviour* 79:108–125.

Coan, E. 1983. The Eastern Pacific Donacidae. *Veliger* 25:273–298.

Coates, A. G., and J.B.C. Jackson. 1985. Morphological themes in the evolution of clonal and aclonal marine invertebrates. In Jackson et al. (eds.), 1985, pp. 67–106.

Coates, A. G., and W. A. Oliver, Jr. 1973. Coloniality in zoantharian corals. In R. S. Boardman, A. H. Cheetham, and W. A. Oliver, Jr. (eds.), *Animal Colonies*, pp. 3–27. Dowden, Hutchinson, and Ross, Stroudsburg, Pennsylvania.

Coe, W. R., and D. L. Fox. 1942. Biology of the California Sea-Mussel (*Mytilus californianus*). I. Influence of temperature, food supply, sex and age on the rate of growth. *J. Exp. Zool.* 90:1–30.

Coen, L. D. 1985. Shear resistance in two bivalve molluscs: Role of hinges and interdigitating margins. *J. Zool. London* (A) 205:479–487.

Colbath, S. L. 1985. Gastropod predation and depositional environments of two molluscan communities from the Miocene Astoria Formation at Beverly Beach State Park, Oregon. *J. Paleont.* 59:849–869.

Coley, P. D. 1980. Effects of leaf age and plant life history patterns on herbivory. *Nature* 284:245–246.

———. 1983. Herbivory and defensive characteristics of tree species in a lowland tropical forest. *Ecol. Monogrs.* 53:209–233.

Colgan, M. W. 1984. The Cretaceous coral *Helipora* (Octocorallia, Coenothecalia)—a common Indo-Pacific reef builder. In Eldredge and Stanley (eds.), 1984, pp. 266–271.

Collette, B. B. 1983. Two new species of coral toadfishes, family Batrachoididae, genus *Sanopus*, from Yucatan, Mexico, and Belize. *Proc. Biol. Soc. Washington* 96:719–724.

Collette, B. B., and J. L. Russo. 1981. A revision of the scaly toadfishes genus *Batrachoides*, with descriptions of two new species from the Eastern Pacific. *Bull. Mar. Sci.* 31:197–233.

Collins, D., and D. M. Rudkin. 1981. *Priscansermarinus barnetti*, a probable lepadomorph barnacle from the Middle Cambrian Burgess Shale of British Columbia. *J. Paleont.* 55:1006–1015.

Commito, J. A. 1982. Effects of *Lunatia heros* predation on the population dynamics of *Mya arenaria* and *Macoma balthica* in Maine, U.S.A. *Mar. Biol.* 69:187–193.

Congdon, J. D., L. J. Vitt, and W. W. King. 1974. Geckos: Adaptive significance and energetics of tail autotomy. *Science* 184:1379–1380.

Connell, J. H. 1961a. Effects of competition, predation by *Thais lapillus*, and other factors on natural populations of the barnacle *Balanus balanoides*. *Ecol. Monogrs.* 31:61–104.

———. 1961b. The influence of interspecific competition and other factors on the distribution of the barnacle *Chthamalus stellatus*. *Ecology* 42:710–723.

———. 1970. A predator-prey system in the rocky intertidal region. I: *Balanus glandula* and several predatory species of *Thais*. *Ecol. Monogrs.* 40:49–78.

———. 1972. On the role of natural enemies in preventing competitive exclusion in some marine animals and in rain forest trees. *Proc. Adv. Stud. Inst. Dynamics Numbers Populations Oosterbeek* 1970:298–312.

Conway Morris, S. 1979. The Burgess Shale (Middle Cambrian) fauna. *Ann. Rev. Ecol. Syst.* 10:327–349.

———. 1985. The Middle Cambrian metazoan *Wiwaxia corrugata* (Matthew) from the Burgess Shale and *Ogygopsis* Shale, British Columbia, Canada. *Phil. Trans. Roy. Soc. London* (B) 307:507–582.

Cook, A. 1985. The organisation of feeding in the carnivorous snail *Euglandina rosea*. *Malacologia* 26:183–189.

Cook, P. J., and J. H. Shergold. 1984. Phosphorus, phosphorites and skeletal evolution at the Precambrian-Cambrian boundary. *Nature* 308:231–236.

Cook, P. L. 1968. Observations on living Bryozoa. *Atti Soc. Ital. Sci. Nat. Mus. Civ. St. Nat. Milano* 108:155–160.

Cook, P. L., and P. J. Chimonides. 1983. A short history of the lunulite Bryozoa. *Bull. Mar. Sci.* 33:566–581.

Coombs, W. P., Jr. 1978a. Theoretical aspects of cursorial adaptations in dinosaurs. *Quart. Rev. Biol.* 53:393–418.

———. 1978b. The families of the ornithischian dinosaur order Ankylosauria. *Palaeontology* 21:143–170.

Cooper, M. R. 1982. A Mid-Permian to earliest Jurassic tetrapod biostratigraphy and its significance. *Arnoldia Zimbabwe* 9:77–104.

Copper, P. 1974. Structure and development of Early Paleozoic reefs. *Proc. Second Intern. Coral Reef Symp.* 1:365–386.

———. 1977. Paleolatitudes in the Devonian of Brazil and the Frasnian-Famennian mass extinction. *Palaeogeogr., Palaeoclimatol., Palaeoecol.* 21:165–207.

Corbet, P. S. 1961. The food of non-cichlid fishes in the Lake Victoria basin with remarks on their evolution and adaptation to lacustrine conditions. *Proc. Zool. Soc. London* 136:1–101.

Corliss, B. H. 1979. Response of deep-sea benthic Foraminifera to development of the psychrosphere near the Eocene-Oligocene boundary. *Nature* 282:63–65.

Corliss, B. H., M.-P. Aubry, W. A. Berggren, J. M. Fenner, L. D. Keigwin, Jr., and G. Keller. 1984. The Eocene/Oligocene boundary event in the deep sea. *Science* 226:806–810.

Cowen, R. 1983. Algal symbiosis and its recognition in the fossil record. In Tevesz and McCall (eds.), 1983, pp. 431–478.

Crepet, W. L. 1984. Advanced (constant) insect pollination mechanisms: Pattern of evolution and implications vis-à-vis angiosperm diversity. *Ann. Missouri Bot. Garden* 71:607–630.

Crick, R. E. 1981. Diversity and evolutionary rates of Cambro-Ordovician nautiloids. *Paleobiology* 7:216–229.

Crimes, T. P., and M. M. Anderson. 1985. Trace fossils from Late Precambrian-Early Cambrian strata of southeastern Newfoundland (Canada): Temporal and environmental implications. *J. Paleont.* 59:310–343.

Croat, T. B., and P. Busey. 1975. Geographical affinities of the Barro Colorado Island flora. *Biotropica* 27:127–135.

Cronin, T. M. 1985. Speciation and stasis in marine Ostracoda: Climatic modulation of evolution. *Science* 227:60–63.

Cronk, Q.C.B. 1980. Extinction and survival in the endemic vascular flora of Ascension Island. *Biol. Conserv.* 17:207–219.

Crowell, J. C. 1982. Continental glaciation through geological times. In Berger and Crowell (eds.), 1982, pp. 77–82.

Crowson, R. A. 1981. *The Biology of the Coleoptera.* Academic Press, New York.

Crush, P. J. 1984. A Late Upper Triassic sphenosuchid crocodilian from Wales. *Palaeontology* 27:131–157.

Culver, D. C. 1982. *Cave Life: Evolution and Ecology.* Harvard University Press, Cambridge.

Cunningham, P. N., and R. N. Hughes. 1984. Learning of predaceous skills by shorecrabs *Carcinus maenas* feeding on mussels and dogwhelks. *Mar. Ecol. Progr. Ser.* 16:21–26.

Currey, J. D. 1977. Mechanical properties of mother of pearl in tension. *Proc. Roy. Soc. London* (B) 196:443–463.

Currey, J. D., and R. N. Hughes. 1982. Strength of the dogwhelk *Nucella lapillus* and the winkle *Littorina littorea* from different habitats. *J. Anim. Ecol.* 41:47–56.

Currey, J. D., A. Nash, and W. Bonfield. 1982. Calcified cuticle in the stomatopod smashing limb. *J. Materials Sci.* 17:1939–1944.

Cutress, C., D. M. Ross, and L. Sutton. 1970. The association of *Calliactis tricolor* with its pagurid, calappid and majid partners in the Caribbean. *Canad. J. Zool.* 48:371–376.

Dahlberg, M. D., and J. C. Conyers. 1973. An ecological study of *Gobiosoma bosci* and *G. ginsburgi* (Pisces, Blenniidae) on the Georgia coast. *Fish. Bull.* 71:279–287.

Dalrymple, G. H. 1979. On the jaw mechanism of the snail-crushing lizards, *Dracaena* Daudin, 1802 (Reptilia, Lacertilia, Teiidae). *J. Herpetol.* 13:303–311.

Dana, T. F. 1975. Development of contemporary Eastern Pacific coral reefs. *Mar. Biol.* 33:355–374.

Daniel, T. L. 1984. Unsteady aspects of aquatic locomotion. *Amer. Zool.* 24:121–134.

Daniels, C. B. 1985. The effect of tail autotomy on the exercise capacity of the water skink, *Sphenomorphus quoyii. Copeia* 1985:1074–1077.

Darnell, R. M. 1961. Trophic spectrum of an estuarine community, based on studies of Lake Pontchartrain, Louisiana. *Ecology* 42:553–568.

Darragh, T. A., and G. W. Kendrick. 1980. Eocene bivalves from the Palliup Siltstone near Walpole, Western Australia. *J. Roy. Soc. Western Australia* 63:5–20.

Darwin, C. 1872. *The Origin of Species by Natural Selection or the Preservation of Favored Races in the Struggle for Life.* 6th ed., reprinted 1962. Colliers, New York.

―――. 1876. *The Movements and Habits of Climbing Plants.* 2d ed., revised. Appleton, New York.

Davenport, J., S. A. Munks, and P. J. Oxford. 1984. A comparison of the swimming of marine and freshwater turtles. *Proc. Roy. Soc. London* (B) 220:447–475.

Davis, G. M. 1979. The origin and evolution of the gastropod family Pomatiopsidae, with emphasis on the Mekong River Triculinae. *Monogr. Acad. Nat. Sci. Philadelphia* 20:120 P.

Dawkins, R., and J. R. Krebs. 1979. Arms races between and within species. *Proc. Roy. Soc. London* (B) 205:489–511.

Dayton, P. K. 1971. Competition, disturbance, and community organization: The provision and subsequent utilization of space in a rocky intertidal community. *Ecol. Monogrs.* 41:351–389.

―――. 1973. Two cases of resource partitioning in an intertidal community: Making the right prediction for the wrong reason. *Amer. Nat.* 107:662–670.

Dayton, P. K., R. J. Rosenthal, L. C. Mahen, and T. Antezana. 1977. Population structure and foraging biology of the predaceous Chilean asteroid *Meyenaster gelatinosus* and the escape biology of its prey. *Mar. Biol.* 39:361–370.

Debach, P. 1974. *Biological Control by Natural Enemies.* Cambridge University Press, London.

Debrenne, F., A. Y. Rozanov, and G. F. Webers. 1984. Upper Cambrian Archaeocyatha from Antarctica. *Geol. Mag.* 121:291–299.

Decary, R. 1926. Observations sur l'*Olios coenobita* Fage et le *Nemoscolus waterloti* Verland. *Arch. Zool. Exper. Gen.* 65:18–21.

Degens, E. T., and P. Stoffers. 1976. Stratified waters as a key to the past. *Nature* 263:22–27.

DeMar, R. 1966. The phylogenetic and functional implications of the armor of the Dissorophidae. *Fieldiana Geology* 16:55–88.

———. 1968. The Permian labyrinthodont amphibian *Dissorophus multicinctus*, and adaptation and phylogeny of the family Dissorophidae. *J. Paleont.* 42:1210–1242.

De Martini, E. E. 1969. A correlative study of the ecology and comparative feeding mechanism morphology of the Embiotocidae (surffishes) as evidence of the family's adaptive radiation into available ecological niches. *Wasmann J. Biol.* 27:177–247.

Denison, R. 1978. *Placodermi*. Handbook of Paleoichthyology. Vol. 2. Gustav Fischer, Stuttgart.

———. 1979. *Acanthodii*. Handbook of Paleoichthyology. Vol. 5. Gustav Fischer, Stuttgart.

Denton, E. J. 1974. On buoyancy and the lives of modern and fossil cephalopods. *Proc. Roy. Soc. London* (B) 185:273–299.

Dial, B. E., and L. C. Fitzpatrick. 1984. Predator escape success in tailed versus tailless *Scincella lateralis* (Sauria: Scincidae). *Anim. Behav.* 32:301–302.

Diamond, J. M. 1984. Historic extinctions: A Rosetta Stone for understanding prehistoric extinctions. In Martin and Klein (eds.), 1984, pp. 824–862.

Diefenbach, C. O. da C. 1979. Ampullarid gastropod—staple food for *Caiman latirostris*? *Copeia* 1979:162–163.

Dingle, H., and R. L. Caldwell. 1978. Ecology and morphology of feeding and agonistic behavior in mudflat stomatopods (Squillidae). *Biol. Bull.* 155:134–149.

Dingus, L., and M. Sadler. 1982. The effects of stratigraphic completeness on estimates of evolutionary rate. *Syst. Zool.* 31:400–412.

Ditadi, A.S.F. 1976. A preliminary note on the mode of life of *Lissomyema exilii* (F. Müller, 1883) Echiura. *Proc. Intern. Symp. Biol. Sipuncula and Echiura* 2:143–146.

Dobzhansky, T. 1968. Adaptedness and fitness. In R. C. Lewontin (ed.), *Population Biology and Evolution*, pp. 109–121. Syracuse University Press, Syracuse.

Dockery, D. T., II. 1977. Mollusca of the Moodys Branch Formation, Mississippi. *Bull. Mississippi Geol., Econ. Topogr. Surv.* (Jackson) 120:212 P.

Domning, D. P. 1976. An ecological model of Late Tertiary sirenian evolution in the North Pacific Ocean. *Syst. Zool.* 25:352–362.

———. 1982. Evolution of manatees: A speculative history. *J. Paleont.* 56:599–619.

Donovan, D. T. 1977. Evolution of the dibranchiate Cephalopoda. *Symp. Zool. Soc. London* 38:15–48.

Douvillé, H. 1931. Symbiose ou parasitisme. *Geol. Soc. France Bull.* (5) 1:391–396.

Dreisig, H. 1981. The rate of predation and its temperature dependence in a tiger beetle, *Cicindela hybrida*. *Oikos* 36:196–202.

Drewes, R. C., and B. Roth. 1981. Snail-eating frogs from the Ethiopian highlands: A new anuran specialization. *Zool. J. Linn. Soc.* 73:267–287.

Drinnan, R. E. 1957. The winter feeding of the oystercatcher (*Haematopus ostralegus*) on the edible cockle (*Cardium edule*). *J. Anim. Ecol.* 26:441–469.

Dudley, E. C., and E. C. Dudley. 1980. Drilling predation on some Miocene marine mollusks. *Nautilus* 94:63–66.

Dudley, E. C., and G. J. Vermeij. 1978. Predation in time and space: Drilling in the gastropod *Turritella*. *Paleobiology* 4:436–441.

Duncan, S. de B., and R. N. Hughes. 1984. Behavioural components of prey-selection by dogwhelks, *Nucella lapillus* (L.), feeding on barnacles, *Semibalanus balanoides* (L.), in the laboratory. *J. Exp. Mar. Biol. Ecol.* 79:91–103.

Dungan, M. L. 1985. Competition and the morphology, ecology, and evolution of acorn barnacles: An experimental test. *Paleobiology* 11:165–173.

Dunn, D. F. 1981. The clownfish sea anemones: Stichodactylidae (Coelenterata: Actinaria) and other sea anemones symbiotic with pomacentrid fishes. *Trans. Amer. Phil. Soc.* 71:115 P.

Dunn, D. F., D. M. Devaney, and B. Roth. 1980. *Stylobates*: A shell-forming sea anemone (Coelenterata, Anthozoa, Actiniidae). *Pacific Sci.* 34:379–388.

Dzik, J. 1981a. Origin of the Cephalopoda. *Acta Palaeont. Polon.* 26:161–189.

———. 1981b. Evolutionary relationships of the Early Palaeozoic "cyclostomatous" Bryozoa. *Palaeontology* 24:827–861.

———. 1984. Phylogeny of the Nautiloidea. *Palaeont. Polon.* 45:219 P.

Eales, N. B. 1949. The food of the dogfish, *Scyliorhinus caniculus* L. *J. Mar. Biol. Assoc. U.K.* 26:791–793.

Ebert, T. A. 1968. Growth rates of a sea urchin *Strongylocentrotus purpuratus* related to food availability and spine abrasion. *Ecology* 59:1075–1991.

Ebling, F. J., J. A. Kitching, L. Muntz, and C. M. Taylor. 1964. The ecology of Lough Ine. XIII. Experimental observations of the destruction of *Mytilus edulis* and *Nucella lapillus* by crabs. *J. Anim. Ecol.* 33:73–83.

Ebling, F. J., A. D. Hawkins, J. A. Kitching, L. Muntz, and V. M. Pratt. 1966. The ecology of Lough Ine. XVI. Predation and diurnal migration in the *Paracentrotus* community. *J. Anim. Ecol.* 35:559–566.

Edmund, A. G. 1985. The armor of fossil giant armadillos (Pampatheriidae, Xenarthra, Mammalia). *Texas Memorial Mus., Univ. Texas, Austin, Pearce-Sellards Ser.* 40:1–20.

Edwards, D. 1980. Early land floras. In Panchen (ed.), 1980, pp. 55–85.

Edwards, D. C. 1969. Predators on *Olivella biplicata*, including a species-specific predator avoidance response. *Veliger* 11:326–333.

Edwards, P. J., and S. D. Wratten. 1982. Wound-induced changes in palatability in birch (*Betula pubescens* Ehrh. ssp. *pubescens*). *Amer. Nat.* 120:816–818.

Edwards, R.R.C., and J. H. Steele. 1968. The ecology of O-group plaice and common dabs at Loch Ewe. I: Population and food. *J. Exp. Mar. Biol. Ecol.* 2:215–238.

Ehrlich, P. R., and P. H. Raven. 1964. Butterflies and plants: A study in coevolution. *Evolution* 18:586–608.

———. 1969. Differentiation of populations. *Science* 165:1228–1231.

Eisenberg, J. F. 1981. *The Mammalian Radiations.* University of Chicago Press, Chicago.

Eisner, T., and J. A. Davis. 1967. Mongoose throwing and smashing millipedes. *Science* 155:577–579.

Eldredge, N., and S. J. Gould. 1972. Punctuated equilibria: An alternative to phyletic gradualism. In T.J.M. Schopf (ed.), *Models in Paleobiology*, pp. 82–115. Freeman, Cooper, and Company, San Francisco.

Eldredge, N., and S. M. Stanley (eds.). 1984. *Living Fossils.* Springer, New York.

Elliott, A. B., and L. K. Karunakaran. 1974. Diet of *Rana cancrivora* in fresh water and brackish water environments. *J. Zool. London* 174:203–215.

Elner, R. W. 1978. The mechanics of predation by the shore crab, *Carcinus maenas* (L.), on the edible mussel, *Mytilus edulis* L. *Oecologia* (Berlin) 36:333–344.

Elner, R. W., and A. Campbell. 1981. Force, function and mechanical advantage in the chelae of the American lobster *Homarus americanus* (Decapoda: Crustacea). *J. Zool. London* 193:269–286.

Elner, R. W., and G. S. Jamieson. 1979. Predation of sea scallops, *Placopecten magellanicus*, by the rock crab, *Cancer irroratus*, and the American lobster, *Homarus americanus*. *J. Fish. Res. Bd. Canada* 36:537–543.

Elner, R. W., and D. G. Raffaelli. 1980. Interactions between two marine snails, *Littorina rudis* Maton and *Littorina nigrolineata* Gray, a predator, *Carcinus maenas* (L.), and a parasite, *Microphallus similis* Jägerskiold. *J. Exp. Mar. Biol. Ecol.* 43:151–160.

Elwood, R. W., A. McLean, and L. Webb. 1979. The development of shell preferences by the hermit crab *Pagurus bernhardus*. *Anim. Behav.* 27:940–946.

Emmons, L. H., and A. H. Gentry. 1983. Tropical forest structure and the distribution of gliding and prehensile-tailed vertebrates. *Amer. Nat.* 121:513–524.

Enders, F. 1975. The influence of hunting manner on prey size, particularly in spiders with long attack distances (Araneidae, Linyphiidae, and Salticidae). *Amer. Nat.* 109:737–763.

Endler, J. A. 1978. A predator's view of animal color patterns. *Evol. Biol.* 11:319–364.

Enequist, P. 1949. Studies on the soft-bottom amphipods of the Skagerak. *Zool. Bidr. Uppsala* 28:297–492.

Erickson, R. 1984. Chelonivorous habits of a Paleocene crocodile *Leidyosuchus formidabilis. Sci. Publ. Sci. Mus. Minnesota*, n.s. 5 (4):3–9.

Estes, R. 1983. *Sauria Terrestria, Amphisbaenia. Encyclopedia of Palaeoherpetology.* Vol. 10-A. Gustav Fischer, Stuttgart.

Estes, R., and E. E. Williams. 1984. Ontogenetic variation in the molariform teeth of lizards. *J. Vert. Paleont.* 4:96–107.

Evans, J. W. 1969. Borers in the shell of the sea scallop, *Placopecten magellanicus. Amer. Zool.* 9:775–782.

Evans, M.E.G., and T. G. Forsythe. 1984. A comparison of adaptations to running, pushing and burrowing in some adult Coleoptera: Especially Carabidae. *J. Zool. London* 202:515–534.

Evans, R. A. 1980–81. *Octopus* predation on *Cypraea spadicea. Of Sea and Shore* 11:225–226.

Fage, L. 1926. Sur quelques araignées de Madagascar, nouvelles ou peu connues, et sur leur curieuse industrie. *Arch. Zool. Exper. Gen.* 65:5–17.

Fairweather, P. G., and A. J. Underwood. 1983. The apparent diet of predators and biases due to different handling times of their prey. *Oecologia* (Berlin) 56:169–179.

Farlow, J. O. 1981. Estimates of dinosaur speeds from a new trackway site in Texas. *Nature* 294:747–748.

Farlow, J. O., and P. Dodson. 1975. The behavioral significance of frill and horn morphology in ceratopsian dinosaurs. *Evolution* 29:351–361.

Farmer, W. M. 1970. Swimming gastropods (Opisthobranchia and Prosobranchia). *Veliger* 13:73–89.

Farrow, G. E. 1971. Back-reef and lagoonal environments of Aldabra Atoll distinguished by their crustacean burrows. *Symp. Zool. Soc. London* 28:455–500.

Fauchald, K., and P. A. Jumars. 1979. The diet of worms: A study of polychaete feeding guilds. *Oceanogr. Mar. Biol. Ann. Rev.* 17:193–284.

Fawcett, M. H. 1984. Local and latitudinal variation in predation on an herbivorous marine snail. *Ecology* 65:1214–1230.

Feare, C. J. 1971. Predation of limpets and dogwhelks by oystercatchers. *Bird Study* 18:121–129.

Feder, H. M. 1955. On the methods used by the starfish *Pisaster ochraceus* in opening three types of bivalve molluscs. *Ecology* 36:764–767.

———. 1981. Aspects of the feeding biology of the brittle star *Ophiura texturata. Ophelia* 20:215–235.

Fedonkin, M. A. 1982. Precambrian soft-bodied fauna and the earliest radiation of invertebrates. *Third North Amer. Paleont. Conv. Proc.* 1:165–167.

Feduccia, A. 1973. Dinosaurs as reptiles. *Evolution* 27:166–169.

Feduccia, A., and H. B. Tordoff. 1979. Feathers of *Archaeopteryx*: Asymmetric veins indicate aerodynamic function. *Science* 203:1021–1023.

Feifarek, B. 1986. Spines and epifauna as antipredator defenses in the thorny oyster *Spondylus americanus* Hermann. *J. Exp. Mar. Biol. Ecol.*, in press.

Fenton, C. L., and M. A. Fenton. 1931. Some snail borings of Paleozoic age. *Amer. Midland Nat.* 12:522–528.

Fischer, A. G. 1960. Latitudinal variations in organic diversity. *Evolution* 14:64–81.

———. 1984. Biological innovations and the sedimentary record. In Holland and Trendall (eds.), 1984, pp. 145–157.

———. 1985. The two Phanerozoic supercycles. In W. A. Berggren and J. A. Van Couvering (eds.), *Catastrophes in Earth History: The New Uniformitarianism*, pp. 129–150. Princeton University Press, Princeton.

Fischer, A. G., and M. A. Arthur. 1977. Secular variations in pelagic realm. In H. E. Cook and P. Enos (eds.), *Deep-Water Carbonate Environments. Soc. Econ. Petrol. Mineral. Spec. Publ.* 25, pp. 19-50.

Fischer, J.-C., and B. Riou. 1982. Les teuthoîdes (Cephalopoda, Dibranchiata) du Callovien Inférieur de la Vulte-sur-Rhône (Ardèche, France). *Ann. Paleont.* (*Vert.-Invert.*) 68:295–325.

Fischer, P. H. 1962. Perforations de fossiles pré-Tertiaires attribuées à des gastéropodes prédateurs. *J. Conchyliol.* 102:68–78.

———. 1966. Perforations de fossiles Tertiaires par des gastéropodes prédateurs. *J. Conchyliol.* 105:66–96.

Fischer, R. 1981. Bioerosion of basalt of the Pacific coast of Costa Rica. *Senckenbergiana Maritima* 13:1–41.

Fisher, D. C. 1977. Functional significance of spines in the Pennsylvanian horseshoe crab *Euproops danae*. *Paleobiology* 3:175–195.

———. 1986. Progress in organismal design. In Raup and Jablonski (eds.), 1986, pp. 99–117.

Fisher, R. A. 1958. *The Genetical Theory of Natural Selection*. 2d ed. Dover, New York.

Fishlin, D. A., and D. W. Phillips. 1980. Chemical camouflaging and behavioral defenses against a predatory seastar by three species of gastropods from the surfgrass *Phyllospadix* community. *Biol. Bull.* 158:34–48.

Fleischer, R. C., and R. F. Johnston. 1982. Natural selection on body size and proportions in house sparrows. *Nature* 298:747–749.

Flessa, K. W., and D. Jablonski. 1985. Declining Phanerozoic background extinction rates: Effect of taxonomic structure? *Nature* 313:216–218.

Florey, E., and G. Hoyle. 1976. The effects of temperature on a nerve-muscle system of the Hawaiian ghost crab *Ocypode ceratophthalma* (Pallas). *J. Comp. Physiol.* 110:51–64.

Flower, R. H. 1955. Saltations in nautiloid coiling. *Evolution* 9:244–260.

———. 1976. The Ordovician cephalopod faunas and their role in correlation. In M. G. Bassett (ed.), *The Ordovician System*, pp. 523–552. University of Wales Press, Cardiff.

Flügel, E., and G. D. Stanley, Jr. 1984. Reorganization, development and evolution of post-Permian reefs and reef organisms. *Palaeontographica Americana* 54:177–186.

Focke, J. W. 1977. The effect of a potentially reef-building vermetid-coralline algal community on an eroding limestone coast, Curaçao, Netherlands Antilles. *Proc. Third Intern. Coral Reef Symp.* 1:239–245.

Ford, E. B. 1965. *Ecological Genetics*. 2d ed. Methuen, London.

Forest, J., M. de Saint Laurent, and F. A. Chace, Jr. 1976. *Neoglyphea inopinata*: A crustacean "living fossil" from the Philippines. *Science* 192:884.

Forester, A. J. 1979. The association between the sponge *Halichondria pan-*

icea (Pallas) and scallop *Chlamys varia* (L.): A commensal-protective mutualism. *J. Exp. Mar. Biol. Ecol.* 36:1–10.

Förster, R. 1967. Die reptanten Dekapoden der Trias. *Neues Jb. Geol. Paläont. Abh.* 128:136–194.

——. 1968. *Paranecrocarcinus libanoticus* n. sp. (Decapoda) und die Entwicklung der Calappidae in der Kreide. *Mitt. Bayer. Staatssamml. Paläont. Hist. Geol.* 8:167–195.

——. 1973. Untersuchungen an Oberjurassischen Palinuridae (Crustacea, Decapoda). *Mitt. Bayer. Staatssamml. Paläont. Hist. Geol.* 13:31–46.

——. 1977. Untersuchungen an Jurassischen Thalassinoidea (Crustacea, Decapoda). *Mitt. Bayer. Staatssamml. Paläont. Hist. Geol.* 17:137–156.

——. 1979. *Eocarcinus praecursor* Withers (Decapoda, Brachyura) from the Lower Pliensbachian of Yorkshire and the early crabs. *Neues Jb. Geol. Paläont. M.H.* 1:15–27.

——. 1982. Heuschreckenkrebse (Crustacea, Stomatopoda) aus dem Alttertiär von Helmstedt und Handorf (Niedersachsen) und der Oberkreide von Nigeria. *Neues Jb. Geol. Paläont. M.H.* 6:321–335.

——. 1984. Bärenkrebse (Crustacea, Decapoda) aus dem Cenoman des Libanon und dem Eozän Italiens. *Mitt. Bayer. Staatssamml. Paläont. Hist. Geol.* 24:57–66.

Fortey, R. A., and R. M. Owens. 1979. Enrollment in the classification of trilobites. *Lethaia* 12:219–226.

Foster, S. A. 1985. Wound healing: A possible role of cleaning stations. *Copeia* 1985:875–880.

Fotheringham, N. 1974. Trophic complexity in a littoral boulder field. *Limnol. Oceanogr.* 19:84–91.

Fox, S. F., and M. A. Rostker. 1982. Social cost of tail loss in *Uta stansburiana. Science* 218:692–693.

Fratt, D. B., and J. H. Dearborn. 1984. Feeding biology of the Antarctic brittle star *Ophionotus victoriae* (Echinodermata: Ophiuroidea). *Polar Biol.* 3:127–139.

Frazzetta, T. H. 1970. From hopeful monsters to bolyerine snakes? *Amer. Nat.* 104:55–71.

Freeman, G., and J. W. Lundelius. 1982. The developmental genetics of dextrality and sinistrality in the gastropod *Lymnaea peregra. Wilhelm Roux's Arch.* 191:69–83.

Fretter, V. 1984. The functional anatomy of the neritacean limpet *Phenacolepas omanensis* Biggs and some comparison with *Septaria. J. Molluscan Stud.* 50:8–18.

Frey, R. W. (ed.). 1975. *The Study of Trace Fossils.* Springer, New York.

Frey, R. W., H. A. Curran, and S. G. Pemberton. 1984. Tracemaking activities of crabs and their environmental significance: The ichnogenus *Psilonichnus. J. Paleont.* 58:333–350.

Fryer, G., and T. D. Iles. 1972. *The Cichlid Fishes of the Great Lakes of Africa: Their Biology and Evolution.* Oliver and Boyd, Edinburgh.

Fürsich, F. T. 1977. Corallian (Upper Jurassic) marine benthic associations from England and Normandy. *Palaeontology* 20:337–385.

Fürsich, F. T., and D. Jablonski. 1984. Late Triassic naticid drillholes: Carnivorous gastropods gain a major adaptation but fail to radiate. *Science* 224:78–80.

Fürsich, F. T., and J. Wendt. 1977. Biostratinomy and paleoecology of the Cassian Formation (Triassic) of the southern Alps. *Palaeogeogr., Palaeoclimatol., Paleoecol.* 22:257–323.

Futuyma, D. J., and M. Slatkin (eds.). 1983. *Coevolution.* Sinauer, Sunderland, Massachusetts.

Gall, L. F., and B. H. Tiffney. 1983. A fossil noctuid moth egg from the Late Cretaceous of eastern North America. *Science* 219:507–509.

Galleni, L., P. Tongiorgi, E. Ferrero, and U. Salghetti. 1980. *Stylochus mediterraneus* (Turbellaria: Polycladida), predator on the mussel *Mytilus galloprovincialis. Mar. Biol.* 55:317–326.

Galton, P. M. 1971. A primitive dome-headed dinosaur (Ornithischia: Pachycephalosauridae) from Lower Cretaceous of England and the function of the dome of pachycephalosaurids. *J. Paleont.* 45:40–47.

———. 1983. Armored dinosaurs (Ornithischia: Ankylosauria) from the Middle and Upper Jurassic of Europe. *Palaeontographica Abt.* (A) 182:1–25.

———. 1985. Diet of prosauropod dinosaurs from the Late Triassic and Early Jurassic. *Lethaia* 18:105–123.

Gardiner, B. G. 1984. Sturgeons as living fossils. In Eldredge and Stanley (eds.), 1984, pp. 148–152.

Garland, T., Jr. 1983. The relation between mammal running speed and body mass in terrestrial mammals. *J. Zool. London* 199:157–170.

Garrett, P. 1970. Phanerozoic stromatolites: Restriction by grazing and burrowing animals. *Science* 169:171–173.

Garrity, S. D. 1984. Some adaptations of gastropods to physical stress on a tropical rocky shore. *Ecology* 65:559–574.

Garrity, S. D., and S. C. Levings. 1981. A predator-prey interaction between two physically and biologically constrained tropical rocky shore gastropods: Direct, indirect and community effects. *Ecol. Monogrs.* 51:267–286.

———. 1983. Homing scars as a defense against predators in the pulmonate limpet *Siphonaria gigas* (Gastropoda). *Mar. Biol.* 72:319–324.

Gasiorowski, S. M. 1973. Les rhyncholites. *Geobios* 6:127–196.

Gauthier, H. 1941. Sur l'éthologie d'un amphipode qui vit dans une coquille. *Bull. Soc. Hist. Nat. Afr. Nord* 32:245–266.

Geist, V. 1966. The evolution of horn-like organs. *Behaviour* 27:175–214.

———. 1967. On fighting injuries and dermal shields of mountain goats. *J. Wildlife Management* 31:192–194.

———. 1978a. *Life Strategies, Human Evolution, Developmental Design; Toward a Biological Theory of Health.* Springer, New York.

———. 1978b. On weapons, combat, and ecology. In L. Krames, P. Pliner, and T. Alloway (eds.), *Advances in the Study of Communication and Effect.*

Vol. 4: *Aggression, Dominance and Individual Spacing*, pp. 1–30. Plenum, New York.

———. 1983. On the evolution of ice age mammals and its significance to an understanding of speciations. *Assoc. Southeastern Biologists Bull.* 30:109–133.

George, R. W., and A. R. Main. 1968. The evolution of spiny lobsters (Palinuridae): A study of evolution in the marine environment. *Evolution* 22:803–820.

Ghiold, J. 1984. Adaptive shifts in clypeasteroid evolution—feeding strategies in the soft-bottom realm. *Neues Jb. Geol. Paläont. Abh.* 169:41–73.

Gibb, J. 1956. Food, feeding habits and territoriality of the rock pipit *Anthus spinoletta*. *Ibis* 98:506–530.

Gibson, J. S. 1970. The function of the operculum of *Thais lapillus* (L.) in relation to desiccation and predation. *J. Anim. Ecol.* 39:159–168.

Gilbert, J. J. 1966. Rotifer ecology and embryological induction. *Science* 151:1234–1237.

Gilbert, L. E. 1983. Coevolution and mimicry. In Futuyma and Slatkin (eds.), 1983, pp. 263–281.

Gilchrist, S. 1984. Specificity of hermit crab attraction to gastropod predation sites. *J. Chem. Ecol.* 10:569–582.

Gill, D. E. 1974. Intrinsic rate of increase, saturation density, and competitive ability. *Amer. Nat.* 108:103–116.

———. 1978. The metapopulation ecology of the red-spotted newt, *Notophthalmus viridescens* (Rafinesque). *Ecol. Monogrs.* 48:145–166.

Gill, G. A., and A. G. Coates. 1977. Mobility, growth patterns and substrate in some fossil and Recent corals. *Lethaia* 10:119–134.

Gillespie, W. H., G. W. Rothwell, and S. E. Scheckler. 1981. The earliest seeds. *Nature* 293:462–464.

Gilmour, T.H.J. 1967. The defensive adaptations of *Lima hians* (Mollusca: Bivalvia). *J. Mar. Biol. Assoc. U.K.* 47:209–221.

Gingerich, P. D. 1977. Patterns of evolution in the mammalian fossil record. In Hallam (ed.), 1977, pp. 469–500.

Givnish, T. J., and G. J. Vermeij. 1976. Sizes and shapes of liane leaves. *Amer. Nat.* 110:743–788.

Glaessner, M. F. 1969. Decapoda. In R. C. Moore (ed.), 1969, pp. R399-R533.

Glynn, P. W. 1974. Rolling stones among the Scleractinia: Mobile coralliths in the Gulf of Panama. *Proc. Second Intern. Coral Reef Symp.* 2:183–198.

———. 1976. Some physical and biological determinants of coral community structure in the Eastern Pacific. *Ecol. Monogrs.* 46:431–456.

———. 1977. Interactions between *Acanthaster* and *Hymenocera* in the field and laboratory. *Proc. Third Intern. Coral Reef Symp.* 1:109–215.

———. 1982a. Coral communities and their modifications relative to past and prospective Central American seaways. *Adv. Mar. Biol.* 19:91–132.

———. 1982b. Individual recognition and phenotypic variability in *Acanthaster planci* (Echinodermata: Asteroidea). *Coral Reefs* 1:89–94.

———. 1983. Crustacean symbionts and the defense of corals: Coevolution on the reef? In Nitecki (ed.), 1983, pp. 111–178.

Glynn, P. W., and G. M. Wellington. 1983. *Corals and Coral Reefs of the Galápagos Islands, with an Annotated List of the Scleractinian Corals of the Galápagos by John W. Wells.* University of California Press, Berkeley.

Glynn, P. W., H. H. Stewart, and J. E. McCosker. 1972. Pacific coral reefs of Panama: Structure, distribution and predators. *Geol. Rundschau* 61:481–519.

Glynn, P. W., E. M. Druffel, and R. B. Dunbar. 1983. A dead Central American coral reef tract: Possible link with the Little Ice Age. *J. Mar. Res.* 41:605–637.

Gohar, H.A.F., and G. N. Soliman. 1963. On the biology of three coralliophilids boring in living corals. *Publ. Mar. Biol. Sta. Al-Ghardaqa* (Red Sea) 12:99–126.

Golikov, A. M., and Y. I. Starobogatov. 1975. Systematics of prosobranch gastropods. *Malacologia* 15:105–232.

Golubic, S., R. D. Perkins, and K. J. Lukas. 1975. Boring microorganisms and microborings in carbonate substrates. In Frey (ed.), 1975, pp. 229–259.

Gonor, J. J. 1966. Escape responses of North Borneo strombid gastropods elicited by predatory prosobranchs *Aulica vespertilio* and *Conus marmoreus. Veliger* 8:226–230.

Gooday, A. 1984. Records of deep-sea rhizopod tests inhabited by metazoans in the north-east Atlantic. *Sarsia* 69:45–53.

Gordon, D. P. 1972. Biological relationships of an intertidal bryozoan population. *J. Nat. Hist.* 6:503–514.

Goreau, T. F., and C. M. Yonge. 1968. Coral community on muddy sand. *Nature* 217:421–423.

Gosliner, T. M. 1980. Systematics and phylogeny of the Aglajidae (Opisthobranchia: Mollusca). *Zool. J. Linn. Soc.* 68:325–360.

Gould, S. J. 1968. Ontogeny and the explanation of form: An allometric analysis. *J. Paleont.* 42, *Paleont. Soc. Mem.* 1. Part II of 2:81–98.

———. 1969. Character variation in two land snails from the Dutch Leeward Islands: Geography, environment, and evolution. *Syst. Zool.* 18:185–200.

———. 1977. *Ontogeny and Phylogeny.* Belknap Press of Harvard University, Cambridge.

———. 1982. Darwinism and the expansion of evolutionary theory. *Science* 216:380–387.

———. 1984. Covariance sets and ordered geographic variation in *Cerion* from Aruba, Bonaire, and Curaçao: A way of studying nonadaptation. *Syst. Zool.* 33:217–237.

———. 1985. The paradox of the first tier: An agenda for paleobiology. *Paleobiology* 11:2–12.

Gould, S. J., and C. B. Calloway. 1980. Clams and brachiopods—ships that pass in the night. *Paleobiology* 6:383–396.

Gould, S. J., and R. C. Lewontin. 1979. The spandrels of San Marco and the

Panglossian paradigm: A critique of the adaptationist programme. *Proc. Roy. Soc. London* (B) 205:581–598.

Gould, S. J., and E. S. Vrba. 1982. Exaptation—a missing term in the science of form. *Paleobiology* 8:4–15.

Goulden, C. E., and L. L. Hornig. 1980. Population oscillations and energy reserves in planktonic Cladocera and their consequences to competition. *Proc. Nat. Acad. Sci. U.S.A.* 77:1716–1720.

Graciansky, P. C. de, G. Deroo, J. P. Herbin, L. Montadert, C. Müller, A. Schaaf, and J. Sigal. 1984. Ocean-wide stagnation episode in the Late Cretaceous. *Nature* 308:346–349.

Graham, J. B., R. H. Rosenblatt, and C. Gans. 1978. Vertebrate air breathing arose in fresh waters and not in the oceans. *Evolution* 32:459–463.

Grant, J.W.G., and I.A.E. Bailey. 1981. Predator induction of crests in morphs of the *Daphnia carinata* King complex. *Limnol. Oceanogr.* 26:201–218.

Grant, P. R. 1972. Centripetal selection and the house sparrow. *Syst. Zool.* 21:23–30.

Grant, R. E. 1966. Spine arrangement and life habits of the productid brachiopod *Waagenoconcha*. *J. Paleont.* 40:1063–1069.

———. 1968. Structural adaptation in two Permian brachiopod genera, Salt Range, West Pakistan. *J. Paleont.* 42:1–32.

———. 1980. Koskinoid perforations in brachiopod shells: Function and mode of formation. *Lethaia* 13:313–319.

Gray, J., and A. J. Boucot (eds.). 1979. *Historical Biogeography, Plate Tectonics, and the Changing Environment*. Oregon State University Press, Corvallis, Oregon.

Gray, J., D. Massa, and A. J. Boucot. 1982. Caradocian land plant microfossils from Libya. *Geology* 10:197–201.

Greene, A. 1975. Biology of the five species of Cychrini (Coeloptera: Carabidae) in the steppe region of southeastern Washington. *Melanderia* 19:1–43.

Greene, H. W. 1973. Defensive tail display by snakes and amphisbaenians. *J. Herpetol.* 7:143–161.

Greenwood, P. H. 1974. The cichlid fishes of Lake Victoria, East Africa: Biology and distribution of a species flock. *Bull. Brit. Mus. (Nat. Hist.), Zool.* 6 (Suppl.): 1–134.

———. 1984. *Polypterus* and *Erpetoichthys*: Anachronistic osteichthyans. In Eldredge and Stanley (eds.), 1984, pp. 143–147.

Griffiths, C. L., and J. L. Seiderer. 1980. Rock-lobsters and mussels—limitations on preferences in a predator-prey interaction. *J. Exp. Mar. Biol. Ecol.* 44:95–109.

Griffiths, D. 1980. Foraging costs and relative prey size. *Amer. Nat.* 116:743–752.

Grime, J. P., and M. A. Mowforth. 1982. Variation in genome size—an ecological interpretation. *Nature* 299:151–153.

Groot, S. J. de. 1971. On the interrelationships between morphology of the alimentary tract, food and feeding behaviour in flatfishes (Pisces: Pleuro-nectiformes). *Netherlands J. Sea Res.* 5:121–196.

Gross, W. 1967. Über das Gebiss der Acanthodier und Placodermen. *J. Linn. Soc. (Zool.)* 47:121–130.

Grünebaum, H., G. Bergman, D. P. Abbott, and J. C. Ogden. 1978. Intraspe-cific agonistic behavior in the rock-boring sea urchin *Echinometra lucun-ter* (L.) (Echinodermata: Echinoidea). *Bull. Mar. Sci.* 28:181–188.

Gubbay, S. 1983. Compressive and adhesive strengths of a variety of British barnacles. *J. Mar. Biol. Assoc. U.K.* 63:541–555.

Guex, J. 1967. Contribution à l'étude des blessures chez les ammonites. *Bull. Soc. Vaud. Sci. Nat.* 69:323–338.

Guinot, D. 1968. Recherches préliminaires sur les groupements naturels chez les crustacés brachyures. VI. Les Carpilinae. *Bull. Mus. Nat. Hist. Na-tur.* (2) 40:320–334.

Gutmann, W. F. 1967. Das Dermalskelett der fossilen "Panzerfische" funk-tionell und phylogenetisch interpretiert. *Senckenbergiana Lethaea* 48:277–283.

Gutmann, W. F., K. Vogel, and H. Zorn. 1978. Brachiopods: Biomechanical interdependences governing their origins and phylogeny. *Science* 199:890–893.

Guṭu, M. 1981. A new contribution to the systematics and phylogeny of the Suborder Monokophora (Crustacea: Tanaidacea). *Trav. Mus. Hist. Nat. Grigore Antipa* 23:81–108.

Hallam, A. 1975. Evolutionary size increase and longevity in Jurassic bi-valves and ammonites. *Nature* 258:493–496.

———. 1976. Stratigraphic distribution and ecology of European Jurassic bi-valves. *Lethaia* 9:245–259.

———. 1977. Jurassic bivalve biogeography. *Paleobiology* 3:58–73.

———. (ed.). 1977. *Patterns of Evolution as Illustrated by the Fossil Record.* Elsevier, Amsterdam.

———. 1978. How rare is phyletic gradualism and what is its evolutionary significance? Evidence from Jurassic bivalves. *Paleobiology* 4:16–25.

———. 1981a. Relative importance of plate movements, eustasy, and cli-mate in controlling major biogeographical changes since the Early Meso-zoic. In G. Nelson and D. E. Rosen (eds.), *Vicariance Biogeography: A Cri-tique*, pp. 303–340. Columbia University Press, New York.

———. 1981b. *Facies Interpretation and the Stratigraphic Record.* Freeman, Oxford.

———. 1981c. The end-Triassic bivalve extinction event. *Palaeogeogr., Pa-laeoclimatol., Palaeoecol.* 35:1–44.

———. 1982. Patterns of speciation in Jurassic *Gryphaea*. *Paleobiology* 8:354–366.

———. 1984. Pre-Quaternary sea-level changes. *Rev. Earth Planet. Sci.* 12:205–243.

———. 1986. The Pliensbachian and Tithonian extinction events. *Nature* 319:765–768.

Hamai, I. 1937. Some notes on relative growth with special reference to the growth of limpets. *Sci. Rep. Tohoku Imp. Univ. Biol.* (4) 12:71–95.

Hamajima, F., T. Fujino, and M. Koga. 1976. Studies on the host-parasite relationship of *Paragonimus westermani* (Kerbert, 1878). IV. Predatory habits of some fresh-water crabs and crayfish on the snail, *Semisulcospira libertina* (Gould, 1859). *Ann. Zool. Japan* 49:274–278.

Hamilton, P. V. 1976. Predation on *Littorina irrorata* (Mollusca: Gastropoda) by *Callinectes sapidus* (Crustacea: Portunidae). *Bull. Mar. Sci.* 26:403–409.

Hamilton, W. J., III., R. Buskirk, and W. H. Buskirk. 1977. Intersexual dominance and differential mortality of gemsbok *Oryx gazella* at Namib Desert waterholes. *Madoqua* 10:5–19.

Hamner, W. M. 1978. Intraspecific competition in *Tridacna crocea*, a burrowing bivalve. *Oecologia* (Berlin) 34:267–281.

Hancock, D. A. 1974. Some aspects of the biology of the sunstar *Crossaster papposus* (L.). *Ophelia* 13:1–30.

Hanlin, H. G. 1978. Food habits of the greater siren, *Siren lacertina*, in an Alabama coastal plain pond. *Copeia* 1978:358–360.

Hannibal, J. T., and R. M. Feldmann. 1981. Systematics and functional morphology of oniscomorph millipedes (Arthropoda: Diplopoda) from the Carboniferous of North America. *J. Paleont.* 55:730–746.

Hansen, T. A. 1978. Larval dispersal and species longevity in Lower Tertiary gastropods. *Science* 199:885–887.

———. 1980. Influence of larval dispersal and geographic distribution on species longevity in neogastropods. *Paleobiology* 6:193–207.

———. 1982. Modes of larval development in Early Tertiary neogastropods. *Paleobiology* 8:367–377.

Harland, W. B., A. V. Cox, P. G. Llewellyn, C.A.G. Pickton, A. G. Smith, and R. Walters. 1982. *A Geologic Time Scale*. Cambridge University Press, Cambridge.

Harless, M., and H. Morlock (eds.). 1979. *Turtles: Perspectives and Research*. John Wiley, New York.

Harper, J. A., and H. B. Rollins. 1982. Recognition of Monoplacophora and Gastropoda in the fossil record: A functional morphologic look at the bellerophont controversy. *Third North Amer. Paleont. Conv. Proc.* 1:227–232.

———. 1985. Infaunal or semi-infaunal bellerophont gastropods: Analysis of *Euphemites* and functionally related taxa. *Lethaia* 18:21–37.

Harris, M. P. 1965. The food of some *Larus* gulls. *Ibis* 107:43–51.

Harry, H. W. 1964. The anatomy of *Chilina fluctuosa* Gray re-examined, with prolegomena on the phylogeny of the higher limnic Basommatophora. *Malacologia* 1:355–385.

Hart, J. L. 1973. Pacific fishes of Canada. *Bull. Fish. Res. Bd. Canada* 180:1–740.

Hartman, W. D. 1979. A new sclerosponge from the Bahamas and its relationship to Mesozoic stromatoporoids. *Colloq. Intern. Centre Nat. Rech. Sci.* 291:467–474.

Hartman, W. D., J. W. Wendt, and F. Wiedenmayer. 1980. Living and fossil sponges: Notes for a short course. *Sedimenta* 8:1–274.

Hartnoll, R. G. 1983. Strategies of crustacean growth. *Aust. Mus. Mem.* 18:121–131.

Hartwick, B., L. Tulloch, and S. Macdonald. 1981. Feeding and growth of *Octopus dofleini* (Wülker), *Veliger* 24:129–138.

Harvell, C. D. 1984. Predator-induced defense in a marine bryozoan. *Science* 224:1357–1359.

Hay, M. E. 1981a. The functional morphology of turf-forming seaweeds: Persistence in stressful marine habitats. *Ecology* 62:739–750.

———. 1981b. Herbivory, algal distribution, and the maintenance of between-habitat diversity on a tropical fringing reef. *Amer. Nat.* 118:520–540.

Hay, M. E., T. Colburn, and D. Downing. 1983. Spatial and temporal patterns in herbivory on a Caribbean fringing reef: The effects on plant distribution. *Oecologia* (Berlin) 58:299–308.

Hayami, I., and Y. Kanie. 1980. Mode of life of a giant capulid gastropod from the Upper Cretaceous of Saghalien and Japan. *Palaeontology* 23:689–698.

Heard, R. W. 1982. Observations on the food and food habits of clapper rails (*Rallus longirostris* Boddaert) from tidal marshes along the east and Gulf coasts of the United States. *Gulf Res. Reps.* 7:125–135.

Heck, K. L., Jr., and E. D. McCoy. 1978. Long-distance dispersal and the reef-building corals of the Eastern Pacific. *Mar. Biol.* 48:349–356.

Heinrich, B. 1979. Foraging strategies of caterpillars: Leaf damage and possible predator avoidance strategies. *Oecologia* (Berlin) 42:325–337.

Heinrich, B., and G. A. Bartholomew. 1979. Roles of endothermy and size in inter- and intraspecific competition for elephant dung in an African dung beetle, *Scarabaeus laevistriatus*. *Physiol. Zool.* 52:484–496.

Heinrich, B., and S. L. Collins. 1983. Caterpillar leaf damage, and the game of hide-and-seek with birds. *Ecology* 64:592–602.

Heinrich, B., and T. P. Mommsen. 1985. Flight of winter moths near $0 \pm C$. *Science* 228:177–179.

Heller, J. 1979. Visual versus non-visual selection of shell colour in an Israeli freshwater snail. *Oecologia* (Berlin) 44:98–104.

Henderson, R. A. 1984. A muscle attachment proposal for septal function in Mesozoic ammonites. *Palaeontology* 27:461–486.

Hendrix, G. Y. 1975. A review of the genus *Pascolion* (Sipuncula) with the descriptions of two new species from the western Atlantic. *Proc. Intern. Symp. Biol. Sipuncula Echiura* 1:117–137.

Henry, J.-L., and E.N.K. Clarkson. 1975. Enrollment and coaptations in some species of the Ordovician trilobite genus *Placoparia*. In Martinsson (ed.), 1975, pp. 87–96.

Heppleston, B. P. 1971. Feeding techniques of the oystercatcher. *Bird Study* 18:15–20.

Hertz, P. E., R. B. Huey, and E. Nevo. 1982. Fight versus flight: Body temperature influences defensive responses of lizards. *Anim. Behav.* 30:676–679.

———. 1983. Homage to Santa Anita: Thermal sensitivity of sprint speed in agamid lizards. *Evolution* 37:1075–1084.

Hewitt, R. A., and R. Watkins. 1980. Cephalopod ecology across a Late Silurian shelf tract. *Neues Jb. Geol. Paläont. Abh.* 160:96–117.

Heymons, R., and H. von Lengerken. 1932. Studien über die Lebenserscheinungen der Silphinae (Coleopt.). VIII. *Ablattaria laevigata* F. *Z. Morph. Ökol. Tiere* 24:259–286.

Heymons, R., H. von Lengerken, and M. Bayer. 1927. Studien über die Lebenserscheinungen der Silphinae (Coleopt.). II. *Phosphuga atrata* L. *Z. Morph. Ökol. Tiere* 19:272–303.

Hiatt, R. W. 1948. The biology of the lined shore crab, *Pachygrapsus crassipes* Forskal. *Pacific Sci.* 2:133–213.

Hiatt, R. W., and D. W. Strasburg. 1960. Ecological relationships of the fish fauna on coral reefs of the Marshall Islands. *Ecol. Monogrs.* 30:65–127.

Hickman, C. S. 1983. Radular patterns, systematics, diversity, and ecology of deep-sea limpets. *Veliger* 26:73–92.

———. 1984. *Pleurotomaria*: Pedigreed perseverance. In Eldredge and Stanley (eds.), 1984, pp. 225–231.

Highsmith, R. C. 1980a. Geographic patterns of coral bioerosion: A productivity hypothesis. *J. Exp. Mar. Biol. Ecol.* 46:177–196.

———. 1980b. Burrowing by the bivalve *Lithophaga curta* in the living reef coral *Montipora berryi* and a hypothesis of reciprocal larval recruitment. *Mar. Biol.* 56:155–162.

———. 1980c. Passive colonization and asexual colony multiplication in the massive coral *Porites lutea* Milne Edwards and Haime. *J. Exp. Mar. Biol. Ecol.* 47:55–67.

———. 1981a. Coral bioerosion at Enewetak: Agents and dynamics. *Int. Rev. Ges. Hydrobiol.* 66:335–375.

———. 1981b. Lime-boring algae in hermatypic coral skeletons. *J. Exp. Mar. Biol. Ecol.* 55:267–281.

———. 1982. Reproduction by fragmentation in corals. *Mar. Ecol. Progr. Ser.* 7:207–226.

Hines, A. H., and T. R. Loughlin. 1980. Observations on sea otters digging for clams at Monterey Harbor, California. *Fish. Bull.* 78:159–163.

Hingston, J. P. 1985. Predation patterns among molluscs in the Victorian Tertiary. *Proc. Roy. Soc. Victoria* 67:49–57.

Hobson, E. S. 1968. Predatory behavior of some shore fishes in the Gulf of California. *U.S. Dept. Interior Bur. Sport Fish. Wildlife Res. Rep.* 73:1–92.

Hochachka, P. W., and G. N. Somero. 1973. *Strategies of Biochemical Adaptation*. Saunders, Philadelphia.

Hoffman, A., and J. Martinell. 1984. Prey selection by naticid gastropods in the Pliocene of Emporda (northeast Spain). *Neues Jb. Geol. Paläont. M.H.* 7:393–399.

Hoffman, A., A. Pisera, and M. Ryszkiewicz. 1974. Predation by muricid and naticid gastropods on the Lower Tortonian mollusks from the Korytnica clays. *Acta Geol. Polon.* 24:249–260.

Holeton, G. F. 1974. Metabolic cold adaptation of polar fish: Fact or artifact? *Physiol. Zool.* 47:137–152.

Holland, H. D. 1984. *The Chemical Evolution of the Atmosphere and Oceans.* Princeton University Press, Princeton.

Holland, H. D., and A. F. Trendall (eds.). 1984. *Patterns of Change in Earth Evolution.* Springer, Berlin.

Holland, M. H. (ed.). 1979. *The Origin of Major Invertebrate Groups.* Academic Press, London.

Holling, C. S. 1966. The functional response of invertebrate predators to prey density. *Mem. Entomol. Soc. Canada* 47:3–86.

Hollmann, R. 1969. Die Entstehung Fossilisationsfähiger Schalen-Frass-reste, dargestellt am Nahrungserwerb von *Homarus gammarus* (Crustacea, Decapoda). *Helgoländer Wissensch. Meeresunters.* 19:401–416.

Holser, W. T. 1977. Catastrophic chemical events in the history of the ocean. *Nature* 267:403–408.

Holthuis, L. B. 1959. Contributions to New Guinea carcinology. III. The occurrence of *Birgus latro* (L.) in Netherlands New Guinea (Crustacea Decapoda, Paguridea). *Nova Guinea*, n.s. 10:303–310.

———. 1963. Contributions to New Guinea carcinology. IV. Further data on the occurrence of *Birgus latro* (L.) in West New Guinea (Crustacea Decapoda, Paguridea). *Nova Guinea* 18:355–359.

———. 1980. *Alphaeus saxidomus* new species, a rock boring snapping shrimp from the Pacific coast of Costa Rica, with notes on *Alphaeus simus* (Guérin-Méneville, 1856). *Zool. Meded.* 55:47–58.

Holthuis, L. B., and R. B. Manning. 1969. Stomatopoda. In R. C. Moore (ed.), 1969, pp. R535–R552.

Hopson, J. A. 1975. The evolution of cranial display structures in hadrosaurian dinosaurs. *Paleobiology* 1:21–43.

———. 1977. Relative brain size and behavior in archosaurian reptiles. *Ann. Rev. Ecol. Syst.* 8:429–448.

Horn, H. S. 1971. *The Adaptive Geometry of Trees.* Princeton University Press, Princeton.

Houbrick, R. S. 1978. The family Cerithiidae in the Indo-Pacific. Part 1: The genera *Rhinoclavis, Pseudovertagus* and *Clavocerithium.* Monogrs. Mar. Mollusca 1:1–130.

———. 1979. Classification and systematic relationships of the Abyssochrysidae, a relict family of bathyal snails (Prosobranchia: Gastropoda). *Smithson. Contribs. Zool.* 290:1–21.

Houbrick, R. S., and V. Fretter. 1969. Some aspects of the functional anatomy and biology of *Cymatium* and *Bursa. Proc. Malacol. Soc. London* 38:415–429.

House, M. R. (ed.). 1979. *The Origin of Major Invertebrate Groups.* Academic Press, London.

House, M. R. 1985. Correlation of mid-Palaeozoic ammonoid evolutionary events with global sedimentary perturbations. *Nature* 313:17–22.

Hsu, K. J., Q. He, J. A. McKenzie, H. Weissert, K. Perch-Nielsen, H. Oberhänsli, K. Kelts, J. LaBrecque, L. Tauxe, U. Krähenbühl, S. F. Percival, Jr., R. Wright, A. M. Karpoff, N. Petersen, P. Tucker, R. Z. Poore, A. M. Gombos, K. Pisciotto, M. F. Carman, Jr., and E. Schreiber. 1982. Mass mortality and its environmental and evolutionary consequences. *Science* 216:249–256.

Hsu, K. J., H. Oberhänsli, J. Y. Gao, S. Shu, C. Haihong, and U. Krähenbühl. 1985. "Strangelove Ocean" before the Cambrian explosion. *Nature* 316:809–811.

Huey, R. B. 1982. Phylogenetic and ontogenetic determinants of sprint performance in some diurnal Kalahari lizards. *Koedoe* 25:43–48.

Hughes, R. N., and R. W. Elner. 1979. Tactics of a predator, *Carcinus maenas*, and morphological responses of the prey, *Nucella lapillus*. *J. Anim. Ecol.* 48:65–78.

Hughes, R. N., and H.P.I. Hughes. 1981. Morphological and behavioural aspects of feeding in the Cassidae (Tonnacea, Mesogastropoda). *Malacologia* 20:385–402.

Hüne, F. R. von. 1956. Paläontologie und Phylogenie der niederen *Tetrapoden*. Gustav Fischer, Jena.

Hutchinson, G. E. 1959. Homage to Santa Rosalia or why are there so many kinds of animals? *Amer. Nat.* 93:145–159.

———. 1967. *A Treatise on Limnology*. Vol. 2. John Wiley, New York.

Huxley, J. 1953. *Evolution in Action*. Harper, New York.

Hylleberg, J. 1975. On the ecology of the sipunculan *Phascolion strombi* (Montagu). *Proc. Intern. Symposium Biol. Sipuncula and Echiura* 1:241–250.

Iliffe, T. M., C. W. Hart, Jr., and R. B. Manning. 1983. Biogeography and the caves of Bermuda. *Nature* 302:141–142.

Iliffe, T. M., H. Wilkens, J. Parzefall, and D. Williams. 1984. Marine lava cave fauna: Composition, biogeography, and origins. *Science* 225:309–311.

Ingham, R. E., and J. A. Zischke. 1977. Prey preferences of carnivorous intertidal snails in the Florida Keys. *Veliger* 20:49–51.

Ingolfsson, A., and B. T. Estrella. 1978. The development of shell-cracking behavior in herring gulls. *Auk* 95:578–579.

Ingram, W. M. 1942. Food habits of *Haplotrema minimum* Ancey and habits of associated mollusks on the Mills College campus. *Nautilus* 55:98–102.

Inouye, D. W. 1982. The consequences of herbivory: A mixed blessing for *Jurinea mollis* (Asteraceae). *Oikos* 39:269–272.

Jaanusson, V. 1971. Evolution of the brachiopod hinge. *Smithson. Contribs. Paleobiol.* 3:33–46.

Jablonski, D. 1980. Apparent versus real biotic effects of transgression and regression. *Paleobiology* 6:397–407.

———. 1982. Evolutionary rates and modes in Late Cretaceous gastropods: Role of larval ecology. *Third North Amer. Paleont. Conv. Proc.* 1:257–262.

Jablonski, D., and D. J. Bottjer. 1983. Soft-bottom epifaunal suspension-feeding assemblages in the Late Cretaceous: Implications for the evolution of benthic paleocommunities. In Tevesz and McCall (eds.), 1983, pp. 747–812.

Jablonski, D., and J. W. Valentine. 1981. Onshore-offshore gradients in Recent Eastern Pacific shelf faunas and their paleobiogeographic significance. In Scudder and Reveal (eds.), 1981, pp. 441–453.

Jablonski, D., J. J. Sepkoski, Jr., D. J. Bottjer, and P. M. Sheehan. 1983. Onshore-offshore patterns in the evolution of Phanerozoic shelf communities. *Science* 222:1123–1125.

Jackson, J.B.C. 1974. Biogeographic consequences of eurytopy and stenotopy among marine bivalves and their evolutionary significance. *Amer. Nat.* 108:541–560.

———. 1977. Competition on marine hard substrata: The adaptive significance of solitary and colonial strategies. *Amer. Nat.* 111:743–767.

———. 1979. Morphological strategies of sessile animals. In G. Larwood and B. R. Rosen (eds.), *Biology and Systematics of Colonial Organisms*, pp. 499–555. Academic Press, London.

———. 1983. Biological determinants of present and past sessile animal distributions. In Tevesz and McCall (eds.), 1983, pp. 39–120.

———. 1985. Distribution and ecology of clonal and aclonal benthic invertebrates. In Jackson et al. (eds.), 1985, pp. 297–355.

Jackson, J.B.C., and S. R. Palumbi. 1979. Regeneration and partial predation in cryptic coral reef environments: Preliminary experiments on sponges and ectoprocts. *Colloq. Intern. Centre Nat. Rech. Sci.* 291:303–308.

Jackson, J.B.C., L. W. Buss, and R. E. Cook (eds.). 1985. *Population Biology and Evolution of Clonal Organisms.* Yale University Press, New Haven.

Jackson, J.B.C., T. F. Goreau, and W. D. Hartman. 1971. Recent brachiopod-coralline sponge communities and their paleoecological significance. *Science* 173:623–625.

James, N. P., T. R. Kobluk, and S. G. Pemberton. 1977. The oldest macroborers: Lower Cambrian of Labrador. *Science* 197:980–983.

Jangoux, M. 1982. Food and feeding mechanisms: Asteroidea. In Jangoux and Lawrence (eds.), 1982, pp. 117–159.

Jangoux, M., and J. M. Lawrence (eds.). 1982. *Echinoderm Nutrition.* Balkema, Rotterdam.

Janis, C. 1976. The evolutionary strategy of the Equidae and the origins of rumen and cecal digestion. *Evolution* 30:757–774.

———. 1982. Evolution of horns in ungulates: Ecology and paleoecology. *Biol. Revs.* 57:261–317.

———. 1984a. Tapirs as living fossils. In Eldredge and Stanley (eds.), 1984, pp. 80–86.

———. 1984b. Tragulids as living fossils. In Eldredge and Stanley (eds.), 1984, pp. 87–94.

Janzen, D. H. 1974. Tropical blackwater rivers, animals, and mast fruiting in the Dipterocarpaceae. *Biotropica* 6:69–103.

Janzen, D. H. 1976. The depression of reptile biomass by large herbivores. *Amer. Nat.* 110:371–400.

——. 1984. Dispersal of small seeds by big herbivores: Foliage is the fruit. *Amer. Nat.* 123:338–353.

Jeanne, R. L. 1979. A latitudinal gradient in rates of ant predation. *Ecology* 60:1211–1224.

Jeffries, M. J., and J. H. Lawton. 1984. Enemy free space and the structure of ecological communities. *Biol. J. Linn. Soc.* 23:269–286.

Jeletzky, J. A. 1966. Comparative morphology, phylogeny, and classification of fossil Coleoidea. *Univ. Kansas Paleont. Contribs., Mollusca, Article* 7:162 P.

Jell, J. S. 1984. Cambrian cnidarians with mineralized skeletons. *Palaeontographica Americana* 54:105–109.

Jenkyns, H. C. 1980. Cretaceous anoxic events: From continents to oceans. *J. Geol. Soc. London* 137:171–188.

Jerison, H. J. 1973. *Evolution of the Brain and Intelligence.* Academic Press, New York.

Johnston, R. F., and R. C. Fleischer. 1981. Overwinter mortality and sexual size dimorphism in the house sparrow. *Auk* 98:503–511.

Johnston, R. F., and R. K. Selander. 1964. House sparrows: Rapid evolution of races in North America. *Science* 144:548–550.

——. 1971. Evolution in the house sparrow. II. Adaptive differentiation in North American populations. *Evolution* 25:1–28.

——. 1973. Evolution in the house sparrow. III. Variation in size and sexual dimorphism in Europe and North and South America. *Amer. Nat.* 107:373–390.

Johnston, R. F., D. M. Niles, and S. A. Rohwer. 1972. Hermon Bumpus and natural selection in the house sparrow *Passer domesticus. Evolution* 26:20–31.

Jones, B. 1982. Paleobiology of the Upper Silurian brachiopod *Atrypoidea. J. Paleont.* 56:912–923.

Jones, M. L. 1969. Boring of shell by *Caobangia* in freshwater snails of Southeast Asia. *Amer Zool.* 9:829–835.

Just, J. 1977. A new genus and species of corophiid Amphipoda from pteropod shells of the bathyal western Atlantic, with notes on related genera (Crustacea). *Steenstrupia* 4:131–138.

——. 1983. Siphonoecetinae subfam. n. (Crustacea, Amphipoda, Corophiidae) 1. Classification. *Steenstrupia* 9:117–135.

Kalin, R. J. 1984. Observations of a feeding method of the Atlantic ribbon worm, *Cerebratulus lacteus. Estuaries* 7:179–180.

Kanie, Y., Y. Fukuda, H. Nakayama, K. Seki, and M. Hattori. 1980. Implosion of living *Nautilus* under increased pressure. *Paleobiology* 6:44–47.

Karl, D. M., C. O. Wirsen, and H. W. Jannasch. 1980. Deep-sea primary production at the Galapagos hydrothermal vents. *Science* 207:1345–1347.

Karr, J. R. 1982a. Avian extinction on Barro Colorado Island, Panama: A reassessment. *Amer. Nat.* 119:220–239.

————. 1982b. Population variability and extinction in the avifauna of a tropical land bridge island. *Ecology* 63:1975–1978.

Kasigwa, P. F., A. J. Mrema, and J. A. Allen. 1983. Predation by mongooses, rodents and snails on *Sitala jenynsi* (Pfr.), *Achatina fulica* Bowdich and other land snails in coastal Tanzania. *J. East Africa Nat. Hist. Soc. Nat. Mus.* 179:1–10.

Kauffman, E. G. 1972. *Ptychodus* predation upon a Cretaceous *Inoceramus*. *Palaeontology* 15:439–444.

Kauffman, E. G., and R. V. Kesling. 1960. An Upper Cretaceous ammonite bitten by a mosasaur. *Contribs. Mus. Paleont. Univ. Michigan* 15:193–248.

Kauffman, E. G., and N. F. Sohl. 1974. Structure and evolution of Antillean Cretaceous rudist frameworks. *Verh. Naturforsch. Ges. Basel* 84:399–467.

Kaufman, L. S., and K. F. Liem. 1982. Fishes of the suborder Labroidei (Pisces: Perciformes): Phylogeny, ecology, and evolutionary significance. *Breviora* 472:1–19.

Kay, E. A. 1968. Review of the bivalved gastropods and a discussion of evolution within the Sacoglossa. *Symp. Zool. Soc. London* 22:109–134.

Kay, R. F. 1981. The nut-crackers—a new theory of the adaptations of the Ramapithecinae. *Amer. J. Phys. Anthropol.* 55:141–151.

Kaźmierczak, J. 1984. Favositid tabulates: Evidence for poriferan affinity. *Science* 225:835–837.

Keigwin, L. D., Jr. 1978. Pliocene closing of the Isthmus of Panama, based on biostratigraphic evidence from nearby Pacific Ocean and Caribbean Sea cores. *Geology* 6:630–634.

————. 1980. Palaeoceanographic change in the Pacific at the Eocene-Oligocene boundary. *Nature* 287:722–725.

————. 1982. Isotopic paleoceanography of the Caribbean and East Pacific: Role of Panama uplift in Late Neogene time. *Science* 217:350–353.

Kemp, P., and M. D. Bertness. 1984. Snail shape and growth rates: Evidence for plastic shell allometry in *Littorina littorea*. *Proc. Nat. Acad. Sci. U.S.A.* 81:811–813.

Kendrick, B., M. J. Risk, J. Michaelides, and K. Bergman. 1982. Amphibious microborers: Bioeroding fungi isolated from live corals. *Bull. Mar. Sci.* 32:862–867.

Kennedy, V. S., and J. A. Blundon. 1983. Shell strength in *Corbicula* sp. (Bivalvia: Corbiculidae) from the Potomac River, Maryland. *Veliger* 26:22–25.

Kennedy, W. J. 1977. Ammonite evolution. In Hallam (ed.), 1977, pp. 251–304.

Kennedy, W. J., and W. A. Cobban. 1976. Aspects of ammonite biology, biogeography, and biostratigraphy. *Spec. Pap. Palaeontol.* 17:1–94.

Kennedy, W. J., N. J. Morris, and J. D. Taylor. 1970. The shell structure, mineralogy and relationships of the Chamacea (Bivalvia). *Palaeontology* 13:379–413.

Kent, B. W. 1979. Interoceanic and latitudinal patterns in spiny lobster mandible size (Decapoda, Palinuridae). *Crustaceana* 5 (Suppl.): 142–146.

———. 1981a. The behavior of the gastropod *Amphissa columbiana* (Prosobranchia: Columbellidae). *Veliger* 23:275–276.

———. 1981b. Feeding and food preferences of the muricid gastropod *Ceratostoma foliatum. Nautilus* 95:38–42.

———. 1981c. Prey dropped by herring gulls (*Larus argentatus*) on soft sediments. *Auk* 98:350–354.

———. 1983a. Patterns of coexistence in busyconine whelks. *J. Exp. Mar. Biol. Ecol.* 66:257–283.

———. 1983b. Diet expansion of *Busycon contrarium* in the absence of *Triplofusus giganteus* (Gastropoda: Buccinacea). *Nautilus* 97:103–104.

———. 1983c. Natural history observations on the busyconine whelks *Busycon contrarium* (Conrad) and *Busycotypus spiratum* (Lamarck). *J. Molluscan Stud.* 49:37–42.

———. 1986. *Ecphora* whelks: Specialized predators of large thick-shelled bivalves. *Paleobiology*, submitted.

Kent, R.M.L. 1981. The effect of *Polydora ciliata* on the shell strength of *Mytilus edulis. J. Conseil Int. Explor. Mer* 39:252–255.

Keough, M. J. 1984. Dynamics of the epifauna of the bivalve *Pinna bicolor*: Interactions among recruitment, predation, and competition. *Ecology* 65:677–688.

Kerfoot, W. C. 1977. Implications of copepod predation. *Limnol. Oceanogr.* 22:316–325.

———. 1978. Combat between predatory copepods and their prey: *Cyclops, Epischura,* and *Bosmina. Limnol. Oceanogr.* 23:1089–1102.

Kerfoot, W. C., D. L. Kellogg, Jr., and J. R. Strickler. 1980. Visual observations of live zooplankters: Evasion, escape, and chemical defenses. In W. C. Kerfoot (ed.), *Evolution and Ecology of Zooplankton Communities*, pp. 10-27. University Press of New England, Hanover, New Hampshire.

Kessel, E. 1938. Der Gelbrand als Schneckenfresser. *Natur u. Volk* 68:572–574.

Kevan, P. G., W. G. Chaloner, and D.B.O. Savile. 1975. Interrelationships of early terrestrial arthropods and plants. *Palaeontology* 18:391–417.

Kielan-Jaworowska, Z. 1968. Scolecodonts versus jaw apparatuses. *Lethaia* 1:39–49.

Kier, P. M. 1965. Evolutionary trends in Paleozoic echinoids. *J. Paleont.* 39:436–465.

———. 1973. The echinoderms and Permian-Triassic time. *Canad. Soc. Petroleum Geol. Mem.* 2:622–629.

———. 1974. Evolutionary trends and their functional significance in the post-Paleozoic echinoids. *J. Paleont.* 42, *Paleont. Soc. Mem.* 5:1–95.

———. 1981. A bored Cretaceous echinoid. *J. Paleont.* 55:656–659.

———. 1982. Rapid evolution in echinoids. *Palaeontology* 25:1–9.

Kiltie, R. A. 1981. The function of interlocking canines in rain forest peccaries (Tayassuidae). *J. Mammol.* 62:459–469.

————. 1982. Bite force as a basis for niche differentiation between rain forest peccaries (*Tayassu tajacu* and *T. pecari*). *Biotropica* 14:188–195.

Kimura, M. 1983. *The Neutral Theory of Molecular Evolution*. Cambridge University Press, Cambridge.

Kingsolver, J. G., and M.A.R. Koehl. 1985. Aerodynamics, thermoregulation, and the evolution of insect wings: Differential scaling and evolutionary change. *Evolution* 39:488–504.

Kitchell, J. A., and D. Pena. 1984. Periodicity of extinctions in the geologic past: Deterministic versus stochastic explanations. *Science* 226:689–692.

Kitchell, J. A., C. H. Boggs, J. F. Kitchell, and J. A. Rice. 1981. Prey selection by naticid gastropods: Experimental tests and application to the fossil record. *Paleobiology* 7:533–542.

Kitchell, J. A., C. H. Boggs, J. A. Rice, J. F. Kitchell, A. Hoffman, and J. Martinell. 1986. Anomalies in naticid predatory behavior: A critique and experimental observations. *Malacologia*, in press.

Kitching, J. A., L. Muntz, and F. J. Ebling. 1966. The ecology of Lough Ine. XV. The ecological significance of shell and body forms in *Nucella*. *J. Anim. Ecol.* 35:113–126.

Kjellesvig-Waering, E. N. 1961. The eurypterids of the Devonian Holland Quarry Shale of Ohio. *Fieldiana Geol.* 14:79–98.

Kleemann, K. 1984. Lebensspuren von *Upogebia operculata* (Crustacea, Decapoda) in Karibischen Steinkorallen (Madreporaria, Anthozoa). *Beitr. Paläont. Österreich* 11:35–57.

Klein, R. G. 1984. Mammalian extinctions and stone age people in Africa. In Martin and Klein (eds.), 1984, pp. 553–573.

Knoll, A. H. 1984. Patterns of extinction in the fossil record of vascular plants. In M. H. Nitecki (ed.), *Extinctions*, pp. 21-68. University of Chicago Press, Chicago.

Knoll, A. H., K. J. Niklas, and B. H. Tiffney. 1979. Phanerozoic land plant diversity in North America. *Science* 206:1400–1402.

Knoll, A. H., K. J. Niklas, P. G. Gensel, and B. H. Tiffney. 1984. Character diversification and patterns of evolution in early vascular plants. *Paleobiology* 10:34–47.

Knutson, R. M. 1974. Heat production and temperature regulation in Eastern Skunk Cabbage. *Science* 186:746–747.

Kobluk, D. R., and M. J. Risk. 1976. Algal borings and framboidal pyrite in Upper Ordovician brachiopods. *Lethaia* 10:135–143.

————. 1977. Rate and nature of infestation of a carbonate substratum by a boring alga. *J. Exp. Mar. Biol. Ecol.* 27:107–115.

Kobluk, D. R., N. P. James, and S. G. Pemberton. 1978. Initial diversification of macroboring ichnofossils and exploitation of the macroboring niche in the Lower Paleozoic. *Paleobiology* 4:163–170.

Koch, C. F. 1980. Bivalve species duration, areal extent and population size in a Cretaceous sea. *Paleobiology* 6:184–192.

Koch, C. F., and N. F. Sohl. 1983. Preservational effects in paleoecological studies: Cretaceous mollusc examples. *Paleobiology* 9:26–34.

Kohlmeyer, J., and E. Kohlmeyer. 1979. *Marine Mycology: The Higher Fungi.* Academic Press, New York.

Kohn, A. J. 1956. Piscivorous gastropods of the genus *Conus. Proc. Nat. Acad. Sci. U.S.A.* 42:168–171.

———. 1959. The ecology of *Conus* in Hawaii. *Ecol. Monogrs.* 29:47–90.

———. 1978. The Conidae (Mollusca: Gastropoda) of India. *J. Nat. Hist.* 12:299–335.

Kohn, A. J., and J. W. Nybakken. 1975. Ecology of *Conus* on eastern Indian Ocean fringing reefs. *Mar. Biol.* 29:211–234.

Kohn, A. J., and V. Waters. 1966. Escape responses of three herbivorous gastropods to the predatory gastropod *Conus textile. Anim. Behav.* 14:340–345.

Kohn, A. J., E. R. Myers, and V. R. Meenakshi. 1979. Interior remodeling of the shell by a gastropod mollusc. *Proc. Nat. Acad. Sci. U.S.A.* 76:3406–3410.

Kojumdjieva, E. 1974. Les gastéropodes perceurs et leurs victimes du Miocene de Bulgarie du nord-oueste. *Bulg. Acad. Sci. Bull. Geol. Inst. (Ser. Paleont.)* 25:5–24.

Kollmann, H. A. 1979. Distribution patterns and evolution of gastropods around the Cretaceous-Tertiary boundary. In W. Kegel Christensen and T. Birkelund (eds.), *Cretaceous-Tertiary Boundary Events.* II. *Proceedings,* pp. 83–87. University of Copenhagen Press, Copenhagen.

———. 1982. Gastropoden-Faunen aus der höheren Unterkreide nordwestdeutschlands. *Geol. Jb.* (A) 65:517–551.

Kollmann, H. A., and E. L. Yochelson. 1976. Survey of Paleozoic gastropods possibly belonging to the subclass Opisthobranchia. *Ann. Naturhist. Mus. Wien* 80:207–220.

Krassilov, V. A. 1981. Changes of Mesozoic vegetation and the extinction of dinosaurs. *Palaeogeogr., Palaeoclimatol., Palaeoecol.* 34:207–224.

Kraus, O. 1974. On the morphology of Palaeozoic diplopods. *Symp. Zool. Soc. London* 32:13–22.

Kropp, R. K. 1982. Responses of five holothurian species to attacks by a predatory gastropod, *Tonna perdix. Pacific Sci.* 36:445–452.

Krueger, D. A., and S. I. Dodson. 1981. Embryological induction and predation ecology in *Daphnia pulex. Limnol. Oceanogr.* 26:219–223.

Kruuk, H. 1972. *The Spotted Hyena: A Study of Predation and Social Behavior.* University of Chicago Press, Chicago.

Kukalova-Peck, J. 1978. Origin and evolution of insect wings and their relation to metamorphosis, as documented by the fossil record. *J. Morphol.* 156:53–126.

Kunze, J. C. 1981. The functional morphology of stomatopod Crustacea. *Phil. Trans. Roy. Soc. London* (B) 292:255–328.

Kunze, J. C., and D. T. Anderson. 1979. Functional morphology of the mouthparts and gastric mill in the hermit crabs *Clibanarius taeniatus* (Milne Edwards), *Clibanarius vittatus* (Krauss), *Paguristes squamosus*

McCulloch and *Dardanus setiger* (Milne-Edwards) (Anomura: Paguridae). *Austral. J. Mar. Fresh-W. Res.* 30:683–721.

LaBarbera, M. 1977. Brachiopod orientation to water movement. 1. Theory, laboratory behavior, and field orientations. *Paleobiology* 3:270–287.

———. 1978. Brachiopod orientation to water movement: Functional morphology. *Lethaia* 11:67–79.

———. 1981a. The ecology of Mesozoic *Gryphaea*, *Exogyra*, and *Ilymatogrya* (Bivalvia, Mollusca) in a modern ocean. *Paleobiology* 7:510–526.

———. 1981b. Water flow patterns in and around three species of articulate brachiopods. *J. Exp. Mar. Biol. Ecol.* 55:185–206.

———. 1984. Feeding currents and particle capture mechanisms in suspension feeding animals. *Amer. Zool.* 24:71–84.

Lamotte, M. 1950. Observations sur la sélection par les prédateurs chez *Cepaea nemoralis*. *J. Conchyliol.* 90:180–190.

Lande, R. 1978. Evolutionary mechanisms of limb loss in tetrapods. *Evolution* 32:73–92.

Lande, R., and S. J. Arnold. 1983. The measurement of selection on correlated characters. *Evolution* 37:1210–1226.

Lang, T. G. 1974. Speed, power, and drag measurements of dolphins and porpoises. In T. Y.-T. Wu, C. J. Brokaw, and C. Brennen (eds.), *Swimming and Flying in Nature*, pp. 553–572. Plenum, New York.

Langley, W. 1981. The effect of prey defenses on the attack behavior of the southern grasshopper mouse (*Onychomys torridus*). *Z. Tierpsychol.* 56:115–127.

Larson, D. W., and D. C. Rhoads. 1983. The evolution of infaunal communities and sedimentary fabrics. In Tevesz and McCall (eds.) 1983, pp. 627–648.

Lauder, G. V. 1982. Patterns of evolution in the feeding mechanism of actinopterygian fishes. *Amer. Zool.* 22:275–285.

———. 1983a. Neuromuscular patterns and the origin of trophic specialization in fishes. *Science* 219:1235–1237.

———. 1983b. Functional design and evolution of the pharyngeal jaw apparatus in euteleostean fishes. *Zool. J. Linn. Soc.* 77:1–38.

Lauder, G. V., and K. F. Liem. 1983. The evolution and interrelationships of the actinopterygian fishes. *Bull. Mus. Comp. Zool.* 150:95–197.

Lavoie, M. E. 1956. How sea stars open bivalves. *Biol. Bull.* 111:114–122.

Laws, H. M., and D. F. Laws. 1972. The escape response of *Donacilla angusta* Reeve (Mollusca: Bivalvia) in the presence of a naticid predator. *Veliger* 14:289–290.

Lawton, P., and R. N. Hughes. 1985. Foraging behaviour of the crab *Cancer pagurus* feeding on the gastropods *Nucella lapillus* and *Littorina littorea*: Comparisons with optimal foraging theory. *Mar. Ecol. Progr. Ser.* 27:143–154.

Layzer, D. 1980. Genetic variation and progressive evolution. *Amer. Nat.* 115:809–826.

Lecointre, G. 1929. Symbiose des cellépores et des gastropodes dans le Faluns de Touraine. *Soc. Géol. France Bull.* (4) 29:401–404.

Leggett, J. K. 1980. British Lower Palaeozoic black shales and their palaeo-oceanographic significance. *J. Geol. Soc. London* 137:139–156.

Lehmann, U. 1975. Über Nahrung und Ernährungsweise von Ammoniten. *Paläont. Z.* 49:187–195.

———. 1981a. Ammonite jaw apparatus and soft parts. In M. R. House and J. R. Senior (eds.), *The Ammonoidea*, pp. 275–287. Systematics Association Special Volume 18.

———. 1981b. *The Ammonites: Their Life and Their World.* Cambridge University Press, Cambridge.

Leigh, E. G. 1971. *Adaptation and Diversity.* Freeman and Cooper, San Francisco.

———. 1973. The evolution of mutation rates. *Genetics* 73 (Suppl.): 1-18.

Levin, D. A. 1975. Pest pressure and recombination systems in plants. *Amer. Nat.* 109:437–451.

———. 1976. Alkaloid-bearing plants: An ecogeographic perspective. *Amer. Nat.* 110:261–284.

Levin, D. A., and B. M. York. 1978. The toxicity of plant alkaloids: An ecogeographic perspective. *Biochem. Systematics Ecol.* 6:61–76.

Lewontin, R. C. 1965. Selection for colonizing ability. In H. G. Baker and G. L. Stebbins (eds.), *The Genetics of Colonizing Species*, pp. 77–91. Academic Press, New York.

———. 1970. The units of selection. *Ann. Rev. Ecol. Syst.* 1:1–18.

———. 1974. *The Genetic Basis of Evolutionary Change.* Columbia University Press, New York.

Lewy, Z., and C. Samtleben. 1979. Functional morphology and palaeontological significance of the conchiolin layers in corbulid pelecypods. *Lethaia* 12:341–351.

Liddell, W. D., and C. E. Brett. 1982. Skeletal overgrowths among epizoans from the Silurian (Wenlockian) Waldron Shale. *Paleobiology* 8:67–78.

Lidgard, S. 1985. Zooid and colony growth in encrusting cheilostome bryozoans. *Palaeontology* 28:255–291.

———. 1986. Ontogeny in animal colonies: A persistent trend in the bryozoan fossil record. *Science* 232:230–232.

Liem, K. F. 1973. Evolutionary strategies and morphological innovations: Cichlid pharyngeal jaws. *Syst. Zool.* 22:425–441.

Liljedahl, L. 1984. Silurian silicified bivalves from Gotland. *Sveriges Geol. Unders.* (C) 804:1–82.

Lindberg, D. R., and K. R. Dwyer. 1983. The topography, formation and role of the home depression of *Collisella scabra* (Gould) (Gastropoda: Acmaeidae). *Veliger* 25:229–234.

Linsley, R. M. 1977. Some "laws" of gastropod shell form. *Paleobiology* 3:196–206.

———. 1978a. The Omphalocirridae: A new family of Palaeozoic Gastropoda which exhibits sexual dimorphism. *Mem. Nat. Mus. Victoria* 39:33–54.

———. 1978b. Locomotion rates and shell form in the Gastropoda. *Malacologia* 17:193–206.

———. 1979. Gastropods of the Devonian. *Special Pap. Palaeont.* 23:249–254.

Linsley, R. M., E. L. Yochelson, and D. M. Rohr. 1978. A reinterpretation of the mode of life of some Paleozoic frilled gastropods. *Lethaia* 11:105–112.

Lipps, J. H. 1970. Plankton evolution. *Evolution* 24:1–22.

Lipps, J. H., and E. Mitchell. 1976. Trophic model for the adaptive radiations and extinctions of pelagic marine mammals. *Paleobiology* 2:147–155.

Littler, M. M., and D. S. Littler. 1980. The evolution of thallus form and survival strategies in benthic marine macroalgae: Field and laboratory tests of a functional form model. *Amer. Nat.* 116:25–44.

———. 1984. Relationships between macroalgal functional form groups and substrata stability in a subtropical rocky-intertidal system. *J. Exp. Mar. Biol. Ecol.* 74:13–34.

Littler, M. M., D. S. Littler, and P. R. Taylor. 1983a. Evolutionary strategies in a tropical barrier reef system: Functional-form groups of marine macroalgae. *J. Phycol.* 19:229–247.

Littler, M. M., P. R. Taylor, and D. S. Littler. 1983b. Algal resistance to herbivory on a Caribbean barrier reef. *Coral Reefs* 2:111–118.

Lowell, R. B. 1987. Safety factors of tropical versus temperate limpet shells: Multiple selection pressures on a single structure. *Evolution*, in press.

Lowry, L. F., K. J. Frost, and J. J. Burns. 1980. Feeding of bearded seals in the Bering and Chukchi Seas and trophic interaction with Pacific walruses. *Arctic* 33:330–342.

Lubchenco, J., and J. Cubit. 1980. Heteromorphic life histories of certain marine algae as adaptations to variations in herbivory. *Ecology* 61:677–687.

Lucas, J. R., and H. J. Brockmann. 1981. Predatory interactions between ants and antlions (Hymenoptera: Formicidae and Neuroptera: Myrmeleontidae). *J. Kansas Entomol. Soc.* 54:228–232.

Luckens, P. A. 1975. Predation and intertidal zonation of barnacles at Leigh, New Zealand. *New Zealand J. Mar. Fresh-W. Res.* 9:355–378.

Lyman, C. P., R. C. O'Brien, G. C. Greene, and E. D. Papafrangos. 1981. Hibernation and longevity in the Turkish hamster *Mesocricetus brandti*. *Science* 212:668–670.

MacArthur, R. H. 1965. Patterns of species diversity. *Biol. Revs.* 40:510–533.

———. 1972. *Geographical Ecology: Patterns in the Distribution of Species*. Harper and Row, New York.

MacArthur, R. H., and E. O. Wilson. 1967. *The Theory of Island Biogeography*. Princeton University Press, Princeton.

MacKinnon, D. I. 1982. *Tuarangia paparua* n. gen. and n. sp., a Late Middle Cambrian pelecypod from New Zealand. *J. Paleont.* 56:589–598.

———. 1985. New Zealand Late Middle Cambrian molluscs and the origin of Rostroconchia and Bivalvia. *Alcheringa* 9:65–81.

MacLeod, N. 1982. The first North American occurrence of the Late Cretaceous elasmobranch *Ptychodus rugosus* Dixon with comments on the functional morphology of the dentition and dermal denticles. *J. Paleont.* 56:403–409.

MacLeod, N., and B. H. Slaughter. 1980. A new ptychodontid shark from the Upper Cretaceous of northeast Texas. *Texas J. Sci.* 32:333–335.

Maes, V. O. 1967. Radulae of two species of *Pleuroploca* (Fasciolariidae) from the Indo-Pacific. *Nautilus* 81:48–54.

Maes, V. O., and D. Raeigle. 1975. Systematics and biology of *Thala floridana* (Gastropoda: Vexillidae). *Malacologia* 15:43–67.

Mahmoud, I. Y., and J. Klicka. 1979. Feeding, drinking, and excretion. In Harless and Morlock (eds.), 1979, pp. 229–243.

Maiorana, V. C. 1977. Tail autotomy, functional conflicts and their resolution by a salamander. *Nature* 265:533–535.

——. 1979. Why do adult insects not molt? *Biol. J. Linn. Soc.* 11:253–258.

Maisey, J. G. 1984. Higher elasmobranch phylogeny and biostratigraphy. *Zool. J. Linn. Soc.* 82:33–54.

Major, P. F. 1979. Piscivorous predators and disabled prey. *Copeia* 1979:158–160.

Mallatt, J. 1984. Feeding ecology of the earliest vertebrates. *Zool. J. Linn. Soc.* 82:261–272.

Manooch, C. S., III. 1977. Foods of the red porgy, *Pagrus pagrus* Linnaeus (Pisces: Sparidae), from North Carolina and South Carolina. *Bull. Mar. Sci.* 27:776–787.

Manton, S. M. 1952. The evolution of arthropodan locomotory mechanisms. Part 3: The locomotion of the Chilopoda and Pauropoda. *J. Linn. Soc. London (Zool.)* 42:118–167.

——. 1954. The evolution of arthropodan locomotory mechanisms. Part 4: The structure, habits and evolution of the Diplopoda. *J. Linn. Soc. London (Zool.)* 42:299–368.

——. 1977. *The Arthropoda: Habits, Functional Morphology, and Evolution.* Clarendon Press, Oxford.

Mapes, R. H., and M. C. Hansen. 1984. Pennsylvanian shark-cephalopod predation: A case study. *Lethaia* 17:175–183.

Mapstone, B. D., A. J. Underwood, and R. G. Creese. 1984. Experimental analyses of the relation between intertidal gastropods *Patelloida mufria* and the trochid *Austrocochlea constricta. Mar. Ecol. Progr. Ser.* 17:85–100.

Marche-Marchad, I. 1977. Remarks on the biology, ecology, and systematic of the genus *Cymbium* Röding, 1798 (Gastropoda Prosobranchia). *Conchiglia* 9:3–9.

Marcus, E., and E. Marcus. 1959. Studies on "Olividae." *Bol. Fac. Fil. Cienc. Letr. Sao Paulo* 232 (*Zool.* 22): 99–188.

Marek, L., and E. L. Yochelson. 1976. Aspects of the biology of Hyolitha (Mollusca). *Lethaia* 9:65–82.

Margolin, A. S. 1964. A running response of *Acmaea* to seastars. *Ecology* 45:191–193.

———. 1976. Swimming of the sea cucumber *Parastichopus californicus* (Stimpson) in response to sea stars. *Ophelia* 15:105–114.

Marincovich, L., Jr. 1977. Cenozoic Naticidae (Mollusca: Gastropoda) of the northeastern Pacific. *Bulls. Amer. Paleont.* 70:169–494.

Marsh, J. A., Jr. 1977. Terrestrial inputs of nitrogen and phosphorus on fringing reefs of Guam. *Proc. Third Intern. Coral Reef Symp.* 1:331–336.

Marshall, L. G. 1981. The great American interchange—an invasion-induced crisis for South American mammals. In Nitecki (ed.), 1981, pp. 133–229.

Marshall, L. G., S. D. Webb, J. J. Sepkoski, Jr., and D. M. Raup. 1982. Mammalian evolution and the great American interchange. *Science* 215:1351–1357.

Martens, E. von. 1903. Durchbohrte Schalen von Landschnecken. *Sitzungsber. Ges. Naturforsch. Freunde Berlin* 8:393–399.

Martens, J. 1969. Die Abgrenzung von Biospezies auf biologisch-ethologischer Grundlage am beispiel der Gattung *Ischyropsalis* C. L. Koch 1839 (Opiliones, Ischyropsalididae). *Zool. Jb. f. Syst.* 96:133–264.

Martin, P. S. 1984. Prehistoric overkill: The global model. In Martin and Klein (eds.), 1984, pp. 354–403.

Martin, P. S., and R. G. Klein (eds.). 1984. *Quaternary Extinctions: A Prehistoric Revolution.* University of Arizona Press, Tucson.

Martinell, J., and E.M.J. Marquina. 1978. Señales de depredación en los Gastropoda procedentes de un yacimiento Pliocénico de Molins de Rei (Barcelona). Implicaciones paleoecológicas. *Acta Geol. Hispan.* 13:125–128.

Martinsson, A. (ed.). 1975. *Fossils and Strata, no. 4: Evolution and Morphology of the Trilobita, Trilobitoidea and Merostomata.* Universitetsforlaget, Oslo.

Massé, H. 1975. Etude de l'alimentation de *Astropecten arantiacus* Linné. *Cah. Biol. Mar.* 16:495–510.

Mather, J. 1982. Choice and competition: Their effects on occupancy of shell homes by *Octopus joubini. Mar. Behav. Physiol.* 8:285–293.

Matsukuma, A. 1977. Notes on *Genkaimurex varicosa* (Kuroda, 1953) (Prosobranchia: Neogastropoda). *Venus* 36:81–88.

———. 1978. Fossil boreholes made by shell-boring predators or commensals. *Venus* 37:29–45.

Matthews, H. R. 1968. Mollusks found in the digestive tract of the fish *Amphichthys cryptocentrus* (Valenciennes, 1837). *Proc. Malacol. Soc. London* 38:247–250.

Mauzey, K. P., C. Birkeland, and P. K. Dayton. 1968. Feeding behavior of asteroids and escape responses of their prey in the Puget Sound region. *Ecology* 49:603–619.

Mazin, J.-M. 1983. *Omphalosaurus nisseri* (Wiman, 1910), un ichthyoptérygien à denture broyeuse du Trias Moyen du Spitsberg. *Bull. Mus. Natn. Hist. Natur. Paris* (4) 5:243–263.

McAlester, A. L. 1962. Upper Devonian pelecypods of the New York Chemung Stage. *Yale Univ. Peabody Mus. Nat. Hist. Bull.* 16:88 P.

McCall, P. L., and M.J.S. Tevesz. 1982. The effects of benthos on physical properties of freshwater sediments. In McCall and Tevesz (eds.), 1982, pp. 105–176.

———— (eds.). 1982. *Animal-Sediment Relations: The Biogenic Alteration of Sediments.* Plenum, New York.

McClintock, J. B., and J. M. Lawrence. 1981. An optimization study on the feeding behavior of *Luidia clathrata* Say (Echinodermata: Asteroidea). *Mar. Behav. Physiol.* 7:263–275.

McDermott, J. J., and P. Roe. 1985. Food, feeding behavior and feeding ecology of nemerteans. *Amer. Zool.* 25:113–125.

McGhee, G. R., Jr. 1982. The Frasnian-Famennian extinction event: A preliminary analysis of Appalachian marine ecosystems. *Geol. Soc. Amer. Special Pap.* 190:491–500.

McGowan, C. 1979. Selection pressure for high body temperatures: Implications for dinosaurs. *Paleobiology* 5:285–295.

McKey, D. 1979. The distribution of secondary compounds within plants. In G. A. Rosenthal and D. H. Janzen (eds.), *Herbivores: Their Interaction with Secondary Plant Metabolites*, pp. 55–133. Academic Press, New York.

McKinney, F. K. 1983. Asexual colony multiplication by fragmentation: An important mode of genetic longevity in the Carboniferous bryozoan *Archimedes. Paleobiology* 9:35–43.

————. 1984. Feeding currents of gymnolaemate bryozoans: Better organization with higher colonial integration. *Bull. Mar. Sci.* 34:315–319.

McKinney, M. L. 1985. Distinguishing patterns of evolution from patterns of deposition. *J. Paleont.* 59:561–567.

McLachlan, A., and Y. Young. 1982. Effects of low temperature on the burrowing rates of four sandy beach molluscs. *J. Exp. Mar. Biol. Ecol.* 65:275–284.

McLaren, D. J. 1982. Frasnian-Famennian extinctions. *Geol. Soc. Amer. Special Pap.* 190:467–484.

McLaughlin, P. A. 1983. Hermit crabs—are they really polyphyletic? *J. Crustacean Biol.* 3:608–621.

McLean, J. H. 1981. The Galapagos rift limpet *Neomphalus*: Relevance to understanding the evolution of a major Paleozoic-Mesozoic radiation. *Malacologia* 21:291–336.

————. 1984a. A case for derivation of the Fissurellidae from the Bellerophontacea. *Malacologia* 25:3–20.

————. 1984b. Shell reduction and loss in fissurellids: A review of genera and species in the *Fissurellidea* group. *Amer. Malacol. Bull.* 2:21–34.

McLean, R. B. 1974. Direct shell acquisition by hermit crabs from gastropods. *Experientia* 30:206–208.

————. 1983. Gastropod shells: A dynamic resource that helps shape benthic community structure. *J. Exp. Mar. Biol. Ecol.* 69:151–174.

McLeod, J. C. 1978. The oldest bryozoans: New evidence from the Early Ordovician. *Science* 200:771–773.

McMahon, R. F., and C. O. McMahon. 1983. Leaping and swimming as protective escape responses in the jackknife clam, *Ensis minor* Dall (Bivalvia: Pharellidae). *Nautilus* 97:55–58.

McNab, B. K. 1978. The evolution of endothermy in the phylogeny of mammals. *Amer. Nat.* 112:1–21.

———. 1980. Food habits, energetics, and the population biology of mammals. *Amer. Nat.* 116:106–124.

———. 1983. Energetics, body size, and the limits to endothermy. *J. Zool. London* 199:1–29.

McNab, B. K., and W. Auffenberg. 1976. The effect of large body size on the temperature regulation of the Komodo dragon, *Varanus komodoensis. Comp. Biochem. Physiol.* 55A:345–350.

McNamara, K. J. 1978. Symbiosis between gastropods and bryozoans in the Late Ordovician of Cumbria, England. *Lethaia* 11:25–40.

———. 1982. Heterochrony and phylogenetic trends. *Paleobiology* 8:130–142.

Mead, A. R. 1961. *The Giant African Snail: A Problem in Economic Malacology.* University of Chicago Press, Chicago.

———. 1963. A flatworm predator of the giant African snail *Achatina fulica* in Hawaii. *Malacologia* 1:305–311.

Mech, L. D. 1966. *The Wolves of Isle Royale.* Government Printing Office, Washington, D.C.

Menge, B. A. 1982. Effects of feeding on the environment: Asteroidea. In Jangoux and Lawrence (eds.), 1982, pp. 521–551.

Menge, B. A., and J. Lubchenco. 1981. Community organization in temperate and tropical rocky intertidal habitats: Prey refuges in relation to consumer pressure gradients. *Ecol. Monogrs.* 51:429–450.

Menge, J. L. 1974. Prey selection and foraging period of the predaceous rocky intertidal snail, *Acanthina punctulata. Oecologia* (Berlin) 17:293–316.

Meyer, D. L. 1985. Evolutionary implications of predation on Recent comatulid crinoids from the Great Barrier Reef. *Paleobiology* 11:154–164.

Meyer, D. L., and D. B. Macurda. 1977. Adaptive radiation of the comatulid crinoids. *Paleobiology* 3:74–82.

Meyer, E. R. 1984. Crocodilians as living fossils. In Eldredge and Stanley (eds.), 1984, pp. 105–131.

Mienis, H. K. 1971. *Theba pisana* in pellets of an Israelian owl. *Basteria* 35:73–75.

Mikulska, I. 1961. Parental care in a rare spider *Pellenes nigrociliatus* (L. Koch) var. *bilunulata* Simon. *Nature* 190:365–366.

Miller, B. A. 1975. The biology of *Terebra gouldi* Deshayes, 1859, and a discussion of life history similarities among other terebrids of similar proboscis type. *Pacific Sci.* 29:227–241.

Miller, K. 1982. Effect of temperature on sprint performance in the frog *Xen-*

opus laevis and the salamander *Necturus maculosus*. *Copeia*, no. 3:695–698.

Miller, M. F. 1984. Distribution of biogenic structures in Paleozoic nonmarine and marine-margin sequences: An actualistic model. *J. Paleont.* 58:550–570.

Miller, M. F., and C. W. Byers. 1984. Abundant and diverse Early Paleozoic infauna indicated by the stratigraphic record. *Geology* 12:40–43.

Miller, R. H., and F. A. Sundberg. 1984. Boring Late Cambrian organisms. *Lethaia* 17:185–190.

Miller, R. S. 1967. Pattern and process in competition. *Adv. Ecol. Res.* 4:1–74.

Miller, S. L. 1974. Adaptive design of locomotion and foot form in prosobranch gastropods. *J. Mar. Biol. Ecol.* 14:99–156.

Miller, W., III. 1983. Biogenic shell damage in the small gastropod *Odostomia impressa* (Say). *Tulane Studies Geol. Paleont.* 17:105–116.

Milner, A. R. 1980. The tetrapod assemblage from Nyrany, Czechoslovakia. In Panchen (ed.), 1980, pp. 439–496.

Minchin, D. 1983. Predation on young *Pecten maximus* (L.) (Bivalvia), by the anemone *Anthopleura bollii* (Cocks). *J. Molluscan Stud.* 49:228–231.

Missarzhevskiy, V. V. 1983. Subdivision and correlation of the Precambrian-Cambrian boundary sequences based on certain ancient groups of skeletal organisms. *Intern. Geol. Rev.* 25:745–759.

Mitchell-Tapping, H. J. 1983. Experimental fracturing of various corals in the reef environment. *Carib. J. Sci.* 19:49–52.

Miyake, S. 1978. *The Crustacean Anomura of Sagami Bay*. Biological Laboratory, Imperial Household, Tokyo.

Mladenov, P. V. 1983. Rate of arm regeneration and potential causes of arm loss in the feather star *Florometra serratissima* (Echinodermata: Crinoidea). *Canad. J. Zool.* 61:2873–2879.

Moffitt, J. 1941. Notes on the food of the California clapper rail. *Condor* 43:270–273.

Moitoza, D. J., and D. W. Phillips. 1979. Prey defense, predator preference, and nonrandom diet: The interactions between *Pycnopodia helianthoides* and two species of sea urchins. *Mar. Biol.* 53:299–304.

Moore, I. A., and J. W. Moore. 1974. Food of shorthorn sculpin, *Myxocephalus scorpius*, in the Cumberland Sound area of Baffin Island. *J. Fish. Res. Bd. Canada* 31:355–359.

Moore, R. C. (ed.). 1969. *Treatise on Invertebrate Paleontology*. Part R: *Arthropoda* 4, 2. University of Kansas Press, Lawrence, Kansas.

Moran, D. P., and M. L. Reaka. 1985. Bioerosion of coral rubble and the availability of shelters for stomatopod crustaceans. *Mar. Ecol. Prog. Ser.*, submitted.

Mordan, P. B. 1977. Factors affecting the distribution and abundance of *Aegopinella* and *Nesovitrea* (Pulmonata: Zonitidae) at Monks Wood National Nature Reserve, Huntingdonshire. *Biol. J. Linn. Soc.* 9:59–72.

Morris, N. J. 1978. The infaunal descendants of the Cycloconchidae: An outline of the evolutionary history and taxonomy of the Heteroconchia, superfamilies Cycloconchacea to Cahamacea. *Phil. Trans. Roy. Soc. London* (B) 284:259–275.

Morris, S., and R. D. Purchon. 1981. The marine shelled Mollusca of West Malaysia and Singapore. Part 3: Bivalvia. *J. Molluscan Stud.* 47:322–327.

Morse, D. H. 1974. Niche breadth as a function of social dominance. *Amer. Nat.* 108:818–830.

———. 1979. Prey capture by the crab spider *Misumena calycina* (Araneae, Thomisidae). *Oecologia* (Berlin) 39:309–319.

———. 1980. *Behavioral Mechanisms in Ecology*. Harvard University Press, Cambridge.

———. 1981. Prey capture by the crab spider *Misumena vatia* (Clerck) (Thomisidae) on three common native flowers. *Amer. Midland Nat.* 105:358–367.

Mortimer, J. A. 1982. Feeding ecology of sea turtles. In K. A. Bjorndal (ed.), *The Biology and Conservation of Sea Turtles*, pp. 103–109. Smithsonian Institution Press, Washington, D.C.

Morton, B. S. 1973. The biology and functional morphology of *Galeomma (Paralepida) takii* (Bivalvia: Leptonacea). *J. Zool. London* 169:133–150.

———. 1974. Some aspects of the biology and functional morphology of *Cleidothaerus maorianus* Finlay (Bivalvia: Anomalodesmata: Pandoracea). *Proc. Malacol. Soc. London* 41:201–222.

———. 1975. Dynamic display in *Galeomma polita* Deshayes (Bivalvia: Leptonacea). *J. Conchol.* 28:365–369.

———. 1978. The biology and functional morphology of *Claudiconcha japonica* (Bivalvia: Veneracea). *J. Zool. London* 184:35–52.

———. 1979. A comparison of lip structure and function correlated with other aspects of the functional morphology of *Lima lima, Limaria (Platylimaria) fragilis,* and *Limaria (Platylimaria) hongkongensis* sp. nov. (Bivalvia: Limacea). *Canad. J. Zool.* 57:728–742.

———. 1980a. The mode of life and functional morphology of *Gregariella coralliophaga* (Gmelin 1791) (Bivalvia: Mytilacea) with a discussion on the evolution of the boring Lithophaginae and adaptive radiation in the Mytilidae. In B. S. Morton and C. K. Tseng (eds.), *Proc. First Intern. Mar. Biol. Workshop: The Marine Flora and Fauna of Hong Kong and Southern China*, pp. 875–895. Hong Kong University Press, Hong Kong.

———. 1980b. Swimming in *Amusium pleuronectes* (Bivalvia: Pectinidae). *J. Zool. London* 190:375–404.

———. 1980c. Selective site segregation in *Patelloida (Chiazacmea) pygmaea* (Dunker) and *P. (C.) lampanicola* Habe (Gastropoda: Patellacea) on a Hong Kong shore. *J. Exp. Mar. Biol. Ecol.* 47:149–171.

———. 1981a. The Anomalodesmata. *Malacologia* 21:35–60.

———. 1981b. Prey capture by the carnivorous septibranch *Poromya granulata* (Bivalvia: Anomalodesmata: Poromyacea). *Sarsia* 66:241–256.

———. 1981c. The mode of life and function of the shell buttress in *Cucullaea concamerata* (Martini) (Bivalvia: Arcacea). *J. Conchol.* 30:295–301.

Morton, B. S. 1982a. The biology, functional and taxonomic status of *Fluviolanatus subtorta* (Bivalvia: Trapeziidae), a heteromyarian bivalve possessing "zooxanthellae." *J. Malacol. Soc. Australia* 5:113–140.

———. 1982b. The functional morphology of *Bathyarca pectunculoides* (Bivalvia: Arcacea) from a deep Norwegian fjord with a discussion of the mantle margin in the Arcoida. *Sarsia* 67:269–282.

———. 1984. The biology and functional morphology of *Clavagella australis* (Bivalvia: Anomalodesmata). *J. Zool. London* 202:489–511.

———. 1985. Prey preference, capture and ration in *Hemifusus tuba* (Gmelin) (Prosobranchia: Melongenidae). *J. Exp. Mar. Biol. Ecol.* 94:191–210.

Morton, B. S., and P.J.B. Scott. 1980. Morphological and functional specializations of the shell, musculature and pallial glands in the Lithophaginae (Mollusca: Bivalvia). *J. Zool. London* 192:179–203.

Morton, J. E. 1955. The functional morphology of *Otina otis*, a primitive marine pulmonate. *J. Mar. Biol. Assoc. U.K.* 34:113–150.

Morton, N. 1983. Pathologically deformed *Graphoceras (Ammonitina)* from the Jurassic of Skye, Scotland. *Palaeontology* 26:443–453.

Moss, S. A. 1977. Feeding mechanisms in sharks. *Amer. Zool.* 17:355–364.

Moy-Thomas, J. A., and R. S. Miles. 1971. *Palaeozoic Fishes*. Chapman and Hall, London.

Muirhead, A., P. A. Tyler, and M. A. Thurston. 1986. Reproductive biology and growth of the genus *Epizoanthus* (Zoantharia) from the north-east Atlantic. *J. Mar. Biol. Assoc. U.K.* 66:131–143.

Muller, J. 1970. Palynological evidence of early differentiation of angiosperms. *Biol. Revs.* 45:417–450.

Muntz, L., F. J. Ebling, and J. A. Kitching. 1965. The ecology of Lough Ine. XIV. Predatory activity of large crabs. *J. Anim. Ecol.* 34:315–329.

Murdock, G. O., and J. D. Currey. 1978. Strength and design of shells of the two ecologically distinct barnacles, *Balanus balanus* and *Semibalanus balanoides* (Cirripedia). *Biol. Bull.* 155:169–192.

Murina, G.-V. V. 1984. Ecology of Sipuncula. *Mar. Ecol. Progr. Ser.* 17:1–7.

Myers, G. S. 1960. The endemic fish fauna of Lake Lanao and the evolution of higher taxonomic categories. *Evolution* 14:323–333.

Nachtigall, W. 1968. *Insects in Flight: A Glimpse Behind the Scenes in Biophysical Research*. Trans. H. Oldroyd, R. H. Abbott, and M. Biedemann. McGraw Hill, New York.

Nakazawa, K., and B. Runnegar. 1973. The Permian-Triassic boundary: A crisis for bivalves? *Canad. Soc. Petroleum Geol. Mem.* 2:608–621.

Nations, J. D. 1975. The genus *Cancer* (Crustacea: Brachyura): Systematics, biogeography and fossil record. *Nat. Hist. Mus. Los Angeles County Sci. Bull.* 23:1–104.

Neville, A. C. 1975. *Biology of the Arthropod Cuticle*. Springer, Berlin.

Newell, C. R., and H. Hidu. 1982. The effects of sediment type on growth rate and shell allometry in the soft shelled clam *Mya arenaria* L. *J. Exp. Mar. Biol. Ecol.* 65:285–295.

Newell, N. D. 1937. Late Paleozoic pelecypods: Pectinacea. *State Geol. Surv. Kansas* 10:1–123.

———. 1971. An outline history of tropical organic reefs. *Novitates* 2465:1–37.

Newell, N. D., and D. W. Boyd. 1970. Oyster-like Permian Bivalvia. *Bull. Amer. Mus. Nat. Hist.* 143:219–281.

———. 1975. Parallel evolution in early trigoniacean bivalves. *Bull. Amer. Mus. Nat. Hist.* 154:53–162.

Newman, W. A. 1979. A new scalpellid (Cirripedia); a Mesozoic relic living near an abyssal hydrothermal spring. *Trans. San Diego Soc. Nat. Hist.* 19:153–167.

———. 1982. Evolution within the Crustacea. Part 3: Cirripedia. In Abele (ed.), 1982, pp. 197–221.

Newman, W. A., and H. S. Ladd. 1974. Origin of coral-inhabiting balanoids (Cirripedia, Thoracica). *Verh. Naturforsch. Ges. Basel* 84:381–396.

Newman, W. A., and A. Ross. 1976. Revision of the balanomorph barnacles; including a catalog of the species. *Mem. San Diego Soc. Nat. Hist.* 9:1–108.

———. 1977. A living *Tesseropora* (Cirripedia: Balanomorpha) from Bermuda and the Azores: First record from the Atlantic since the Oligocene. *Trans. San Diego Soc. Nat. Hist.* 18:207–216.

Newman, W. A., and S. M. Stanley. 1981. Barnacle ecology: Is competition important? Competition wins out overall: Reply to Paine. *Paleobiology* 7:561–569.

Newman, W. A., V. A. Zullo, and T. H. Withers. 1969. Cirripedia. In R. C. Moore (ed.), 1969, pp. R206–R295.

Ng, K. L., and L. Tan. 1984. The shell-peeling structure of the box crab, *Calappa philargeus* (Linn.) and other crabs in relation to mollusc shell architecture. *J. Singapore Nat. Acad. Sci.* 13:195–199.

Nicol, D. 1958. A survey of inequivalve pelecypods. *J. Washington Acad. Sci.* 48:56–62.

———. 1965. Ecological implications of living pelecypods with calcareous spines. *Nautilus* 78:109–116.

———. 1967. Some characteristics of cold-water marine pelecypods. *J. Paleont.* 41:1330–1340.

Nicolson, S. W., G. A. Bartholomew, and M. K. Seely. 1984. Ecological correlates of locomotion speed, morphometrics and body temperature in three Namib Desert tenebrionid beetles. *South Afr. J. Zool.* 19:131–134.

Nielsen, C. 1975. Observations on *Buccinum undatum* L. attacking bivalves and on prey responses, with a short review of attack methods of other prosobranchs. *Ophelia* 13:87–108.

Niklas, K. J. (ed.). 1981. *Paleobotany, Paleoecology, and Evolution.* Vol. 1. Praeger, New York.

———. 1982. Computer simulations of early land plant morphologies: Canalization of patterns during evolution? *Paleobiology* 8:196–210.

———. 1984. Size-related changes in the primary xylem anatomy of some early tracheophytes. *Paleobiology* 10:487–506.

Niklas, K. J. 1985. The evolution of tracheid diameter in early vascular plants and its implications on the hydraulic conductance of the primary xylem strand. *Evolution* 39:1110–1122.

Niklas, K. J., and V. Smocovitis. 1983. Evidence for a conducting strand in Early Silurian (Llandoverian) plants: Implications for the evolution of the land plants. *Paleobiology* 9:126–137.

Niklas, K. J., B. H. Tiffney, and A. H. Knoll. 1980. Apparent changes in the diversity of fossil plants: A preliminary assessment. *Evol. Biol.* 12:1–89.

———. 1983. Patterns in vascular land plant diversification. *Nature* 303:614–616.

Nitecki, M. H. (ed.). 1981. *Biotic Crises in Ecological and Evolutionary Time*. Academic Press, New York.

——— (ed.). 1983. *Coevolution*. University of Chicago Press, Chicago.

Nixon, M. 1979. Hole-boring in shells by *Octopus vulgaris* Cuvier in the Mediterranean. *Malacologia* 18:431–443.

Nordsieck, H. 1982. Die Evolution des Verschlussapparats der Schliessmundschnecken (Gastropoda: Clausiliidae). *Arch. Molluskenkunde* 112:27–43.

Norman, D. B., and D. Weishampel. 1985. Ornithopod feeding mechanisms: Their bearing on the evolution of herbivory. *Amer. Nat.* 126:151–164.

Norris, K. S., and B. Mohl. 1983. Can odontocetes debilitate prey with sound? *Amer. Nat.* 122:85–104.

Norton-Griffiths, M. 1967. Some ecological aspects of the feeding behaviour of the oystercatcher *Haematopus ostralegus* on the Edible Mussel *Mytilus edulis. Ibis* 109:412–424.

Nowlan, G. S., G. M. Narbonne, and W. H. Fritz. 1985. Small shelly fossils and trace fossils near the Precambrian-Cambrian boundary in the Yukon Territory, Canada. *Lethaia* 18:233–256.

Odin, G. S., D. Curry, N. H. Gale, and W. J. Kennedy. 1982. The Phanerozoic time scale in 1981. In G. S. Odin (ed.), *Numerical Dating in Stratigraphy.* Part II, pp. 957–960. Wiley, Chichester.

Odin, G. S., N. H. Gale, B. Auvrary, M. Bielski, F. Doré, J.-R. Lancelot, and P. Pasteels. 1983. Numerical dating of Precambrian-Cambrian boundary. *Nature* 301:21–23.

O'Donald, P. 1973. A further analysis of Bumpus' data: The intensity of natural selection. *Evolution* 27:398–404.

Officer, C. B., and C. L. Drake. 1985. Terminal Cretaceous environmental events. *Science* 227:1161–1167.

Okutani, T. 1961. Description of *Solariella nectonica* sp. nov. with special reference to its swimming behavior. *Venus* 21:304–308.

Oliver, J. S., P. M. Slattery, E. F. O'Connor, and L. F. Lowry. 1983a. Walrus, *Odobenus rosmarus*, feeding in the Bering Sea: A benthic perspective. *Fish. Bull.* 81:501–512.

Oliver, J. S., P. M. Slattery, M. A. Silberstein, and E. F. O'Connor. 1983b. A comparison of gray whale, *Eschrichtius robustus*, feeding in the Bering Sea and Baja California. *Fish. Bull.* 81:513–522.

Olson, E. C. 1975. Permo-Carboniferous paleoecology and morphotypic series. *Amer. Zool.* 15:371–389.

———. 1976. The exploitation of land by early tetrapods. *Linn. Soc. Symp. Ser.* 3:1–30.

———. 1983. Coevolution or coadaptation: Permo-Carboniferous vertebrate chronofauna. In Nitecki (ed.), 1983, pp. 307–338.

Olson, S. L., and A. Feduccia. 1980. *Presbyornis* and the origin of the Anseriformes (Aves: Charadriomorphae). *Smithson. Contribs. Zool.* 323:1–24.

Olson, S. L., and H. F. James. 1982. Fossil birds from the Hawaiian Islands: Evidence for wholesale extinction by man before Western contact. *Science* 217:633–635.

Olson, S. L., and D. W. Steadman. 1978. The fossil record of the Glareolidae and Haematopodidae (Aves: Charadriiformes). *Proc. Biol. Soc. Washington* 91:972–981.

Orians, G. H., and O. T. Solbrig. 1977. A cost-income model of leaves and roots with special reference to arid and semiarid areas. *Amer. Nat.* 111:677–690.

Orth, R. J. 1975. Destruction of eelgrass, *Zostera marina*, by the cownosed ray, *Rhinoptera bonasus*, in the Chesapeake Bay. *Chesapeake Sci.* 16:205–208.

Osse, J.W.M. 1985. Jaw protrusion, an optimization of the feeding apparatus of teleosts? *Acts Biotheoretica* 34:219–232.

Ott, J. C., B. Fuchs, R. Fuchs, and A. Malasek. 1976. Observations on the biology of *Callianassa stebbingi* Borrodaille and *Upogebia littoralis* Risso and their effect upon the sediment. *Senckenbergiana Maritima* 8:61–79.

Otter, G. W. 1937. Rock-destroying organisms in relation to coral reefs. *Brit. Mus. (Nat. Hist.) Great Barrier Reef Exped. 1928-29, Sci. Rep.* 1:323–352.

Otto, C., and B. S. Svensson. 1980. The significance of case material selection for the survival of caddis larvae. *J. Anim. Ecol.* 49:855–865.

Owens, R. M. 1983. A review of Permian trilobite genera. *Special Pap. Palaeont.* 30:15–41.

Pabst, W. 1953. Zur Biologie der Mitteleuropäischen Troguliden. *Zool. Jb. f. Syst.* 82:1–46.

Packard, A. 1972. Cephalopods and fish: The limits of convergence. *Biol. Revs.* 47:241–307.

Packer, C. 1983. Sexual dimorphism: The horns of African antelopes. *Science* 221:1191–1193.

Padian, K. 1983. A functional analysis of flying and walking in pterosaurs. *Paleobiology* 9:218–239.

Page, L. M. 1983. *Handbook of Darters.* TFH Publications, Neptune City, New Jersey.

Paine, R. T. 1963a. Trophic relationships of eight sympatric predatory gastropods. *Ecology* 44:63–73.

———. 1963b. Food recognition and predation on opisthobranchs by *Navanax inermis* (Gastropoda: Opisthobranchia). *Veliger* 6:1–9.

Paine, R. T. 1966. Function of labial spines, composition of diet, and size of certain marine gastropods. *Veliger* 9:17–24.

———. 1976. Biological observations on a subtidal *Mytilus californianus* bed. *Veliger* 19:125–130.

———. 1980. Food webs: Linkage, interaction strength and community infrastructure. *J. Anim. Ecol.* 49:667–685.

———. 1981. Barnacle ecology: Is competition important? The forgotten roles of disturbance and predation. *Paleobiology* 7:553–560.

Paine, R. T., and A. R. Palmer. 1978. *Sycyases sanguineus*: A unique trophic generalist from the Chilean intertidal zone. *Copeia* 1978:75–80.

Palmer, A. 1984. The biomere problem: Evolution of an idea. *J. Paleont.* 58:599–611.

Palmer, A. R. 1977. Function of shell sculpture in marine gastropods: Hydrodynamic destabilization in *Ceratostoma foliatum*. *Science* 197:1293–1295.

———. 1979. Fish predation and the evolution of gastropod shell sculpture: Experimental and geographic evidence. *Evolution* 33:697–713.

———. 1980. Locomotion rates and shell form in the Gastropoda: A re-evaluation. *Malacologia* 19:289–296.

———. 1981. Do carbonate skeletons limit the rate of body growth? *Nature* 292:150–152.

———. 1982. Predation and parallel evolution: Recurrent parietal plate reduction in balanomorph barnacles. *Paleobiology* 8:31–45.

———. 1983a. Relative cost of producing skeletal organic matrix versus calcification: Evidence from marine gastropods. *Mar. Biol.* 75:287–292.

———. 1983b. Growth rate as a measure of food value in thaidid gastropods: Assumptions and implications of prey morphology and distribution. *J. Exp. Mar. Biol. Ecol.* 73:95–124.

Palmer, A. R., J. Szymanska, and L. Thomas. 1982. Prolonged withdrawal: A possible predator evasion behavior in *Balanus glandula* (Crustacea: Cirripedia). *Mar. Biol.* 67:51–55.

Palmer, T. J. 1982. Cambrian to Cretaceous changes in hardground communities. *Lethaia* 15:309–323.

Palmer, T. J., and C. D. Hancock. 1973. Symbiotic relationships between ectoprocts and gastropods, and ectoprocts and hermit crabs in the French Jurassic. *Palaeontology* 16:563–566.

Palumbi, S. R., and J.B.C. Jackson. 1982. Ecology of cryptic coral reef communities. II. Recovery of small disturbance events by encrusting Bryozoa: The influence of "host" species and lesion size. *J. Exp. Mar. Biol. Ecol.* 64:103–116.

Panchen, A. L. (ed). 1980. *The Terrestrial Environment and the Origin of Land Vertebrates*. Academic Press, London.

Papp, A., H. Zapfe, F. Bachmayer, and A. F. Tauber. 1947. Lebensspuren mariner Krebse. *Königl. Akad. Wissensch. Wien, Mathem. Naturwiss. Klasse, Sitzber.* 155:281–317.

Patterson, C. 1964. A review of Mesozoic acanthopterygian fishes, with special reference to those of the English Chalk. *Phil. Trans. Roy. Soc. London* (B) 247:213–482.

Paul, C.R.C. 1977. Evolution of primitive echinoderms. In Hallam (ed.), 1977, pp. 123–157.

Paul, C.R.C., and A. B. Smith. 1984. The early radiation and phylogeny of echinoderms. *Biol. Revs.* 59:443–481.

Pawson, D. L. 1980. Holothuroidea. In T. W. Broadhead and J. A. Waters (eds.), *Echinoderms: Notes for a Short Course. Univ. Tennessee Dept. Geol. Sci., Stud. Geol.* 3:175–189.

Pedder, A.E.H. 1982. The rugose coral record across the Frasnian-Famennian boundary. *Geol. Soc. Amer. Special Pap.* 190:485–489.

Peel, J. S. 1975. A new Silurian gastropod from Wisconsin and the ecology of uncoiling in Paleozoic gastropods. *Bull. Geol. Soc. Denmark* 24:211–221.

———. 1984. Attempted predation and shell repair in *Euomphalopterus* (Gastropoda) from the Silurian of Gotland. *Bull. Geol. Soc. Denmark* 32:163–168.

Pemberton, G. S., M. J. Risk, and D. E. Buckley. 1976. Supershrimp: Deep bioturbation in the Strait of Canso, Nova Scotia. *Science* 192:790–791.

Pennington, B. J., and J. D. Currey. 1984. A mathematical model for the mechanical properties of scallop shells. *J. Zool. London* 202:239–263.

Percival, S. F., Jr., and A. G. Fischer. 1977. Changes in the calcareous nanoplankton in the Cretaceous-Tertiary biotic crisis at Zumaya, Spain. *Evol. Theory* 2:1–35.

Perry, D. M. 1985. Function of the shell spine in the predaceous rocky intertidal snail *Acanthina spirata* (Prosobranchia: Muricacea). *Mar. Biol.* 88:51–58.

Peters, J. A. 1960. The snakes of the subfamily Dipsadinae. *Misc. Publ. Mus. Zool. Univ. Michigan* 114:1–224.

Peterson, C. H. 1979. Predation, competitive exclusion, and diversity in the soft-sediment benthic communities of estuaries and lagoons. In R. J. Livingston (ed.), *Ecological Processes in Coastal and Marine Systems*, pp. 233–264. Plenum, New York.

———. 1982. Clam predation by whelks (*Busycon* spp.): Experimental tests of the importance of prey size, prey density, and seagrass cover. *Mar. Biol.* 66:159–170.

Peterson, C. H., and M. L. Quammen. 1982. Siphon nipping: Its importance to small fishes and its impact on growth of the bivalve *Protothaca staminea* (Conrad). *J. Exp. Mar. Biol. Ecol.* 63:249–268.

Pettitt, C. 1975. A review of the predators of *Littorea*, especially those of *L. saxatilis* (Olivi) (Gastropoda: Prosobranchia). *J. Conchol.* 28:343–357.

Petuch, E. J. 1982. Geographical heterochrony: Contemporaneous coexistence of Neogene and Recent molluscan faunas in the Americas. *Palaeogeogr., Palaeoclimatol., Palaeoecol.* 37:277–312.

Phillips, D. W., and M. L. Chiarappa. 1980. Defensive responses of gastropods to the predatory flatworms *Freemania litoricola* (Heath and McGregor) and *Notoplana acticola* (Leone). *J. Exp. Mar. Biol. Ecol.* 47:179–189.

Phillips, R. R. 1977. The relationship between social behaviour and the use of space in the benthic fish *Chasmodes bosquianus* Lacepede (Teleostei: Blenniidae). IV. Effects of topography on habitat selection and shelter choice. *Behaviour* 60:1–27.

Pickerill, R. K. 1984. Comment and reply on "Abundant and diverse Early Paleozoic infauna indicated by the stratigraphic record." *Geology* 12:567–568.

Pickerill, R. K., and T. L. Harland. 1984. Middle Ordovician microborings of probable sponge origin from eastern Canada and southern Norway. *J. Paleont.* 58:885–891.

Pierce, S. K. 1971. Volume regulation and valve movements by marine mussels. *Comp. Biochem. Physiol.* 39A:103–117.

Poag, C. W. 1971. Notes on the morphology and habit of *Vasiglobulina alabamensis* (Foraminiferida). *J. Paleont.* 45:961–962.

Pocock, K. J. 1974. A unique case of teratology in trilobite segmentation. *Lethaia* 7:63–66.

Pohlo, R. 1982. Evolution of the Tellinacea (Bivalvia). *J. Molluscan Stud.* 48:245–256.

Pohowsky, R. A. 1978. The boring ctenostomate Bryozoa: Taxonomy and paleobiology based on cavities in calcareous substrata. *Bulls. Amer. Paleont.* 73:1–92.

Pojeta, J., Jr. 1978. The origin and early taxonomic diversification of pelecypods. *Phil. Trans. Roy. Soc. London* (B) 284:225–246.

Pojeta, J., Jr., and T. J. Palmer. 1976. The origin of rock boring in mytilacean pelecypods. *Alcheringa* 1:167–179.

Pojeta, J., Jr., and B. Runnegar. 1976. The paleontology of rostroconch mollusks and the early history of the phylum Mollusca. *U.S. Geol. Surv. Prof. Pap.* 968:1–88.

Pollard, E. 1975. Aspects of the ecology of *Helix pomatia* L. *J. Anim. Ecol.* 44:305–329.

Ponder, W. F. 1970. The morphology of *Alcithoe arabica* (Gastropoda: Volutidae). *Malacol. Revs.* 3:127–165.

———. 1983. Xenophoridae of the world. *Aust. Mus. Mem.* 17:1–126.

Ponder, W. F., and R. U. Gooding. 1978. Four new eulimid gastropods associated with shallow-water diadematid echinoids in the Western Pacific. *Pacific Sci.* 32:157–181.

Pough, F. H. 1980. The advantages of ectothermy for tetrapods. *Amer. Nat.* 115:92–112.

Pratt, B. R. 1982. Stromatolite decline—a reconsideration. *Geology* 10:512–515.

Pregill, G. 1984. Durophagous feeding adaptations in an amphisbaenid. *J. Herpetol.* 18:186–191.

Price, H. A. 1982. An analysis of factors determining seasonal variation in the byssal attachment strength of *Mytilus edulis*. *J. Mar. Biol. Assoc. U.K.* 62:147–155.

Price, P. W. 1980. *Evolutionary Biology of Parasites*. Princeton University Press, Princeton.

Pritchard, P.C.H. 1979. Taxonomy, evolution, and zoogeography. In Harless and Morlock (eds.), 1979, pp. 1–42.

Prothero, D. R. 1985. Mid-Oligocene extinction event in North American land mammals. *Science* 229:550–551.

Provenzano, A. J., Jr. 1971. Rediscovery of *Munidopagurus macrocheles* (A. Milne-Edwards, 1880) (Crustacea, Decapoda, Paguridae), with a description of the first zoeal stage. *Bull. Mar. Sci.* 21:256–266.

Punzo, F. 1982. Tail autotomy and running speed in the lizards *Cophosaurus texanus* and *Uma notata*. *J. Herpetol.* 16:329–331.

Purchon, R. D. 1955a. The structure and function of the British Pholadidae (rock-boring Lamellibranchia). *Proc. Zool. Soc. London* 124:859–911.

———. 1955b. The functional morphology of the rock-boring lamellibranch *Petricola pholadiformis* Lamarck. *J. Mar. Biol. Assoc. U.K.* 34:257–278.

Quast, J. C. 1968. Observations on the food of the kelp-bed fishes. *California Dept. Fish. and Game Bull.* 139:109–142.

Radinsky, L. 1978. Evolution of brain size in carnivores and ungulates. *Amer. Nat.* 112:815–831.

Raffaelli, D. G. 1978. The relationship between shell injuries, shell thickness and habitat characteristics of the intertidal snail *Littorina rudis* Maton. *J. Molluscan Stud.* 44:166–170.

Raffi, S., S. M. Stanley, and R. Marasti. 1985. Biogeographic patterns and Plio-Pleistocene extinction of Bivalvia in the Mediterranean and southern North Sea. *Paleobiology* 11:368–388.

Rampino, M. R., and R. B. Stothers. 1984. Terrestrial mass extinctions, cometary impacts, and the sun's motion perpendicular to the galactic plane. *Nature* 308:709–712.

Rand, A. S. 1954. Variation and predator pressure in an island and a mainland population of lizards. *Copeia* 1954:260–262.

Randall, J. E. 1964. Contribution to the biology of the queen conch, *Strombus gigas*. *Bull. Mar. Sci. Gulf Caribbean* 14:246–295.

———. 1967. Food habits of reef fishes of the West Indies. *Stud. Trop. Oceanogr. Inst. Mar. Sci. Univ. Miami* 5:665–847.

———. 1974. The effect of fishes on coral reefs. *Proc. Second Intern. Coral Reef Symp.* 1:159–166.

Rasmussen, E. 1973. Systematics and ecology of the Isefjord marine fauna (Denmark). *Ophelia* 11:1–495.

Raup, D. M. 1961. The geometry of coiling in gastropods. *Proc. Nat. Acad. Sci. U.S.A.* 47:602–609.

———. 1966. Geometric analysis of shell coiling: General problems. *J. Paleont.* 40:1178–1190.

———. 1967. Geometric analysis of shell coiling: Coiling in ammonoids. *J. Paleont.* 41:43–65.

———. 1972. Taxonomic diversity during the Phanerozoic. *Science* 177:1065–1071.

Raup, D. M. 1976a. Species diversity in the Phanerozoic: A tabulation. *Paleobiology* 2:279–288.

———. 1976b. Species diversity in the Phanerozoic: An interpretation. *Paleobiology* 2:289–297.

———. 1979. Biases in the fossil record of species and genera. *Bull. Carnegie Mus. Nat. Hist.* 13:85–91.

Raup, D. M., and R. E. Crick. 1981. Evolution of single characters in the Jurassic ammonite *Kosmoceras*. *Paleobiology* 7:200–215.

Raup, D. M., and S. J. Gould. 1974. Stochastic simulation and evolution of morphology—towards a nomothetic paleontology. *Syst. Zool.* 23:305–322.

Raup, D. M., and D. Jablonski (eds.). 1986. *Patterns and Processes in the History of Life*. Springer, Berlin.

Raup, D. M., and J. J. Sepkoski, Jr. 1982. Mass extinctions in the marine fossil record. *Science* 215:1501–1503.

———. 1984. Periodicity of extinctions in the geologic past. *Proc. Nat. Acad. Sci. U.S.A.* 81:801–805.

Ray, C. E. 1976. Geography of phocid evolution. *Syst. Zool.* 25:391–406.

Reaka, M. L., and R. B. Manning. 1981. The behavior of stomatopod Crustacea, and its relationships to rates of evolution. *J. Crustacean Biol.* 1:309–327.

Reid, D. G. 1986. Predation by crabs on tree snails in a mangrove habitat in northern Queensland. Manuscript.

Reid, R.E.H. 1968. Bathymetric distributions of Calcarea and Hexactinellida in the present and the past. *Geol. Mag.* 105:546–559.

Reif, W.-E. 1976. Morphogenesis, pattern formation and function of the dentition of *Heterodontus* (Selachii). *Zoomorphologie* 83:1–47.

———. 1978. Plicae and cardinal-crurae in pectinids: Protective devices against starfish predation? *Neues Jb. Geol. Paläont. Abh.* 157:115–118.

———. 1980. Tooth enameloid as a taxonomic criterion. 3. A new primitive shark family from the Lower Keuper. *Neues Jb. Geol. Paläont. Abh.* 160:61–72.

———. 1982. Evolution of the dermal skeleton and dentition in vertebrates: The odontode regulation theory. *Evol. Biol.* 15:287–368.

Reimchen, T. E. 1979. Substratum heterogeneity, crypsis, and colour polymorphism in an intertidal snail (*Littorina mariae*). *Canad. J. Zool.* 57:1070–1085.

———. 1983. Structural relationships between spines and lateral plates in threespined stickleback (*Gasterosteus aculeatus*). *Evolution* 37:931–946.

Reise, K. 1978. Experiments on epibenthic predation in the Wadden Sea. *Helgol. Wissensch. Meeresunters.* 31:55–101.

Reise, K., and P. Ax. 1979. A meiofaunal "thiobios" limited to the anaerobic sulfide system of marine sand does not exist. *Mar. Biol.* 54:225–237.

Reitner, J. 1978. Ein Teuthiden-Rest aus dem Obernor (Kössener-Schichten) der Lahnewies-Neidernachmulde bei Garmisch-Partenkirchen (Bayern). *Paläont. Z.* 52:205–212.

Reitner, J., and T. Engeser. 1982. Phylogenetic trends in phragmocone-bearing coleoids (Belemnomorpha). *Neues Jb. Geol. Paläont. Abh.* 164:156–162.

———. 1983. Contributions to the systematics and the paleoecology of the family Acanthochaetetidae Fischer, 1870 (order Tabulospongida, class Sclerospongiae). *Geobios* 16:773–779.

Repenning, C. A. 1976a. *Enhydra* and *Enhydriodon* from the Pacific coast of North America. *J. Res. U.S. Geol. Surv.* 4:305–315.

———. 1976b. Adaptive evolution of sea lions and walruses. *Syst. Zool.* 25:375–390.

Repetski, J. E. 1978. A fish from the Upper Cambrian of North America. *Science* 200:529–531.

Retallack, G., and D. L. Dilcher. 1981. A coastal hypothesis for the dispersal and rise to dominance of flowering plants. In Niklas (ed.), 1981, pp. 27–77.

Rex, M. A. 1973. Deep-sea species diversity: Decreased gastropod diversity at abyssal depths. *Science* 181:1051–1053.

Rex, M. A., and K. J. Boss. 1976. Open coiling in Recent gastropods. *Malacologia* 15:289–297.

Reyne, A. 1938. On the distribution of *Birgus latro* L. in the Dutch East Indies. *Arch. Neerl. Zool.* 3 (suppl.): 239–247.

———. 1939. On the food habits of the coconut crab (*Birgus latro* L.), with notes on its distribution. *Arch. Neerl. Zool.* 3:283–320.

Rhoads, D. C. 1970. Mass properties, stability, and ecology of marine muds related to burrowing activity. In T. P. Crimes and J. C. Harper (eds.), *Trace Fossils. Geol. J. Special Issue* 3:391–406.

Rhoads, D. C., and L. F. Boyer. 1982. The effects of marine benthos on physical properties of sediments: A successional perspective. In McCall and Tevesz (eds.), 1982, pp. 3–52.

Rhoads, D. C., and R. A. Lutz (eds.). 1980. *Skeletal Growth of Aquatic Organisms: Biological Records of Environmental Change.* Plenum, New York.

Rhoads, D. C., and J. W. Morse. 1971. Evolutionary and ecologic significance of oxygen-deficient marine basins. *Lethaia* 4:413–428.

Rhoads, D. C., and D. K. Young. 1970. The influence of deposit-feeding organisms on sediment stability and community trophic structure. *J. Mar. Res.* 28:150–178.

Ribbink, A. J., B. A. Marsh, A. C. Marsh, A. C. Ribbink, and B. J. Sharp. 1983. A preliminary survey of the cichlid fishes of rocky habitats in Lake Malawi. *South Afr. J. Zool.* 18:149–310.

Rice, A. L. 1962. The food of the sea scorpion *Acanthocottus bubalis* (Teleosti-Scleroparei) from Manx waters. *Proc. Zool. Soc. London* 138:295–303.

Rice, M. E. 1975. Survey of the Sipuncula of the coral and beach-rock communities of the Caribbean Sea. *Proc. Intern. Symp. Biol. Sipuncula and Echiura* 1:35–49.

Richards, R. P. 1972. Autecology of Richmondian brachiopods (Late Ordovician of Indiana and Ohio). *J. Paleont.* 46:386–405.

Richardson, B. J., P. M. Rogers, and G. M. Hewitt. 1980. Ecological genetics of the wild rabbit in Australia. II. Protein variation in British, French and Australian rabbits and the geographical distribution of the variation in Australia. *Aust. J. Biol. Sci.* 33:371–383.

Ricklefs, R. E. 1979. Adaptation, constraint, and compromise in avian postnatal development. *Biol. Revs.* 54:269–290.

Ridder, C. de, and J. M. Lawrence. 1982. Food and feeding mechanisms: Echinoidea. In Jangoux and Lawrence (eds.), 1982, pp. 57–115.

Riedl, R. 1978. *Order in Living Organisms.* Trans. R.P.S. Jeffries. Wiley, New York.

Rieppel, I., and L. Labhardt. 1979. Mandibular mechanics in *Varanus niloticus* (Reptilia: Lacertilia). *Herpetologica* 35:158–163.

Riggs, S. R. 1984. Paleoceanographic model of Neogene phosphorite deposition, U.S. Atlantic continental margin. *Science* 223:123–131.

Robba, E., and F. Ostinelli. 1975. Studi paleoecologici sul Pliocene Ligure I. Testimonianze di predazione sui molluschi Pliocenici di Albenga. *Riv. Ital. Paleont.* 81:309–372.

Robilliard, G. A. 1971. Predation by the nudibranch *Dirona albolineata* on three species of prosobranchs. *Pacific Sci.* 25:429–435.

Robinson, M. H., and B. Robinson. 1973. Ecology and behavior of the giant wood spider *Nephila maculata* (Fabricius) in New Guinea. *Smithson. Contribs. Zool.* 149:1–76.

Robinson, M. H., L. G. Abele, and B. Robinson. 1970. Attack autotomy: A defense against predators. *Science* 169:300–301.

Rodriguez, J., and R. C. Gutschick. 1977. Barnacle borings in live and dead hosts from the Louisiana Limestone (Famennian) of Missouri. *J. Paleont.* 51:718–724.

Roger, J., and E. Buge. 1947. L'association cellépore-gastéropode dans les Faluns de la Touraine. *Géol. Soc. France Bull.* (5) 17:461–470.

Rohr, D. M. 1976. Silurian predator borings in the brachiopod *Dicaelosia* from the Canadian Arctic. *J. Paleont.* 50:1175–1179.

Rolfe, W.D.I. 1980. Early invertebrate terrestrial faunas. In Panchen (ed.), 1980, pp. 117–157.

———. 1985. Form and function in Thylacocephala, Conchyliocarida and Concavicarida (? Crustacea). *Trans. Roy. Soc. Edinburgh* 76:391–399.

Rome, L. C. 1983. Effect of long-term exposure to different temperatures on the mechanical performance of frog muscle. *Physiol. Zool.* 56:33–40.

Rome, L. C., P. T. Loughna, and J. Goldspink. 1985. Temperature acclimation: Improved sustained swimming performance in carp at low temperatures. *Science* 228:194–196.

Romer, A. S. 1933. Eurypterid influence on vertebrate history. *Science* 78:114–117.

———. 1966. *Vertebrate Paleontology.* 3d ed. University of Chicago Press, Chicago.

Root, A. 1963. Notes on feeding habits of the openbilled stork *Anastomus lamelligerus. Ibis* 105:399–400.

Rosewater, J. 1965. The family Tridacnidae in the Indo-Pacific. *Indo-Pacific Mollusca* 1:347–394.

Ross, A., and W. A. Newman. 1973. Revision of the coral-inhabiting barnacles (Cirripedia: Balanidae). *Trans. San Diego Soc. Nat. Hist.* 17:137–174.

Ross, D. M. 1971. Protection of hermit crabs (*Dardanus* spp.) from *Octopus* by commensal sea anemones (*Calliactis* spp.) *Nature* 230:401–402.

———. 1974. Evolutionary aspects of associations between crabs and sea anemones. In W. B. Vernberg (ed.), *Symbiosis in the Sea*, pp. 111–125. University of South Carolina Press, Columbia.

Roughgarden, J. 1979. *Theory of Population Genetics and Evolutionary Ecology: An Introduction.* MacMillan, New York.

Rowland, S. M. 1984. Were there framework reefs in the Cambrian? *Geology* 12:181–183.

Rudkin, D. M. 1979. Healed injuries in *Ogygopsis klotzi* (Trilobita) from the Middle Cambrian of British Columbia. *Roy. Ontario Mus. Life Sci. Occ. Pap.* 32:1–8.

Rudman, W. B. 1972a. Structure and functioning of the gut in the Bullomorpha (Opisthobranchia). Part 3: Philinidae. *J. Nat. Hist.* 6:459–474.

———. 1972b. Structure and functioning of the gut in the Bullomorpha (Opisthobranchia). Part 4: Aglajidae. *J. Nat. Hist.* 6:547–560.

———. 1981. Further studies on the anatomy and ecology of opisthobranch molluscs feeding on the scleractinian coral *Porites. Zool. J. Linn. Soc.* 71:373–412.

———. 1982. A new species of *Phestilla;* the first record of a corallivorous aeolid nudibranch from tropical America. *J. Zool. London* 192:465–471.

Rudwick, M.J.S. 1959. The growth and form of brachiopod shells. *Geol. Mag.* 96:1–24.

———. 1961. The feeding mechanism of the brachiopod *Prorichthofenia. Palaeontology* 3:450–471.

———. 1962. Notes on the ecology of brachiopods in New Zealand. *Trans. Roy. Soc. New Zealand* 25:327–335.

———. 1970. *Living and Fossil Brachiopods.* Hutchinson, London.

Rudwick, M.J.S., and R. Cowen. 1967. The functional morphology of some aberrant strophomenide brachiopods from the Permian of Sicily. *Boll. Soc. Paleont. Ital.* 6:113–176.

Runham, N. W., and P. J. Hunter. 1970. *Terrestrial Slugs.* Hutchinson, London.

Runnegar, B. 1974. Evolutionary history of the bivalve subclass Anomalodesmata. *J. Paleont.* 48:904–940.

———. 1978. Origin and evolution of the class Rostroconchia. *Phil. Trans. Roy. Soc. London* (B) 284:319–333.

———. 1981. Muscle scars, shell form and torsion in Cambrian and Ordovician univalved molluscs. *Lethaia* 14:311–322.

———. 1982a. Oxygen requirements, biology and phylogenetic significance of the Late Precambrian worm *Dickinsonia*, and the evolution of the burrowing habit. *Alcheringa* 6:223–239.

———. 1982b. The Cambrian explosion: Animals or fossils? *J. Geol. Soc. Australia* 29:395–411.

———. 1983. Molluscan phylogeny revisited. *Mem. Assoc. Australas. Palaeontols.* 1:121–144.

Runnegar, B., and C. Bentley. 1983. Anatomy, ecology and affinities of the Australian Early Cambrian bivalve *Pojetaia runnegari* Jell. *J. Paleont.* 57:73–92.

Runnegar, B., and P. A. Jell. 1976. Australian Middle Cambrian molluscs and their bearing on early molluscan evolution. *Alcheringa* 1:109–138.

Runnegar, B., and J. Pojeta, Jr. 1974. Molluscan phylogeny: The paleontological viewpoint. *Science* 186:311–317.

Runnegar, B., J. Pojeta, Jr., M. E. Taylor, and D. Collins. 1979. New species of the Cambrian and Ordovician chitons *Mathevia* and *Chelodes* from Wisconsin and Queensland: Evidence for the early history of polyplacophoran mollusks. *J. Paleont.* 53:1374–1394.

Russell, B. C. 1983. The food and feeding habits of rocky reef fish of northeastern New Zealand. *New Zealand J. Mar. Fresh-W. Res.* 17:121–145.

Russell, D. A., and P. Beland. 1976. Running dinosaurs. *Nature* 264:486.

Rützler, K. 1971. Bredin-Archbold-Smithsonian biological survey of Dominica: Burrowing sponges, *Siphonodictyon* Bergquist, from the Caribbean. *Smithson. Contribs. Zool.* 77:1–17.

———. 1974. The burrowing sponges of Bermuda. *Smithson. Contribs. Zool.* 165:1–32.

———. 1975. The role of burrowing sponges in bioerosion. *Oecologia* (Berlin) 19:203–216.

Ryland, J. S. 1970. *Bryozoans*. Hutchinson, London.

Rypstra, A. L. 1984. A relative measure of predation on web-spiders in temperate and tropical forests. *Oikos* 43:129–142.

Sadler, P. M. 1981. Sediment accumulation rates and the completeness of stratigraphic sections. *J. Geol.* 89:569–584.

Safriel, U. 1966. Recent vermetid formation on the Mediterranean shore of Israel. *Proc. Malacol. Soc. London* 37:27–34.

———. 1974. Vermetid gastropods and intertidal reefs in Israel and Bermuda. *Science* 186:1113–1115.

———. 1975. The role of vermetid gastropods in the formation of Mediterranean and Atlantic reefs. *Oecologia* (Berlin) 20:85–101.

Samuelsen, T. J. 1970. The biology of six species of Anomura (Crustacea, Decapoda) from Raunefjorden, western Norway. *Sarsia* 45:25–52.

Sander, F., and C. M. Lalli. 1982. A comparative study of the mollusk communities on the shelf-slope margin of Barbados, West Indies. *Veliger* 24:309–318.

Sanders, H. L. 1968. Marine benthic diversity: A comparative study. *Amer. Nat.* 102:243–282.

Sars, G. O. 1895. *An Account of the Crustacea of Norway, With Short De-*

scriptions and Figures of All the Species. Vol. 1: *Amphipoda*. Cammer-meyers, Christiania.

Saunders, W. B. 1981. The species of living *Nautilus* and their distribution. *Veliger* 24:8–17.

Saunders, W. B., and A.R.H. Swan. 1984. Morphology and morphologic diversity of mid-Carboniferous (Namurian) ammonoids in time and space. *Paleobiology* 10:195–228.

Saunders, W. B., C. Spinosa, C. Teichert, and R. C. Banks. 1978. The jaw apparatus of Recent *Nautilus* and its palaeontological implications. *Palaeontology* 21:129–141.

Savitzky, A. H. 1983. Coadapted character complexes among snakes: Fossoriality, piscivory, and durophagy. *Amer. Zool.* 23:397–409.

Schaeffer, B. 1973. Fishes and the Permian-Triassic boundary. *Canad. Soc. Petroleum Geol. Mem.* 2:493–497.

Schäfer, W. 1954. Form und Funktion der Brachyuren Scheere. *Abh. Senckenberg. Naturforsch. Ges.* 489:1–66.

———. 1972. *Ecology and Palaeontology of Marine Environments*. Trans. A. Oertel. University of Chicago Press, Chicago.

Schaffer, W. M. 1968. Intraspecific combat and the evolution of the Caprini. *Evolution* 22:817–825.

Schaller, G. B. 1972. *The Serengeti Lion: A Study of Predator-Prey Relations*. University of Chicago Press, Chicago.

Scheltema, R. S. 1978. On the relationship between dispersal of pelagic veliger larvae and the evolution of marine prosobranch gastropods. In B. Battaglia and J. A. Beardmore (eds.), *Marine Organisms*, pp. 303–322. Plenum, New York.

Schembri, P. J. 1982. Feeding behaviour of fifteen species of hermit crabs (Crustacea: Decapoda: Anomura) from the Otago region, southeastern New Zealand. *J. Nat. Hist.* 16:859–878.

Schenck, H. 1892. Beiträge zur Biologie der Lianen, im Besonderen der in Brasilien einheimischen Arten. I. Beiträge zur Biologie der Lianen. *Bot. Mitt. Trop.* 4:1–248.

Schindel, D. E. 1980. Microstratigraphic sampling and the limits of paleontologic resolution. *Paleobiology* 6:408–426.

———. 1982. Resolution analysis: A new approach to the gaps in the fossil record. *Paleobiology* 8:340–353.

———. 1986. Morphometrics of gastropod coiling: New procedures, morphological integration and architectural constraints. Submitted.

Schindel, D. E., G. J. Vermeij, and E. Zipser. 1982. Frequencies of repaired shell fractures among the Pennsylvanian gastropods of north-central Texas. *J. Paleont.* 56:729–740.

Schmalfuss, H. 1978. Constructional morphology of cuticular terraces in trilobites, with conclusions on synecological evolution. *Neues Jb. Geol. Paläont. Abh.* 157:164–168.

Schmid, G. 1933. Die Käferlarve *Silpha* als Schneckenräuber. *Natur und Museum* 6:193–198.

Schmitt, R. J. 1982. Consequences of dissimilar defenses against predation in a subtidal marine community. *Ecology* 63:1588–1601.

Schmitt, R. J., C. W. Osenberg, and M. G. Bercovitch. 1983. Mechanisms and consequences of shell fouling in the kelp snail, *Norrisia norrisi* (Sowerby) (Trochidae): Indirect effects of *Octopus* drilling. *J. Exp. Mar. Biol. Ecol.* 69:267–281.

Schnell, G. D. 1974. Flight speeds and wingbeat frequencies of the magnificent frigatebird. *Auk* 91:564–570.

Schnell, G. D., and J. J. Hellack. 1979. Bird flight speeds in nature: Optimized or a compromise? *Amer. Nat.* 113:53–66.

Schoch, R. M. 1983. The function of the horns of titanotheres (Mammalia, Perissodactyla). *Evol. Theory* 6:274.

Schoener, A., and T. W. Schoener. 1981. The dynamics of the species-area relation in marine fouling systems. 1. Biological correlates of changes in the species-area slope. *Amer. Nat.* 118:339–360.

Schoener, T. W. 1983. Field experiments on interspecific competition. *Amer. Nat.* 122:240–285.

Scholander, P. F., W. Flagg, V. Walters, and L. Irving. 1953. Climatic adaptation in Arctic and tropical poikilotherms. *Physiol. Zool.* 26:67–92.

Schopf, T.J.M. 1974. Permo-Triassic extinctions: Relation to sea-floor spreading. *J. Geol.* 82:129–143.

———. 1980. *Paleoceanography*. Harvard University Press, Cambridge.

———. 1981. Punctuated equilibrium and evolutionary stasis. *Paleobiology* 7:156–166.

Schram, F. R. 1977. Paleozoogeography of Late Paleozoic and Triassic Malacostraca. *Syst. Zool.* 26:367–379.

———. 1979. The genus *Archaeocaris*, and a general review of the Palaeostomatopoda (Hoplocarida: Malacostraca). *San Diego Soc. Nat. Hist. Trans.* 19:57–66.

———. 1982. The fossil record and the evolution of Crustacea. In Abele (ed.), 1982, pp. 93–147.

———. (ed.). 1983. *Crustacean Phylogeny*. Balkema, Rotterdam.

Schram, F. R., and R. R. Hessler. 1984. Anaspidid Syncarida. In Eldredge and Stanley (eds.), 1984, pp. 192–195.

Schram, F. R., and W. A. Newman. 1980. *Verruca withersi* n. sp. (Crustacea: Cirripedia) from the middle of the Cretaceous of Colombia. *J. Paleont.* 54:229–233.

Scott, A. C., and T. N. Taylor. 1983. Plant/animal interactions during the Upper Carboniferous. *Bot. Rev.* 49:259–307.

Scrutton, C. T. 1979. Early fossil cnidarians. In House (ed.), 1979, pp. 161–207.

Scudder, G.G.E., and J. L. Reveal (eds.). 1981. *Evolution Today: Proceedings of the Second International Congress of Systematic and Evolutionary Biology*. Hunt Institute of Botanical Documentation, Carnegie-Mellon University, Pittsburgh.

Scully, E. P. 1979. The effects of gastropod shell availability and habitat characteristics on shell utilization in the intertidal hermit crab *Pagurus longicarpus* Say. *J. Exp. Mar. Biol. Ecol.* 37:139–152.

———. 1983. The behavioral ecology of competition and resource utilization among hermit crabs. In S. Rebach and D. W. Dunham (eds.), *Studies in Adaptation: The Behavior of Higher Crustacea*, pp. 23–55. Wiley, New York.

Sebens, K. P. 1976. The ecology of Caribbean sea anemones in Panama: Utilization of space on a coral reef. In G. O. Mackie (ed.), *Coelenterate Ecology and Behavior*, pp. 67–77. Plenum, New York.

Secretan, S. 1985. Conchyliocarida, a class of fossil crustaceans: Relationships to Malacostraca and postulated behaviour. *Trans. Roy. Soc. Edinburgh* 76:381–389.

Sedberry, G. R., and J. A. Musick. 1978. Feeding strategies of some demersal fishes of the continental slope and rise off the Mid-Atlantic coast of the USA. *Mar. Biol.* 44:357–375.

Seed, R. 1968. Factors influencing shell shape in the mussel *Mytilus edulis*. *J. Mar. Biol. Assoc. U.K.* 48:561–584.

———. 1969. The ecology of *Mytilus edulis* L. (Eulamellibranchia) on exposed rocky shores. II. Growth and mortality. *Oecologia* (Berlin) 3:317–350.

Seilacher, A. 1977. Evolution of trace fossil communities. In Hallam (ed.), 1977, pp. 357–376.

———. 1979. Constructional morphology of sand dollars. *Paleobiology* 5:191–221.

———. 1984. Late Precambrian and Early Cambrian Metazoa: Preservational or real extinctions? In Holland and Trendall (eds.), 1984, pp. 159–168.

Selden, P. A. 1984. Autecology of Silurian eurypterids. *Special Pap. Palaeontol.* 32:39–54.

Sepkoski, J. J., Jr. 1979. A kinetic model of Phanerozoic taxonomic diversity. II. Early Phanerozoic families and multiple equilibria. *Paleobiology* 5:222–251.

———. 1981. A factor analytic description of the Phanerozoic marine fossil record. *Paleobiology* 7:36–53.

———. 1982a. Mass extinctions in the Phanerozoic oceans: A review. *Geol. Soc. Amer. Special Pap.* 190:283–289.

———. 1982b. A compendium of fossil marine families. *Milwaukee Publ. Mus. Contribs. Biol. Geol.* 51:125 P.

———. 1982c. Flat-pebble conglomerates, storm deposits, and the Cambrian bottom fauna. In G. Einsele and A. Seilacher (eds.), *Cyclic and Event Stratification*, pp. 371–385. Springer, Berlin.

———. 1984. A kinetic model of Phanerozoic taxonomic diversity. III. Post-Paleozoic families and mass extinctions. *Paleobiology* 10:246–267.

———. 1986. Phanerozoic overview of mass extinctions. In Raup and Jablonski (eds.), 1986, pp. 277–295.

Sepkoski, J. J., Jr., and P. M. Sheehan. 1983. Diversification, faunal change, and community replacement during the Ordovician radiations. In Tevesz and McCall (eds.), 1983, pp. 673–717.

Sepkoski, J. J., Jr., R. K. Bambach, D. M. Raup, and J. W. Valentine. 1981. Phanerozoic marine diversity and the fossil record. *Nature* 293:435–437.

Shachak, M., U. N. Safriel, and R. Hunum. 1981. An exceptional event of predation on desert snails by migratory thrushes in the Negev Desert, Israel. *Ecology* 62:1441–1449.

Shaffer, H. B. 1978. Relative predation pressure on salamanders (Caudata: Plethodontidae) along an altitudinal transect in Guatemala. *Copeia* 1978:268–272.

Shear, W. A., W. M. Bonamo, J. D. Grierson, W.D.I. Rolfe, E. L. Smith, and R. A. Norton. 1984. Early land animals in North America: Evidence from Devonian age arthropods from Gilboa, New York. *Science* 224:492–494.

Sheehan, P. M. 1975. Brachiopod synecology in a time of crisis (Late Ordovician-Early Silurian). *Paleobiology* 1:205–212.

———. 1982. Brachiopod macroevolution at the Ordovician-Silurian boundary. *Third North Amer. Paleont. Conv. Proc.* 2:477–481.

———. 1985. Reefs are not so different—they follow the evolutionary pattern of level-bottom communities. *Geology* 13:46–49.

Sheehan, P. M., and T.A. Hansen. 1986. Detritus feeding as a buffer to extinction at the end of the Cretaceous. *Geology,* in press.

Sheehan, P. M., and P. J. Lespérance. 1978. Effect of predation on the population dynamics of a Devonian brachiopod. *J. Paleont.* 52:812–817.

Sheehan, P. M., and D.R.J. Schiefelbein. 1984. The trace fossil *Thalassinoides* from the Upper Ordovician of the eastern Great Basin: Deep burrowing in the Early Paleozoic. *J. Paleont.* 58:440–447.

Sheldon, R. P. 1981. Ancient marine phosphorites. *Ann. Rev. Earth Planet. Sci.* 9:251–284.

Shepherd, S. A. 1973. Studies on Southern Australian abalone (genus *Haliotis*). I. Ecology of the five sympatric species. *Austral. J. Mar. Fresh-W. Res.* 24:217–257.

Sheppard, C.R.C. 1982. "Reach" of aggressively interacting corals, and relative importance of interactions at different depths. *Proc. Fourth Intern. Coral Reef Symp.* 2:363–368.

Shimek, R. L. 1984. The diets of Alaskan *Neptunea. Veliger* 26:274–281.

Shivji, M., M. J. Parker, B. Hartwick, M. J. Smith, and N. A. Sloan. 1983. Feeding and distribution study of the sunflower sea star *Pycnopodia helianthoides* (Brandt, 1835). *Pacific Sci.* 37:133–140.

Shoup, J. B. 1968. Shell opening by crabs of the genus *Calappa. Science* 160:887–888.

Shumway, S. E. 1982. Oxygen consumption in brachiopods and the possible role of punctae. *J. Exp. Mar. Biol. Ecol.* 58:207–220.

Sieg, J. 1983. Evolution of Tanaidacea. In Schram (ed.), 1983, pp. 229–256.

Siegfried, W. R. 1977. Mussel-dropping behaviour of kelp gulls. *South Afr. J. Sci.* 73:337–341.

Signor, P. W., III. 1982a. Growth-related surficial resorption of the penultimate whorl in *Terebra dimidiata* (Linnaeus 1758) and other marine prosobranch gastropods. *Veliger* 25:79–82.

———. 1982b. Resolution of life habits using multiple morphologic criteria: Shell form and life mode in turritelliform gastropods. *Paleobiology* 8:378–388.

———. 1982c. Influence of shell shape on burrowing rates in infaunal turritelliform snails. *Third North Amer. Paleont. Conv. Proc.* 2:483–487.

———. 1983. Burrowing and the functional significance of ratchet sculpture in turritelliform gastropods. *Malacologia* 23:313–320.

———. 1985. The role of shell geometry as a deterrent to predation in terebrid gastropods. *Veliger* 28(2):179–185.

Signor, P. W., III, and C. E. Brett. 1984. The mid-Paleozoic precursor to the Mesozoic marine revolution. *Paleobiology* 10:229–245.

Signor, P. W., III, and P. W. Kat. 1984. Functional significance of columellar folds in turritelliform gastropods. *J. Paleont.* 58:210–216.

Sih, A. 1982. Optimal patch use: Variation in selective pressure for efficient foraging. *Amer. Nat.* 120:666–685.

———. 1985. Evolution, predator avoidance, and unsuccessful predation. *Amer. Nat.* 125:153–157.

Simberloff, D. S. 1978. Using island biogeographic distributions to determine if colonization is stochastic. *Amer. Nat.* 112:713–726.

———. 1981. Community effects of introduced species. In Nitecki (ed.), 1981, pp. 53–81.

Simberloff, D., B. J. Brown, and S. Lowrie. 1978. Isopod and insect wood borers may benefit Florida mangroves. *Science* 201:630–632.

Simpson, R. D. 1976. Physical and biotic factors limiting the distribution and abundance of littoral molluscs on MacQuarie Island (sub-Antarctic). *J. Exp. Mar. Biol. Ecol.* 21:11–49.

Sims, R. W. 1955. The morphology of the head of the hawfinch (*Coccothrostes coccothrostes*) with special reference to the myology of the jaw. *Brit. Mus. Nat. Hist. Bull. (Zool.)* 2:369–393.

Sinclair, M. E. 1977. Agonistic behaviour of the stone crab, *Menippe mercenaria* (Say). *Anim. Behav.* 25:193–207.

Skelton, P. W. 1978. The evolution of functional design in rudists (Hippuritacea) and its taxonomic implications. *Phil. Trans. Roy. Soc. London* (B) 284:305–318.

Slatkin, M. 1983. Genetic background. In Futuyma and Slatkin (eds.), 1983, pp. 14–32.

Sleder, J. 1981. *Acanthina punctulata* (Neogastropoda: Muricacea): Its distribution, activity, diet, and predatory behavior. *Veliger* 24:172–180.

Sloan, N. A., and S.M.C. Robinson. 1983. Winter feeding by asteroids on a subtidal sandbed in British Columbia. *Ophelia* 22:125–140.

Sloan, R. E., K. J. Rigby, Jr., L. Van Valen, and D. Gabriel. 1986. Gradual dinosaur extinction and simultaneous ungulate radiation in the Hell Creek Formation. *Science* 232:629–633.

Slocum, C. J. 1980. Differential susceptibility to grazers in two phases of an intertidal alga: Advantages of heteromorphic generations. *J. Exp. Mar. Biol. Ecol.* 46:99–110.

Smart, J., and N. F. Hughes. 1973. The insect and the plant: Progressive paleoecological integration. *Symp. Roy. Entomol. Soc. London* 6:143–156.

Smit, J. 1982. Extinction and evolution of planktonic Foraminifera after a major impact at the Cretaceous/Tertiary boundary. *Geol. Soc. Amer. Special Paper* 190:329–352.

Smith, A. B. 1984. *Echinoid Paleobiology*. Allen and Unwin, London.

Smith, D.A.S. 1975. Polymorphism and selective predation in *Donax faba* Gmelin (Bivalvia: Tellinacea), *J. Exp. Mar. Biol. Ecol.* 17:205–219.

Smith, O. R. 1953. Notes on the ability of the horseshoe crab, *Limulus polyphemus*, to locate soft-shelled clams, *Mya arenaria. Ecology* 34:636–637.

Smith, S. A., C. W. Thayer, and C. E. Brett. 1985. Predation in the Paleozoic: Gastropod-like drillholes in Devonian brachiopods. *Science* 230:1033–1035.

Smith-Gill, S. J., and K. A. Berven. 1979. Predicting amphibian metamorphosis. *Amer. Nat.* 113:563–585.

Snyder, N.F.R., and H. A. Snyder. 1969. A comparative study of mollusc predation by limpkins, Everglade kites, and boat-tailed grackles. *Living Bird* 8:177–223.

———. 1971. Defenses of the Florida apple snail *Pomacea paludosa. Behaviour* 40:175–215.

Sober, E. 1984. *The Nature of Selection: Evolutionary Theory in Philosophical Focus*. MIT Press, Cambridge.

Sohl, N. F. 1964. Neogastropoda, Opisthobranchia and Basommatophora from the Ripley, Owl Creek, and Prairie Bluff Formations. *U.S. Geol. Surv. Prof. Paper* 331-B:153–344.

———. 1969. The fossil record of shell boring by snails. *Amer. Zool.* 9:725–734.

Sohl, N. F., and H. A. Kollmann. 1985. Cretaceous acteonellid gastropods from the western hemisphere. *U.S. Geol. Surv. Prof. Paper* 1304:1–104.

Solem, A. 1972. Microarmature and barriers in the aperture of land snails. *Veliger* 15:81–87.

———. 1974. Patterns of radular tooth structure in carnivorous land snails. *Veliger* 17:81–88.

———. 1976. *Endodontid Land Snails from Pacific Islands (Mollusca: Pulmonata: Sigmurethra). Part I: Family Endodontidae*. Field Museum of Natural History, Chicago.

———. 1978. Classification of the land Mollusca. In V. Fretter and J. Peake (eds.), *Pulmonates*, Vol. 2A: *Systematics, Evolution and Ecology*, pp. 49–97. Academic Press, London.

———. 1979. Biogeographic significance of land snails, Paleozoic to Recent. In Gray and Boucot (eds.), 1979, pp. 277–287.

Solem, A., and E. L. Yochelson. 1979. North American Paleozoic land snails, with a summary of other Paleozoic nonmarine snails. *U.S. Geol. Surv. Prof. Paper* 1072:1–42.

Solem, A., S. Tillier, and P. Mordan. 1984. Pseudo-operculate pulmonate land snails from New Caledonia. *Veliger* 27:193–199.

Sousa, W. P., S. C. Schroeter, and S. D. Gaines. 1979. Latitudinal variation in intertidal algal community structure: The influence of grazing and vegetative propagation. *Oecologia* (Berlin) 48:297–307.

Sparks, D. K., R. D. Hoare, and R. D. Kesling. 1980. Epizoans on the brachiopod *Paraspirifer bownockeri* (Stewart) from the Middle Devonian of Ohio. *Papers on Paleontology, Univ. Michigan Mus. Paleont.* 23:1–105.

Speden, I. G. 1970. The type Fox Hills Formation, Cretaceous (Maastrichtian), South Dakota. Part 2: Systematics of the Bivalvia. *Yale Univ. Peabody Mus. Nat. Hist. Bull.* 33:222 P.

———. 1971. Notes on New Zealand fossil Mollusca—2. Predation on New Zealand Cretaceous species of *Inoceramus* (Bivalvia). *N.Z. J. Geol. Geophys.* 14:56–70.

Spight, T. M. 1977. Availability and use of shells by intertidal hermit crabs. *Biol. Bull.* 152:120–133.

———. 1981. Why small hermit crabs have large shells. *Ecosynthesis* 1:225–254.

Stachowitsch, M. 1977. The hermit crab microbiocoenosis—the role of mobile secondary hard bottom elements in a north Adriatic benthic community. In B. F. Keegan, P. O. Ceidigh, and P.J.S. Boaden (eds.), *Biology of Benthic Organisms; 11th Eur. Symp. Mar. Biol. (Galway, Oct., 1976)*, pp. 549–558. Pergamon, Oxford.

———. 1980. The epibiotic and endolithic species associated with the gastropod shells inhabited by the hermit crabs *Paguristes oculatus* and *Pagurus cuanensis. Mar. Ecol.* 1:73–101.

Stanley, G. D., Jr. 1979. Paleoecology, structure, and distribution of Triassic coral build-ups in Western North Ameria. *Univ. Kansas Paleont. Contribs.* 65:1–58.

Stanley, S. M. 1968. Post-Paleozoic adaptive radiation of infaunal bivalve molluscs—a consequence of mantle fusion and siphon formation. *J. Paleont.* 42:214–229.

———. 1969. Bivalve mollusk borrowing aided by discordant shell ornamentation. *Science* 166:634–635.

———. 1970. Relation of shell form to life habits of the Bivalvia (Mollusca). *Geol. Soc. Amer. Mem.* 125:1–296.

———. 1972. Functional morphology and evolution of byssally attached bivalve molluscs. *J. Paleont.* 46:165–212.

———. 1973. An explanation of Cope's Rule. *Evolution* 27:1–26.

———. 1974. Relative growth of the titanothere horn: A new approach to an old problem. *Evolution* 28:447–457.

———. 1975a. A theory of evolution above the species level. *Proc. Nat. Acad. Sci. U.S.A.* 72:646–650.

———. 1975b. Why clams have the shape they have: An experimental analysis of burrowing. *Paleobiology* 1:48–58.

———. 1976. Fossil data and the Precambrian-Cambrian evolutionary transition. *Amer. J. Sci.* 276:56–76.

Stanley, S. M. 1977a. Trends, rates, and patterns of evolution in the Bivalvia. In Hallam (ed.), 1977, pp. 209–250.

———. 1977b. Coadaptation in the Trigoniidae, a remarkable family of burrowing bivalves. *Palaeontology* 20:869–899.

———. 1979. *Macroevolution: Pattern and Process*. Freeman, San Francisco.

———. 1981. Infaunal survival: Alternate functions of shell ornamentation in the Bivalvia (Mollusca). *Paleobiology* 7:384–393.

———. 1982a. Gastropod torsion: Predation and the opercular imperative. *Neues Jb. Geol. Paläont. Abh.* 164:95–107.

———. 1982b. Species selection involving alternative character states: Approach to macroevolutionary analysis. *Third North Amer. Paleont. Conv. Proc.* 2:505–510.

———. 1984. Temperature and biotic crises in the marine realm. *Geology* 12:205–208.

———. 1986. Population size, extinction, and speciation: The fission effect in Neogene Bivalvia. *Paleobiology* 12:89–110.

Stanley, S. M., and L. D. Campbell. 1981. Neogene mass extinction of Western Atlantic molluscs. *Nature* 293:457–459.

Stanley, S. M., and W. A. Newman. 1980. Competitive exclusion in evolutionary time: The case of the acorn barnacles. *Paleobiology* 6:173–183.

Stanley, S. M., B. Van Valkenburgh, and R. S. Steneck. 1983. Coevolution and the fossil record. In Futuyma and Slatkin (eds.), 1983, pp. 328–349.

Stasek, C. R. 1967. Autotomy in the Mollusca. *Occ. Pap. Calif. Acad. Sci.* 61:1–44.

———. 1972. The molluscan framework. In M. Florkin and B. T. Sheer (eds.), *Chemical Zoology*, Vol. 7: *Mollusca*, pp. 1–44. Academic Press, New York.

Steadman, D. W., and P. S. Martin. 1984. Extinction of birds in the Late Pleistocene of North America. In Martin and Klein (eds.), 1984, pp. 466–477.

Stearns, S. C. 1976. Life-history tactics: A review of the ideas. *Quart. Rev. Biol.* 51:3–47.

———. 1984. How much of the phenotype is necessary to understand evolution at the level of the gene? In K. Hörmann and V. Loeschcke (eds.), *Population Biology and Evolution*, pp. 31–45. Springer, Berlin.

———. 1986. Natural selection and fitness, adaptation and constraint. In Raup and Jablonski (eds.), 1986, pp. 23–44.

Stebbins, G. L. 1967. Adaptive radiation and trends of evolution in higher plants. *Evol. Biol.* 1:101–142.

———. 1969. *The Basis of Progressive Evolution*. University of North Carolina Press, Chapel Hill.

Stebbins, G. L., and G.J.C. Hill, 1980. Did multicellular plants invade the land? *Amer. Nat.* 115:343–353.

Steel, R. 1973. *Crocodylia*. Encyclopedia of Palaeoherpetology. Part 16. Gustav Fischer, Stuttgart.

Steele-Petrović, H. M. 1975. An explanation for the tolerance of brachiopods and relative intolerance of filter-feeding bivalves for soft muddy bottoms. *J. Paleont.* 49:552–556.

———. 1979. The physiological differences between articulate brachiopods and filter-feeding bivalves as a factor in the evolution of marine level-bottom communities. *Palaeontology* 22:101–134.

Stehli, F. G., R. D. Douglas, and N. D. Newell. 1969. Generation and maintenance of gradients in taxonomic diversity. *Science* 164:947–949.

Stein, R. A., C. G. Goodman, and E. A. Marschall. 1984. Using time and energetic measures of cost in estimating prey value for fish predators. *Ecology* 65:702–715.

Stein, R. A., J. F. Kitchell, and B. Knezević. 1975. Selective predation by carp (*Cyprinus carpio* L.) on benthic molluscs in Skadar Lake, Yugoslavia. *J. Fish. Biol.* 7:391–399.

Steneck, R. S. 1982. A limpet-coralline alga association: Adaptations and defenses between a selective herbivore and its prey. *Ecology* 63:507–522.

———. 1983a. Escalating herbivory and resulting adaptive trends in calcareous algal crusts. *Paleobiology* 9:44–61.

———. 1983b. Quantifying herbivory on coral reefs: Just scratching the surface and still biting off more than we can chew. In M. L. Reaka (ed.), *The Ecology of Deep and Shallow Coral Reefs*, Symposia Series for Undersea Research, NOAA's Undersea Research Program, Vol. 1 (1), pp. 103–111.

———. 1985. Adaptations of crustose coralline algae to herbivory: Patterns in space and time. In D. F. Toomey and M. H. Nitecki (eds.), *Paleoalgology: Contemporary Research and Applications*, pp. 352–366. Springer, Berlin.

Steneck, R. S., and L. Watling. 1982. Feeding capabilities and limitations of herbivorous molluscs: A functional group approach. *Mar. Biol.* 68:299–319.

Ştevcić, Z. 1971. The main features of brachyuran evolution. *Syst. Zool.* 20:331–340.

Stimson, J. 1970. Territorial behavior of the owl limpet, *Lottia gigantea*. *Ecology* 51:113–118.

———. 1973. The role of the territory in the ecology of the intertidal limpet *Lottia gigantea* (Gray). *Ecology* 54:1020–1030.

Stitt, J. H. 1971. Repeating evolutionary pattern in Late Cambrian trilobite biomeres. *J. Paleont.* 45:178–181.

———. 1976. Functional morphology and life habits of the Late Cambrian trilobite *Stenopilus pronus* Raymond. *J. Paleont.* 50:561–576.

Störmer, L. 1976. Arthropods from the Lower Devonian (Lower Emsian) of Alken an der Mosel (Germany). Part 5: Myriapoda and additional forms, with general remarks on fauna and problems regarding invasion of land by arthropods. *Senckenbergiana Lethaea* 57:87–183.

———. 1977. Arthropod invasion of land during Late Silurian and Devonian times. *Science* 197:1362–1364.

Strathmann, R. R. 1978a. The evolution and loss of feeding larval stages of marine invertebrates. *Evolution* 32:894–906.

———. 1978b. Progressive vacating of adaptive types during the Phanerozoic. *Evolution* 32:907–914.

Strathmann, R. R. 1981. The role of spines in preventing structural damage to echinoid tests. *Paleobiology* 7:400–406.

Strathmann, R. R., and E. S. Branscomb. 1979. Adequacy of cues to favorable sites used by settling larvae of two intertidal barnacles. In S. Stancyk (ed.), *Reproductive Ecology of Marine Invertebrates*, pp. 77–89. University of South Carolina Press, Columbia.

Strathmann, R. R., E. S. Branscomb, and K. Vedder. 1981. Fatal errors in set as a cost of dispersal and the influence of intertidal flora on set of barnacles. *Oecologia* (Berlin) 48:13–18.

Strauch, F. 1970. Die Thule-Landbrücke als Wanderweg und Faunenscheide zwischen Atlantik und Skandik im Tertiär. *Geol. Rundschau* 60:381–417.

———. 1972. Phylogenese, Adaptation und Migration einiger nordischer mariner Molluskengenera (*Neptunea, Panomya, Cyrtodaria* und *Mya*). *Abh. Senckenberg. Naturforsch. Ges.* 531:1–211.

Stridsberg, S. 1981. Apertural constrictions in some oncocerid cephalopods. *Lethaia* 14:269–276.

———. 1984. Aptychopsid plates—jaw elements or protective operculum. *Lethaia* 17:93–98.

Strong, D. R., L. A. Szyska, and D. S. Simberloff. 1979. Tests of community-wide character displacement against null hypotheses. *Evolution* 33:897–913.

Struhsaker, T. T., and L. Leland. 1977. Palm-nut smashing by *Cebus a. apella* in Colombia. *Biotropica* 9:124–126.

Suchaneck, T. H. 1978. The ecology of *Mytilus edulis* L. in exposed rocky intertidal communities. *J. Exp. Mar. Biol. Ecol.* 31:105–120.

———. 1983. Control of seagrass communities and sediment distribution by *Callianassa* (Crustacea, Thalassinidea) bioturbation. *J. Mar. Res.* 41:281–298.

Surlyk, F., and M. B. Johansen. 1984. End-Cretaceous brachiopod extinctions in the chalk of Denmark. *Science* 223:1174–1177.

Sutton, S. L. 1972. *Woodlice*. Ginn, London.

Swain, T. 1978. Plant-animal coevolution: A synoptic view of the Paleozoic and Mesozoic. In J. B. Harborne (ed.), *Biochemical Aspects of Plant and Animal Coevolution*, pp. 3–19. Academic Press, London.

Swain, T., and G. Cooper-Driver. 1981. Biochemical evolution in early land plants. In Niklas (ed.), 1981, pp. 103–134.

Tappan, H. 1968. Primary production, isotopes, extinctions and the atmosphere. *Palaeogeogr., Palaeoclimatol., Palaeoecol.* 4:187–210.

———. 1970. Phytoplankton abundance and Late Paleozoic extinctions: A reply. *Palaeogeogr., Palaeoclimatol., Palaeoecol.* 8:49–66.

———. 1982. Extinction or survival: Selectivity and causes of Phanerozoic crises. *Geol. Soc. Amer. Special Pap.* 190:265–276.

Tappan, H., and A. R. Loeblich, Jr. 1973a. Smaller protistan evidence and explanation of the Permian-Triassic crisis. *Canad. Soc. Petrol. Geol. Mem.* 2:465–489.

———. 1973b. Evolution of the oceanic plankton. *Earth Sci. Revs.* 9:207–240.

Targett, T. E. 1978. Food resource partitioning by the pufferfishes *Sphoeroides spengleri* and *S. testudineus* from Biscayne Bay, Florida. *Mar. Biol.* 49:83–91.

Taylor, D. W. 1966. A remarkable snail fauna from Coahuila, Mexico. *Veliger* 9:152–228.

Taylor, E. C. 1983. Lake Tanganyika shell dwellers. *Trop. Fish Hobbyist* (October, 1983): 8–17.

Taylor, J. D. 1970. Feeding habits of predatory gastropods in a Tertiary (Eocene) molluscan assemblage from the Paris Basin. *Palaeontology* 13:255–260.

———. 1976. Habitat, abundance and diets of muricacean gastropods at Aldabra Atoll. *Zool. J. Linn. Soc.* 59:155–193.

———. 1978. Habitats and diet of predatory gastropods at Addu Atoll, Maldives. *J. Exp. Mar. Biol. Ecol.* 31:83–103.

———. 1983. The food of coral-reef *Drupa* (Gastropoda). *Zool. J. Linn. Soc.* 78:209–316.

———. 1984. A partial food web involving predatory gastropods on a Pacific fringing reef. *J. Exp. Mar. Biol. Ecol.* 74:273–290.

Taylor, J. D., and D. G. Reid. 1984. The abundance and trophic classification of molluscs upon coral reefs in the Sudanese Red Sea. *J. Nat. Hist.* 18:175–209.

Taylor, J. D., and C. N. Taylor. 1977. Latitudinal distribution of predatory gastropods on the Eastern Atlantic shelf. *J. Biogeogr.* 4:73–81.

Taylor, J. D., R. J. Cleevely, and N. J. Morris. 1983. Predatory gastropods and their activities in the Blackdown Greensand (Albian) of England. *Palaeontology* 26:521–553.

Taylor, J. D., N. J. Morris, and C. N. Taylor. 1980. Food specialization and the evolution of predatory prosobranch gastropods. *Palaeontology* 23:375–409.

Taylor, P. D. 1979. Palaeoecology of the encrusting epifauna of some British Jurassic bivalves. *Palaeogeogr., Palaeoclimatol., Palaeoecol.* 28:241–262.

Taylor, P. D., and P. L. Cook. 1981. *Hippoporidra edax* (Busk 1859) and a revision of some fossil and living *Hippoporidra* (Bryozoa). *Bull. Brit. Mus. Nat. Hist. (Geol.)* 35:243–251.

Temple, S. A. 1977. Plant-animal mutualism: Coevolution with dodo leads to near extinction of plant. *Science* 197:885–886.

Tevesz, M.J.S., and P. L. McCall (eds.). 1983. *Biotic Interactions in Recent and Fossil Benthic Communities*. Plenum, New York.

Thayer, C. W. 1972. Adaptive features of swimming monomyarian bivalves (Mollusca). *Forma et Functio* 5:1–32.

———. 1975a. Morphologic adaptations of benthic invertebrates to soft substrata. *J. Mar. Res.* 33:177–189.

———. 1975b. Strength of pedicle attachment in articulate brachiopods: Ecologic and paleoecologic significance. *Paleobiology* 1:388–399.

Thayer, C. W. 1979. Biological bulldozers and the evolution of marine benthic communities. *Science* 203:458–461.

———. 1983. Sediment-mediated biological disturbance and the evolution of marine benthos. In Tevesz and McCall (eds.), 1983, pp. 479–625.

———. 1985. Brachiopods versus mussels: Competition, predation, and palatability. *Science* 228:1527–1528.

———. 1986. Are brachiopods better than bivalves? Mechanisms of turbidity tolerance and their interaction with feeding in articulates. *Paleobiology* 12:161–174.

Thien, L. B., P. Bernhardt, G. W. Gibbs, O. Pellmyr, G. Bergström, I. Groth, and G. McPherson. 1985. The pollination of *Zygogynum* (Winteraceae) by a moth, *Sabatinca* (Micropterygidae): An ancient association? *Science* 227:540–541.

Thierstein, H. R., and W. H. Berger. 1978. Injection events and ocean history. *Nature* 276:461–466.

Thomas, R.D.K. 1978a. Shell form and the ecological range of living and extinct Arcoida. *Paleobiology* 4:181–194.

———. 1978b. Limits to opportunism in the evolution of the Arcoida (Bivalvia). *Phil. Trans. Roy. Soc. London* (B) 284:335–344.

———. 1984. When and how did plants and animals take to the land? *Paleobiology* 10:1–8.

Thomerson, J. E. 1969. Variation and relationship of the studfishes, *Fundulus catenatus* and *Fundulus stellifer* (Cyprinodontidae, Pisces). *Tulane Stud. Zool. Bot.* 16:1–21.

Thompson, D. W. 1942. *On Growth and Form*. Cambridge University Press, London.

Thompson, F. G. 1980. Proserpinoid land snails and their relationships within the Archaeogastropoda. *Malacologia* 20:1–33.

Thomson, K. S. 1969. The biology of the lobe-finned fishes. *Biol. Revs.* 44:91–154.

———. 1976. Explanation of large scale extinction of lower vertebrates. *Nature* 261:578–580.

———. 1977. The pattern of diversification among fishes. In Hallam (ed.), 1977, pp. 376–404.

Thorson, G. 1950. Reproductive and larval ecology of marine bottom invertebrates. *Biol. Revs.* 25:1–45.

Thulborn, R. A. 1981. Estimated speed of a giant bipedal dinosaur. *Nature* 292:273–274.

Thulborn, R. A., and M. Wade. 1979. Dinosaur stampede in the Cretaceous of Queensland. *Lethaia* 12:275–279.

Thunell, R. C. 1981. Cenozoic paleotemperature changes and planktonic foraminiferal speciation. *Nature* 289:670–672.

Thurmond, J. T. 1974. Lower vertebrate faunas of the Trinity Division in north-central Texas. *Geoscience and Man* 8:103–129.

Tiffney, B. H. 1981. Diversity and major events in the evolution of land plants. In Niklas (ed.), 1981, pp. 193–230.

———. 1984. Seed size, dispersal syndromes, and the rise of the angiosperms: Evidence and hypothesis. *Ann. Missouri Bot. Garden* 71:551–576.

———. 1986. Evolution of seed dispersal syndromes according to the fossil record. In D. R. Murray (ed.), *Seed Dispersal*. Academic Press, in press.

Tillier, S. 1984. Relationships of gymnomorph gastropods (Mollusca: Gastropoda). *Zool. J. Linn. Soc.* 82:345–362.

Tomlinson, J. T. 1969. The burrowing barnacles (Cirripedia: order Acrothoracica), *U.S. Nat. Mus. Bull.* 296:1–162.

Townsend, C. R., and T. K. McCarthy. 1980. On the defense strategy of *Physa fontinalis* (L.), a freshwater pulmonate snail. *Oecologia* (Berlin) 46:75–79.

Trevallion, A., R.R.C. Edwards, and J. H. Steele. 1970. Dynamics of a benthic bivalve. In J. H. Steele (ed.), *Marine Food Chains*, pp. 285–295. University of California Press, Los Angeles.

Trott, L. B. 1970. Contributions to the biology of carapid fishes (Paracanthopterygii: Gadiformes). *Univ. California Publ. Zool.* 89:1–60.

Trueman, E. R. 1968. The burrowing activities of bivalves. *Symp. Zool. Soc. London* 22:167–186.

Trueman, E. R., and and A. Packard. 1968. Motor performances of some cephalopods. *J. Exp. Biol.* 49:495–507.

Trueman, E. R., A. R. Brand, and P. Davis. 1966. The effect of substrate and shell shape on the burrowing of some common bivalves. *Proc. Malacol. Soc. London* 37:97–109.

Tucker, M. E., and M. J. Benton. 1982. Triassic environments, climates and reptile evolution. *Palaeogeogr., Palaeoclimatol., Palaeoecol.* 40:361–389.

Tucker, M. E., and T. P. Burchette. 1977. Triassic dinosaur footprints from South Wales: Their context and preservation. *Palaeogeogr., Palaeoclimatol., Palaeoecol.* 22:195–208.

Tucker, V. A., and K. Schmidt-Koenig. 1971. Flight speeds of birds in relation to energetics and wind directions. *Auk* 88:97–107.

Tunnicliffe, V. 1979. The role of boring sponges in coral fracture. *Colloq. Intern. Centre Nat. Rech. Sci.* 291:309–315.

Tuomi, J., P. Niemela, E. Haukioja, S. Siren, and S. Neuvonen. 1984. Nutrient stress: An explanation for plant anti-herbivore responses to defoliation. *Oecologia* (Berlin) 61:208–210.

Turner, H. J., Jr. 1955. How clam drills capture razor clams. *Nautilus* 69:20–22.

Turner, J.R.G. 1981. Adaptation and evolution in *Heliconius*: A defense of NeoDarwinism. *Ann. Rev. Ecol. Syst.* 12:99–121.

———. 1983. Mimetic butterflies and punctuated equilibria: Some old light on a new paradigm. *Biol. J. Linn. Soc.* 20:277–300.

Turner, R. D., and T. R. Roberts. 1978. Mollusks as prey of pariid catfish in the Fly River. *Bull. Amer. Malacol. Union* 1978:33–40.

Tyler, J. C. 1980. Osteology, phylogeny, and higher classification of the fishes of the order Plectognathi (Tetraodontiformes). *NOAA Techn. Rep. NMFS Circular* 434:1–422.

Tynan, M. C. 1983. Coral-like microfossils from the Lower Cambrian of California. *J. Paleont.* 57:1188–1211.

Vacelet, J. 1979. Description et affinités d'une éponge sphinctozoaire actuelle. *Colloq. Intern. Centre Nat. Rech. Sci.* 291:483–493.

———. 1981. Eponges hypercalcifiées ("pharétronides," "sclérosponges") des cavités des récifs coralliens de Nouvelle-Calédonie. *Bull. Mus. Natn. Hist. Nat. Paris* (4) 3:313–351.

———. 1982. Algal-sponge symbioses in the coral reefs of New Caledonia: A morphological study. *Proc. Fourth Intern. Coral Reef Symp.* 2:713–719.

Valentine, J. W. 1968. Climatic regulation of species diversification and extinction. *Geol. Soc. Amer. Bull.* 79:273–276.

———. 1969. Niche diversity and niche size pattern in marine fossils. *J. Paleont.* 43:905–915.

Valentine, J. W., and D. Jablonski. 1982. Major determinants of the biogeographic pattern of the shallow-sea fauna. *Bull. Soc. Géol. France* (7) 24:893–899.

———. 1983a. Larval adaptations and patterns of brachiopod diversity in space and time. *Evolution* 37:1052–1061.

———. 1983b. Speciation in the shallow sea: General patterns and biogeographic controls. In R. W. Sims, J. H. Price, and P.E.S. Whalley (eds.), *Evolution, Time and Space: The Emergence of the Biosphere*, pp. 201–226. Academic Press, London.

Valentine, J. W., and E. M. Moores. 1970. Plate-tectonic regulation of faunal diversity and sea level: A model. *Nature* 228:657–659.

Vance, R. R. 1978. A mutualistic interaction between a sessile marine clam and its epibionts. *Ecology* 59:679–685.

Vandel, A. 1943. Essai sur l'origine, l'évolution et la classification des Oniscoidea (isopodes terrestres). *Bull. Biol. France Belg. Suppl.* 30:1–136.

———. 1948. Les isopodes volvationnels exoantennés et la genèse de leurs coaptations. *Bull. Biol. France Belg.* 82:388–428.

———. 1959. Nouvelles recherches sur les isopodes volvationnels exoantennés et la genèse de leurs coaptations. *Bull. Biol. France Belg.* 93:121–139.

Van Valen, L. 1973. A new evolutionary law. *Evol. Theory* 1:1–18.

———. 1975. Group selection, sex, and fossils. *Evolution* 29:87–93.

———. 1976. Energy and evolution. *Evol. Theory* 1:179–229.

———. 1983. How pervasive is coevolution? In Nitecki (ed.), 1983, pp. 1–19.

———. 1984. A resetting of Phanerozoic community evolution. *Nature* 307:50–52.

———. 1985a. How constant is extinction? *Evol. Theory* 7:93–106.

———. 1985b. A theory of origination and extinction. *Evol. Theory* 7:133–142.

Van Valen, L., and V. C. Maiorana. 1985. Patterns of origination. *Evol. Theory* 7:107–125.

Van Valen, L., and R. E. Sloan. 1977. Ecology and the extinction of the dinosaurs. *Evol. Theory* 2:37–64.

Van Veldhuizen, H. D., and D. W. Phillips. 1978. Prey capture by *Pisaster brevispinus* (Asteroidea: Echinodermata) on soft substrate. *Mar. Biol.* 48:89–97.

Vermeij, G. J. 1970. Adaptive versatility and skeleton construction. *Amer. Nat.* 104:253–260.

———. 1971. Gastropod evolution and morphological diversity in relation to shell geometry. *J. Zool. London* 163:15–23.

———. 1972. Intraspecific shore-level size gradients in intertidal molluscs. *Ecology* 53:693–700.

———. 1973a. Morphological patterns in high intertidal gastropods: Adaptive strategies and their limitations. *Mar. Biol.* 20:319–346.

———. 1973b. Molluscs in mangrove swamps: Physiognomy, diversity, and regional differences. *Syst. Zool.* 22:609–624.

———. 1974a. Regional variations in tropical high intertidal gastropod assemblages. *J. Mar. Res.* 32:343–357.

———. 1974b. Marine faunal dominance and molluscan shell form. *Evolution* 28:656–664.

———. 1975. Evolution and distribution of left-handed and planispiral coiling in snails. *Nature* 254:419–420.

———. 1976. Interoceanic differences in vulnerability of shelled prey to crab predation. *Nature* 260:135–136.

———. 1977a. Patterns in crab claw size: The geography of crushing. *Syst. Zool.* 26:138–151.

———. 1977b. The Mesozoic marine revolution: Evidence from snails, predators and grazers. *Paleobiology* 3:245–258.

———. 1978. *Biogeography and Adaptation: Patterns of Marine Life.* Harvard University Press, Cambridge.

———. 1979a. Shell architecture and causes of death in Micronesian reef snails. *Evolution* 33:686–696.

———. 1979b. The architectural geography of some gastropods. In Gray and Boucot (eds.), 1979, pp. 427–433.

———. 1980a. Drilling predation of bivalves in Guam: Some paleoecological implications. *Malacologia* 19:329–334.

———. 1980b. Drilling predation in a population of the edible bivalve *Anadara granosa. Nautilus* 94:123–125.

———. 1980c. Gastropod growth rate, allometry, and adult size: Environmental implications. In Rhoads and Lutz (eds.), 1980, pp. 379–394.

———. 1981. Apertural form in gastropods. *Lethaia* 14:104.

———. 1982a. Environmental change and the evolutionary history of the periwinkle *Littorina littorea* in North America. *Evolution* 36:561–580.

———. 1982b. Gastropod shell form, repair, and breakage in relation to breakage by the crab *Calappa. Malacologia* 23:1–12.

———. 1982c. Phenotypic evolution in a poorly dispersing snail after arrival of a predator. *Nature* 299:349–350.

———. 1982d. Unsuccessful predation and evolution. *Amer. Nat.* 120:701–720.

Vermeij, G. J. 1983a. Intimate associations and coevolution in the sea. In Futuyma and Slatkin (eds.), 1983, pp. 311–327.

———. 1983b. Traces and trends in predation, with special reference to bivalved animals. *Palaeontology* 26:455–465.

———. 1983c. Shell-breaking predation through time. In Tevesz and McCall (eds.), 1983, pp. 649–669.

———. 1985. Aptations, effects and fortuitous survival: Comment on a paper by A. Sih. *Amer. Nat.* 125:470–472.

———. 1986a. The biology of human-caused extinction. In B. G. Norton (ed.), *The Preservation of Species*, pp. 28–49. Princeton University Press, Princeton.

———. 1986b. Survival during biotic crises: The properties and evolutionary significance of refuges. In D. K. Elliott (ed.), *Dynamics of Extinction*, pp. 231–246. Wiley, New York.

Vermeij, G. J., and A. P. Covich. 1978. Coevolution of freshwater gastropods and their predators. *Amer. Nat.* 112:833–843.

Vermeij, G. J., and J. D. Currey. 1980. Geographical variation in the strength of thaidid snail shells. *Biol. Bull.* 158:383–389.

Vermeij, G. J., and E. C. Dudley. 1982. Shell repair and drilling in some gastropods from the Ripley Formation (Upper Cretaceous) of the south-eastern U.S.A. *Cretaceous Res.* 3:397–403.

———. 1985. Distribution of adaptations: A comparison between the functional shell morphology of freshwater and marine pelecypods. In E. R. Trueman (ed.), *Biology of the Mollusca*, Vol. 10: *Evolution*, pp. 461–478. Academic Press, London.

Vermeij, G. J., and E. J. Petuch. 1986. Differential extinction in tropical American molluscs: Endemism, architecture, and the Panama land bridge. *Malacologia* 27:29–41.

Vermeij, G. J., and J. A. Veil. 1978. A latitudinal pattern in bivalve shell gaping. *Malacologia* 17:57–61.

Vermeij, G. J., and E. Zipser. 1986a. A short-term study of growth and death in a population of the gastropod *Strombus gibberulus* in Guam. *Veliger* 28:314–317.

———. 1986b. Burrowing performance of some tropical Pacific gastropods. *Veliger* 29:200–206.

Vermeij, G. J., E. A. Kay, and L. G. Eldredge. 1984. Molluscs of the Northern Mariana Islands, with special reference to the selectivity of open-ocean dispersal barriers. *Micronesica* 19:27–55.

Vermeij, G. J., D. E. Schindel, and E. Zipser. 1981. Predation through geological time: Evidence from gastropod shell repair. *Science* 214:1024–1026.

Vermeij, G. J., E. Zipser, and E. C. Dudley. 1980. Predation in time and space: Peeling and drilling in terebrid gastropods. *Paleobiology* 6:352–364.

Vermeij, G. J., E. Zipser, and R. Zardini. 1982. Breakage-induced shell repair in some gastropods from the Upper Triassic of Italy. *J. Paleont.* 56:233–235.

Vidal, G., and A. H. Knoll. 1982. Radiations and extinctions of plankton in the Late Proterozoic and Early Cambrian. *Nature* 297:57–60.

Vine, P. J. 1974. Effects of algal grazing and aggressive behaviour of the fishes *Pomacentrus lividus* and *Acanthurus sohal* on coral-reef ecology. *Mar. Biol.* 24:131–136.

Vitt, L. J., and R. E. Ballinger. 1982. The adaptive significance of a complex caudal adaptation in the tropical gekkonid lizard *Lygodactylus klugei*. *Canad. J. Zool.* 60:2582–2587.

Vitt, L. J., J. D. Congdon, and N. A. Dickson. 1977. Adaptive strategies and energetics of tail autotomy in lizards. *Ecology* 58:326–337.

Vitt, L. J., J. D. Congdon, A. C. Hulse, and J. E. Platz. 1974. Territorial aggressive encounters and tail breaks in the lizard *Sceloporus magister*. *Copeia* 1974:990–993.

Vivien, M. L. 1973. Contribution à la connaissance de l'éthologie alimentaire de l'icthyofaune du platier interne des récifs coralliens de Tulear (Madagascar). *Téthys Suppl.* 5:221–308.

Voigt, E. 1975. Tunnelbaue Rezenter und fossiler Phoronidea. *Paläont. Z.* 49:135–167.

Vrba, E. S. 1980. Evolution, species and fossils: How does life evolve? *South Afr. J. Sci.* 76:61–84.

———. 1983. Macroevolutionary trends: New perspectives on the roles of adaptation and incidental effect. *Science* 221:387–389.

———. 1984. Evolutionary pattern and process in the sister-group Alcelaphini-Aepycerotini (Mammalia: Bovidae). In Eldredge and Stanley (eds.), 1984, pp. 62–79.

Vrba, E. S., and N. Eldredge. 1984. Individuals, hierarchies and processes: Towards a more complete evolutionary theory. *Paleobiology* 10:146–171.

Wächtler, W. 1927. Zur Biologie der Raublungenschnecke *Poiretia (Glandina) algira* Brug. *Zool. Anz.* 72:191–197.

Wade, B. A. 1967. Studies on the biology of the West Indian beach clam, *Donax denticulatus* Linné. I. Ecology. *Bull. Mar. Sci.* 17:149–174.

Wahle, C. M. 1983. Regeneration of injuries among Jamaican gorgonians: The roles of colony physiology and environment. *Biol. Bull.* 165:778–790.

Walker, A.J.M. 1972. Introduction to the ecology of the Antarctic limpet *Patinigera polaris* (Hombron and Jacquinot) at Signy Island, South Orkney Islands. *Brit. Antarctic Surv. Bull.* 28:49–69.

Walker, M. H. 1978. Food and feeding habits of *Lethrinus chrysostomus* Richardson (Pisces: Perciformes) and other lethrinids on the Great Barrier Reef. *Austral. J. Mar. Fresh-W. Res.* 29:623–630.

Waller, T. R. 1969. The evolution of the *Argopecten gibbus* stock (Mollusca: Bivalvia) with emphasis on the Tertiary and Quaternary species of eastern North America. *J. Paleont.* 43, Part II of 2, *Paleont. Soc. Mem.* 3:1–125.

———. 1972. The functional significance of some shell microstructures in the Pectinacea (Mollusca: Bivalvia). *Intern. Geol. Congr., 24th Session, Montreal, Canada, Sec. 7, Paleont.*: 48–56.

———. 1978. Morphology, morphoclines and a new classification of the

Pteriomorphia (Mollusca: Bivalvia). *Phil. Trans. Roy. Soc. London* (B) 284:345–365.

———. 1984. The ctenolium of scallop shells: Functional morphology and evolution of a key family-level character in the Pectinacea (Mollusca: Bivalvia). *Malacologia* 25:203–219.

Walter, M. R., and G. R. Heys. 1985. Links between the rise of the Metazoa and the decline of stromatolites. *J. Precambrian Res.* 29:159–174.

Ward, P. D. 1979. Functional morphology of Cretaceous helically coiled ammonite shells. *Paleobiology* 5:415–422.

———. 1980. Shell shape distributions in Jurassic-Cretaceous ammonites and Jurassic-Tertiary nautilids. *Paleobiology* 6:32–43.

———. 1981. Shell sculpture as a defensive adaptation in ammonoids. *Paleobiology* 7:96–100.

———. 1982. The relationship of siphuncle size to emptying rates in chambered cephalopods: Implications for cephalopod paleobiology. *Paleobiology* 8:426–433.

———. 1986a. Cretaceous ammonite shell shapes. *Malacologia* 27:3–28.

———. 1986b. *The Natural History of Nautilus*. Allen and Unwin, London.

Ward, P., and P. W. Signor, III. 1983. Evolutionary tempo in Jurassic and Cretaceous ammonites. *Paleobiology* 9:183–198.

Ward, P., and G.E.G. Westermann. 1977. First occurrence, systematics, and functional morphology of *Nipponites* (Cretaceous Lytoceratina) from the Americas. *J. Paleont.* 51:367–372.

Ward, P., and M. K. Wicksten. 1980. Food sources and feeding behavior of *Nautilus macromphalus*. *Veliger* 23:119–124.

Ward, P., R. Stone, G.E.G. Westermann, and A. Martin. 1977. Notes on shell weight, cameral fluids, swimming speed, and color polymorphism of the cephalopod *Nautilus pompilius* in the Fiji Islands. *Paleobiology* 3:377–388.

Warme, J. E. 1975. Borings as trace fossils, and the processes of marine bioerosion. In Frey (ed.), 1975, pp. 181–227.

Warmke, G. L., and D. S. Erdman. 1963. Records of mollusks eaten by bonefish in Puerto Rican waters. *Nautilus* 76:115–120.

Warner, G. 1982. Food and feeding mechanisms: Ophiuroidea. In Jangoux and Lawrence (eds.), 1982, pp. 161–181.

Warner, G. F., and A. R. Jones. 1976. Leverage and muscle types in crab chelae (Crustacea: Brachyura). *J. Zool. London* 180:57–68.

Warner, G. F., D. Chapman, N. Hawkey, and D. G. Waring. 1982. Structure and function of the chelae and chela closer muscles of the shore crab *Carcinus maenas* (Crustacea: Brachyura). *J. Zool. London* 196:431–438.

Wassersug, R. J., and K. Hoff. 1979. A comparative study of the buccal pumping mechanism of tadpoles. *Biol. J. Linn. Soc.* 12:225–259.

Watanabe, J. M. 1983. Anti-predator defenses of three kelp forest gastropods: Contrasting adaptations of closely-related prey species. *J. Exp. Mar. Biol. Ecol.* 73:257–270.

Watkins, R. 1974. Molluscan paleobiology of the Miocene Wimer Formation, Del Norte County, California. *J. Paleont.* 48:1264–1282.

———. 1979. Benthic community organization in the Ludlow Series of the Welsh Borderland. *Bull. Brit. Mus. Nat. Hist. (Geol.)* 31:175–280.

Weaver, P.P.E., and P. J. Schultheis. 1983. Vertical open burrows in deep-sea sediments 2 m in length. *Nature* 301:329–331.

Webb, P. W. 1982. Locomotor patterns in the evolution of actinopterygian fishes. *Amer. Zool.* 22:329–342.

———. 1984. Body form, locomotion and foraging in aquatic vertebrates. *Amer. Zool.* 24:107–120.

Webb, S. D. 1977. A history of savanna vertebrates in the New World. Part 1: North America. *Ann. Rev. Ecol. Syst.* 8:355–380.

———. 1984. Ten million years of mammal extinctions in North America. In Martin and Klein (eds.), 1984, pp. 189–210.

Weishampel, D. B. 1984. Interactions between Mesozoic plants and vertebrates: Fructifications and seed predation. *Neues Jb. Geol. Paläont. Abh.* 167:224–259.

Wellington, J. M., and R. K. Trench. 1985. Persistence and coexistence of a nonsymbiotic coral in open reef environments. *Proc. Nat. Acad. Sci. U.S.A.* 82:2432–2436.

Wells, H. W. 1958. Predation of pelecypods and gastropods by *Fasciolaria hunteria* (Perry). *Bull. Mar. Sci. Gulf Caribbean* 8:152–166.

———. 1969. Hydroid and sponge commensals of *Cantharus cancellarius* with a "false shell." *Nautilus* 82:93–102.

Wells, R. A. 1980. Activity pattern as a mechanism of predator avoidance in two species of acmaeid limpet. *J. Exp. Mar. Biol. Ecol.* 48:151–168.

Wenner, C. A., and J. A. Musick. 1975. Food habits and seasonal abundance of the American eel, *Anguilla rostrata*, from the Lower Chesapeake Bay. *Chesapeake Sci.* 16:62–66.

West-Eberhard, M. J. 1983. Sexual selection, social competition, and speciation. *Quart. Rev. Biol.* 58:155–183.

Westermann, G.E.G. 1971. Form, structure and function of shell and siphuncle in coiled Mesozoic ammonoids. *Life Sci. Contribs. Roy. Ontario Mus.*, no. 78:1–39.

———. 1975. Architecture and buoyancy of simple cephalopod phragmocones and remarks on ammonoids. *Paläont. Z.* 49:221–234.

———. 1977. Form and function of orthoconic cephalopod shells with concave septa. *Paleobiology* 3:300–321.

Westermann, G.E.G., and P. Ward. 1980. Septum morphology and bathymetry in cephalopods. *Paleobiology* 6:48–50.

Westoll, T. S. 1949. On the evolution of the Dipnoi. In G. L. Jepsen, E. Mayr, and G. G. Simpson (eds.), *Genetics, Paleontology, and Evolution*, pp. 121–184. Princeton University Press, Princeton.

Westphal, F. 1975. Bauprinzipien im Panzer der Placodonten (Reptilia Triadica). *Paläont. Z.* 49:197–225.

Westrop, S. R. 1983. The life habits of the Ordovician illaenine trilobite *Bumastoides*. *Lethaia* 16:15–24.

Weyl, P. K. 1968. The role of the oceans in climatic change: A theory of the ice ages. *Meteorol. Monogrs.* 8:37–62.

Whitaker, J. O., Jr., and T. W. French. 1984. Foods of six species of sympatric shrews from New Brunswick. *Canad. J. Zool.* 62:622–626.

Whitney, R. R. 1961. *Bairdiella, Bairdiella icistius* (Jordan and Gilbert). *Calif. Dept. Fish. and Game Bull.* 113:105–151.

Whittington, H. B. 1979. Early arthropods, their appendages and relationships. In M. R. Holland (ed.), 1979, pp. 253–268.

Whittington, H. B., and D.E.G. Briggs. 1985. The largest Cambrian animal, *Anomalocaris*, Burgess Shale, British Columbia. *Phil. Trans. Roy. Soc. London* (B) 309:569–609.

Wickler, W., and U. Seibt. 1970. Das Verhalten von *Hymenocera picta* Dana, einer Seesterne fressenden Garnele (Decapoda, Natantia, Gnathophyllidae). *Z. Tierpsychol.* 27:352–368.

Wilber, T. P., Jr., and W. Herrnkind. 1982. Rate of new shell acquisition by hermit crabs in a salt marsh habitat. *J. Crustacean Biol.* 2:587–592.

———. 1984. Predaceous gastropods regulate new-shell supply to salt marsh hermit crabs. *Mar. Biol.* 79:145–150.

Wilde, P., and W.B.N. Berry. 1984. Destabilization of the oceanic density structure and its significance to marine "extinction" events. *Palaeogeogr. Palaeoclimatol., Palaeoecol.* 48:143–162.

Wiley, E. O. 1976. The phylogeny and biogeography of fossil and Recent gars (Actinopterygii: Lepisosteidae). *Univ. Kansas Mus. Nat. Hist. Misc. Publ.* 64:1–111.

Williams, A., and J. M. Hurst. 1977. Brachiopod evolution. In Hallam (ed.), 1977, pp. 79–121.

Williams, D. D., A. T. Read, and K. A. Moore. 1983. The biology and zoogeography of *Helicopsyche borealis* (Tricoptera: Helicopsychidae): A Nearctic representative of a tropical genus. *Canad. J. Zool.* 61:2288–2289.

Williams, F. X. 1951. Life-history studies of East African *Achatina* snails. *Bull. Mus. Comp. Zool. Harvard Univ.* 105:295–317.

Williams, G. C. 1966. *Adaptation and Natural Selection: A Critique of Some Current Evolutionary Thought*. Princeton University Press, Princeton.

———. 1975. *Sex and Evolution*. Princeton University Press, Princeton.

Williams, M. J. 1978. Opening of bivalve shells by the mud crab *Scylla serrata* Forskal. *Austral. J. Mar. Fresh-W. Res.* 29:699–702.

Williamson, P. G. 1981. Palaeontological documentation of speciation in Cenozoic molluscs from Turkana Basin. *Nature* 293:437–443.

Willis, L., S. T. Threlkeld, and C. C. Carpenter. 1982. Tail loss patterns in *Thamnophis* (Reptilia: Colubridae) and the probable fate of injured individuals. *Copeia* 1982:98–101.

Wilson, B. R., and R. Tait. 1984. Systematics, anatomy and boring mechanisms of the rock-boring mytilid bivalve *Botula*. *Proc. Roy. Soc. Victoria* 96:113–125.

Wilson, E. O. 1971. *The Insect Societies*. Belknap Press of Harvard University, Cambridge.

Wilson, J. L. 1975. *Carbonate Facies in Geologic History*. Springer, New York.

Wilson, M. A. 1985. Disturbance and ecologic succession in an Upper Ordovician cobble-dwelling hardground community. *Science* 228:575–577.

Wiltse, W. I. 1980. Predation by juvenile *Polinices duplicatus* (Say) on *Gemma gemma* (Totten). *J. Exp. Mar. Biol. Ecol.* 42:187–199.

Witman, J. D., and T. H. Suchaneck. 1984. Mussels in flow: Drag and dislodgment by epizoans. *Mar. Ecol. Progr. Ser.* 16:259–268.

Wnuk, C., and H. W. Pfefferkorn. 1984. The life habits and paleoecology of Middle Pennsylvanian pteridosperms based on an *in situ* assemblage from the Bernice Basin (Sullivan County, Pennsylvania U.S.A.). *Rev. Paleobot. Palynol.* 41:329–351.

Wodinsky, J. 1969. Penetration of the shell and feeding from gastropods by *Octopus*. *Amer. Zool.* 9:997–1010.

———. 1973. Mechanism of hole boring in *Octopus vulgaris*. *J. Gen. Psychol.* 88:179–183.

Wolbach, W. S., R. S. Lewis, and E. Anders. 1985. Cretaceous extinctions: Evidence for wildfires and search for meteoritic material. *Science* 230:167–170.

Wolff, T. 1961. Description of a remarkable deep-sea hermit crab, with notes on the evolution of the Paguridea. *Galathea Rep.* 4:11–32.

Wonders, J.B.W. 1977. The role of benthic algae in the shallow reef of Curaçao (Netherlands Antilles). III. The significance of grazing. *Aquat. Bot.* 3:357–390.

Woodin, S. A. 1978. Refuges, disturbance, and community structure: A marine soft-bottom example. *Ecology* 59:274–284.

Wright, A. D. 1979. Brachiopod radiation. In House (ed.), 1979, pp. 235–252.

Wright, C. W., and J.S.H. Collins. 1972. British Cretaceous crabs. *Palaeontogr. Soc. Monogrs.* 126:1–114.

Yaldwyn, J. C. 1965. Antarctic and subantarctic decapod Crustacea. In P. van Oye and J. van Mieghem (eds.), *Biogeography and Ecology in Antarctica*, pp. 324–332. Junk, The Hague.

Yamaoka, K. 1978. Pharyngeal jaw structure in labrid fish. *Publ. Seto Mar. Biol. Lab.* 14:409–426.

Yancey, T. E., and D. W. Boyd. 1983. Revision of the Alatoconchidae: A remarkable family of Permian bivalves. *Palaeontology* 26:497–520.

Yin, H. F. 1985. Bivalves near the Permian-Triassic boundary in South China. *J. Paleont.* 59:572–600.

Yin, H. F., and E. L. Yochelson. 1983. Middle Triassic Gastropoda from Qingyan, Ghizhou Province, China: 3—Euomphalacea and Loxonematacea. *J. Paleont.* 57:1098–1127.

Yingst, J. Y., and D. C. Rhoads. 1980. The role of bioturbation in the enhancement of bacterial growth rates in marine sediments. In K. R. Tenore and B. C. Coull (eds.), *Marine Benthic Dynamics*, pp. 407–421. University of South Carolina Press, Columbia.

Yochelson, E. L. 1971. A new Late Devonian gastropod and its bearing on problems of open coiling and septation. *Smithson. Contribs. Paleobiol.* 3:231–241.

———. 1979. Gastropod opercula as objects of paleobiogeographic study. In Gray and Boucot (eds.), 1979, pp. 37–43.

———. 1984a. Historic and current considerations for revision of Paleozoic gastropod classification. *J. Paleont.* 58:259–269.

———. 1984b. Speculative functional morphology and morphology that could not function: The example of *Hyolithes* and *Biconulites. Malacologia* 25:255–264.

Yochelson, E. L., and H. Yin. 1985. Redescription of *Bellerophon asiaticus* Wirth (Early Triassic: Gastropoda) from China, and a survey of Triassic Bellerophontacea. *J. Paleont.* 59:1305–1319.

Yochelson, E. L., D. Dockery, and H. Wolf. 1983. Predation on sub-Holocene scaphopod mollusks from southern Louisiana. *U.S. Geol. Surv. Prof. Pap.* 1282:1–13.

Yom-Tov, Y. 1970. The effect of predation on population densities of some desert land snails. *Ecology* 51:907–911.

Yonge, C. M. 1939. The protobranchiate Mollusca: A functional interpretation of their structure and evolution. *Phil. Trans. Roy. Soc. London* (B) 230:79–147.

———. 1946. On the habits and adaptations of *Aloidis (Corbula) gibba. J. Mar. Biol. Assoc. U.K.* 26:358–376.

———. 1951a. Studies on Pacific coast mollusks. II. Structure and adaptations of rock boring in *Platyodon cancellatus* (Conrad). *Univ. Calif. Publ. Zool.* 55:401–407.

———. 1951b. Studies on Pacific coast mollusks. III. Observations on *Hinnites multirugosus* (Gale). *Univ. Calif. Publ. Zool.* 55:409–420.

———. 1953. Form and habit in *Pinna carnea* Gmelin. *Phil. Trans. Roy. Soc. London* (B) 237:355–374.

———. 1955. Adaptation to rock boring in *Botula* and *Lithophaga* (Lamellibranchia, Mytilidae) with a discussion of the evolution of this habit. *Quart. J. Microscop. Sci.* 96:383–410.

———. 1962. On *Etheria elliptica* Lam. and the course of evolution, including assumption of monomyarianism, in the family of Etheriidae (Bivalvia: Unionacea). *Phil. Trans. Roy. Soc. London* (B) 244:423–458.

———. 1967. Form, habit and evolution in the Chamidae (Bivalvia) with reference to conditions in the rudists (Hippuritacea). *Phil. Trans. Roy. Soc. London* (B) 252:49–105.

———. 1978a. On the monomyarian, *Acostaea rivoli* and evolution in the family Etheriidae (Bivalvia: Unionacea). *J. Zool. London* 184:429–448.

———. 1978b. On the Dimyidae (Mollusca: Bivalvia) with special reference to *Dimya corrugata* Hedley and *Basiliomya goreaui* Bayer. *J. Molluscan Stud.* 44:357–375.

Young, D. K. 1969. *Okadaia elegans*, a tube-boring nudibranch mollusc from the Central and West Pacific. *Amer. Zool.* 9:903–907.

Zach, R. 1978. Selection and dropping of whelks by Northwestern crows. *Behaviour* 67:134–148.

Zajac, R. N. 1985. The effects of sublethal predation on reproduction by the spionid polychaete *Polydora ligni* Webster. *J. Exp. Mar. Biol. Ecol.* 88:1–19.

Zangerl, R. 1981. *Chondrichthyes*. I. *Paleozoic Elasmobranchii. Handbook of Paleoichthyology*. Vol. 3. Gustav Fischer, Stuttgart.

Ziegelmeier, E. 1954. Beobachtungen über den Nahrungserwerb bei der Naticide *Lunatia nitida* Donovan (Gastropoda Prosobranchia). *Helgoländer Wissensch. Meeresunters.* 5:1–33.

Zimmermann, M. H. 1983. *Xylem Structure and the Ascent of Sap*. Springer, Berlin.

Zinner, H. 1978. Observations on little owls feeding on snails in the Negev. *Argamon, Israel J. Malacol.* 6:57–60.

Zipser, E., and G. J. Vermeij. 1978. Crushing behavior of tropical and temperate crabs. *J. Exp. Mar. Biol. Ecol.* 31:155–172.

———. 1980. Survival after non-lethal shell damage in the gastropod *Conus sponsalis. Micronesica* 16:229–234.

Zucker, W. V. 1983. Tannins: Does structure determine function? *Amer. Nat.* 121:335–365.

Zusi, R. L. 1984. A functional and evolutionary analysis of rhynchokinesis in birds. *Smithson. Contribs. Zool.* 395:1–40.

Index

Library of Congress Cataloging-in-Publication Data

Vermeij, Geerat J., 1946–
Evolution and escalation.

Bibliography: p.   Includes index.
1. Evolution. 2. Ecology. I. Title.
QH371.V5 1987   575   86-25389
ISBN 0-691-08446-7